Conservation of Freshwater Fishes

Freshwater fishes are one of the most diverse groups of vertebrates, but are also amongst the most threatened. With contributions from leaders in the field, this is the first assessment of the global state of freshwater fish diversity, synthesising the opportunities, challenges and barriers facing the conservation of freshwater fish biodiversity.

The book includes the first global assessment of the number, type and distribution of threatened freshwater fish species, discussing the features of freshwater fish biology and ecology that render so many species vulnerable to extinction. Introductory chapters on why freshwater fish are so sensitive to environmental change and disturbance lead into chapters providing detailed reviews of the key threatening processes and potential solutions. A concluding chapter summarises the key issues and looks to the future for opportunities and challenges for the conservation and management of freshwater fish.

GERARD P. CLOSS is Associate Professor and Head of the Department of Zoology at the University of Otago. He has published over 100 papers exploring various freshwater ecological topics and has advised government agencies on catchment and river protection, conservation of native fish species, and management of introduced fish.

MARTIN KRKOSEK is Assistant Professor and Alfred P. Sloan Research Fellow in Ocean Science at the University of Toronto. His research focuses on the ecology, epidemiology and conservation of fish populations and coastal ecosystems, and is represented in over 50 scientific papers. He is also an advisor to government agencies and conservation organisations.

JULIAN D. OLDEN is H. Mason Keeler Endowed Professor in the School of Aquatic and Fishery Sciences, University of Washington, and Adjunct Research Fellow at the Australian Rivers Institute, Griffith University. His research focuses on the conservation of freshwater ecosystems and is represented by over 160 scientific papers.

Conservation Biology

This series aims to present internationally significant contributions from leading researchers in particularly active areas of conservation biology. It focuses on topics where basic theory is strong and where there are pressing problems for practical conservation. The series includes both authored and edited volumes and adopts a direct and accessible style targeted at interested undergraduates, postgraduates, researchers and university teachers.

Conservation of Freshwater Fishes

Edited by

GERARD P. CLOSS
University of Otago, New Zealand

MARTIN KRKOSEK
University of Toronto, Canada

JULIAN D. OLDEN
University of Washington, USA

CAMBRIDGE
UNIVERSITY PRESS

CAMBRIDGE
UNIVERSITY PRESS

University Printing House, Cambridge CB2 8BS, United Kingdom

Cambridge University Press is part of the University of Cambridge.

It furthers the University's mission by disseminating knowledge in the pursuit of education, learning and research at the highest international levels of excellence.

www.cambridge.org
Information on this title: www.cambridge.org/9781107040113

© Cambridge University Press 2016

First published 2016

Printed in the United Kingdom by TJ International Ltd. Padstow Cornwall

A catalogue record for this publication is available from the British Library

Library of Congress Cataloguing in Publication data
Conservation of freshwater fishes / edited by Gerard P. Closs, University of Otago,
New Zealand, Martin Krkosek, University of Toronto, Canada, Julian D. Olden,
University of Washington, USA, and Griffith University, Australia.
  pages cm
Includes bibliographical references and index.
ISBN 978-1-107-04011-3 (hbk) – ISBN 978-1-107-61609-7 (pbk)
1. Freshwater fishes–Conservation. I. Closs, Gerry. II. Krkosek, Martin.
III. Olden, Julian D.
QL624.C658 2015
597.176–dc23   2015023048
ISBN 978-1-107-04011-3 Hardback
ISBN 978-1-107-61609-7 Paperback

Contents

Contributors

PAUL L. ANGERMEIER U.S. Geological Survey, Virginia Cooperative Fish and Wildlife Research Unit, Virginia Tech

ROBERT ARLINGHAUS Faculty of Agriculture and Horticulture, Humboldt University of Berlin and Department of Biology and Ecology of Fishes, Leibniz Institute of Freshwater Ecology and Inland Fisheries

STEPHEN R. BALCOMBE Australian Rivers Institute, Griffith University

PIER GIORGIO BIANCO Department of Biology, University of Naples

EVAN CHILDRESS Center for Limnology, University of Wisconsin

GERARD P. CLOSS Department of Zoology, University of Otago

MARIA J. COLLARES-PEREIRA Centre for Ecology, Evolution and Environmental Changes, University of Lisbon

STEVEN J. COOKE Department of Biology, Carleton University

EDWIN AGUDELO CÓRDOBA Instituto Amazónico de Investigaciones Científicas SINCHI

IAN G. COWX Hull International Fisheries Institute, University of Hull

WILLIAM R. T. DARWALL Global Species Programme, IUCN (International Union for the Conservation of Nature)

JOHN M. DETTMERS Great Lakes Fishery Commission

JAAKKO ERKINARO Natural Resources Institute, Finland

JOHN M. EPIFANIO Illinois Natural History Survey

JÖRG FREYHOF German Centre for Integrative Biodiversity Research (iDiv)

KEITH B. GIDO Division of Biology, Kansas State University

ELLEN J. HAMANN Center for Limnology, University of Wisconsin

JANI HEINO Finnish Environment Institute, Natural Environment Centre, University of Oulu

VIRGILIO HERMOSO Australian Rivers Institute, Griffith University

J. DEREK HOGAN Department of Life Sciences, Texas A&M University – Corpus Christi

PAUL HUMPHRIES School of Environmental Sciences, Charles Sturt University

ARI HUUSKO Natural Resources Institute, Finland

STEPHANIE R. JANUCHOWSKI-HARTLEY Center for Limnology, University of Wisconsin and School of Life Sciences, Texas A&M University – Corpus Christi

MARK J. KENNARD National Environmental Research Program – Northern Australia Hub and Australian Rivers Institute, Griffith University

VALERIO KETMAIER Department of Biology and Biotechnology, University of Rome; Institute of Biochemistry and Biology, University of Potsdam, and Leibniz Institute for Research in Evolution and Biodiversity, Humboldt University

AARON A. KONING Center for Limnology, University of Wisconsin

MARTIN KRKOSEK Department of Ecology and Evolutionary Biology, University of Toronto

KATHARINA LANGE Center for Ecology, Evolution and Biochemistry, Department of Fish Ecology and Evolution, EAWAG Swiss Federal Institute of Aquatic Science and Technology

NICOLAS W. R. LAPOINTE The Nature Conservancy of Canada

SIMON LINKE National Environmental Research Program – Northern Australia Hub and Australian Rivers Institute, Griffith University

MISKA LUOTO Department of Geosciences and Geography, University of Helsinki

MICHAEL P. MARCHETTI Department of Biology, St. Mary's College of California

CHRISTOPH D. MATTHAEI Department of Zoology, University of Otago

JENIFER K. MCINTYRE Washington State University

PETER B. MCINTYRE Center for Limnology, University of Wisconsin

THOMAS M. NEESON Center for Limnology, University of Wisconsin

VIVIAN M. NGUYEN Department of Biology, Carleton University

DANIEL L. OELE Center for Limnology, University of Wisconsin

JULIAN D. OLDEN School of Aquatic and Fishery Sciences, University of Washington

JOSHUAH S. PERKIN Department of Biology, Tennessee Tech University

MARCELA PORTOCARRERO-AYA Instituto de Investigación de Recursos Naturales Alexander von Humboldt

ROBERT POULIN Department of Zoology, University of Otago

BRENDA M. PRACHEIL Environmental Sciences Division, Oak Ridge National Laboratory

BRADLEY J. PUSEY Centre of Excellence in Natural Resource Management, University of Western Australia

MICHAEL C. QUIST Idaho Cooperative Fish and Wildlife Research Unit, Department of Fish and Wildlife Sciences, University of Idaho

RAJEEV RAGHAVAN Conservation Research Group, Department of Fisheries, St. Albert's College

CATHERINE REIDY LIERMANN Center for Limnology, University of Wisconsin

JOHN S. RICHARDSON Department of Forest & Conservation Sciences, University of British Columbia

NATHANIEL L. SCHOLZ NOAA Fisheries, Northwest Fisheries Science Center

PAUL H. SKELTON South African Institute for Aquatic Biodiversity and Department of Ichthyology and Fisheries Science, Rhodes University

THOMAS F. TURNER Museum of Southwestern Biology, University of New Mexico

DENIS TWEDDLE South African Institute of Aquatic Biodiversity

M. JAKE VANDER ZANDEN Center for Limnology, University of Wisconsin

ROBIN S. WAPLES NOAA Fisheries, Northwest Fisheries Science Center

OLAF L. F. WEYL South African Institute of Aquatic Biodiversity

JAMES E. WHITNEY Department of Biology, Pittsburgh State University

KIRK O. WINEMILLER Department of Wildlife and Fisheries Sciences, Texas A&M University

MARK S. WIPFLI US Geological Survey, Alaska Cooperative Fish & Wildlife Research Unit, Institute of Arctic Biology, University of Alaska Fairbanks

Preface

Do unto those downstream as you would have those upstream do unto you.

Wendell Berry

Early naturalists Comte de Buffon, Alfred Russel Wallace and Charles Darwin inspired famous ideas of change in life, but also revealed the range of life distributed across the Earth. Some species are known from only a single location, whereas others occur high to low, east to west, or poles to the tropics. Freshwater fishes exemplify this phenomenon. High diversity and endemism stem largely from the fact that fresh waters are embedded within a terrestrial landscape that limits large-distance movement among drainage basins. This constrained geography is at least partially responsible for the fantastic diversity of freshwater fishes seen around the world, and the magnificent differences in regional faunas ranging from the Neotropics to the Palearctic.

Equally impressive is how humanity's migrations across, and subsequent modification of, the landscape have had innumerable effects on the many organisms with which we share this world. Quite simply, the human enterprise is nothing short of gigantic, and as a result, freshwaters are subject to a plethora of anthropogenic threats, including habitat loss and modification, dams and fragmentation, climate change, overexploitation, pollution and the spread of invasive species. It is quite telling that humans now appropriate over half the available freshwater run-off, reservoirs trap a quarter of the global sediment load before it reaches the oceans, river systems have been fragmented by over one million dams globally, and many inland waters are polluted and their fisheries persistently over-harvested. The damage of these and many other insults is that freshwater fishes are among the most imperilled faunas worldwide.

So, have fishes had their chips? Scientific experts respond in the chapters that follow with an empathetic and resounding 'NO'! Despite the enormity of the challenge, the pages of this book speak to hope and not despair. No matter how daunting the prospects, the Blue Planet crisis merely provides more opportunities for conservation success. We hope that the writings in this book inspire both current and future generations of scientists to roll up their sleeves and get to work to ensure a sustainable future for our freshwater fishes.

Books such as this one become a reality only with the support of many people. We would like to thank Alan Crowden, Megan Waddington and the staff of Cambridge University Press for their encouragement throughout the publication process, and acknowledge the fantastic efforts of Susan Harvey and Gerry Barbalich in formatting the final chapters. We thank Jeremy Monroe and Dave Herasimtschuk from Freshwaters Illustrated for providing an inspirational photo for the book cover. Numerous anonymous referees provided critical feedback on the chapters; to them we extend our deepest thanks. Our thanks also go to Matt Jarvis for his excellent work on the index.

All of us have had the benefit of talkin' fish with many colleagues over the years and we wish to especially thank the following individuals for generously sharing their ideas and perspectives on a range of topics: Paul Angermeier, Tobias Bickel, Bruno David, Kurt Fausch, Keith Gido, Eric Hansen, Andy Hicks, Don Jackson, Peter Jones, Mark Kennard, Esben Kristensen, Bob McDowall, Jeremy Monroe, Peter Moyle, LeRoy Poff, Rick Stoffels, Jake Vander Zanden, Manna Warburton and Kirk Winemiller.

Finally, we offer thanks to our home institutions for providing logistic and financial support while we worked on this book. In particular, we acknowledge funding from the Performance Based Research Fund (GC), the H. Mason Keeler Endowed Professorship from School of Aquatic and Fishery Sciences – University of Washington (JDO) and the Natural Sciences and Engineering Research Council of Canada (MK).

Gerry Closs, Julian Olden and Martin Krkosek

Lost fishes, who is counting? The extent of the threat to freshwater fish biodiversity

WILLIAM R. T. DARWALL
AND JÖRG FREYHOF

1.1 INTRODUCTION

Freshwater rivers, lakes and wetlands are among the most threatened ecosystems on the planet, facing growing pressures from an expanding human population and increased socioeconomic development (Ormerod *et al.*, 2010; Vörösmarty *et al.*, 2010; Carpenter *et al.*, 2011). This pressure on freshwater ecosystems is accompanied by correspondingly high levels of threat to freshwater biodiversity (Dudgeon *et al.*, 2006; WWF, 2010; Thieme *et al.*, 2011; Collen *et al.*, 2014), as is clearly demonstrated by the high species extinction rates and levels of threat recorded on the IUCN Red List of Threatened Species (www.iucnredlist.org), hereafter referred to as the IUCN Red List. North American freshwater bivalves are, for example, notable for having the greatest proportion of extinct species worldwide. In Europe, freshwater species top the IUCN Red List with the highest proportion of threatened species. Remarkable twenty-first-century extinctions such as the baiji (*Lipotes vexillifer*), the golden toad (*Incilius periglenes*) and the Alaotra grebe (*Tachybaptus rufolavatus*), just to name a few, are all freshwater species. However, no-one has yet comprehensively assessed the level of threat and extinction rates for freshwater fishes at the global scale – and it is likely that many fishes are disappearing without record.

The twenty-first century is a critical time for the future of freshwater fishes. Human actions have a serious impact on freshwater ecosystems around the world and the freshwater fish species face ever increasing risks. Unless actions are taken rapidly to reduce the multiple threats facing freshwater fishes, many species will be lost. A sobering example,

Conservation of Freshwater Fishes, eds G. P. Closs, M. Krkosek and J. D. Olden. Published by Cambridge University Press. © Cambridge University Press 2016.

which serves well to demonstrate the severity of the threat, is the perilous state of the world's sturgeons and paddlefishes (Acipenseriformes). Sturgeon have survived on this Earth for 250 million years, but now face the serious possibility of becoming extinct in this century as a direct result of human activities. Illegal fishing, overfishing, obstructions to migratory routes and pollution has resulted in 23 of the 27 sturgeon species being assessed as threatened on the IUCN Red List. Of these, 17 species are Critically Endangered and four are possibly Extinct, including the Chinese paddlefish (*Psephurus gladius*), the world's longest freshwater fish for which only two adult specimens (both females) have been recorded since 2002. Human exploitation of freshwater ecosystems and the fishes within them must operate within sustainable limits, and critical sites for freshwater species must be identified and protected before it is too late for many species. Much greater attention to conservation action on the ground is required if we are to improve the status of freshwater fishes.

At the time of writing there were 15,750 described species of freshwater fish (Eschmeyer & Fong, unpubl. data), where the definition of 'freshwater' refers to those species that spend a significant part of their life history in freshwater. Diadromous species, which undergo obligatory migrations between marine and freshwater habitats, are included under this definition. This number of species represents 48% of the global diversity of all described fishes (Eschmeyer & Fong, 2013) and approximately 25% of all vertebrate diversity. Freshwater fishes fulfil a wide range of ecological roles essential to the long-term functioning of freshwater ecosystems across the world. They are also of considerable interest to scientists, hobbyists and recreational fishers and, most significantly, provide 33% of the world's small-scale fish catch and employment for an estimated 60 million people (UNEP, 2010). The supply of freshwater fish is critically important for human nutrition, especially in Africa and parts of Asia. Over 200 million of Africa's 1 billion people consume fish and nearly half of this comes from inland fisheries (UNEP, 2010). However, despite the great significance of freshwater fishes, there has been no globally comprehensive assessment of their conservation status and of the major threats to their survival. IUCN (International Union for the Conservation of Nature) and its partners have been actively working to address this information gap since 2002 by collating available information to map species' distributions and assess extinction risk. The work is ongoing and this chapter presents an assessment of the state of knowledge and condition of global freshwater fish diversity in 2013.

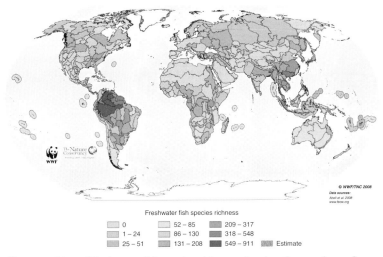

Figure 1.1 Map of freshwater fish species richness showing the numbers of species present in each ecoregion.

Source: Abell *et al.* (2008); downloaded 21 June 2014 from: www.feow.org/maps/biodiversity/freshwater_fish_species_richness

1.2 GLOBAL DISTRIBUTION OF FRESHWATER FISHES

Our knowledge of the global distribution of freshwater fish species is improving but is still incomplete, with an average of two new freshwater fish species described every three days between 2002 and 2012 (Nicolas Bailey, pers. comm.). Many regions are still poorly surveyed. For example, in the 9 years following publication of the *Checklist of the Freshwater Fishes of South and Central America* (CLOFFSCA) (Reis *et al.*, 2003), an average of one new species was described every 3.5 days, and from Mediterranean freshwaters alone, 99 new species have been described since the year 2000 (representing 22% of the fauna recognised in 2000) (Geiger *et al.*, 2014). The global pattern of distributions is well described at the scales of biogeographic regions (Berra, 2001; Lévêque *et al.*, 2008), freshwater ecoregions (Abell *et al.*, 2008) (Figure 1.1) and river basins (Oberdorff *et al.*, 2011). At the ecoregion scale, more than 6900 of the 13,400 species considered as freshwater-dependent are endemic to a single ecoregion (Abell *et al.*, 2008). Following Abell *et al.* (2008) and previous global assessments such as McAllister *et al.* (1997), Groombridge and Jenkins (1998) and Revenga and Kura (2003), outstanding areas for both fish species richness and endemism include

Box 1.1. The IUCN Red List Categories and Criteria

The IUCN Red List Categories and Criteria are intended to be an easily and widely understood system for classifying species risk of global extinction. Species are assigned to one of nine possible categories: Extinct (EX); Extinct in the Wild (EW); Critically Endangered (CR); Endangered (EN); Vulnerable (VU); Near Threatened (NT); Least Concern (LC); Data Deficient (DD), or; Not Evaluated (NE). Species assessed as Critically Endangered, Endangered, or Vulnerable are considered to be Threatened.

The percentage of the freshwater fishes threatened was calculated as a mid-point estimate (i.e. assuming the DD species are threatened in the same proportion as the species for which there are sufficient data) as follows:

$$\% \text{ threat} = \frac{(CR + EN + VU)}{(\text{total assessed - EX } - \text{ EW } - \text{ DD})} \times 100$$

large portions of Africa's Congo basin, the Southern Gulf of Guinea drainages and Lakes Malawi, Tanganyika, and Victoria; Asia's Zhu Jiang (Pearl River) basin and neighbouring systems; and large portions of South America's Amazon and Orinoco basins. Other areas confirmed for globally high species richness include Asia's Brahmaputra, Ganges and Yangtze basins, as well as large portions of the Mekong, Chao Phraya and Sitang and Irrawaddy; Africa's lower Guinea; and South America's Paraná and Orinoco. When species richness is adjusted for ecoregion area, additional systems such as the Tennessee, Cumberland, Mobile Bay, Apalachicola and Ozark highlands in the Southeastern United States; portions of Africa's Niger River Basin; the islands of New Caledonia, Vanuatu and Fiji; China's Hainan Island; and large parts of Sumatra and Borneo, among many other areas, are also noteworthy (Abell *et al.*, 2008).

1.3 STATE OF KNOWLEDGE ON THE EXTENT AND DISTRIBUTION OF THREATS

The IUCN Red List provides the only globally consistent measure of threat to individual species, assessed as the risk of an individual species going extinct. Each species' assessment for the IUCN Red List requires: (i) a map of the species distributions and (ii) an assessment of its risk of global extinction (see Box 1.1). As of 2013, 46% (7301 species) of all described species of freshwater fishes had been mapped to river

Conservation Status of Freshwater Fishes

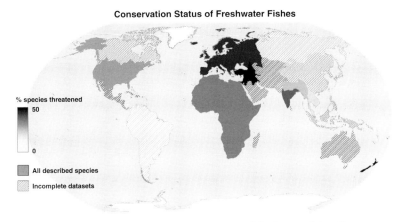

Figure 1.2 Map of global progress in assessment of freshwater fishes for the IUCN Red List of Threatened Species. The level of threat is shown as the percentage of species assessed that are threatened (excluding all species assessed as 'Extinct', 'Extinct in the Wild' or 'Data Deficient'). Those regions depicted with solid shading have been comprehensively assessed, meaning that all described species have been assessed. For those regions with hatched shading, not all species have been assessed such that the overall level of threat may not be representative of all species, representing only a small proportion of species in some regions.

and lake sub-basins, major threats identified and their risk of extinction assessed (IUCN, 2013), according to the IUCN Red List Categories and Criteria (Mace & Lande, 1991) (Figure 1.2). Comprehensive assessments have been completed for all described species in many regions including: Europe, Africa, India, Indo-Burma, United States of America, New Zealand, Oceania and the Middle East. For other regions, only a subset of species have been assessed, leaving significant information gaps in particular for South America, large parts of Northern and Eastern Asia and Indonesia. In these regions, a number of species have been assessed representing the combined results of a random sample (Collen *et al.*, 2014) and ad hoc assessments. As such, the sample of species is potentially biased by a tendency to first assess those species thought most likely to be at risk. For these regions we also draw on additional information from National Red List Assessments and other sources. However, National Assessments often differ in their application of the IUCN Red List Categories and Criteria such that direct regional comparison of numbers of threatened species should be made with caution.

1.3.1 Extent of threat

Of the 7300 freshwater fish species assessed for the IUCN Red List by 2013, 31% are threatened with extinction (classified as Critically Endangered, Endangered, or Vulnerable). This percentage figure excludes the 1571 species for which insufficient information was available to assess the extinction risk (classified as Data Deficient) and the 69 species assessed as either Extinct or Extinct in the Wild. This level of threat is relatively high as compared with other globally comprehensive assessments, which find 13% of birds, 20% of mammals and 34% of amphibians threatened (IUCN, 2013).

Burkhead (2012) applied his own criteria to estimate the background extinction rate for freshwater fishes in the twentieth century as nearly twice that of other vertebrate groups. He used an extrapolation from the North American proportion of extinct species to suggest that around 410 species, 3% of all described freshwater fishes, are extinct globally. Alternatively, Harrison and Stiassny (1999) estimated 245 species (2% of freshwater species) to be potentially extinct or seriously threatened, and Freyhof and Brooks (2011) estimated 2% of European species as being extinct. However, it still remains difficult to definitively assess a species as being extinct. For example, species not recorded for many decades are occasionally rediscovered in poorly known areas. As a result, many fish experts hesitate to declare a species as extinct.

1.3.2 Causes of threat

Why are so many freshwater fishes threatened? The major threats to freshwater ecosystems as summarised by Thieme et al. (2011) include: (i) the position of freshwater ecosystems in the landscape as sinks for terrestrial run-off that includes pollutants, pesticides and heavy sediment loads; (ii) intense competition with humans for use of water for various purposes such as hydropower, irrigation, livelihoods and transport; (iii) regionally intense harvest pressures; and (iv) the high degree of connectivity within and between freshwater ecosystems that often facilitates rapid and widespread dispersal of pollutants and invasive alien species. This summary is supported by an analysis of the threat categories assigned to all freshwater fish on the IUCN Red List reflecting the nature and scale of past and present development activities, and highlighting habitat degradation and loss, water abstraction and flow modifications, invasive species, pollution and overexploitation as major threats (Figure 1.3). The inherent features of freshwater fishes

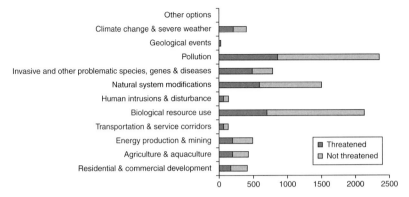

Figure 1.3 The numbers of species classified to each of the main categories of threat as coded in the IUCN Red List.

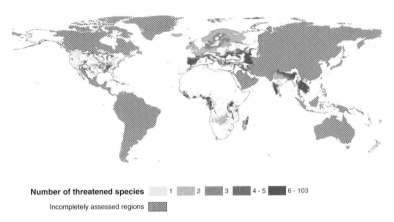

Figure 1.4 Map showing the distribution of all freshwater fish species assessed as threatened showing the number of threatened species within individual river and lake subcatchments. Data are presented for those regions of the world where all described species have been assessed.

that make them so susceptible to these types of threat are addressed in Chapter 2.

1.3.3 Regional comparisons

Patterns of richness for threatened species are displayed for all comprehensively assessed regions (Figure 1.4). The extent and distribution of threats varies significantly at the regional scale, thus warranting an overview of each region covering patterns of species distributions and the levels and types of threat faced.

1.3.3.1 Europe

The most recent assessment of European freshwater fishes identifies 531 valid species native to the region (Freyhof & Brooks, 2011). Almost 42% (200 species) are threatened with global extinction and 15 species are globally Extinct or Extinct in the Wild (Kottelat & Freyhof, 2007). The geographic distribution of threatened species is shown in Figure 1.4.

The 12 species known to be Extinct in Europe include 6 species of *Coregonus* and 2 *Salvelinus*. All were strongly impacted by modifications to their lake environments. The Lake Constance whitefish *Coregonus gutturosus* and the charr *Salvelinus profundus* were endemic to Lake Constance where they formed the basis of a commercial fishery until the 1960s, but disappeared as the lake became highly eutrophic. Eutrophication of Central European lakes is believed to have wiped out a significant proportion of the endemic fish fauna (Vonlanthen *et al.*, 2012).

Of particular concern are the fishes of Mediterranean rivers that are subject to water abstraction, dam construction and alien species invasion. This region is inhabited by many endemic species, most of which are highly vulnerable to a range of threats. Despite the Mediterranean being highlighted as a global biodiversity hotspot (Myers, 1988; Mittermeier *et al.*, 1999), its freshwater ecosystems receive limited conservation support. Another highly threatened group are the diadromous migrants, especially salmonids and sturgeons. Six of the seven sturgeons have been assessed as Critically Endangered, an optimistic view given that many populations only persist through artificial stocking.

The single most important threat to European freshwater fishes is the over-abstraction of ground and surface water from rivers and lakes. In the Mediterranean region this is a major concern for many species, especially as the extraction is often illegal and poorly policed. This situation is becoming increasingly serious as drought events increase and the rates of water consumption rise. Many of those countries already suffering from limited water supplies also support high numbers of endemic and threatened species.

Invasive species are recorded as a threat to about 55% of species in Europe. Kottelat and Freyhof (2007) recorded 28 established alien species introduced from outside of Europe and an additional 77 species native to Europe but expanding beyond their natural range. Once established, their spread is often rapid, as demonstrated for the topmouth gudgeon (*Pseudorasbora parva*), a Chinese cyprinid that spread across much of Europe in 40 years.

The increasing numbers of dams across Europe are also of concern as water flows are disrupted, habitats transformed and fish migration routes disrupted. Few rivers in Europe remain free of hydropower or irrigation dams. In many cases migratory species have lost all former spawning areas. For example, the beloribitsa (*Stenodus leucichthys*), a migratory species of the Caspian Sea, is now considered to be Extinct in the Wild as it is unable to complete its 3000-km upriver migration into the Volga to spawn following construction of the Volgograd dam in 1959 (Freyhof & Kottelat, 2008).

1.3.3.2 *Western Asia*

Western Asia, or the Middle East, is biogeographically similar to Europe with many fish genera shared between the two regions and the conservation concerns are similar to those for the Mediterranean region. Western Asia covers the countries south of the Caucasus such as Georgia and some adjacent parts of Russia, Azerbaijan, Armenia, Turkey (only the Asian part), Cyprus, Syria, Jordan, Lebanon, Israel, Iraq, Kuwait, the countries of the Gulf and Iran. The freshwater fish diversity of Western Asia has never been reviewed comprehensively and only rough estimates of species numbers are known. Approximately 450 described species occur, and of these, around 30 established species are alien to the region. Some of these alien species are from within the region but have been translocated outside of their natural ranges. Western Asia is still poorly explored when it comes to freshwater biodiversity including fishes; at least 100 species, maybe considerably more, are awaiting description. Of an estimated 520 native species (including both described and undescribed species), 320 (approximately 60%) are endemic. The centres of endemism are Western and Central Anatolia and Central Iran, where many species with highly restricted ranges occur, particularly within the genera *Aphanius*, *Pseudophoxinus* and *Cobitis*. The largest river system, the Euphrates and Tigris drainage, is inhabited by an almost entirely endemic fish fauna of around 100 species including 5 species of cave fish and one of the largest cypriniform fishes in the world, the pike barbel, (*Luciobarbus esocinus*) growing to more than 2 m in length. Preliminary results of an assessment for the IUCN Red List clearly demonstrate that the fish fauna is subject to high threat levels across this mostly arid region that supports a dense human population. At least 13 species are already thought to be extinct. The habitats of the long-spine bream

(*Acanthobrama centisquama*) (Lake Amiq, Turkey) and *Mirogrex hulensis* (Lake Hula, Israel) were both drained. Introduced alien species caused the extinction of a number of other lacustrine endemics in Anatolia.

The main threat to freshwater fishes, as in other arid or semi-arid areas across the world, stems from the intense competition between people and biodiversity for a limited water resource. Across the region, dams for irrigation and hydropower production have been constructed, and many more are planned or are under construction. In many, possibly the majority, of cases, no water flows out of the reservoirs during dry periods. In large rivers such as the Euphrates, large reservoirs lead to significant loss of water through evaporation which, when combined with water withdrawals for irrigation, leave a considerably reduced supply downstream. The impacts of dams are not only restricted to the dry parts of Western Asia. For example, almost all rivers flowing to the Southern Caspian Sea are blocked shortly before they enter the sea such that all populations of migratory fishes are now almost entirely conservation-dependent. For example, sturgeons (*Huso huso, Acipenser* spp.) are no longer able to spawn naturally. Even highly valued migratory fishes such as the Black Sea roach (*Rutilus frisii*) rely almost entirely on artificial breeding. Below the dams, poaching levels are so high that the rivers from which broodstock are taken for artificial reproduction are now protected by armed guards.

Surface and ground water are abstracted in huge quantities throughout the dry parts of Western Asia and abstraction is rarely sustainable. Recently, Voss *et al.* (2013) found that large parts of Western Asia are losing groundwater reserves at an alarming rate. The overall rate of freshwater loss from the region during the seven-year study period is one of the highest in the world, second only to water loss in India. Climate change scenarios that predict reduced rainfall in this region (Chenoweth *et al.*, 2011) suggest a bleak future for freshwater fishes in large parts of Western Asia. Many areas have already dried out and many fish species, once widespread, are now restricted to small refuges. For example, the once extensive spring areas at Ras al Ain in Northern Syria have almost completely dried out, as has the famous spring of the Barada River near Damascus, along with almost the entire Damascus hydrological basin.

1.3.3.3 Continental Africa

Africa has a diverse fish fauna of almost 3000 described species of which 27% are threatened (Snoeks *et al.*, 2011). The distribution of threatened freshwater fish species (Figure 1.4) is largely focused in a band that runs along the coast of Western Africa and the Lower Guinea province from Senegal to Republic of Congo, throughout the Zambezi/Okavango basins in the northern part of southern Africa, through the river basins and lakes of the Rift Valley of eastern Africa, also including some of the coastal basins of eastern Africa, and some basins in the eastern and southern parts of South Africa. There are pockets of threatened species along the Uele River in central Africa, in the region of Lake Tana in northeastern Africa, and in the Atlantic and Mediterranean northwest Africa freshwater ecoregions in the Maghreb region of northern Africa. Three species are listed as being Extinct (*Aplocheilichthys* sp. Nov. 'Naivasha'; *Barbus microbarbis*; *Salmo pallaryi*); all three had restricted distributions and their apparent extinctions are attributed, at least in part, to introductions of alien species.

The following account of regional patterns of threat is based on that given by Snoeks *et al.* (2011). The highest number of threatened species occurs in Lake Malawi, which is well known for its exceptionally rich and unique fish fauna (e.g. Fryer & Iles, 1972; Ribbink, 1988; Turner, 1994, 1996; Côté & Reynolds, 1998; Snoeks, 2004). With an estimated 500–1000 fish species, most of which are endemic to the lake, Lake Malawi has the most species-rich fish fauna of any lake globally (Snoeks, 2000). Approximately 99% of the fish species are cichlids, of which only some 300 species have been formally described, the rest undescribed or identified by informal names (Snoeks, 2000). Of these cichlid species, 99% are endemic to the lake. Of the 358 species assessed for the IUCN Red List, 105 are recorded as threatened (29% of the total number of species assessed for this lake). In this case it is mainly the highly restricted ranges and low numbers of offspring of the cichlids which make them vulnerable to extinction. The main threat is increasing sedimentation along the lake margins where many of these species live, sometimes restricted to only a few hundred metres of shoreline.

Lake Victoria, with an estimated 545 species, a large proportion of which are endemic to the lake (Snoeks, 2000), has the next highest number of threatened species (81 species; 44% of the assessed species in the lake), resulting from a combination of well-documented threats, including: introduction of the piscivorous predator, Nile perch (*Lates niloticus*), and the alien invasive water hyacinth (*Eichornia crassipes*), which has reduced light and oxygen levels in the lake's waters; overfishing and use of fish poisons; and habitat deterioration and eutrophication resulting from increasing lakeside agriculture, urbanisation and deforestation. However, the full extent of the threat to these species is still not fully understood and extensive surveys are required throughout the lake to fully assess the conservation status of all cichlid species present.

Lake Tanganyika also supports an exceptional diversity of fish species (Snoeks, 2000). Of the estimated 325 fish species in the lake, around 250 are endemic cichlids (Snoeks, 2000); 12 of the 245 species assessed are threatened (5% of the total assessed). The proportion of threatened species (relative to the total number) is lower than for Lakes Victoria and Malawi because the threats are generally more localised (Cohen *et al.*, 1996) (especially compared to Lake Victoria), and the Tanganyikan species tend to have a wider distribution, extending into less-populated areas where there are fewer threats.

Outside the African Rift Valley, high numbers of threatened species occur in and around the rapids in the Lower Congo and in the coastal basins of Western Africa and Lower Guinea. The Lower Congo has up to 24 threatened species just upstream of Inga, and has one Critically Endangered species of cichlid, *Teleogramma brichardi*, apparently restricted to the Kinsuka rapids near Kinshasa. This species is increasingly threatened by the impacts of urbanisation at Kinshasa and Brazzaville (Stiassny *et al.*, 2011). Several other Endangered species are recorded, especially in the vicinity of Malebo Pool.

At least 13 threatened species are found in the delta region of the Niger River, including two Critically Endangered species that are threatened by the impacts of oil exploration, the distichodontid *Neolebias powelli* and the killifish *Fundulopanchax powelli*. Six threatened species are recorded nearby in the species-rich lower Ogun River at Lagos lagoon. These species are threatened mainly by deforestation (e.g. *Brycinus brevis*) as well as agricultural and urban development (e.g. the mormyrid *Marcusenius brucii*); however, the small red-eyed tetra, an alestid, *Arnoldichthys spilopterus*, is threatened by an extensive harvesting for the aquarium fish trade. Just north of the Niger Delta, in a tributary of the Benue River on the Bauchi plateau, the cyprinid *Garra trewavasae* is Critically Endangered due to the impacts of tin mining.

At least 10 threatened species are found in the coastal drainages of Sierra Leone and Liberia; these include some Critically Endangered species (e.g. *Labeo curriei, Barbus carcharhinoides, Epiplatys ruhkopfi, Tilapia cessiana* and *T. coffea*), and several Endangered species, especially in the vicinity of the St. Paul and Lofa rivers. These species are threatened by habitat degradation caused by deforestation and mining. In the Konkouré River in Guinea, the catfish *Synodontis dekimpei* is Critically Endangered for the same reasons. In the Fouta-Djalon ecoregion of

Guinea, the Kindia killifish (*Scriptaphyosemion cauveti*), a Critically Endangered species from a tributary to the Kolenté River, is threatened by expansion of the nearby city of Kindia.

Twenty-six threatened species are recorded from the Western Equatorial Crater Lakes freshwater ecoregion and the river drainages nearby, at the border of Southwest Cameroon and Nigeria. Many of these species are Endangered or Critically Endangered, and the majority are cichlids endemic to crater lakes, although there are also several killifishes and some cyprinids and catfishes. Within Lower Guinea, high numbers of threatened species (12–14 species) are found in sections of the Ivindo, Bouniandjé and Nouna (a tributary to the upper Ivindo) systems.

In Eastern and Southern Africa, the number of threatened species is low for most basins (excluding the Rift Valley lakes). The greatest numbers of threatened species in eastern Africa are found in the small Ruvu River, a coastal basin near Dar es Salaam that harbours nine such species. Although other basins in eastern Africa have fewer numbers of threatened species, some include one or more Critically Endangered species, most of which are cichlids.

In Southern Africa, the Olifants River in the southwest has the greatest number of threatened species, with seven species (70–75% of all assessed species in the Olifants basin), including two Critically Endangered species, the Twee River redfin (*Barbus erubescens*) and an undescribed species of *Pseudobarbus*. Both are threatened by introduced invasive species, plus deterioration in habitat and water abstraction caused by intensive farming. Slightly to the south, in the Tradou catchment of the Breede River system, the Tradou redfin (*Pseudobarbus burchelli*) is Critically Endangered due to introduced species and pollution. A number of other species are assessed as Endangered in the southern part of the country and in Lesotho (e.g. the small scale redfin, *Pseudobarbus asper* and the Maloti minnow, *Pseudobarbus quathlambae*), where they face similar threats to the above-mentioned species. South Africa and Mozambique also harbour some undescribed but Critically Endangered species of *Pseudobarbus*, *Kneria* and *Barbus*. Two Critically Endangered species are found in the Karstveld Sinkholes ecoregion of central Namibia: one is *Tilapia guinasana*, which occurs naturally only in Lake Guinas where it is threatened by groundwater extraction, as well as competition and predation from, as well as possible hybridisation with, introduced tilapiines; and the other is the cave catfish (*Clarias cavernicola*) known only from a single tiny lake (18 m × 2.5 m) in the Aigamas Cave, near the town of Otavi, which is threatened by water abstraction and possibly collections made for the aquarist trade. There are three Critically Endangered species in the Zambezi River basin. In the upper Zambezi floodplain ecoregion the banded neolebias (*Neolebias lozii*) is restricted to the Sianda River that has been canalised to aid drainage for agriculture. Finally, unlike most other Critically Endangered species with restricted distributions, the Kariba tilapia (*Oreochromis mortimeri*) is widely distributed in the Middle Zambezi-Luangwa ecoregion and parts of the Zambezian Highveld ecoregion. This species is threatened by the widespread introduction of the Nile tilapia (*Oreochromis niloticus*), which is displacing it throughout much of its range.

1.3.3.4 Madagascar

An assessment of the 98 endemic fish species in 2004 (IUCN, 2004) found almost 85% of species for which we have sufficient data (55 species) to be threatened. This represents the highest level of extinction risk yet recorded for a regional assemblage of fishes. Given the rapid rate of environmental degradation in the period since the 2004 assessment, a reassessment is considered a priority because the situation is likely to have deteriorated even further. As an example, the Mangarahara cichlid (*Ptychochromis insolitus*) represents a species, yet to be formally assessed for the IUCN Red List, driven to the brink of extinction. The species was only recently described (Stiassny & Sparks, 2006) and was until very recently thought to be Extinct in the Wild because its only known habitat dried out following construction of a dam on the Mangarahara River. The recent discovery of a small population in the wild brings new hope for the survival of this species for which there were thought to be only three known captive individuals.

In addition to dams, the main threats identified include: (i) deforestation of associated catchments; (ii) introduced alien invasive fish species; and (iii) conversion of wetland habitats for farming, particularly for rice fields. A number of taxa (e.g. *Pantanodon* spp.) are dependent upon wetland habitats such as marshlands, and the loss of these habitats has had a clearly negative impact on these dependent taxa.

1.3.3.5 Siberia and Central Asia

While Siberia is a vast area dominated by some of the largest rivers of the world, freshwater fish diversity is remarkably low, as might be expected for such cold and high northern-latitude rivers. An estimated 275 species of freshwater fishes are native to Siberia. An estimated 60 species are found in Lake Baikal, 30 of which comprise the large and very diverse adaptive radiation of sculpins (Cottidae) endemic to the lake. In contrast, Central Asia hosts huge endorheic basins including Lake Balkhash, Lakes Issyk Kul and Ala-Kul, and the Aral Sea. All these basins include rivers, usually with a largely mountainous flow regime, and large, often brackish, lakes. The native fish fauna of the smaller lake basins is poor and dominated by cypriniform fishes. In most cases there are a few endemics, often nemacheilid loaches and cyprinids from the *Rhynchocypris* or *Schizothorax* species groups.

The freshwater biodiversity of Siberia and Central Asia has, and continues to be, impacted by human activities. At local and regional scales, rivers are dammed and polluted. On a wider scale, vast areas in Siberia have been clear-felled; the effects of this massive habitat change across the 'Taiga' forests are largely unstudied, but significant impacts on freshwater habitats and the nutrient and flow regime have to be expected. Due to the high precipitation levels in Siberia, water abstraction is not a significant problem, but it is a serious problem in the warmer, semi-arid areas of Central Asia. All the larger lake basins are impacted by water retention along their inflowing rivers. The worst example is the Aral Sea where, in the 1960s, the Soviet Union undertook a major water diversion project on the arid plains of Kazakhstan, Uzbekistan and Turkmenistan. The region's two major rivers, fed by snowmelt and precipitation in distant mountains, were used to transform the desert into farms for cotton and other crops. Prior to the project the Syr Darya and the Amu Darya rivers flowed down from the mountains and pooled together in the lowest part of the basin to make the Aral Sea, which was once the fourth largest lake in the world (http://earthobservatory.nasa.gov/Features/WorldOfChange/aral_sea.php). Today, the Aral Sea is no longer inhabited by freshwater fishes and the lower reaches of its inflowing rivers still suffer heavily from over-abstraction of water and from the effects of alien species introduced from across the world, especially from China. The basin, including the headwaters of the inflowing rivers, was previously inhabited by around 50 native species of freshwater fish, the best known of which are the scurrile dwarf sturgeons of the genus *Pseudoscaphirhynchus*. However, the area is poorly studied and fieldwork is urgently needed to better understand its fish diversity. Now only a few introduced marine species exist in the Aral Sea itself and no long-distance migrants remain. As for most anadromous species, a component of the population does not migrate and non-migratory populations of *Luciobarbus brachycephalus*, *L. capito*, *Alburnus aralensis* and *Rutilus heckelii* still survive in the rivers. Others, such as the pike asp (*Aspiolucius esocinus*) and the dwarf sturgeons are now very close to extinction. The desiccation of the Aral Sea is widely recognised as an ecological disaster, and there seems to be little chance of restoring the sea to its former state.

1.3.3.6 Eastern Himalaya

A report on the status of freshwater fishes of the Eastern Himalaya region in 2010, encompassing the Ganges–Brahmaputra, the Chinwin–Irrawaddy and the Kaladan/Kolodyne catchments, found records for 520 valid species representing relatively high species richness within the region (Allen *et al.*, 2010). The centres of richness are the Tista, Kameng, Dikrong, Subansiri and Siang basins of the Ganges–Brahmaputra system, although a lack of sampling effort may explain the relatively low diversity reported for some areas such as the Gangetic Plains. Of the 379 species for which there was sufficient information to assess extinction risk, 70 (18.5%) are threatened, the other 141 species being assessed as Data Deficient. No species is recorded to have gone extinct. The majority of threatened species are in the Chadwin basin in Manipur. Five species are Critically Endangered including three species of sawfish that venture into freshwaters. These sawfishes are primarily threatened through overfishing in the marine parts of their ranges. The other two Critically Endangered species are snow trouts (genus *Schizothorax*), both endemic to Lake Rara in Nepal, where they are threatened by overfishing, pollution and siltation.

The main threats to fishes across the region are habitat loss and degradation, overexploitation, hydrological modifications and water pollution. Hydropower dams block the migration routes for a number of species (Dudgeon, 2005), including the hilsa (*Tenualosa ilisha*), an anadromous species of major commercial importance (Rahman, 2006). The Ganges is also subject to intense water abstraction leading to its ranking by WWF as one of the 10 most endangered rivers in the world (Wong *et al.*, 2007). Some stretches of the Ganges are almost devoid of fish due to extreme pollution (Edds *et al.*, 2002; Gopal & Agarwal, 2003). Consistent with the rest of the Indian subcontinent, intensive harvesting of fish for food is a major threat to the freshwater fishes of the region, with a lack of effective legislation compounding the problem (see Chapter 15).

1.3.3.7 Peninsular India

India is rich in freshwater fish diversity (Kottelat & Whitten, 1996; Shaji *et al.*, 2000; Dahanukar *et al.*, 2004). An assessment of all known freshwater fishes in peninsular India recorded 290 described species (Molur *et al.*, 2011) with 37% of assessed species (97 species) threatened. No species are known to have gone extinct in the recent past. The centre of species diversity, endemism and of threatened species is the Western Ghats. The Periyar, Chalakkudy and Pambar rivers have the highest numbers of threatened species (Figure 1.4). The Western Ghats holds the highest number of Critically Endangered species (seven species), all of which are restricted to Kerala State. Of the 96 threatened species endemic to peninsula India, 50 are endemic to the Western Ghats region.

Deforestation and drainage basin alteration, river regulation, pollution, over-harvesting and invasive alien species are threatening freshwater ecosystems throughout India (Dahanukar *et al.*, 2004; Raghavan *et al.*, 2008, 2011). Many Asian streams and rivers are grossly polluted and among the most degraded globally (Dudgeon, 2000). Over-harvesting for food and for the aquarium trade is also a problem. Overfishing is widespread, and the unmanaged collection and trade of endemic species for the aquarium trade is an emerging conservation issue in the Kerala part of the Western Ghats (Raghavan, 2010). For example, red-lined torpedo barbs *Sahyadria denisonii* and *S. chalakkudiensis*, both of which are very popular in the aquarium trade, are overexploited in at least three of the rivers in the region (Raghavan *et al.*, 2013). Thirteen species of alien fishes are currently distributed across river basins in the Kerala part of the Western Ghats (Lowe *et al.*, 2000; Raghavan *et al.*, 2008; Krishnakumar *et al.*, 2009).

1.3.3.8 Indo-Burma

There are 1178 recognised freshwater species in the Indo-Burma region (Kottelat *et al.*, 2012) including 151 species from the Salween drainage, 328 from the Mae Khlong–Chao Phraya drainages, 500 from the Mekong drainage, 253 from the Red River drainage, 160 from the streams draining the eastern slope of the Annamite range and 221 from the Malay Peninsula (south of the isthmus of Kra). The Malay Peninsula south of Thailand and those parts of the Salween, Mekong and Red River drainages in China have not been assessed. *Platytropius siamensis*, the Siamese flat-barbelled catfish, is the only species of fish from the region considered to be Extinct. An additional four species are considered Critically Endangered (Possibly Extinct) – Balitoridae: *Schistura tenura*, *Schistura nasifilis*; and Cyprinidae: *Puntius compressiformis, Balantiocheilos ambusticauda*). The dwarf chain loach (*Ambastaia sidthimunki*) is thought to be present in just two small streams in the Mae Khlong catchment but has possibly been extirpated from the wild. Finally, the red-tailed black shark (*Epalzeorhynchos bicolor*) was long considered Extinct in the Wild, but a small population has recently been rediscovered. These last two species are available through the aquarium trade, but their survival in captivity relies entirely on cultivation in only a few farms in Thailand. The wider overview presented below is taken directly from Kottelat *et al.* (2012).

The lower and middle Mekong and Chao Phraya drainages have the most diverse fish fauna. The fish fauna is least diverse in mountainous areas (for example, in Northern Thailand and Northern Laos), but these sites are very remote and hard to access and have received limited sampling. Of the extant species for which sufficient data are available to determine their conservation status, 16.9% (112 species) are threatened, representing a similar degree of threat to that observed in the Eastern Himalaya (18.5% threatened; Allen *et al.*, 2010). However, while high, it contrasts with higher levels of threat for Africa (27%; Darwall *et al.*, 2011), the Western Ghats in India (37%; Molur *et al.*, 2011) and Europe (42%; Kottelat & Freyhof, 2007). Although this may reflect a lower current level of threat in some areas, it may also be a product of the large number of Data Deficient species (43.6% here, versus 27.1% in the East Himalaya region). It is also feared that the level of threat will increase dramatically if plans for many new hydroelectric schemes throughout the Mekong River Basin are approved and implemented (Ziv *et al.*, 2012).

The distribution pattern of threatened species largely parallels that of species richness. This reflects the reality that more studies have been conducted along the main rivers, and that the species identified as being threatened are in most cases species important for fisheries for which there is more available information. The reported absence of threatened species in most of the Red River drainage may be due to lack of information, with most species assessed as Data Deficient.

The main threats identified are agricultural and forestry effluents, affecting 402 species (including 63 threatened species), dams and other modifications of aquatic habitats, affecting 298 species (including 62 threatened species) and the over-harvesting of natural resources, affecting 313 species (including 59 threatened species) (Kottelat *et al.*, 2012). The most notorious species in the Mekong River Basin, on account of their large size, include the giant Mekong catfish (*Pangasianodon gigas*), the giant freshwater whipray (*Himantura polylepis*) and the giant carp

(*Catlocarpio siamensis*), all of which are highly threatened through a combination of overfishing, pollution and habitat destruction (see Chapter 12). Given that even these high-profile commercially valued species are already threatened, we might question what can be done to raise awareness and conservation action for the large number of lesser-known threatened species. The situation in the Mekong demonstrates the scale of the challenge facing freshwater fish conservation globally. Even the potential risk posed to the world's largest inland capture fishery in the Tonlé Sap, also one of the most species-rich lakes in the world (Campbell *et al.*, 2006), appears insufficient to outweigh perceived potential gains from building mainstream dams for hydropower on the Mekong (Orr *et al.*, 2012).

1.3.3.9 China

The freshwater fish fauna of China has been reviewed by Kang *et al.* (2013) and this following section is largely based on their findings. The country supports an estimated 920 fish species, of which 613 are endemic. Yunnan Province, including the upper reaches of the Yangtze, Red, Mekong and Salween rivers, holds the highest number of species (373) and highest number of country-endemic species (216), with many species specialised to the high-altitude streams and lakes of this part of the world. The Qinghai Tibetan Plateau Region is also characterised by high levels of species endemism, and includes a number of species in the genera *Schizothorax* and *Triplophysa* specialised to cold-water high-altitude torrent rivers. The northwestern part of China (Xinjiang Province) is dominated by species suited to cold, clear waters, including the threatened Siberian taimen (*Hucho taimen*). The southern part of China is characterised by species preferring warm waters. The upper part of the Yangtze River, which is Asia's longest river, is species-rich with 330 species reported, many of which are endemic to the region. The Middle–Lower Yangtze Plain Region contains mostly Cyprinidae in addition to some of the world's most threatened species such as the Chinese sturgeon (*Acipenser sinensis*) and the Chinese paddlefish (*Psephurus gladius*), both of which are Critically Endangered following decades of overfishing and more recent obstruction to their migration routes by the Gezhouba and Three Gorges Dams. The paddlefish has just recently been reported to be extinct, although surveys still continue.

Only 545 of the estimated 920 species documented for China (Kang *et al.*, 2013) have been assessed for the IUCN Red List (IUCN, 2013). Of these, 197 are Data Deficient, thus indicating a significant need for additional research. Of the remaining 348 species, 78 (22%) are either threatened or Extinct. Two species are thought to have become extinct, *Anabarilius macrolepis* and *Cyprinus yilongensis*, when Yilong Lake dried up as a result of water abstraction for agriculture in 1981.

The main threats identified are pollution, overexploitation and hydrological alterations. One in three species assessed in China are impacted by overexploitation. The clearest example for hydrological alteration comes from the South–North Transfer project reviewed by WWF (Pittock *et al.*, 2009). Since the late 1970s, annual water withdrawals on the North China Plain have exceeded the limits of the annual renewable supplies. So severe was the level of water abstraction that in the 1990s the Yellow River, China's second largest river, failed to reach the sea on a number of occasions.

The scale of impact relating to the planned construction of dams throughout China is perhaps best demonstrated through the case of hydropower developments on the Yangtze River (Qiu, 2012). In 2011, the central government solidified plans to increase China's reliance on non-fossil fuel energy from the 2010 level of 8% to 15% of the energy mix by 2020. With nearly two-thirds of this target coming from hydropower, ecologists are concerned that the already degraded ecosystems of the Upper Yangtze will be threatened (Qiu, 2012). The Upper Yangtze River Rare and Endemic Fish Reserve, established to help offset impacts of the Three Gorges Dam, is a critical habitat for an estimated 190 fish species, yet in 2005 officials reduced this 500 km long reserve by 150 km to make way for the Xiangjiaba and Xiluodu dams (Qiu, 2012). Now the Xiaonanhai Dam will further erode the size of the reserve through creation of a 100-km long reservoir in the heart of the reserve itself.

1.3.3.10 Japan

Most of the rivers of Japan are short and steep, draining relatively small catchment areas. There are an estimated 297 species of freshwater fishes, of which an estimated 42% are endemic to Japan (Yuma *et al.*, 1998), some with highly restricted ranges. For example, unique species complexes have evolved within Lake Biwa, which is about 4 million years old. According to the national Red List of Japanese freshwater fishes (Ministry of Environment, Government of Japan, 2007), which largely follows the methodology for applying the IUCN Red List Categories and Criteria, around half of Japan's freshwater fishes are threatened. Four species are thought to now be Extinct, 61 species are Critically

Endangered, 48 are Endangered, 35 are Vulnerable and a further 26 species are Near Threatened.

The major threats reported include: (i) the destruction and fragmentation of habitat; (ii) overexploitation; (iii) modification of paddy field habitats; (iv) invasive alien species; and (v) pollution (Hosoya, 2008). Dam construction is thought to have heavily impacted the migrations of many sea-run fishes such as the Sakhalin green sturgeon (*Acipenser medirostris*), now thought to be extirpated in Japan. As in so many regions of the world, alien invasive species are also impacting the local fish fauna, e.g. many native species are predated by the introduced largemouth bass (*Micropterus salmoides*) and smallmouth bass (*M. dolomieu*). Finally, the heavy use of pesticides in the 1960s appears to have coincided with the start of a marked reduction in fish populations within paddy fields, and is thought to have been partly responsible for the local extirpation of the southern nine-spined stickleback (*Pungitius sinensis*) from Kyoto and Hyogo. Overall, the consolidation of paddy fields and alien invasive species are thought to pose the most significant threats.

A more recent overview of Japan's biodiversity in inland waters (Ministry of Environment, Government of Japan, 2010) concluded there had been significant losses of biodiversity during the period of assessment (1950s–2010) and this was likely to continue. Habitat degradation and loss was again highlighted as a major problem with river bed excavation of sand and gravels, reclamation of lakes, marshes and wetlands, combined with poor water quality, leading to an overall loss of available habitat for freshwater fishes. It was estimated that around 60% of wetlands have been lost from 1900 to the 1990s. The number of lakes also declined significantly, with 15% of major lakes being reclaimed between 1945 and the 1980s, and domestic and industrial/agricultural effluents being discharged into many lakes and rivers. Many dams have been constructed since the 1950s. By the 1980s, 46 (41%) of the country's 113 major rivers, along which species such as the cherry salmon (*Oncorhynchus masou*) and Japanese charr (*Salvelinus leucomaenis*) migrate, were passable for less than 50% of their length. For species such as gobies, with lower upstream swimming capabilities, the situation is thought to be even more serious. An additional major impact has been the artificial modification of river and lake shorelines as part of flood control measures. For example, in Lake Biwa, shoreline reed beds were reduced by about 50% between the 1950s and the 1990s, resulting in a huge loss of habitat for many fish species.

1.3.3.11 Indonesia

Indonesia harbours a vast diversity of freshwater fishes. It is ranked third in the world in terms of numbers of freshwater fish species, with 1189 native species, 125 of them endemic to the country, and 20 introduced species (Froese & Pauly, 2014). This high diversity of species is all the more impressive given that Indonesia is only around 20% of the size (in terms of surface land cover) of these other countries. The freshwater fish fauna is poorly documented, with many additional species awaiting discovery.

Most remarkable is the extremely high diversity of small and very small fishes; indeed, the smallest fish in the world (*Paedocypris progenetica*) is from the Indonesian islands of Sumatra and Bintan. Numerous popular aquarium fishes such as small barbs, rainbow fishes and gouramis originate from Indonesia, and the export and farming of ornamental fishes is a major industry. Most important in terms of the numbers exported is the clown loach (*Chromobotia macracanthus*) endemic to Sumatra and Borneo. Sumatra and Borneo are also the home of the highly prized arowana (*Scleropages formosus*) that is also widely distributed in Southeast Asia. The arowana is strictly protected in the wild, and artificial selection has led to several strains which can sell for more than 10,000 Euro.

Sadly, this incredible diversity of fishes is massively impacted by the country's efforts to be a major exporter of wood pulp, plywood, rubber and palm oil. Extensive areas of the country have been logged and rainforests continue to be harvested and converted into oil palm plantations. Indonesia has one of the largest biodiversity treasures in the world, yet it is also being destroyed at a rate that exceeds most other countries. From a freshwater fish perspective, the most heavily impacted habitats are the peat swamps that support a tremendous, often locally endemic, fauna including many miniature fishes. Posa *et al.* (2011) and Ng *et al.* (1994) emphasise the very high rate of peat swamp forest loss. Regionally, mining for minerals is also having an impact but, as mentioned above, the overwhelming destruction comes from the massively expanding oil palm industry, which transforms diverse rainforests into homogenous plantations throughout much of Indonesia.

The individual conservation status of all the species of the mega-diverse fish fauna of Indonesia remains to be assessed. Most likely, Indonesia's natural capital is already massively damaged and many species are expected to have gone extinct in the past 30 years. Although no systematic assessment has yet been completed for

Indonesia, around 389 species have been assessed for the IUCN Red List, of which around 24% are threatened. No species on the IUCN Red List has been assessed to be Extinct in this region, but the assessment is far from complete.

1.3.3.12 *Pacific Islands of Oceania*

An assessment of 167 species of freshwater fishes in the Polynesia–Micronesia region, focusing mainly on the endemic species (Pippard, 2012), found most (91 species) to be widely distributed and at a low risk of extinction. A further 63 species were lacking sufficient information to assess their status. Of the remaining 104 species, only 12 (8%) were considered threatened. Three of these species are Critically Endangered, eight Endangered and one Vulnerable. Around 5% are thought to be declining, 20% are considered stable, and no populations were increasing.

Deforestation and the conversion of land to agriculture have led to increased levels of sedimentation impacting many fish species, such as the gobies, that require clear rocky substrates. The leaching of pesticides, such as for production of taro (on Wallis and Fortuna) and for Sakau (in the Federated States of Micronesia), can also lead to injury or mortality in fish species. Some rivers, such as those in New Caledonia, are impacted by mining effluents leading to siltation and degradation of important fish habitats. As with many regions, an increasing number of dams are obstructing migration routes for many of Indonesia's fishes. Over-harvesting is not a major problem at present, although there is potential for over-harvesting of some of the very beautiful gobies for aquaria. Finally, although alien invasive species are not currently a major problem, tilapia (*Oreochromis* spp.) have now reached these islands and it is only a matter of time before they spread and impact the local fish fauna, as has occurred elsewhere (Canonico *et al.*, 2005).

1.3.3.13 *Australia*

Although relatively species-poor by world standards, with 256 native species and 37 alien species, the Australian freshwater ichthyofauna has long been recognised as highly distinctive, with an estimated 74% of native species endemic to the country (Lintermans, 2013). Many taxa are more diverse and with higher levels of endemism than elsewhere in the world, including the Galaxiidae and the lungfish (*Neoceratodus forsteri*). The status of Australia's freshwater fishes has been reviewed most thoroughly by Lintermans (2013).

No known species has been recorded as extinct in Australia since 1788 (Lintermans, 2013), but some species may have disappeared unnoticed (Hammer *et al.*, 2009). There have also been some close calls and many species are now clearly on the verge of extinction. The Lake Eacham rainbowfish (*Melanotaenia eachamensis*) disappeared from Lake Eacham on the Atherton Tablelands in Queensland (its only known habitat at the time) in the 1980s, following the introduction of alien fishes (Hammer *et al.*, 2009). It was rediscovered in nearby streams, but the distinctive lake population has been lost. There are many other species clinging to small pockets of ever-dwindling habitat (Hammer *et al.*, 2009).

Of 73 fish taxa listed as threatened under state or national legislation, the Galaxiidae has the most threatened taxa (18 of 23 described taxa) (Lintermans, 2011). The Australian freshwater fish fauna has not yet been comprehensively assessed for the IUCN Red List but, of the 169 species assessed up to 2013, 23% are threatened (IUCN, 2013). On current trends, extinctions are predicted within the next 30 years (Lintermans, 2013). The effects of land clearance and agriculture have been especially severe on aquatic ecosystems, with over-abstraction of water, increased turbidity and pollution from agricultural run-off. Other issues include increasing salinisation, especially in streams and wetlands dependent on groundwater inflows, as in Southwestern Australia (Beatty *et al.*, 2011). Industrial and agricultural chemicals, elevated nutrient levels (leading to eutrophication), acidification through oxidation of sulfidic soils, and water cooling and hypoxic conditions resulting from low-level offtakes below dams are all potentially problematic for fishes. Australia's biodiversity is still responding and adapting to these relatively recent land use changes (Bates *et al.*, 2010). Flow regulation through impoundments has also had a major impact on Australian rivers, with over 446 large dams recorded in 2000 (Arthington & Pusey, 2003). River corridors in Australia have been transformed for urban and agricultural land use to the extent that many floodplain ecosystems are functionally extinct (Tockner *et al.*, 2008). Most rivers in Eastern Australia are now regulated by dams, weirs and other structures, changing flow and temperature regimes downstream (Kingsford, 2000; Arthington & Pusey, 2003).

Negative interactions with the 37 established non-indigenous species are considered a significant risk to over three-quarters of native species (Lintermans, 2004). Following the initial period of sport fish introductions the aquarium trade is now considered the main source of new introductions (Lintermans, 2004; Corfield *et al.*, 2008). Common carp (*Cyprinus carpio*) have been reported to cause an estimated $16 million damage annually, primarily resulting from the environmental impacts of their bottom-feeding habits that lead to an increase in suspended sediments (McLeod, 2004). Introduced salmonids have fragmented and greatly reduced the distribution and abundance of multiple species of *Galaxias* (McDowall, 2006).

1.3.3.14 New Zealand

New Zealand's long isolation has led to the development of a distinctive fish fauna including 41 species, nearly half of which are in the family Galaxiidae. Thirty-three of these species are endemic to New Zealand. Preliminary findings indicate 49% of species are threatened, including 18 endemic species (Gibson, pers. comm.). Most of the threatened species are non-diadromous galaxiids for which introduced salmonids represent the major threat. Other threats include drainage of wetlands, pollution of waterways, removal of riparian vegetation and placement of barriers in waterways. For example, an estimated 85–90% of New Zealand's wetlands have been lost in the last 100 years, mainly through drainage for agriculture (Smith *et al.*, 1997). As around half of New Zealand's fish species are diadromous, they require unimpeded passage through their freshwater habitats. Many barriers block, or partially block, the migration and recruitment of these species, such as 'the large galaxiids' (*Galaxias postvectis, G. argenteus, G. brevipinnis, G. fasciatus*) and the long-fin eel (*Anguilla dieffenbachii*). The possible overharvesting of 'whitebait', which includes juveniles of five migratory galaxiids, combined with the destruction of spawning grounds, is also thought to have impacted galaxiid populations for many years now. A notable extinction is the New Zealand grayling (*Prototroctes oxyrhynchus*), with no individuals sighted since the 1940s. The precise reason for extinction is uncertain, but salmonid introduction almost certainly contributed (McDowall, 1990).

1.3.3.15 North America

North America, geographically defined as Canada, the United States and Mexico, has an estimated 1213 species of freshwater fishes (Burkhead, 2012). Burkhead (2012) described an increase in species richness per drainage from Canada southward, with over two-thirds of species occurring east of the Great Continental Divide, especially in the Interior and Appalachian Highlands. Catchments in the Western United States and Mexico exhibit relatively lower species richness but generally have higher endemism and disproportionately higher numbers of extinct species

(Minckley & Douglas, 1991; Burr & Mayden, 1992; Contreras-MacBeath et al., 2014).

The conservation status of freshwater fishes in North America has been assessed three times by the American Fisheries Society. The latest assessment (Jelks et al., 2008) reported around 39% of described species as imperilled (the equivalent term to 'threatened' species on the IUCN Red List). Of those previously assessed as imperilled in their 1989 assessment, most (89%) were found to be in the same condition or worse; only 8 taxa had improved their status. Habitat degradation and alien invasive species were listed as the main threats. Burkhead (2012) determined there to be 60 extinct taxa in North America in recent times. He attributed the post-1950s increase in extinction to urbanisation, construction of dams, alteration of natural water bodies (e.g. through channelisation and pollution) and the impacts of industrial expansion. A recent assessment of the majority of the United States component of the fish fauna (756 species) for the IUCN Red List found that 22% are threatened and 18 species are reported as Extinct. The difference between these two estimates for the number of extinct species reflects differences in approach in classifying a species as extinct. The 2013 assessment of United States fishes for the IUCN Red List identified hydrological modifications (such as dams), pollution and invasive alien species as the main threats.

The Southeastern United States has many threatened fishes, especially the minnows, ictalurid catfishes and darters; the mid-Pacific coast has many threatened lampreys, salmonids, sticklebacks and minnows; and the lower Rio Grande and coastal and endorheic basins of Mexico have many threatened minnows, characids, goodeids, silversides, pupfishes and livebearers (Burkhead, 2012; Contreras-MacBeath et al., 2014). The Tennessee River ecoregion has the greatest number of threatened fishes with 58 listed taxa (Burkhead, 2012).

Canada has around 200 species of freshwater fishes, of which an estimated 28 species are endemic. According to the Committee on the Status of Endangered Wildlife in Canada (COSEWIC) 20 species, or populations of species, are classified as 'Endangered', 17 as 'Threatened' and a further 29 as 'Special Concern' (COSEWIC, 2012). Dextrase and Mandrak (2006) reported habitat degradation or loss and alien species as the greatest threats, along with overfishing, pollution, climate change and interactions between wild and farmed species (CESCC, 2006). Although a comprehensive assessment of Canada's freshwater fish is yet to be conducted for the IUCN Red List, 167 species have been assessed (IUCN, 2013), of which 3 species are listed as Extinct, and 11 threatened.

As for most countries, Mexico suffers from the impacts of unsustainable development and its wetland ecosystems are suffering badly. The extinction risk for each of the 624 species (43% of which are endemic) has been assessed (Contreras-MacBeath, pers. comm.) and preliminary results suggest that approximately 37% of species are threatened. The majority of threatened species are within the families Cyprinidae (56 species), Goodeidae (38 species), and Poeciliidae, Atherinopsidae and Cyprinodontidae with 23 threatened species in each. Twenty species are Extinct or Extinct in the Wild, of which 15 are endemic to Mexico, representing a high proportion of all known fish extinctions worldwide. Among the species facing the highest risk of extinction are those with highly restricted distributions (such as the Critically Endangered Tequila splitfin *Zoogoneticus tequila*), as well as a number of species of pupfish. The main threats are over-abstraction of water (the country being predominantly arid or semi-arid), overexploitation, habitat disruption and fragmentation, pollution and competition with growing numbers of alien and native invasive species. Saltwater intrusions into rivers, such as the Rio Bravo del Norte (Rio Grande), are of particular concern because freshwater species are being replaced by marine species.

1.3.3.16 South and Central America

The freshwater fish fauna of South America is the most diverse of all continents, with an estimated number of species exceeding 4000 (Reis, 2013). When the entire Neotropical region is considered, the number of species jumps to above 5000, representing almost 10% of all known vertebrates (Lundberg *et al.*, 2000; Albert *et al.*, 2011). In the 9 years following publication of the CLOFFSCA (Reis *et al.*, 2003), more than 900 additional freshwater fish species have been described from the region, at an average of one new species every 3.5 days, thus elevating the number of known species to around 5400 in 2012. At this pace of species descriptions, the final number of fish species in the Neotropical region may exceed the 8000 estimated by Schaefer (1998).

The high diversity of freshwater habitats includes many of the largest rivers in the world, extensive marshes such as the Llanos of the Orinoco River and the Pantanal of the Paraguay River, and high-altitude lakes and rivers. Despite this great diversity and extent of freshwater habitats, the level of threat to their long-term survival is rising. The primary threats include ongoing and extensive deforestation, water abstraction or diversion for agriculture, alluvial gold mining and hydroelectric power (dams). At the time of writing, 151 new dams were proposed for construction in five of the six major Andean tributaries of the Amazon (Finer & Jenkins, 2012). These dams will compromise the migratory requirements and habitat for many species. Some species are also threatened by overexploitation and invasive species, although currently to a lesser degree than in many other parts of the world. However, the newly formed reservoirs associated with the hydropower developments will increase the risk of introduction and spread of alien species.

A comprehensive assessment of the conservation status of these species is urgently required but, at the time of writing, only 332 species had been assessed for the IUCN Red List, of which 31 are threatened, mainly due to overexploitation, pollution and invasive species – clearly, there is much work to be done. A number of national and international initiatives are under way to help document the conservation status and distribution of South and Central America's freshwater fish fauna. Until these assessments are completed, it is difficult to quantify the level of threat to this impressive diversity of species. One ongoing assessment of the Lower Plata River Basin in Argentina, Paraguay and Brazil (Cappato & Yanosky, 2009) has already recorded 3 Endangered, 8 Vulnerable and 3 Near Threatened species among the 184 assessed species (Baigún et al., 2012). At a regional scale, almost 81% of fishes in the state of Santa Catarina in southern Brazil are threatened, including 11 Critically Endangered, 3 Endangered, 20 Vulnerable, 8 Least Concern and 47 Data Deficient species.

Given the extraordinary diversity of freshwater fish species, the current and emerging threats and the great importance of this species diversity to local livelihoods and national economies, a comprehensive assessment of their distribution and status is a priority for underpinning future efforts at developing a conservation action plan for the region. The remaining gap in our knowledge of species distributions and their conservation requirements in South America is truly a cause for concern.

1.4 SUMMARY AND CONCLUSIONS

The scale of the problems facing freshwater fishes around the globe is clear. We know that approximately one out of every three freshwater fish species is threatened (Table 1.1). This means that, should the current threats to their survival continue unabated, they are predicted to go extinct in the not too distant future. The main threats include loss and degradation of habitat, pollution, over-abstraction of water and invasive alien species. Not surprisingly, the threat is most acute in: (i) the more heavily developed regions of the world, such as Europe, where almost

Table 1.1 *Summary of the level of threat and state of knowledge in 2013 for freshwater fishes in all regions of the world. Regions are loosely ordered according to their state of coverage for the IUCN Red List.*

Region	Estimated no. valid species described for the region	Estimated no. species endemic to the region	No. species globally assessed for IUCN Red List	No. species assessed as Data Deficient on the IUCN Red List	No. species globally threatened	% species assessed for the IUCN Red List globally threatened	Number of recorded global extinctions in the wild	State of coverage for the IUCN Red List
Europe	525	426	525	34	200	41	15	Good
Peninsular India	290	189	290	26	97	37	0	Good
Eastern Himalayas	520	??	520	141	70	19	0	Good
Indo-Burma	1178	~630	1178	514	112	17	1 (4 possibly)	Good
New Zealand**	41	33	41	4	20	49	1	Good
Continental Africa	2892	2788	2892	517	642	27	5	Good
United States	790	440	790	19	169	22	19	Good
Canada	~200	~28	164	1	11	7	3	Good
Western Asia	~300	~245	245	12	105	46	7	Good
Mexico**	624	273	616	55	223	41	20	Good
Japan – National RL	~297	~125	??	??	144	~48	4	Good
Japan – IUCN RL	~297	~125	129	27	11	11	4	Medium
Madagascar*	155	98	136	31	54	53	3	Medium
Australia	256	~190	169	28	32	23	0	Medium
Pacific Islands	??	??	167	63	12	12	0	Medium
Siberia	~275	??	??	??	??	??	??	Poor
Central Asia	??	??	82	4	17	22	1	Poor
China	920	613	545	197	76	22	2	Poor
Indonesia	1189	125	389	85	72	24	0	Poor
South & Central America	>5400	>3100	537	94	123	29	21	Poor
Global	**15,750**	**n/a**	**7300**	**1571**	**1988**	**31**	**69**	

*Needs updating; **unpublished data; ?? numbers unknown.

one in two species are threatened; and (ii) those regions suffering from extreme water stress, such as Western Asia, where once again almost one in two species are threatened. Madagascar's fish fauna is thought to be most at risk, with a staggering 85% of endemic species threatened, mainly due to land conversion for agriculture, deforestation and introduction of alien invasive species. This probably represents the highest recorded level of threat to any taxonomic group in a single country, and is even more notable given the distinctive isolated island fauna that it represents. We also find that particular taxonomic groups of fishes are more susceptible than others. For example, a recent study has shown that the freshwater sharks and rays are subject to considerable threats from habitat degradation, such that over one-third (36%) of the 90 obligate and euryhaline freshwater species are threatened (Dulvy *et al.*, 2014). In general, the findings summarised in this chapter support earlier proposals (e.g. Reynolds *et al.*, 2005; Olden *et al.*, 2007; Dudgeon, 2011) that the largest, smallest and migratory species are disproportionately threatened. For example, 21 of the world's 27 sturgeons are Critically Endangered, as are a number of the large pangasid catfishes in Southeast Asia. The proliferation of large dams is a major cause for concern as many of these species are now unable to reach their spawning grounds, the migration routes being blocked. Although we know less about the smaller, less conspicuous species, many of which probably remain undiscovered and undescribed, a number, such as those found in the peat swamps of Southeast Asia, are also highly threatened.

Those countries where coverage on the IUCN Red List remains poorest and for which we suspect a severe threat to freshwater fishes include: (i) China, where a recent survey reported almost half of the rivers previously mapped to no longer exist due to over-abstraction and diversion of water (www.theatlantic.com/china/archive/2013/04/28-000-rivers-disappeared-in-china-what-happened/275365/); and (ii) Indonesia, where some river fish faunas are reported to now comprise almost entirely non-native species.

With more than one in five species that have been described and assessed for the IUCN Red List classified as Data Deficient, and a conservatively estimated 7000 described species still to be assessed, the scale of the knowledge gap is clear. In addition to the Data Deficient species, many more are yet to be discovered and described. In Europe alone, considered one of the better-known of the world freshwater fish faunas, 5–10 new species are described each year, with some suggesting the total number of species may stabilise at around 700–800 species, indicating that we are yet to discover and describe an additional 150–200 species in Europe. Large areas of the Indo-Burma region are still in the exploration and discovery phase, with 434 (37%) of the 1178 species described after 1989 and 285 of those (24%) after 1999 (Kottelat *et al.*, 2012).

The high level of threat to freshwater fishes combined with our lack of knowledge for so many species is a tragedy from the conservation perspective but, with more than 3 billion people dependent upon fish as a primary protein source (FAO, 2010), these predicted species losses will have a major impact on many of the world's poorest people. We should not assume that fisheries will continue to be productive and resilient if we lose a large part of the species base supporting those fisheries.

In conclusion, the future looks very uncertain for freshwater fishes. No region fully escapes the growing multitude of threats to freshwater and wetland habitats. As the human population rises towards a projected peak of 9 billion people in 2050, the forecast is for a 55% increase in water demands, an 80% increase in energy needs, and a 60% increase in food demands by 2050 (OECD, 2012). Freshwater fishes are part of the environmental collateral damage in this race to meet the growing needs of the human population (Chapter 2). If freshwater fishes are to survive in their natural environment, global development has to adopt new technologies and management approaches that will allow fishes to survive where their environment is currently becoming more hostile by the day. For example, dams must be designed that allow fish to pass freely (both upstream and downstream); river flows must be sufficient in quality, quantity and timing to enable the freshwater ecosystems to function sustainably (Chapter 4); the spread of invasive species must be controlled (Chapter 8); protected areas must be sited and designed to be effective for freshwater species (Chapter 14); and fisheries must be managed at sustainable levels (Chapter 15). The survival of freshwater fishes is also dependent upon a radical change in attitudes such that fish species are valued as legitimate conservation targets and not purely as exploitable and replaceable food resources (Chapters 2 and 18). Let us hope that we can rise to this challenge and create a better future for freshwater fishes.

1.5 DISCUSSION QUESTIONS

1. Given the increasing and competing demands on water by humans, what might be considered a realistic target for the conservation of freshwater fish species at the global scale?

2. How can we effectively communicate the problems facing freshwater fishes given that people rarely see most species? Freshwater fishes are largely out of sight and out of mind.

3. How might we change attitudes so that people value fish as species to be conserved (as we do birds, for example) rather than as replaceable agricultural products?

4. Can Protected Areas, one of the main tools for conserving species, work effectively for freshwater fishes?

5. An overriding threat to many freshwater fishes is climate change: where habitats become unsuitable for species survival under future climate change, should we conduct assisted migrations for species unable to move to more suitable environments due to barriers to natural migration?

1.6 REFERENCES

Abell, R., Thieme, M. L., Revenga, C., *et al.* (2008). Freshwater ecoregions of the world: a new map of biogeographic units for freshwater biodiversity conservation. *Bioscience*, 58, 403–414.

Albert, J. S., Petry, P. & Reis, R. E. (2011). Major biogeographic and phylogenetic patterns. In *Historical Biogeography of Neotropical Freshwater Fishes*. Albert, J. S. & Reis, R. E. (Eds). Berkeley, CA: University of California Press, pp. 21–58.

Allen, D. J., Molur, S. & Daniel, B. A. (compilers) (2010). *The Status and Distribution of Freshwater Biodiversity in the Eastern Himalaya*. Cambridge, UK and Gland, Switzerland: IUCN, and Coimbatore, India: Zoo Outreach Organisation.

Arthington, A. H. & Pusey, B. J. (2003). Flow restoration and protection in Australian rivers. *River Research and Applications* 19, 377–395.

Baigún, C. R. M., Colautti, D., López, H. L., Van Damme, P.A. & Reis, R. E. (2012). Application of extinction risk and conservation criteria for assessing fish species in the lower La Plata River basin, South America. *Aquatic Conservation: Marine and Freshwater Ecosystems*, 22, 181–197.

Bates, B., Bunn, S., Baker, P., *et al.* (2010). *National Climate Change Adaptation Research Plan: Freshwater Biodiversity*. NCCARF, Griffith University, Australia.

Beatty, S. J., Morgan, D. L., Rashnavadi, M. & Lymbery, A. J. (2011). Salinity tolerances of endemic freshwater fishes of south-western Australia: implications for conservation in a biodiversity hotspot. *Marine and Freshwater Research*, 62, 91–100.

Berra, T. M. (2001). *Freshwater Fish Distribution*. London: Academic Press.

Burkhead, N. M. (2012). Extinction rates in North American freshwater fishes, 1900–2010. *BioScience*, 62, 798–808.

Burr, B. M. & Mayden, R. L. (1992). Phylogenetics and North American freshwater fishes. In *Systematics, Historical Ecology, and North American Freshwater Fishes*. Mayden, R. L. (Ed.). Stanford, CA: Stanford University Press, pp. 18–75.

Campbell, I. C., Poole, C., Giesen, W. & Valbo-Jorgensen, J. (2006). Species diversity and ecology of Tonle Sap Great Lake, Cambodia. *Aquatic Sciences*, 68, 355–373.

Canonico, G. C., Arthington, A., McCrary, J. K. & Thieme, M. L. (2005). The effect of introduced tilapias on native biodiversity. *Aquatic Conservation: Marine and Freshwater Ecosystems*, 15, 463–483.

Cappato, J. & Yanosky, A. (2009): *Uso Sostenible de Peces en la Cuenca del Plata*. Argentina, Brazil & Paraguy: UICN-Guyra-Proteger.

Carpenter, S. R., Stanley, E. H. & Vander Zanden, M. J. (2011). State of the world's freshwater ecosystems: physical, chemical, and biological changes. *Annual Review of Environment and Resources*, 36, 75–99.

CESCC. (2006). Wild species 2005: the general status of species in Canada [online]. Government of Canada. www.wildspecies.ca/wildspecies2005/index.cfm?lang=e&sec=49&view=10 (accessed 7 August 2013).

Chenoweth, J., Hadjinicolaou, P., Bruggeman, A., *et al.* (2011). Impact of climate change on the water resources of the eastern Mediterranean and Middle East region: modeled 21st century changes and implications. *Water Resources Research*, 47, W06506.

Cohen, A., Kaufman, L. S. & Ogutu-Ohwayo, R. (1996). Anthropogenic threats, impacts, and conservation strategies in the African Great Lakes: A review. In: *The Limnology, Climatology, and Paleoclimatology of the East African Lakes.* Johnson, T. C. & Odada, E. O. (Eds). Amsterdam: Gordon and Breech Publishers, pp. 575–624.

Collen, B., Whitton, F., Dyer, E. E., *et al.* (2014). Global patterns of freshwater species diversity, threat and endemism. *Global Ecology and Biogeography*, 23, 40–51.

Contreras-MacBeath T., Rodrigues M. B., Sorani V., Goldspink, C. & McGregor Reid , G. (2014). Richness and endemism of the freshwater fishes of Mexico. *Journal of Threatened Taxa*, 6, 5421–5433.

Corfield, J., Diggles, B., Jubb, C., *et al.* (2008). *Review of the Impacts of introduced Ornamental Fish Species that have Established Wild Populations in Australia.* Canberra: Australian Government, Department of the Environment, Water, Heritage and the Arts.

COSEWIC. (2012). Canadian Wildlife Species at Risk. Committee on the Status of Endangered Wildlife in Canada. www.cosewic.gc.ca/eng/sct0/rpt/rpt_csar_e.cfm (accessed 15 October 2012).

Côté, I. M. & Reynolds, J. D. (1998). Tropical fish: explosions and extinctions. *Trends in Ecology and Evolution*, 13, 475–476.

Dahanukar, N., Raut, R. & Bhat, A. (2004). Distribution, endemism and threat status of freshwater fishes in the Western Ghats of India. *Journal of Biogeography*, 31, 123–136.

Darwall, W.R.T., Smith K.G., Allen D.J., *et al.* (Eds). (2011) *The Diversity of Life in African Freshwaters: Under Water, Under Threat. An Analysis of the Status and Distribution of Freshwater Species Throughout Mainland Africa.* Gland, Switzerland and Cambridge, UK: IUCN,.

Dextrase, A. J. & Mandrak, N. E. (2006). Impacts of alien invasive species on freshwater fauna at risk in Canada. *Biological Invasions*, 8, 13–24.

Dudgeon, D. (2000). The ecology of tropical Asian rivers and streams in relation to biodiversity conservation. *Annual Reviews of Ecology and Systematics*, 31, 239–263.

Dudgeon, D. (2005). River rehabilitation for conservation of freshwater biodiversity in monsoonal Asia. *Ecology and Society*, 10, 15. www.ecologyandsociety.org/vol10/iss2/ art15/.

Dudgeon, D. (2011). Asian river fishes in the Anthropocene: threats and conservation challenges in an era of rapid environmental change. *Journal of Fish Biology*, 79, 1487–1524.

Dudgeon, D., Arthington, A. H., Gessner, M. O., *et al.* (2006). Freshwater biodiversity: importance, threats, status and conservation challenges. *Biological Reviews*, 81, 163–182.

Dulvy, N. K., Fowler, S. L., Musick, J. A., *et al.* (2014). Extinction risk and conservation of the world's sharks and rays. *eLife*, 3, e00590.

Edds, D. R., Gillette, D. P., Maskey, T. M. & Mahato, M. (2002). Hot-soda process paper mill effluent effects on fishes and macroinvertebrates in the Narayani River, Nepal. *Journal of Freshwater Ecology*, 17, 543–554.

Eschmeyer, W. N. & Fong, J. D. (2013). Species by family/subfamily. http://research.calacademy.org/research/ichthyology/catalog/SpeciesByFamily.asp (accessed 29 July 2013).

FAO. (2010). *The State of World Fisheries and Aquaculture.* Rome: Food and Agriculture Organisation of the United Nations.

Finer, M. & Jenkins, C. N. (2012). Proliferation of hydroelectric dams in the Andean Amazon and implications for Andes–Amazon connectivity. *PLoS ONE*, 7, e35126.

Freyhof, J. & Brooks, E. (2011). *European Red List of Freshwater Fishes.* Luxembourg: Publications Office of the European Union.

Freyhof, J. & Kottelat, M. (2008). *Stenodus leucichthys.* In: IUCN 2013. IUCN Red List of Threatened Species. Version 2013.1. www.iucnredlist.org (downloaded 31 July 2013).

Froese, R. & Pauly, D. (Eds) (2014). FishBase. www.fishbase.org, version 06/2014.

Fryer, G. & Iles, T. D. (1972). *The Cichlid Fishes of the Great Lakes of Africa.* London: Oliver and Boyd.

Geiger, M. F., Herder, F., Monaghan, M. T., *et al.* (2014). Spatial heterogeneity in the Mediterranean Biodiversity Hotspot affects barcoding accuracy of its freshwater fishes. *Molecular Ecology Resources*, 14, 1210–1221.

Gopal, K. & Agarwal, A. (2003). *River Pollution in India and Its Management*. New Delhi: APH Publishing.

Groombridge, B. & Jenkins, M. (1998). *Freshwater Biodiversity: A Preliminary Global Assessment*. Cambridge: World Conservation Monitoring Centre.

Hammer, M., Wedderburn, S. & van Weenen, J. (2009). *Action Plan for South Australian Freshwater Fishes*. Adelaide: Native Fish Australia (SA) Inc.

Harrison, I. J. & Stiassny, M. L. J. (1999). The quiet crisis: a preliminary listing of the freshwater fishes of the world that are extinct or 'missing in action'. In *Extinctions in Near Time*. MacPhee, R. D. E. (Ed.). New York: Kluwer Academic/Plenum, pp. 271–331.

Hosoya, K. (2008). Circumstance of protection for threatened freshwater fishes in Japan. *Korean Journal of Ichthyology*, 20, 133–138.

IUCN. (2004). Red List Assessment of Madagascar's Freshwater Fishes. http://intranet.iucn.org/webfiles/doc/SpeciesProg/FBU/IUCN_Madagascar_freshwater_fish_2004.pdf

IUCN. (2013). IUCN Red List of Threatened Species. Version 2013.1. www.iucnredlist.org (downloaded 23 July 2013).

Jelks, H. L., Walsh, S. J., Burkhead, N. M., *et al.* (2008). Conservation status of imperiled North American freshwater and diadromous fishes. *Fisheries*, 33, 372–407.

Kang, B., Deng, J., Wu, Y., *et al.* (2013). Mapping China's freshwater fishes: diversity and biogeography. *Fish and Fisheries*, 15, 209–230.

Kingsford, R. T. (2000). Ecological impact of dams, water diversions and river management on floodplain wetlands in Australia. *Austral Ecology*, 25, 109–127.

Kottelat, M. & Freyhoff, J. (2007). *Handbook of European Freshwater Fishes*. Kottelat, Cornol, Switzerland and Freyhoff, Berlin, Germany.

Kottelat, M. & Whitten, T. (1996). *Freshwater biodiversity in Asia with special reference to fish*. World Bank Technical Paper 343.

Kottelat, M., Baird, I. G., Kullander, S. O., *et al.* (2012). The status and distribution of freshwater fishes of Indo-Burma. In *The Status and Distribution of Freshwater Biodiversity in Indo-Burma*. Allen, D. J., Smith, K. G. & Darwall, W. R. T. (Compilers). Cambridge, UK and Gland, Switzerland: IUCN.

Krishnakumar, K., Raghavan, R., Prasad, G., *et al.* (2009). When pets become pests: exotic aquarium fishes and biological invasions in Kerala, India. *Current Science*, 97, 474–476.

Lévêque, C., Oberdorff T., Paugy, D., Stiassny, M. L. J. & Tedesco, P. A. (2008). Global diversity of fish (Pisces) in freshwater. *Hydrobiologia*, 595, 545–567.

Lintermans, M. (2004). Conservation status of Australian fishes (2004). *Australian Society for Fish Biology Newsletter*, 34, 59–64.

Lintermans, M. (2011). Conservation status of Australian Fishes (2011). *Australian Society for Fish Biology Newsletter*, 41, 94–97.

Lintermans, M. (2013), Conservation and management. In *The Ecology of Australian Freshwater Fish*. Humphries, P. & Walker, K. (Eds) Collingwood: CSIRO Publishing, pp. 283–316.

Lowe, S., Browne, M., Boudjelas, S. & de Poorter, M. (2000). *100 of the World's worst invasive alien species. A selection from the global invasive species database*. Published by the ISSG as a special lift out in *Aliens*, 12 December 2000, 12 pp.

Lundberg, J. G., Kottelat, M., Smith, G. R., Stiassny, M. L. J. & Gill, A. C. (2000). So many fishes, so little time: an overview of recent ichthyological discovery in continental waters. *Annals of the Missouri Botanical Garden*, 87, 26–62.

Mace, G. M. & Lande, R. (1991). Assessing extinction threats: towards a re-evaluation of IUCN threatened species categories. *Conservation Biology*, 5, 148–157.

McAllister, D. E., Hamilton, A. L. & Harvey, B. H. (1997). Global freshwater biodiversity: striving for the integrity of freshwater ecosystems. *Sea Wind*, 11.

McDowall, R. M. (1990) *New Zealand Freshwater Fishes: A Natural History and Guide*. Auckland: Heinemann Reed.

McDowall, R. M. (2006). Crying wolf, crying foul, or crying shame: alien salmonids and a biodiversity crisis in the southern cool-temperate galaxioid fishes? *Reviews in Fish Biology and Fisheries*, 16, 233–422.

McLeod, R. (2004). *Counting the Cost: Impact of Invasive Animals in Australia*. Canberra: Cooperative Research Centre for Pest Animal Control.

Minckley, W. L. & Douglas, M. E. (1991). Discovery and extinction of western fishes: a blink of eye in geologic time. In *Battle Against Extinction: Native Fish Management in the American West*. Minckley, W. L. & Deacon, J. E. (Eds). Tucson, AZ: University of Arizona Press.

Ministry of Environment, Government of Japan. (2007). Red List of freshwater and brackish water fish. www.env.go.jp/press/file_view.php?serial=9944&hou_id=8648

Ministry of Environment, Government of Japan. (2010). *Report of Comprehensive Assessment of Biodiversity in Japan (Japan Biodiversity Outlook)*. Nature Conservation Bureau, Ministry of the Environment, Japan.

Mittermeier, R. A., Myers, N., Mittermeier, C. G. & Robles-Gil, P. (1999). *Hotspots: Earth's Biologically Richest and Most Endangered Terrestrial Ecoregions*. Mexico: CEMEX, SA and Agrupación Sierra Madre, SC.

Molur, S., Smith, K. G., Daniel, B. A. & Darwall, W. R. T. (Compilers). (2011). *The Status and Distribution of Freshwater Biodiversity in the Western Ghats, India*. Cambridge, UK and Gland, Switzerland: IUCN, and Coimbatore, India: Zoo Outreach Organisation.

Myers, N. (1988). Threatened biotas: "hot spots" in tropical forests. *The Environmentalist*, 8, 187–208.

Ng, P. K. L., Tay, J. B. & Lim, K. K. P. (1994). Diversity and conservation of blackwater fishes in Peninsular Malaysia, particularly in the north Selangor peat swamp forest. *Hydrobiologia*, 285, 203–218.

Oberdorff, T., Tedesco, P. A., Hugueny, B., *et al.* (2011). Global and regional patterns in riverine fish species richness: a review. *International Journal of Ecology*. Article ID 967631, 12 pages.

OECD (2012). *OECD Environmental Outlook to 2050. The Consequences of Inaction*. Paris: OECD.

Olden, J. D., Hogan, Z. S. & Vander Zanden, M. J. (2007). Small fish, big fish, red fish, blue fish: size-biased extinction risk of the world's freshwater and marine fishes. *Global Ecology and Biogeography*, 16, 694–701.

Ormerod, S. J., Dobson, M., Hildrew, A. G. & Townsend, C. R. (2010). Multiple stressors in freshwater ecosystems. *Freshwater Biology*, 55 (Suppl. 1), 1–4.

Orr, S., Pittock, J., Chapagain, A. & Dumaresq, D. (2012). Dams on the Mekong River: Lost fish protein and the implications for land and water resources. *Global Environmental Change*, 22, 925–932.

Pippard, H. (2012). *The Current Status and Distribution of Freshwater Fishes, Land Snails and Reptiles in the Pacific Islands of Oceania*. IUCN: Gland, Switzerland.

Pittock, J., Meng J., Geiger M., & Chapagain, A. K. (2009). *Interbasin Water Transfers and Water Scarcity in a Changing World – A Solution or a Pipedream?* Berlin: WWF Germany.

Posa, M. C., Wijedasa, L. S. & Corlett, R. T. (2011). Biodiversity and conservation of tropical peat swamp forests. *BioScience*, 61, 49–57.

Qiu, J. (2012). Trouble on the Yangtze. *Science*, 336, 288–291.

Raghavan, R. (2010). Ornamental fisheries and trade in Kerala. In *Fish Conservation in Kerala*. Benziger, A. & Sonnenschein, L. (Eds). St. Louis, USA: World Aquariums and Oceans Federation, pp. 169–197.

Raghavan, R., Prasad, G., Ali, A. & Pereira, B. (2008). Exotic fishes in a global biodiversity hotspot – a case study from River Chalakudy, part of Western Ghats, Kerala, India. *Biological Invasions*, 10, 37–40.

Raghavan, R., Ali, A., Dahanukar, N. & Rosser, A. (2011). Is the Deccan Mahseer, *Tor khudree* (Sykes, 1839) fishery in the Western Ghats Hotspot sustainable? A participatory approach to stock assessment. *Fisheries Research*, 110, 29–38.

Raghavan, R., Dahanukar, N., Tlusty, M. F., *et al.* (2013). Uncovering an obscure trade: threatened freshwater fishes and the aquarium pet markets. *Biological Conservation*, 164, 158–169.

Rahman, M. J. (2006). Recent advances in the biology and management of Indian shad (*Tenualosa ilisha* Ham.). *SAARC Journal of Agriculture*, 4, 76–98.

Reis, R. E. (2013). Conserving the freshwater fishes of South America. *International Zoo Yearbook*, 47, 1–6.

Reis, R. E., Kullander, S. O. & Ferraris Jr, C. J. (Eds). (2003). *Checklist of the Freshwater Fishes of South and Central America*. Porto Alegre: Edipucrs.

Revenga, C. & Kura, Y. (2003). *Status and Trends of Biodiversity of Inland Water Ecosystems*. Montreal, Secretariat of the Convention on Biological Diversity, Technical Series no. 11.

Reynolds, J. D., Webb, T. J. & Hawkins, L. A. (2005). Life history and ecological correlates of extinction risk in European freshwater fishes. *Canadian Journal of Fisheries and Aquatic Science*, 62, 854–862.

Ribbink, A. J. (1988). Evolution and speciation of African cichlids. In *Biology and Ecology of African Freshwater Fishes*. Lévêque, C., Bruton, M. N. & Ssentongo, G. W. (Eds). Paris: ORSTOM, pp. 35–51.

Schaefer, S. A. (1998). Conflict and resolution: impact of new taxa on phylogenetic studies of the Neotropical cascudinhos (Siluroidei: Loricariidae). In *Phylogeny and Classification of Neotropical Fishes*. Malabarba, L. R., Reis, R. E., Vari, R. P., Lucena, Z. M. S. & Lucena, C. A. S. (Eds). Porto Alegre: Edipucrs, pp. 375–400.

Shaji, C. P., Easa, P. S. & Gopalakrishnan, A. (2000). Freshwater fish diversity of Western Ghats, In: *Endemic Fish Diversity of Western Ghats*. Ponniah, A. G. & Gopalakrishnan, A. (Eds). Lucknow: NBFGR-NATP publication, National Bureau of Fish Genetic Resources, pp. 35–35.

Smith, I., Cochrane, P., Stephenson, B., & Gibbs, N. (1997). *State of New Zealand's Environment 1997*. Wellington, New Zealand: The Ministry for the Environment.

Snoeks, J. (2000). How well known is the ichthyodiversity of the large East African lakes? *Advances in Ecological Research*, 31, 17–38.

Snoeks, J. (2004). *The Cichlid Diversity of Lake Malawi/Nyassa/Niassa: Identification, Distribution and Taxonomy*. El Paso, TX: Cichlid Press.

Snoeks, J., Harrison, I. J. & Stiassny M. L. J. (2011). The status and distribution of freshwater fishes. In *The Diversity of Life in African Freshwaters: Under Water, Under Threat. An Analysis of the Status and Distribution of Freshwater Species Throughout Mainland Africa*. Darwall W. R. T., Smith K. G., Allen D. J., et al. (eds). Gland, Switzerland and Cambridge, UK: IUCN.

Stiassny, M. L. J. & Sparks, J. S. (2006). Phylogeny and taxonomic revision of the endemic Malagasy genus *Ptychochromis* (Teleostei: Cichlidae), with the description of five new species and a diagnosis for *Katria*, new genus. *American Museum Novitates*, 3535, 1–55.

Stiassny, M. L. J., Brummett, R. E., Harrison, I. J., Monsembula, R. & Mamonekene, R. (2011). Status and distribution of freshwater fishes of Central Africa. In *The Status and Distribution of freshwater biodiversity in Central Africa*. Brooks, E. G. E., Allen, D. J. and Darwall, W. R. T. (Eds.), Gland, Switzerland and Cambridge, UK: IUCN.

Thieme, M. L., Turak, E., McIntyre, P., et al. (2011). Freshwater ecosystems under threat – the ultimate hotspot. In *Fresh Water The Essence of Life*. Mittermeier, R. A., Farrell, T., Harrison, I. J., Upgren, A. J. & Brooks, T. (Eds). Boulder, CO: CEMEX and iLCP.

Tockner, K., Bunn S. E., Gordon C., et al. (2008). Flood plains: critically threatened ecosystems. In: *Aquatic Ecosystems*. Polunin, N. (Ed.). Cambridge University Press.

Turner, G. F. (1994). Speciation mechanisms in Lake Malawi cichlids: a critical review. *Archiv für Hydrobiologie – Beiheft Ergebnisse der Limnologie*, 44, 139–160.

Turner, G. F. (1996). *Offshore Cichlids of Lake Malawi*. El Paso, TX: Cichlid Press.

UNEP. (2010). *Blue Harvest: Inland Fisheries as an Ecosystem Service*. Penang, Malaysia: WorldFish Center.

Vonlanthen, P., Bittner, D., Hudson, A. G., et al. (2012). Eutrophication causes speciation reversal in whitefish adaptive radiations. *Nature*, 482, 357–362.

Vörösmarty, C. J., McIntyre, P. B., Gessner, M. O., et al. (2010). Global threats to human water security and river biodiversity. *Nature*, 467, 555–561.

Voss, K. A., Famiglietti J. S., Lo, M., et al. (2013). Groundwater depletion in the Middle East from GRACE with implications for transboundary water management in the Tigris–Euphrates–Western Iran region. *Water Resources Research*, 49, 904–914.

Wong, C. M., Williams, C. E., Pittock, J., Collier, U. & Schelle, P. (2007). *World's Top 10 Rivers at Risk*. Gland, Switzerland: WWF International.

WWF. (2010). *Living Planet Report 2010: Biodiversity, Biocapacity and Development*. London: WWF in Association with Zoological Society of London & Global Footprint Network.

Yuma, M., Hosoya, K. & Nagata, Y. (1998). Distribution of the freshwater fishes of Japan: an historical overview. *Environmental Biology of Fishes*, 52, 97–124.

Ziv, G., Baran, E., Nam, S., Rodríguez-Iturbe, I. & Levin, S. A. (2012). Trading-off fish biodiversity, food security, and hydropower in the Mekong River Basin. *Proceedings of the National Academy of Sciences*, 109, 5609–5614.

Why are freshwater fish so threatened?

GERARD P. CLOSS, PAUL L. ANGERMEIER,
WILLIAM R. T. DARWALL AND STEPHEN R. BALCOMBE

2.1 INTRODUCTION

The huge diversity of freshwater fishes is concentrated into an area of habitat that covers only about 1% of the Earth's surface, and much of this limited area has already been extensively impacted and intensively managed to meet human needs (Dudgeon *et al.*, 2006). As outlined in Chapter 1, the number and proportions of threatened species tend to rise wherever fish diversity coincides with dense human populations, intensive resource use and development pressure. Of particular concern is the substantial proportion of the global diversity of freshwater fishes concentrated within the Mekong and Amazon Basins and west-central Africa (Berra, 2001; Abell *et al.*, 2008; Dudgeon, 2011; Chapter 1) with extensive exploitation of water resources planned to accelerate in future years (Dudgeon, 2011; Chapter 1). If current trends continue, and the social, political and economic models that have been used to develop industrialised regions of the world over the past two centuries prevail, then the future of a significant proportion of global diversity of freshwater fish species is clearly uncertain.

Understanding why so many freshwater fish species are threatened requires some understanding of their biology, diversity, distribution, biogeography and ecology, but also some appreciation of the social, economic and political forces that are causing humans to destroy the natural ecosystems upon which we all ultimately depend. To begin to understand the diversity of freshwater fishes, we first need to consider the processes that generated and continue to sustain the diversity of species we see today. Based on an understanding of how freshwater

Conservation of Freshwater Fishes, eds G. P. Closs, M. Krkosek and J. D. Olden. Published by Cambridge University Press. © Cambridge University Press 2016.

fish diversity is generated and sustained, we consider how vulnerable or resilient various freshwater fishes are to the range of anthropogenic impacts that impinge on freshwater ecosystems. Finally, we discuss how social, political and economic drivers influence human impacts on natural systems, and the changes needed to current models of development that can lead to a sustainable future for humans and the diverse range of freshwater fish species with which we share our planet. The aim of this chapter is to provide an overview of the key issues and threats driving the declines in freshwater fish diversity identified in Chapter 1; subsequent chapters provide more detail on the key issues and address our options for developing a sustainable future for freshwater fishes.

2.2 DRIVERS AND PATTERNS OF FRESHWATER FISH DIVERSITY

The diversity of freshwater fish assemblages is determined by interacting hydrological, geomorphic, ecological and physiological processes that act as a hierarchical series of 'filters' to determine which types (Families, Orders, Genera) and species of fish can survive at any particular location (Figure 2.1; Tonn *et al.*, 1990; Angermeier & Winston, 1999; Unmack, 2001a; Smith & Bermingham, 2005; Filipe *et al.*, 2010; Oberdorff *et al.*, 2011; Tisseuil *et al.*, 2013). At the largest spatial and temporal scales, past and present geological and climatic events, latitude and patterns of biogeographic distribution and dispersal of various taxa determine the pool of species available within a particular region (Tonn *et al.*, 1990; Abell *et al.*, 2008). At finer spatial and temporal scales, contemporary processes filter out which species will survive at any particular location within that region (Tonn *et al.*, 1990). These contemporary processes include the physical and chemical conditions within regions, rivers or lakes, barriers to migration within catchments, and biological interactions between species, life-history stages and individuals (Tonn *et al.*, 1990; Angermeier & Winston, 1999). At these finer scales, the morphological, physiological and behavioural attributes of different species, life stages and individuals interact with the physical and chemical environment to determine microhabitat selection, feeding strategies, access to key resources and the outcomes of various inter- and intraspecific interactions (Figure 2.1; Angermeier & Winston, 1999; Dunham *et al.*, 2008).

An additional element in understanding the distribution of freshwater fish taxa is how the role of biogeography and past events determine distribution. Many freshwater fishes face significant challenges

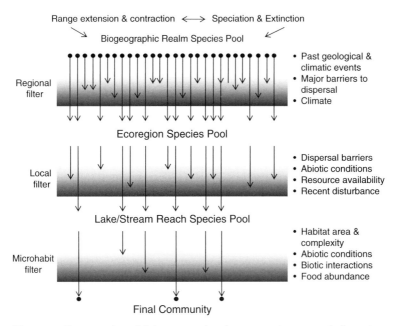

Figure 2.1 Conceptual model demonstrating the progressive removal of species (shown as arrows) as geographic scale decreases. The largest pool of species is available at the scale of the biogeographic realm, then reduced by a series of 'environmental filters' that refine the range of species available to colonise the habitat available at any particular location (modified from Tonn *et al.*, 1990).

dispersing between catchments or watersheds (Tonn *et al.*, 1990; Matthews, 1998; Unmack 2001b). Myers (1938) recognised that some groups of freshwater fish (e.g. Cypriniformes, the most speciose order of freshwater fish) have little or no tolerance of sea water, either now or in their evolutionary history, describing them as 'primary division' freshwater fishes. Hence, for these species, dispersal is only possible through connected fresh waters. Opportunities for natural range extension between catchments are limited to rare events, such as changes to drainage patterns due to mountain-building, volcanism, hydrological connections over drainage divides, or falls in sea level that allow drainages to merge on coastal shelves (Unmack, 2001a). Many families of primary division freshwater fishes have relatively limited distributions, perhaps reflecting limited past or present opportunities for dispersal away from the region in which they first evolved (Berra, 2001). Further, given the limited scope for dispersal, extinction due to adverse events and conditions is always more likely than colonisation, and once a species is lost from a catchment, recolonisation is unlikely (Unmack, 2001a, 2013; Tisseuill *et al.*, 2013). Hence, catastrophic disturbances (e.g. glaciation or volcanism) that eliminate particular types or species of fish from a catchment or region can have long-lasting impacts on diversity and composition of the fish communities of the disturbed area.

'Secondary' and 'peripheral' freshwater fish divisions represent those taxa that either can or have previously dispersed through marine habitats (Myers, 1938). A past or present ability to disperse through marine habitats means the distributions of many secondary and peripheral fish taxa span multiple biogeographic realms, catchment boundaries or oceanic islands (e.g. the freshwater anguillid eels which are found on most continents).

Globally, major differences can be seen in the distribution and composition of freshwater fish diversity. The majority of the 138 freshwater fish families are primary division (93 primary, 15 secondary and 30 peripheral; Table 2.1), and hence have limited past or present capacity to disperse through marine habitats (Berra, 2001). Consequently, strong differences in the taxonomic composition of fish communities can be seen across the six biogeographic realms that support freshwater fishes (Nearctic, Neotropical, Palarctic, Ethiopian, Oriental and Australasian). Overall, 72 primary, 7 secondary and 8 peripheral freshwater fish families are endemic to a single biogeographic realm (Table 2.1; Berra, 2001). The greatest diversity of freshwater fishes occurs in the Neotropical realm, with 51 freshwater fish families, of which 35 are endemic. The Oriental and Ethiopian realms are also very diverse, supporting 41 and 37 families, of which 15 and 19 are endemic, respectively. In contrast, the Australasian realm supports 21 freshwater fish families, of which only 7 are endemic (although most species are endemic).

This high degree of regional endemicity at the family level generates a distinctive fauna dominated by different families within each biogeographic realm (Berra, 2001). The most diverse orders are Cypriniformes (> 2600 species), Siluriformes (> 2300), Perciformes (> 2200), Characiformes (> 1300) and Cyprinodontiformes (> 800), with most other families contributing < 100 species per group (Berra, 2001). The Cypriniformes dominate fish assemblages across much of the Nearctic, Palarctic, Ethiopian and Oriental realms, but are completely absent from the Neotropical and Australasian realms. The fauna of the Neotropical realm is highly diverse, but representatives of the Siluriformes, Characiformes, Cyprinodontiformes and Perciformes (particularly Cichlidae) dominate. In contrast, the Australasian realm exhibits relatively

Table 2.1 *Numbers of primary, secondary and peripheral freshwater fish families and patterns of endemism across the 6 biogeographic realms (after Berra, 2001).*

	Nearctic	Neotropical	Palearctic	Ethiopian	Oriental	Australasian
No. primary	15	35	15	29	27	2
No. secondary	7	8	2	5	3	4
No. peripheral	9	8	13	3	11	14
No. endemic primary	8	32	0	18	13	2
No. endemic secondary	1	2	1	1	0	2
No. endemic peripheral	0	1	2	0	2	3
Total endemic families	9	35	3	19	15	7

low diversity, but is very distinctive with Antheriniformes (particularly Melanotaeniidae) dominating the northern regions, and Osmeriformes (particularly Galaxiidae) and Perciformes (Percichthyidae) dominating the southern regions (Berra, 2001; Unmack, 2001a). The Nearctic region, which has been and continues to be subject to harsh climatic conditions (e.g. extensive ice-age glaciation) is generally characterised by low levels of diversity and endemism (Abell *et al.*, 2008). Nearctic streams are also dominated by secondary division families, particularly Salmonidae, reflecting the ability of these fishes to recolonise following the retreat of ice-age glaciers (Abell *et al.*, 2008)

Relatively few families of freshwater fish are distributed across three or more biogeographic realms. Some primitive primary division families (e.g. Osteoglossidae) have broad distributions, most likely reflecting ancient patterns of continental breakup and drift. However, most widespread groups are either secondary or peripheral division fishes that have a marine ancestry, a marine life-history stage, or are tolerant of salt water (e.g. Anguillidae, Cichlidae, Galaxiidae, Eleotridae). Some very widespread families (e.g. Clupeidae and Gobiidae) are primarily marine but contain some freshwater species (Table 2.2). The Gobiidae, in particular, occur on many isolated oceanic islands, particularly across the Indo-Pacific Region, where species either have a recent marine ancestry or retain a marine larval life-history stage (Keith, 2003). While many freshwater gobiids are widely distributed, others have distributions that are restricted to single islands or catchments on larger islands (Keith, 2003; Maruyama *et al.*, 2004; Lord *et al.*, 2012; Taillebois *et al.*, 2012).

Table 2.2 *Distribution of biogeographically significant and/or particularly diverse orders and families of freshwater fish (defined as those species that are dependent on freshwater, spending either all or a significant proportion of their life cycle in freshwaters). The number of freshwater species (# sp.) in each family is based on Berra (2001). The numbers of known species in many of these groups are underestimates given the description of new species continues. Full spellings of the regions are provided in Table 2.1.*

Order	Family	# sp.	Ne	Neo	Pal	Eth	Or	Au
Acipenseriformes	Acipenseridae	24	x		x		x	
Osteoglossiformes	Osteoglossidae	7		x		x	x	x
	Mormyridae	95				x		
Anguilliformes	Anguillidae	20	x	x	x	x	x	x
Clupeiformes	Clupeidae	50	x	x	x	x	x	x
Cypriniformes	Cyprinidae	2010	x		x	x	x	
	Catostomidae	66	x		x			
	Cobitidae	164			x		x	
	Balitoridae	485			x	x	x	
Characiformes	Characidae	700	x	x				
	Anostomidae	105		x				
	Distichodontidae	90				x		
Siluriformes	Loricariidae	650		x				
	Callichthyidae	174		x				
	Sisoridae	120			x		x	
	Ictaluridae	45	x					
Esociformes	Esocidae	5	x		x			
Osmeriformes	Osmeridae	236	x		x			
	Retropinnidae	6						x

Salmoniformes	Galaxiidae	51	x		x			x
	Salmonidae	66				x	x	x
Atheriniformes	Melanotaeniidae	68	x					x
Cyprinodontiformes	Cyprinodontidae	100			x	x	x	
	Rivulidae	200			x			
	Poeciliidae	300			x		x	x
Perciformes	Percichthyidae	20			x			x
	Centrarchidae	30	x		x			
	Percidae	195				x		
	Cichlidae	2000		x	x		x	
	Eleotridae	150	x	x	x		x	x
	Gobiidae	200		x	x	x	x	x

The large biogeographic realms can be further divided into distinct ecoregions (Figure 1.1); Abell *et al.* (2008) recognised 426 ecoregions, defined by unique assemblages of fish largely delineated by major catchment or watershed boundaries. It is at this smaller ecoregion scale that we most clearly see the interplay between regional and local processes shaping the diversity of freshwater fish communities. Superimposed on the distinctive fauna of each biogeographic realm is the imprint of a variety of long-term, large-scale processes, including continental drift, mountain-building, glaciation, climate, increasing aridity and previous marine transgressions, that may affect all or part of the various realms, thus creating unique evolutionary and environmental histories, and unique assemblages of fishes for each region and river system (Matthews, 1998; Unmack, 2001a; Abell *et al.*, 2008; Olden *et al.*, 2010).

Despite the different evolutionary starting points for the fauna of each ecoregion, many freshwater fish communities and species exhibit similar 'traits' at different locations (Winemiller & Rose, 1992; Angermeier & Winston, 1999; Heino *et al.*, 2013). Traits can be viewed as the morphological, physiological, behavioural and life-history attributes of each species, with different combinations of traits determining where each species and their various life-history stages can live (Winemiller & Rose, 1992; Olden *et al.*, 2008; Chapter 3). Even though they may be taxonomically distinct, fishes living in similar types of habitats (i.e. lakes, slow rivers, mountain streams, etc.) face similar physical and chemical challenges. Hence, similar traits represent similar 'solutions' to common problems, i.e. evolutionary convergence (Winemiller & Rose, 1992; Angermeier & Winston, 1999; Lamouroux *et al.*, 2002). High-elevation, fast-flowing streams in the Northern Hemisphere are dominated by fusiform salmonids, whereas similarly shaped galaxiids dominate in many small streams in the Southern Hemisphere. Slow-flowing rivers and lakes are often dominated by a large biomass of detritivorous fishes. Across Eurasia and Africa, various cyprinids dominate these low-velocity, turbid habitats, whereas in North America cyprinids or catostomids (suckers) dominate. In South America, where cyprinids or catostomids are absent, detritivorous prochilodontids dominate (Taylor *et al.*, 2006) and other species of Characidae occupy a variety of non-benthic niches. Species that possess a set of traits specialised for life in a very specific type of habitat are generally considered to be at greater risk of extinction than more generalised species that possess sets of traits that enable them to live across a broad range of habitats (Angermeier, 1995; Olden *et al.*, 2008).

2.3 LIFE HISTORY, SPECIES TRAITS AND THE RISK OF EXTINCTION

Life histories evolve to cope with the conditions likely to be encountered over an organism's life (Winemiller & Rose, 1992). The life histories of many species exhibit combinations of traits, known as trait synergisms or syndromes, that influence the risk of extinction when challenged by unexpectedly adverse conditions (Figure 2.2; Winemiller & Rose, 1992; O'Grady et al., 2004; Frimpong & Angermeier, 2010; Mims et al., 2010). A tri-lateral continuum model of life-history strategies has been proposed for freshwater fish species (Winemiller & Rose, 1992), with the three end-point strategies defined as periodic, opportunistic and equilibrium. These three strategies enable different species to cope with specific environmental conditions, the predictability of which varies in time and space (Winemiller & Rose, 1992; Mims et al., 2010). Opportunistic life-history strategists are typified by small-bodied species with early maturation and low juvenile survival, and often associated with frequently disturbed habitats. Equilibrium life-history strategists tend to be small to medium-sized, with moderate age at maturation (relative to other freshwater fish species), low fecundity per spawning event and high juvenile survival largely associated with intensive parental care and small clutch size. Such species are likely to survive in stable habitats. Periodic life-history strategists are characterised by large body size, late maturation, high fecundity but low juvenile survivorship, and are often associated with highly periodic (seasonal) environments (Winemiller & Rose, 1992; Mims et al., 2010).

The life-history model proposed by Winemiller and Rose (1992) suggests that certain combinations of traits increase the vulnerability of species to decline and extinction when faced with disturbance. Equilibrium and periodic breeding species are considered to be more prone to extirpation than opportunistic species when confronted with unpredictable anthropogenic impacts, given they require relatively stable environments, and exhibit prolonged growth and late maturity (Mims et al., 2010). Other key traits include extent of distributional range, body morphology, reproductive strategy, and specialisation in habitat, feeding, or life history (Winemiller & Rose, 1992; Angermeier, 1995; Olden et al., 2008). Distributional range has been observed to be a key predictor of extinction risk (Angermeier, 1995; Van Allen et al., 2012); species with limited ecological and geographical ranges tend to be more vulnerable to extinction due to their lower overall population size, a tendency towards specialised habitat requirements, and increased probability that a significant

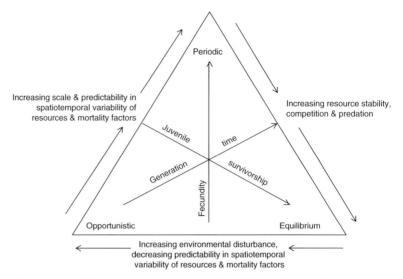

Figure 2.2 Life-history continuum model adapted from Mims *et al.* (2010), Winemiller (2005) and Winemiller and Rose (1992). Inside arrows summarise the key trade-offs defining the three end-point strategies. Outside arrows summarise the selection pressures that favour each strategy in relation to environment.

proportion of their natural range will be affected by threatening processes (Angermeier, 1995; Reynolds *et al.*, 2005). Body size, another trait associated with increased extinction risk, is correlated with growth, which in turn is associated with natural mortality rates, longevity, age at maturity and reproductive output (Winemiller & Rose, 1992; Winemiller, 2005).

A bimodal relationship exists between body size and extinction vulnerability, with small and large species being disproportionately threatened (Reynolds *et al.*, 2005; Olden *et al.*, 2007). Small-bodied species are more diverse, often with limited or fragmented distributions, and limited capacity to disperse due to physiological constraints and body morphology (Olden *et al.*, 2007). Additionally, smaller-bodied fishes are more likely to have adverse interactions with non-native species, being more susceptible to predation and less likely to be competitively dominant (McDowall, 2006). Large-bodied species are often long-lived, with vulnerable extended migration and dispersal pathways, late maturity, and are often subject to fishing pressure (Reynolds *et al.*, 2005; Olden *et al.*, 2007; Stone, 2007). Of course, these results do not mean that all small or large species are at risk of decline and extinction. Small body size can confer a high degree of resilience to adverse impacts, often facilitated by rapid growth to maturity and early reproduction, i.e. an opportunistic life-history strategy (Winemiller & Rose, 1992). Similarly, large body size can confer longevity and high fecundity provided off-spring survival is occasionally high (Winemiller, 2005).

Trait analyses by Olden *et al.* (2008) on lower Colorado River Basin desert fishes highlight how key traits influence survival under varying conditions, and how traits interact to increase extinction risk. Species such as razorback sucker (*Xyrauchen texanus*) that provide little parental care (i.e. invest little energy in reproduction with respect to guarding and nest-building) exhibit reduced population persistence due to low off-spring survival and a consequent lack of population resilience following disturbance. That said, even when fish displayed greater parental care, extinction risk was increased when coupled with a highly specialised diet, highlighting synergistic effects between life-history traits. Olden *et al.* (2008) also found that herbivores and detritivores were three times more widespread compared to other species, indicating that a special-ised diet may increase the risk of local extirpation and extinction. They observed that fishes with a large body size at maturity, or that combined a small size at maturity with low fecundity, had an increased extinction risk. In contrast, small, highly fecund fish were most likely to persist after a disturbance event (Olden *et al.*, 2008). Species with relatively high parental care, a trait typically associated with 'slow', vulnerable life cycles, were buffered from extinction if they were feeding generalists or strong swimmers. Such 'antagonistic' relationships between traits highlight the complex nature of trait synergisms, and explain how different life-history strategies may be successful or fail in the same environment (Olden *et al.*, 2008). Our understanding and prediction of fish species extinctions would be improved by explicitly considering how biological traits com-bine to predispose species to the primary components of the extinction process (Olden *et al.*, 2007). Trait-based analysis is increasingly used to predict species and community responses to (a) habitat and climate alter-ation and (b) species extinction and invasion (Frimpong & Angermeier, 2010). Such analyses are especially useful in elucidating how habitat tem-plates 'filter' species to produce existing patterns or to forecast commu-nity responses to future conditions (e.g. Novak *et al.*, 2011).

2.4 MOVEMENT AND DISPERSAL OF FISH IN DENDRITIC DRAINAGE NETWORKS

Freshwater systems form complex dendritic river and stream networks that are intricately connected to their catchment by the flow of water (Poff *et al.*, 1997; Malmqvist & Rundle, 2002). All freshwater systems, including lakes, are characterised and structured by the constant movement of water (Poff *et al.*, 1997). Water enters as rainfall or groundwater infiltration, and leaves, often after only a few hours or days, as either evaporation, re-infiltration back into groundwater or discharge to the sea. The constant loss of water from freshwater systems requires constant replacement if the system is to avoid desiccation. The predominant flux of materials through a catchment is from upstream to downstream (Malmqvist & Rundle, 2002), but there can also be movement of material upstream; for example, the upstream migration of many species of fish can have important consequences for headwater ecosystems (Schindler *et al.*, 2013). Consequently, freshwater systems are at high risk for the spread of water-borne pollutants (nutrients, silt, toxicants) from upstream to downstream, changes to flow regimes downstream of water abstraction and dams (Poff *et al.*, 1997), and the spread of introduced aquatic organisms (Malmqvist & Rundle, 2002). Further, dendritic drainage systems form long, narrow, linear channels that wind through the landscape, creating an extensive riparian interface with the surrounding terrestrial environment; any change or inputs associated with this riparian interface can have profound impacts on the adjoining aquatic systems (Dudgeon *et al.*, 2006; Chapter 9). Longitudinal connectivity along narrow, linear river systems is also readily disrupted by any number of mechanisms, including the presence of lakes, waterfalls, weirs and dams, low discharge, pollution and habitat alteration (Poff *et al.*, 1997; Malmqvist & Rundle, 2002; Dudgeon *et al.*, 2006). Thus, protecting and isolating key areas from anthropogenic impacts is difficult, particularly in large catchments or watersheds, and especially those located in densely populated or intensively managed lowland areas (Dudgeon *et al.*, 2006).

Natural landscapes are mosaics shaped by a variety of physical and biotic processes operating across different spatial and temporal scales (Dunham & Rieman, 1999). Riverine landscapes in particular exhibit natural patchiness, due to their dendritic structure and the longitudinal variation of habitat from the headwaters to the sea (Fagan *et al.*, 2002). As such, the distribution of organisms is rarely homogenous across a riverine landscape or 'riverscape' (*sensu* Fausch *et al.*, 2002), and populations may be subdivided or patchy due to specialised habitat requirements, local extinctions, and the non-uniform movement of animals between suitable habitat patches (Rieman & Dunham, 2000; Woodford & McIntosh, 2010, 2011). Small populations face a higher risk of extinction through demographic and environmental stochasticity, and reduced genetic diversity (Morita & Yamamoto, 2002). Generally, the more fragmented a species' distribution, the more vulnerable that species is to extinction (Fagan *et al.*, 2002; Morita & Yamamoto, 2002; Chapter 10). This fragmented distribution can generate a 'metapopulation' spatial structure, whereby discrete subpopulations exist in more or less favourable habitats, but long-term persistence may depend on dispersal from or to other subpopulations (Schlosser & Angermeier, 1995; Woodford & McIntosh, 2010, 2011). Subpopulations intermittently go extinct but, provided dispersal rates are adequate, the metapopulation persists. As populations become more distant from each other, they also become more isolated and exchange individuals less frequently, and the area of 'sink' habitat, where persistence is impossible and individuals are lost from the population, increases (Woodford & McIntosh, 2011). Fragmented populations, and the multiple factors affecting the ability of organisms to disperse among populations, makes metapopulation dynamics a challenging but crucial aspect of understanding the processes driving the distribution of fishes within freshwater systems (Schlosser & Angermeier, 1995; Dunham & Rieman, 1999; Woodford & McIntosh, 2011).

The dynamics of movement and dispersal of organisms within complex dendritic riverscapes have distinct properties, not readily predicted from a consideration of the simple linear or two-dimensional frameworks commonly applied to rivers and terrestrial systems, respectively (Fagan, 2002). While a short geographic distance may separate the tips of headwater streams, the actual pathway through the connecting stream network may be considerable, presenting many potential barriers to migration (Figure 2.3). Thus, events not constrained to the dendritic network may impact headwater streams that are within the same local area, but are isolated within the river network due to a confluence many kilometres downstream, or may even drain into completely different river systems, or even different oceans (Fagan, 2002). Hence, populations in different streams may be protected from invasive aquatic species by relatively impermeable geographic catchment boundaries, but may be vulnerable to a single event in the surrounding terrestrial habitat that crosses catchment boundaries. For example, Brown *et al.* (2001) observed extirpations of endangered Gila trout (*Oncorhynchus gilae gilae*) populations that were in geographically close headwater streams subject to a single large wildfire, but were otherwise isolated within river networks.

The location of barriers to movement within dendritic networks may also significantly alter the patterns of isolation and movement

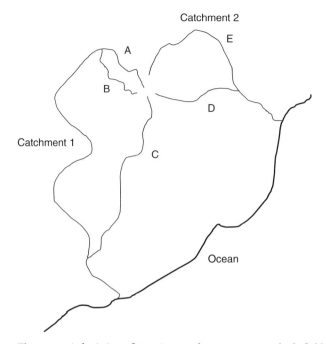

Figure 2.3 A depiction of two river catchments or watersheds fed by five headwater streams. While the geographic distance separating the headwaters of all streams is small, the distances to travel between streams through the dendritic network varies greatly depending on the distance of the confluence downstream. Dispersal between tributaries A and B, and E and D, requires only a relatively short distance downstream. However, individuals in A and B must move a significantly longer distance downstream if they are to enter tributary C. Movement of individuals between catchments 1 and 2 requires long-distance dispersal through freshwater and marine environments, even though the headwaters of each system are in close geographic proximity to each other.

within a network (Figure 2.3). Barriers to movement in trunks will tend to generate larger isolated fragments than barriers nearer the branch tips (Fagan, 2002). However, even seemingly large river fragments can be too small to accommodate certain life-history traits. For example, several cyprinids (e.g. peppered chub, *Macrhybopsis tetranema* and Arkansas River shiner, *Notropis girardi*) of the Great Plains (North America) have been extirpated from river fragments < 136 km long because shorter fragments are too short to allow their drifting eggs and larvae to mature properly (Perkin & Gido, 2011). Large fragments comprising multiple subpopulations that continue to exchange individuals (i.e. metapopulations) tend to be more resilient and resistant to extirpation than small populations that are distant from other populations (Dunham & Rieman, 1999; Woodford & McIntosh, 2010, 2011). In many systems, the pathways between habitat patches may be additionally fragmented by anthropogenic impacts, including introductions of non-native fishes, habitat degradation, dams and altered disturbance regimes, further reducing the likelihood that patches of suitable habitat can be recolonised following stochastic events that might occasionally extirpate local populations (Malmqvist & Rundle, 2002; Dudgeon, 2011; Woodford & McIntosh, 2011). Combinations of natural and anthropogenic barriers to migration and dispersal can have synergistic impacts, exacerbating the isolation of populations within a riverscape and increasing the likelihood of extinction over time (Winston *et al.*, 1991; Reyes-Gavilán *et al.*, 1996; Woodford & McIntosh, 2011). Despite the various risks to fragmented populations, in some situations, isolation and fragmentation can confer a degree of protection from introduced predators and competitors. Numerous isolated populations of threatened galaxiid species persist upstream of barriers to introduced trout in headwater refuges in New Zealand, southern Australia and South Africa (McDowall, 2006) and many barriers are constructed in the western US to prevent the movement of non-native trout upstream into areas containing native trout (Novinger & Rahel, 2003).

Barriers to movement and patterns of fragmentation may also change over time (Fagan, 2002). Unmack (2001b) observed that desert fish populations in separate streams were isolated seasonally when downstream reaches dried out, with movement between streams only possible during floods. Closs and Lake (1996) observed upstream invasion by trout into galaxiid habitat during a series of wet years, resulting in the fragmentation of the native galaxiid population. Subsequently, trout were eliminated in a dry year, allowing galaxiids to repopulate sections of stream from which they had been previously excluded by trout. Fragmentation may also be caused by species' abiotic limitations and barriers created by key environmental parameters, such as temperature, that may change over time. Nakano *et al.* (1996) modelled the potential impact of a long-term increase in temperature of 4°C on Japanese char, and predicted a 90% reduction in distribution, including fragmentation of the presently highly connected population into isolated high-elevation streams, potentially increasing extinction vulnerability.

2.5 COMPETING FOR FRESH WATER: UNDERSTANDING, MANAGING AND MITIGATING HUMAN IMPACTS ON AQUATIC FAUNA

Assemblages of freshwater fish are shaped by, and closely linked to, natural and anthropogenic processes associated with the landscapes and catchments in which they are located (Angermeier & Winston, 1998). Increasingly, human activities associated with the production of food, energy, shelter and wealth are altering natural landscapes and catchments on a global scale (Malmqvist & Rundle, 2002; Dudgeon *et al.*, 2006; Strayer & Dudgeon, 2010; Dudgeon, 2011; Naiman & Dudgeon, 2011). Consequently, human activities are now the single most important determinant of the diversity and abundance of freshwater fishes across a significant proportion of the global landscape (Dudgeon, 2010b, 2011; Naiman & Dudgeon, 2011). Despite the many complexities involved, the anthropogenic factors impacting freshwater fish assemblages are generally well known and understood, as are many of the strategies to avoid or mitigate their impacts (Dudgeon, 2010b, 2011; Naiman & Dudgeon, 2011; Naiman, 2013; Vörösmarty *et al.*, 2013). Hence, the adverse impacts of human activities are not an inevitable outcome of human economic development; rather, they are the (often inadvertent) outcome of particular sociopolitical choices stemming from how society values and treats ecosystems. Depending on the choices and decisions taken, most impacts on freshwaters can be minimised or mitigated significantly, thereby facilitating conservation of their biota and ecosystem services (Naiman & Dudgeon, 2011; Villamagna *et al.*, 2013).

Human populations are generally concentrated, and their impacts greatest, in productive lowland regions, where populations of freshwater fish are often most abundant and diverse (see Abell *et al.*, 2008). However, the impacts associated with the provision of the food, shelter and energy required to sustain modern industrial human societies often extend into the headwaters of catchments; impacts can include the loss of migratory fishes due to barriers to migration, or the construction of dams in steep valleys for water storage and hydroelectric generation, or even the more general impacts of deforestation and climate change (Dudgeon *et al.*, 2006; Chapters 3 and 4). Thus, almost all habitats within a catchment or watershed may be affected by human activities in some way across extensive areas of the planet (Malmqvist & Rundle, 2002; Dudgeon *et al.*, 2006). Although the effects of human economic activity and associated alterations to watercourses are diverse, they can be summarised under four interrelated headings: ecosystem destruction, habitat alteration and degradation, water pollution and altered water chemistry, and species additions and removals (Table 2.3; Malmqvist & Rundle, 2002; Dudgeon *et al.*, 2006).

2.5.1 Environmental destruction

Ecosystem destruction is the most extreme end-point of a continuum of anthropogenic impacts. Deterioration at the broad catchment level often results from a failure to associate the cumulative effects of local actions to larger-scale spatial and temporal processes, with these cumulative impacts ultimately causing ecological collapse across larger areas (Vörösmarty *et al.*, 2013). Intensive urban and agricultural development is often associated with the physical and ecological destruction of natural riverscapes through local drainage, channelisation and piping, or in many cases, the total removal of surface water through abstraction (removal of water from waterway by channelisation or pumping) from multiple, often poorly regulated water intakes (Malmqvist & Rundle, 2002; Wedderburn *et al.*, 2012). The desiccation of the Aral Sea due to diversion of inflowing rivers and the subsequent loss of multiple species is one of the most graphic examples of extensive ecosystem destruction due to water abstraction (Chapter 1). Urban streams, in particular, often suffer severe impacts, with increases in the impervious surface area within the catchment and stormwater management resulting in multiple widespread changes to catchment hydrology, flow regime and sediment dynamics that disconnect streams from the floodplain and result in channel simplification (Bernhardt & Palmer, 2007). In many cases, a stream may be reduced to an 'efficient gutter' by the construction of piped and sealed channels to convey water away as rapidly as possible from urban environments, with no consideration given to the habitat they might provide (Bernhardt & Palmer, 2007). Surface mining can also result in the destruction of watercourses across extensive areas in the effort to gain access to underlying minerals (Palmer *et al.*, 2010). However, the destruction of natural riverscapes is not an inevitable consequence of urban, agricultural or industrial development. Adverse impacts can be mitigated by maintaining in-stream habitat and enhancing infiltration, evaporation, transpiration and storage (Walsh *et al.*, 2005). The destruction of streams from mining and agriculture results from lax environmental policy and enforcement across multiple regions (Dudgeon, 2010a; Palmer *et al.*, 2010). In many semi-arid regions, surface water remains, but ecologically damaging processes, such as secondary salinisation due to the cumulative impact of local

Table 2.3 *Summary of the main anthropogenic impacts impinging on fish assemblages in freshwaters (after Malmqvist & Rundle 2002).*

Ultimate forcing factor	Subfactor	Proximate causes	Abiotic impact	Biotic implications
Ecosystem destruction		Urban and agricultural expansion, water abstraction	Complete ecosystem loss	Extinction of entire assemblage
Habitat alteration	Hydrology	Dams, channelisation, water abstraction, deforestation, water transfer	Alteration to natural flow regime, loss of upstream-downstream linkage	Altered habitat conditions, loss of cues for key life-history stages
	Siltation	Deforestation, agricultural and urban land use	Reduced substratum complexity, reduced hyporheic exchange	Altered habitat conditions
	Altered riparian corridor	Agriculture, urbanisation	Altered energy inputs, temperature regime, woody debris and marginal habitats	Altered habitat conditions and trophic dynamics
Water chemistry	Acidification	Industrial emissions	Reduced pH, increased Al^{3+}	Direct physiological and toxic effects
	Nutrient addition	Agricultural and urban discharge, atmospheric deposition of NO_x	Increased N and P	Altered habitat conditions and trophic dynamics
	Organic pollution	Agricultural and urban waste	Increased suspended solids, reduced O_2	Reduced habitat availability
	Various toxins	Agriculture, urban and industrial	Wide variety of acute and chronic impacts	Direct physiological/toxic effects
Species removal and addition		Fisheries, aquaculture, ornamental, intercontinental and intercatchment transport	Invasive species	Increased/reduced competition and predation, altered trophic dynamics, hybridisation, loss of genetic structure

processes such as over-irrigation and clearing of natural vegetation, render any remaining water unsuitable for most species of freshwater fish, and alter or eliminate freshwater communities irreversibly (Beatty *et al.*, 2011). The combination of increasing human populations and climate change threaten to expand the areas affected by such adverse processes (Bernhardt & Palmer, 2007; Beatty *et al.*, 2011). There is an urgent need to develop sustainable water research and management strategies that integrate and accommodate local, regional and global perspectives and processes, aimed at reducing the cumulative large and small impacts that collectively cause widespread destruction (Vörösmarty *et al.*, 2013).

2.5.2 Habitat degradation and alteration

Habitat degradation and alteration occur before the end-point of 'ecosystem destruction' is reached. Clearly, there is a continuum of development and impacts that can ultimately end in ecological destruction, but determining the appropriate developmental end-points before large-scale deterioration occurs is challenging due to political and social pressures, poor knowledge of ecological system structure and function, and the complex dynamics that result from multiple cross-scale interactions (Gupta *et al.*, 2013; Vörösmarty *et al.*, 2013). Urban and agricultural development degrades streams in a variety of ways by increasing inputs of multiple pollutants, increasing sedimentation, changing the structure of riparian vegetation resulting in the loss of shading and inputs of woody debris, altering flow regimes, eroding and simplifying stream channels, and creating barriers to movement (Bernhardt & Palmer, 2007; Collier *et al.*, 2009; Beatty *et al.*, 2011). These impacts often interact in complex ways to collectively create 'multiple stressor' effects (Wagenhoff *et al.*, 2012; Chapter 6). Channel complexity and habitat structure are often reduced to create space for roads and infrastructure, with limited options for restoration due to the presence of infrastructure on floodplains and stream corridors (Bernhardt & Palmer, 2007). These changes ultimately lead to a reduction in the amount of suitable stream habitat available to fishes and their prey, and altered flow regimes (Bernhardt & Palmer, 2007). The processes that lead to habitat degradation and the measures to mitigate many of the negative impacts of agricultural and urban development on streams and lakes are generally well understood, although the social and political will to address these issues is frequently weak or lacking altogether (Dudgeon, 2010b; Verburg *et al.*, 2013; Wilcock *et al.*, 2013).

Urban and agricultural streams tend to support fewer sensitive fish taxa compared to natural systems, with communities in severely impacted systems often comprising a few resilient, and often non-native, fish species (Genito *et al.*, 2002; Collier *et al.*, 2009). Fish diversity and richness generally decrease with agricultural intensification, and widespread landscape modification usually has an increasingly negative effect on the integrity of aquatic communities (Mensing *et al.*, 1998; Allan, 2004). Barriers to migration and fragmented fish populations may result from built structures such as culverts, weirs and dams, or from the degradation of stream reaches due to habitat loss or pollution (Berhardt & Palmer, 2007; Collier *et al.*, 2009; David *et al.*, 2014; Chapter 4). Globally, river systems have been fragmented by over 50,000 large dams and an unknown number of smaller dams which act as barriers to fish migrations (Richter *et al.*, 2010). Obviously, dams are a major threat to diadromous species such as eels (*Anguilla*), shads (*Alosa*), sturgeons (*Acipenser*), salmonids (e.g. *Salmo, Salvelinus, Hucho, Onchorhynchus, Stenodus*) and many other species, as they inhibit or prevent migration between spawning and foraging grounds. Less obvious is the interruption of migrations of many potamodromous species that migrate within river basins (Wells *et al.*, 2003; Chapter 11). For example, the construction of multiple dams in the Yangtze River basin has been a major contributing factor to the extinction or near-extinction of endemic mega-fishes such as the Chinese paddlefish (*Psephurus gladius*), Chinese sturgeon (*Acipenser sinensis*) and Yangtze sturgeon (*Acipenser dabryanus*) (Dudgeon, 2010a). To some degree, barriers to upstream migration created by large dams and weirs can be overcome using fish passes, although the success of such measures depends on multiple factors such as size of structure, volume and hydraulics of water moving through, flow regimes up- and downstream of the barrier, and species involved (Mallen-Cooper & Brand, 2007; Chapters 4 and 11). Smaller barriers to upstream migration, such as low weirs or culverts, may be overcome through innovative and relatively low-cost measures such as the placement of mussel spat ropes across barriers to reduce water velocities and create refuges for passing fishes (David *et al.*, 2014). However, less attention has been given to downstream migration, and in some cases the options for mitigation may be limited. The downstream drift of eggs and larvae is a feature of the life cycle of many fish species in large tropical river systems, but is blocked by large impoundments, resulting in the death of eggs and larvae (Pompeu *et al.*, 2012).

Water resource development (for direct human use, agriculture, industry and flood protection) creates further habitat degradation through the alteration of natural flow regimes through rivers and lakes

(Poff *et al.*, 1997). Flow is a major determinant of physical habitat, and reduced or altered flow reduces the area and suitability of habitat available to specialised species (Poff *et al.*, 2007). Flow management for human purposes also tends to homogenise flow by reducing peak and increasing minimum flows (Moyle & Mount, 2007; Poff *et al.*, 2007). Many fish species have evolved life-history strategies adapted to predictable but seasonally variable flow regimes, with patterns of flow providing critical cues and conditions for various life-history stages (Humphries *et al.*, 1999; Humphries & Lake, 2000; King *et al.*, 2003). For example, in Appalachian streams, cottid and cyprinid species are adapted to spawn under distinct combinations of flow, temperature and photoperiod (DeHaven *et al.*, 1992). The altered condition of flow-managed rivers often excludes these native specialists in favour of more generalist species (Humphries & Lake, 2000). Indeed, it has been argued that the global homogenisation of river discharge through various forms of water management is also tending to homogenise the global fish fauna, favouring a common and somewhat ubiquitous group of invasive species, including common carp (*Cyprinus carpio*), centrarchid basses (*Micropterus* spp.) and various salmonids (Moyle & Mount, 2007). Further, replacement of native warm-water riverine fish fauna, often by salmonids, is facilitated by the cold 'tail water' release downstream from large reservoirs (Moyle & Mount, 2007). The impacts of dams on tail-water temperatures can be mitigated using a number of effective strategies (Olden & Naiman, 2010), and water releases from dams can be managed to mimic more natural flow regimes (Poff *et al.*, 2003; Chapter 4). However, other downstream impacts, such as changes to habitat resulting from altered sediment-transport dynamics through impoundments, can be complex and difficult to manage (Grant, 2012). The accumulation of sediments in impoundments can also create problems when dams are removed, potentially resulting in the erosion and release of large amounts of sediment (Marks *et al.*, 2010; Chapters 4 and 11).

2.5.3 Water pollution

Water pollution is a complex and diverse topic due to the huge diversity of chemicals that can find their way into freshwater systems, with multiple and varied impacts on freshwater fishes (see Chapters 5 and 6). The topic of pollution can encompass the effects of chemical toxins, nutrients and sediments, acid rain, biological pathogens and natural toxins produced by a variety of freshwater organisms (Chapter 5). Water falling through the atmosphere and draining through a catchment accumulates multiple chemicals, and the susceptibility of organisms to the various changes in water chemistry that occur varies hugely (Allan, 2004). Because rivers and lakes are tightly linked to their surrounding catchments, any significant changes to the condition of a catchment, such as clearing of vegetation, release of pollutants or changes to drainage patterns, will result in changes to the chemical composition of the water draining from that catchment (Allan, 2004; Dudgeon *et al.*, 2006; Dudgeon, 2011; Chapter 5). Although many of these releases are from apparently multiple minor and diffuse cross-catchment sources, cumulatively they can have major impacts, particularly if accumulation of pollutants occurs within ecosystems (Dudgeon, 2011). However, in some cases, widespread degradation within a river system can be traced back to a single point source, as happened following the release of mine tailing waste from the OK Tedi mine into the Fly River system in Papua New Guinea (Storey *et al.*, 2008), highlighting the connectedness of river systems and the distance significant impacts may extend downstream from a major source of pollutants. The development of the OK Tedi mine has had widespread and complex impacts on the fish communities of the Fly River system, although the effects of water pollution and ancillary development remain poorly studied there (Allen *et al.*, 2008).

Despite a longstanding awareness of the risks posed by chemical toxins, they continue to present significant challenges (Carson, 1962; Malmqvist & Rundle, 2002; Dudgeon *et al.*, 2006; Dudgeon, 2011; Chapter 5). A plethora of chemical compounds continue to enter freshwater systems from multiple sources, including the atmosphere, sewage, run-off from urban and agricultural land and toxic leachates from mines (Malmqvist & Rundle, 2002; Dudgeon *et al.*, 2006; Palmer *et al.*, 2010; Chapter 5). The sheer number and diversity of synthetic chemicals in use inevitably results in releases into the environment, and once there, the ecotoxicology of many is poorly understood (see Chapter 5). Diffuse releases of pollutants represent a significant challenge; for example, the widespread release of mercury from fossil fuel burning has resulted in the deposition and accumulation of mercury across sites and regions remote from any significant urban or industrial activity, such as the high Arctic (Lu *et al.*, 2001). The release of environmentally persistent chemicals, including mercury, creates legacy effects that last long after their release due to bioaccumulation or remobilisation following the erosion of contaminated sediments or melting of ice (Ma *et al.*, 2011; Chapter 5). Leachate from abandoned mines can also create legacy effects that are difficult to manage (Palmer *et al.*, 2010). Unfortunately, toxic legacies continue to be created; for example, the use of mercury to extract gold from sediments is a widespread problem associated with small-scale mining operations across much of South America (Ouboter *et al.*, 2012). Increasing levels of mercury have been reported from fishes and top predators from various Amazonian locations, suggesting that bioaccumulation and biomagnification is posing a direct and significant threat to the health of freshwater fishes and their consumers, including humans (Ouboter *et al.*, 2012).

The release of non-toxic contaminants is also a significant pollution issue, causing widespread degradation of water and habitat quality, and often through multiple point sources and diffuse inputs (Allan, 2004). Atmospheric pollution, mostly related to the burning of fossil fuels, has led to the acidification and nitrification of various Northern Hemisphere freshwater systems, with varied and complex impacts on the biota, including fishes (Malmqvist & Rundle, 2002). Large-scale agriculture and urbanisation also tends to produce diffuse inputs of nutrients to waterways, potentially causing dramatic changes to the food webs of aquatic systems, altering the availability of critical resources, and predatory and competitive interactions (Malmqvist & Rundle, 2002; Allan, 2004). Inputs of fine inorganic sediments from urban and agricultural catchments are also a significant pollutant in their own right, but can interact synergistically with nutrients, producing complex 'multiple-stressor impacts' (Wagenhoff *et al.*, 2012; Chapter 6). Acidification has had severe impacts on salmonid populations in some regions of North America and Europe, and although improved management of the issue has resulted in some recovery, acidification of waterways remains a problem in many areas (Battarbee *et al.*, 2014). Unfortunately, acidification of waterways associated with increased industrialisation has also become a problem in parts of Asia, although the extent of impacts on aquatic biota is not generally known (Mant *et al.*, 2013). Many of the problems associated with inputs of non-toxic pollutants relate to a lack of awareness of the connectedness of land–water systems, and a poor understanding of the consequences of individual choices, actions and behaviour on wider ecosystem integrity and health (Angermeier, 2007).

2.5.4 Inland capture fisheries, aquaculture and the introduction of exotic species

Humans change the structure of freshwater fish communities by both removing and adding fish and other species to a wide variety of systems

(Malmqvist & Rundle, 2002). Direct exploitation of freshwater fishes, involving both the capture and release of fish of many species, occurs for food, recreation, ornamental display and aquaculture, with each activity having different impacts and risks (Arlinghaus *et al.*, 2002; Welcomme *et al.*, 2010; Chapters 12, 13 and 15). Understanding of the impacts of the direct exploitation of freshwater fishes is relatively poor for various reasons, including the diversity of types of fishing (e.g. commercial, recreational, subsistence or aquaculture), the variety of impacts that different fishing approaches and rationales create (e.g. direct removal, stocking with native or introduced fishes, disease threats, etc.), the diverse range of stakeholders and management agencies that may have conflicting interests in fisheries management, and the direct impacts of fishing that are often obscured by the other interacting factors that may be affecting freshwater fish populations at any one time (Arlinghaus *et al.*, 2002; Malmqvist & Rundle, 2002; Welcomme *et al.*, 2010; Chapter 15). Commercial fishing continues to be a significant activity in Asia, Africa and South America; however, its significance has declined in Northern America, Europe and Australasia, with fisheries becoming increasingly directed towards recreational fishing, display and conservation (Arlinghaus *et al.*, 2002; Welcomme *et al.*, 2010). Accurate estimates of the economic value of many inland fisheries are lacking, partly due to the small-scale and diffuse nature of many fishing activities, but also due to the difficulties of estimating the value of support industries, including sellers of fishing equipment, bait, accommodation providers, boat builders, etc. (Welcomme *et al.*, 2010). Recreational fishing activity is also increasing rapidly in Asia, Latin America and parts of Africa, with potentially significant economic benefits (Welcomme *et al.*, 2010). It is also important to note that the existence of commercial, recreational and subsistence fisheries, and the use and promotion of fishes for conservation, can also have significant indirect benefits, including incentives to protect waterways and associated ecosystems that sustain fisheries, individual and social benefits, increased economic activity, and a greater awareness of the existence of and threats confronting freshwater fish species and populations (Arlinghaus *et al.*, 2002; Welcomme *et al.*, 2010).

Targeted capture fisheries, both commercial and recreational, tend to preferentially focus on large, piscivorous individuals and species, switching to smaller, less desirable fishes as larger fishes are depleted (Arlinghaus *et al.*, 2002; Castello *et al.*, 2014). Historically, intensive commercial fishing had significant impacts on freshwater populations in temperate regions – for example, in southeast Australia, the extensive stocks of Murray cod (*Maccullochella peelii*) and other favoured native fish species present prior to European settlement were significantly depleted, primarily by commercial fishing, by the late 1800s (Allan *et al.*, 2005; Harris *et al.*, 2013). Currently, overfishing contributes to declines in the abundance of various iconic, often apex predatory fish, including large species in the Mekong River basin such as the giant catfish (*Pangasianodon gigas*), giant barb (*Catlocarpio siamensis*) and giant pangasius (*Pangasius sanitwongsei*), the Amazonian arapaima or pirarucu (*Arapaima gigas*), various species of sturgeon and the New Zealand longfin eel (*Anguilla dieffenbachii*) (Allan *et al.*, 2005; Wright, 2013; Chapter 12). The known impacts of over-harvest on these large iconic species contrasts with reported annual catches of freshwater fish from various regions that are either relatively stable or even increasing (Allan *et al.*, 2005; Welcomme *et al.*, 2010). This suggests that, similar to marine multispecies fisheries (Pauly *et al.*, 1998), a pattern of 'fishing down the food web' also occurs in multispecies freshwater systems. Fishing targets and removes the largest individuals and species first, with the fishery being sustained over time by the targeting of progressively smaller, lower trophic level and less desirable individuals and species (Allan *et al.*, 2005). The insidious effect of a 'shifting baseline' results in each new generation of fishers accepting the now degraded fishery as the 'new normal' (Humphries & Winemiller, 2009). Relative to the effects of habitat degradation and modification, the impacts of fishing (including altered population structure, littering and disturbance to wildlife) are generally considered to be less severe and potentially reversible with appropriate management (Arlinghaus *et al.*, 2002). However, the replacement of preferred species with less desirable fishes, combined with poor monitoring and reporting of catches, obscures a full understanding of the severity and extent of the impact of overharvest on freshwater fish assemblages in many regions (Allan *et al.*, 2005; Welcomme *et al.*, 2010).

Commercial and recreational fishing pressure that targets large piscivorous fishes not only represents a direct threat to those species, but can also increase the risk of other adverse impacts related to habitat degradation and modification (Arlinghaus *et al.*, 2002; Allan *et al.*, 2005). Under certain food web configurations, the loss of large piscivorous fishes may reduce predation on smaller planktivorous fishes, triggering trophic cascades that can exacerbate the effects of nutrient enrichment via 'negative biomanipulation' (Carpenter *et al.*, 1985; Arlinghaus *et al.*, 2002; Allan *et al.*, 2005). The collection of fishes for the ornamental aquarium trade can also impact fish populations, with cichlids in both South America and in African lakes being over-harvested for both recreational and ornamental uses (Andrews, 1990). Aquaculture, although an important feature of many freshwater fisheries, including those supporting the ornamental fish trade, is unlikely to replace wild-caught fisheries given that both activities compete for common resources (water, space, food, etc.), often requires the collection of wild propagules, is also affected by widespread environmental degradation, and tends to focus on high-value large fishes rather than the diverse small fishes that characterise many existing freshwater fisheries (Naylor *et al.*, 2000; Welcomme *et al.*, 2010). Aquaculture can play an important role in supplementing fish populations in high-pressure, high-yield recreational fisheries; however, supplemental stocking of fisheries is also associated with a range of problems, a significant one being the spread of exotic species and disease (Welcomme *et al.*, 2010; Chapters 7 and 8).

Invasive alien species are a major cause of biodiversity loss globally; they are also among the least controlled and least reversible of human impacts on the world's ecosystems (Leprieur *et al.*, 2008; Strayer, 2010; Chapter 8). While some have debated the extent of the threat posed by exotic species (Gido & Brown, 1999), none dispute that the spread of exotic species is homogenising freshwater fish faunas, particularly where human activity is highest (Rahel, 2007; Leprieur *et al.*, 2008), nor that the impact of non-native fishes on some freshwater fish taxa and communities can be very severe, e.g. the negative impacts of salmonids on Southern Hemisphere galaxiids (McDowall, 2006) or Nile perch on cichlids in the East African great lakes (Snoeks, 2000). For most non-native introductions, however, the ecological risk and impact are scarcely understood, in part because standard methods for quantifying impact are lacking (Lapointe *et al.*, 2012). The spread of non-native species other than fishes is also of concern, including zebra mussel (*Dreissena polymorpha*) in North America (Strayer *et al.*, 1999) and the bloom-forming diatom *Didymosphenia geminata* in New Zealand, South America and North America; both species can be accidently spread by aquatic recreational activities (Whitton *et al.*, 2009; Reid *et al.*, 2012). While non-native species have been released to supplement commercial and recreational fisheries, non-native species are also deliberately and accidentally released for a variety of other reasons including ornamental, biocontrol, bait, and aquarium display purposes (Morgan *et al.*, 2004; Rahel, 2007).

A strong association between human economic activity and the numbers of introduced species present in regional fish communities exists for a number of reasons (Leprieur *et al.*, 2008). Regions supporting high levels of economic activity tend to have larger human populations, some of which will be engaged in activities associated with the translocation of non-native species (Rahel, 2007). Dense human populations also tend to result in increased habitat modification, homogenisation and degradation, all of which create conditions favouring certain invasive non-native species (Moyle & Mount, 2007). The spread of such species is also facilitated by water management infrastructure that affects the permeability of biogeographic barriers, e.g. interbasin water transfers for water supply and hydroelectric power generation, high concentrations of shipping traffic that release non-native species from ballast water, and intentional or accidental release of fishes for a variety of reasons (Rahel, 2007; Leprieur *et al.*, 2008). The positive association between economic activity and non-native fish diversity has raised concerns that the rate of spread of non-native species is likely to increase, particularly in regions with diverse natural faunas and with rapidly growing economies, including Asia, Africa and South America, highlighting an urgent need to develop global strategies to combat such trends (Rahel, 2007; Leprieur *et al.*, 2008). Slowing the spread of aquatic non-native species will involve implementing regulations that limit release opportunities, increasing the public's awareness about the consequences of releasing non-native species, and developing technological solutions that prevent the unintentional movement of aquatic organisms, or eliminate them before they become irreversibly established (Rahel, 2007).

2.6 SOCIETAL ATTITUDES AND ECONOMIC DEVELOPMENT

The social, economic and political answers to 'why are freshwater fish so threatened?' are complex and reflect how society values freshwaters and their fishes. Understanding the perceived importance (or unimportance) of freshwater ecosystems to people can help conservationists (a) develop conservation strategies likely to enjoy political support, and (b) identify groups to target for conservation education (Angermeier, 2010). In this section, we sketch the socioeconomic backdrop for how freshwater ecosystems are managed. Because that backdrop is similar for most elements of biodiversity, we do not distinguish imperilment of fishes versus other biota. Our perspective is based largely on experiences in the US, but most patterns apply elsewhere, especially to wealthy nations.

2.6.1 Public knowledge of and attitudes toward biodiversity and freshwaters

A fundamental cause of biodiversity loss is that most people do not view it as a serious problem. Few studies explicitly examine attitudes toward freshwater fishes, or even toward biodiversity generally. However, attitudes toward the environment and threatened or extinct species are more widely studied. Especially instructive are Roper surveys conducted in the 1990s (summarised by Coyle, 2005) and Gallup surveys conducted 2000–2014 (www.gallup.com; accessed 11–12 June 2014). Below, we assume that 'fish conservation' issues are subsumed under 'environmental' issues.

Public ignorance and complacency about environmental quality and biodiversity are pervasive. Although most US adults consider themselves 'environmentalists', few (< 3%) are environmentally literate (Coyle, 2005). The average American adult does not understand essential aspects of environmental science, important causal relationships, or basic ecological processes (Coyle, 2005). Only 40% know what biodiversity is. Lack of understanding of causal connections (among ecosystem components, between human actions and environmental conditions) is especially common. Annual Gallup data since 2000 consistently show that most Americans do not worry about species extinction, with this issue ranking in the bottom half of the issues covered. Other concerns, such as personal health, financial security, crime and terrorism always trump environmentalism. Fewer than half of Americans think environmental laws 'do not go far enough', including laws to protect natural areas and endangered species; majorities held these opinions in 1992 (Coyle, 2005). Apathy regarding biodiversity loss extends to global consumption patterns and limits the success of certification programmes for marine fisheries (Kaiser & Edwards-Jones, 2006) and timber products (Dauvergne & Lister, 2010). Given that freshwater fishes are less loved than birds or mammals (Czech *et al.*, 1998; Kaiser & Edwards-Jones, 2006), we suspect that the lack of public concern for biodiversity loss in general is accentuated for fishes in particular.

Attitudes in the European Union seem more favourable toward biodiversity than those in the US. Gallup's Flash Eurobarometer survey in 2010 found that > 90% of adults viewed biodiversity loss as a serious global problem. Most citizens (70%) said they made personal efforts to protect biodiversity. However, knowledge is localised: only 37% of Europeans felt 'well informed' about biodiversity.

Public support for environmental quality and biodiversity has eroded since the 1990s, especially during economic downturns.

This shift has reduced political traction for environmentalism (Shellenberger & Nordhaus, 2005). The proportion of young adults who expend *no* personal effort to 'help the environment' has tripled across the last three generations (Twenge *et al.*, 2012). Since 2001, Americans worry less about all the environmental issues Gallup measures. Given there is no evidence that biodiversity loss is slowing, this decline in 'worry' reflects reduced concern for biodiversity loss, rather than resolution of the problem. Gallup data show that since 2004, the proportion of Americans favouring economic growth versus environmental protection, when the two conflict, has been similar. However, in the preceding two decades, the proportion of Americans favouring environmental protection was typically twice that of the proportion favouring economic growth.

In contrast to their tepid concern for biodiversity, Americans show strong concern for freshwater. Most say the main reason to protect the environment is to protect human health (Coyle, 2005). Thus, environmental features such as water quality (strong tie to personal health) are widely scrutinised. In annual Gallup surveys, water pollution and water supply consistently rank at the top of Americans' environmental worries. Some form of water pollution has been the top-ranking environmental concern in each survey since 1989. However, that concern does not necessarily reflect accurate knowledge: < 25% of Americans know (a) contaminated water is the leading cause of childhood deaths globally, (b) < 1% of the world's water is fresh and available for use, (c) the term 'non-point source pollution', and (d) run-off is the most common source of water pollution (Coyle, 2005).

The contrast in public concern for biodiversity versus freshwater suggests a promising path forward for fish conservation and a catalytic role for fish biologists as environmental educators (Angermeier, 2007). There is considerable overlap in the tactics used to manage watersheds for clean water, which is highly valued, and for fish conservation. Thus, additional research and outreach to demonstrate the synergies and cost savings produced by programmes that merge these two management goals could be instrumental in making conservation appealing, or at least acceptable, to a broader portion of stakeholders.

2.6.2 Environmental protection versus economic growth

Threats to freshwater fishes reflect the tension between societal goals of economic growth and environmental protection. The tension stems from the facts that most freshwaters are public resources, but freshwater is crucial to many private economic sectors such as agriculture, industry, electric power, transportation and recreation. For some economic activities (e.g. food production) freshwater is a key ingredient; for others (e.g. urban sewage disposal) freshwater is the waste-processing medium. However, many (probably most) anthropogenic impacts on freshwaters are inadvertent, usually avoidable, outcomes of economic activities. Widespread examples include airborne mercury contamination of surface waters by coal-fired power plants, disruption of natural flow regimes of rivers by dams, and introduction of non-native species into surface waters by anglers. About two-thirds of Americans think environmental protection and economic growth are compatible (Coyle, 2005) but in practice, economic development commonly harms freshwater fishes.

Both purposeful and inadvertent economic uses of freshwaters contribute to fish imperilment by shifting environmental conditions away from evolutionary baselines. The main causes of freshwater fish endangerment in the US include water pollution and water projects (e.g. reservoirs and diversions) but these causes stem from the economics of agriculture, mining, logging, industry and urban development (Czech *et al.*, 2000). As these and other economic sectors expand under current macroeconomic policies, the suitable resources available to other biota will diminish, effectively endangering even more species via competitive exclusion (Czech *et al.*, 2000).

The dominant school of economic thought (i.e. neoclassical economics) espouses unfettered markets as the best mechanism to allocate societal resources. However, markets transactions can easily exclude from prices certain environmental and social costs associated with the production of goods; such 'externalities' include adverse biological impacts. Moreover, many components of natural ecosystems (e.g. beauty, biodiversity, life support) benefit people or have great value but cannot be readily traded in markets (Agius, 2001). In general, markets cannot efficiently produce or allocate public resources (Daly & Farley, 2011). Thus, when economic production involves both public and private resources, markets favour production paths that convey benefits (e.g. profit) to the capitalist but distribute costs to the public (e.g. cost of purifying polluted water so the public can use it). If left unregulated by moral or legal codes, this arrangement can lead to severe degradation of public resources (Hardin, 1968). In modern societies, government regulation is the most common lever applied to markets to reduce externalised costs to public resources such as freshwater and fisheries. However, the extent to which government constraints on economic activity are politically acceptable depends on the relative social values of the specific activity and externalised costs, as well as the political power of the stakeholders (Ludwig *et al.*, 1993).

Ecological economics, which explicitly values ecosystems and biodiversity, is an emerging alternative to neoclassical economics (Daly & Farley, 2011). A major difference between these schools of thought is that ecological economists view the economy as a subsystem of the global ecosystem, whereas neoclassical economists view ecosystems simply as components of the economy. This distinction has profound implications for how economic activity can or should proceed. For example, ecological economics distinguishes between natural capital and man-made capital; both are important contributors to human welfare, but they have limited substitutability. Ecological economists assert that economic choices should account for changes in both forms of capital. Further, if the macroeconomy grows so much that the associated loss of ecosystem services (from natural capital) exceeds the value of the economic gain (in man-made capital), that growth is considered 'uneconomic' (Daly & Farley, 2011). Symptoms of uneconomic growth include loss of a landscape's biodiversity or capacity to purify freshwater. In contrast, neoclassical economists posit that man-made capital is readily substitutable for natural capital, which obviates the need for (or desirability of) any caps on economic growth. Incorporating principles of ecological economics into macroeconomic policy would fundamentally alter the dynamics between economic growth and environmental protection, presumably leading to greater conservation of freshwater fishes and the services generated from their habitats.

2.7 CONCLUSIONS

The conservation of freshwater fish diversity presents us with significant challenges. Freshwater fishes are an exceptionally diverse group of organisms living in an environment that also happens to be critical for human development and survival (Dudgeon *et al.*, 2006). Freshwater habitats are inextricably linked to their surrounding catchments, and changes in that terrestrial environment will usually impact the freshwater habitats embedded within them to some extent. Additional human demands for water will deplete surface waters and alter the flow regimes that play a critical role in shaping the abiotic and biotic habitat components, delivering food, and providing cues for behaviour and key life-history events. Given the diversity of forms, life histories, feeding strategies and behaviours exhibited by freshwater fishes, many of which are highly specialised for the conditions likely to be encountered in the natural environment in which they evolved, the pressures exerted by humans almost inevitably have the potential to generate negative impacts on freshwater fish abundance and distribution, and favour a suite of invasive species that are becoming increasingly widespread. Industrialisation increases the capacity of humans to modify environments; hence, accelerating economic growth, particularly in the high-diversity regions of Asia, parts of Africa and tropical South America, has the potential to have hugely negative impacts on the diversity of freshwater fishes (Dudgeon, 2000, 2011; Leprieur *et al.*, 2008; Humphries & Winemiller, 2009). Added to this, the catchments of many river systems transcend jurisdictional boundaries, complicating the development of coherent and coordinated management strategies considerably (Dudgeon, 2011).

Freshwater fishes suffer from a lack of visibility and public/political interest compared to the more visible and often charismatic terrestrial megafauna (although some freshwater fishes have extreme prestige, e.g. salmonids, and hence have spread globally), and many of the adverse impacts on freshwater systems have gone, and continue to go, largely unnoticed (Dudgeon, 2000; Humphries & Winemiller, 2009). Human demand for water, and the impacts that large populations have on river catchments and the systems that drain them, mean some loss of freshwater fish diversity is almost inevitable over the next 20 years. Broad-scale changes in the environment, such as those related to climate change, also have the potential to further intensify the pressures on freshwater fish communities (Malmqvist & Rundle, 2002). If the loss of freshwater fish diversity associated with human activity is not to become catastrophic, then major changes in our approach to managing and mitigating the impacts of development are essential. The degree of resilience of some high-diversity systems in the face of multiple adverse impacts, such as the haplochromine species flocks of Lake Victoria (Witte *et al.*, 2013), gives some hope for the future survival of a significant proportion of the global freshwater fish fauna, provided measures are taken to avoid or ameliorate the worst impacts. The key threats to freshwater fishes are well known: water pollution, habitat degradation and destruction (including altered flow regimes), overexploitation, and invasion by non-native species (Dudgeon *et al.*, 2006). Strategies to minimise these threats, thereby conserving freshwater fishes, to some degree exist (Malmqvist & Rundle, 2002), but are not widely implemented because the prevailing social and economic value systems do not favour strong conservation programmes. Addressing these issues requires public education and strong advocacy by those who understand the threats and challenges, ultimately aiming to build a degree of societal and political will to save whatever remains of global freshwater fish diversity (Dudgeon, 2000, 2010; Dudgeon *et al.*, 2006). Freshwater fish biologists are crucial catalysts in making conservation a reality (Angermeier, 2007). The challenges are great and the problems significant, but not insurmountable. It is only by publicly acknowledging the challenges and problems, and advocating persistently and effectively for change, that we can begin to develop effective strategies to address the key issues.

2.8 DISCUSSION QUESTIONS

1. What is the biogeographical significance of families of primary, secondary and peripheral division freshwater fishes?
2. Discuss how a hierarchical series of 'environmental filters' determines the composition of the fish community at any particular location.
3. Describe an example of where either water pollution, habitat degradation and destruction, altered flow regimes, overexploitation, or invasion by non-native species has driven a species to near-extinction or even extinction.
4. How many native fish species occur in your country? How many are threatened with extinction?
5. Outline a strategy required to save at least one threatened species in your region.

2.9 ACKNOWLEDGEMENTS

We thank Scott Bonar and Eric Hallerman for helpful comments on the manuscript. The Virginia Cooperative Fish and Wildlife Research Unit is jointly sponsored by U.S. Geological Survey, Virginia Tech, Virginia Department of Game and Inland Fisheries, and Wildlife Management Institute.

2.10 REFERENCES

Abell, R., Thieme, M. L., Revenga, C., *et al.* (2008). Freshwater ecoregions of the world: a new map of biogeographic units for freshwater biodiversity conservation. *Bioscience*, 58, 403–414.

Agius, J. (2001). Biodiversity credits: creating missing markets for biodiversity. *Environmental and Planning Law Journal*, 18, 481–504.

Allan, J. D. (2004). Landscapes and riverscapes: the influence of land use on stream ecosystems. *Annual Review of Ecology, Evolution, and Systematics*, 35, 257–284.

Allan, J. D., Abell, R., Hogan, Z., *et al.* (2005). Overfishing of inland waters. *BioScience*, 55, 1041–1051.

Allen, G. R., Storey, A. W. & Yarrao, M. (2008). *Freshwater Fishes of the Fly river Papua New Guinea*. Tabubil, PNG: Ok Tedi Mining.

Andrews, C. (1990). The ornamental fish trade and fish conservation. *Journal of Fish Biology*, 37, 53–59.

Angermeier, P. L. (1995). Ecological attributes of extinction-prone species: loss of freshwater fishes of Virginia. *Conservation Biology*, 9, 143–158.

Angermeier, P. L. (2007). The role of fish biologists in helping society build ecological sustainability. *Fisheries*, 32, 9–20.

Angermeier, P. L. (2010). Preface: Conservation challenges for stream fish ecologists. *American Fisheries Society Symposium*, 73, 303–309.

Angermeier, P. L. & Winston, M. R. (1998). Local vs. regional influences on local diversity in stream fish communities of Virginia. *Ecology*, 79, 911–927.

Angermeier, P. L. & Winston, M. R. (1999). Characterizing fish community diversity across Virginia landscapes: prerequisite for conservation. *Ecological Applications*, 9, 335–349.

Arlinghaus, R., Mehner, T. & Cowx, I. G. (2002). Reconciling traditional inland fisheries management and sustainability in industrialized countries, with emphasis on Europe. *Fish and Fisheries*, 3, 261–316.

Battarbee, R. W., Shilland, E. M., Kernan, M., Monteith, D. T. & Curtis, C. J. (2014). Recovery of acidified surface waters from acidification in the United Kingdom after twenty years of chemical and biological monitoring (1988–2008). *Ecological Indicators*, 37, 267–273.

Beatty, S. J., Morgan, D. L., Rashnavadi, M. & Lymbery, A. J. (2011). Salinity tolerances of endemic freshwater fishes of south-western Australia: implications for conservation in a biodiversity hotspot. *Marine and Freshwater Research*, 62, 91–100.

Bernhardt, E. S. & Palmer, M. A. (2007) Restoring streams in an urbanizing world. *Freshwater Biology*, 52, 738–751.

Berra, T. M. (2001). *Freshwater Fish Distribution*. San Diego, CA: Academic Press.

Brown, D. K., Echelle, A. A., Propst, D. L., Brooks, J. E. & Fisher, W. L. (2001). Catastrophic wildfire and number of populations as factors influencing risk of extinction for Gila trout (*Oncorhynchus gilae*). *Western North American Naturalist*, 61, 139–148.

Carpenter, S. R., Kitchell, J. F. & Hodgson, J. R. (1985) Cascading trophic interactions and lake productivity. *Bioscience*, 35, 634–639.

Carson, R. (1962). *Silent Spring*. Boston, MA: Houghton Mifflin.

Castello, L., Arantes, C. C., McGrath, D. G., Stewart, D. J. & de Sousa, F. S. (2014). Understanding fishing-induced extinctions in the Amazon. *Aquatic Conservation: Maine and Freshwater Ecosystems*. DOI: 10.1002/aqc.2491

Closs, G. P. & Lake, P. S. (1996). Drought, differential mortality and the coexistence of a native and an introduced fish species in a south east Australian intermittent stream. *Environmental Biology of Fishes*, 47, 17–26.

Collier, K. J., Aldridge, B. M. T. A., Hicks, B. J., *et al.* (2009). Ecological values of Hamilton urban streams (North Island, New Zealand): constraints and opportunities for restoration. *New Zealand Journal of Ecology*, 33, 177–189.

Coyle, K. (2005). *Environmental Literacy in America: What Ten Years of NEETF/Roper Research and Related Studies Say about Environmental Literacy in the U.S.* Washington, DC: The National Environmental Education & Training Foundation.

Czech, B. (2000). Economic growth as a limiting factor for wildlife conservation. *Wildlife Society Bulletin*, 28, 4–14.

Czech, B., Krausman, P. R. & Borkhataria, R. (1998). Social construction, political power, and the allocation of benefits to endangered species. *Conservation Biology*, 12, 1103–1112.

Czech, B., Krausman, P. R. & Devers, P. K. (2000). Economic associations among causes of species endangerment in the United States. *BioScience*, 50, 593–601.

Daly, H. E. & Farley, J. (2011). *Ecological Economics: Principles and Applications*, second edition. Washington, DC: Island Press.

Dauvergne, P. & Lister, J. (2010). The prospects and limits of eco-consumerism: shopping our way to less deforestation? *Organization & Environment*, 23, 132–154.

David, B. O., Tonkin, J. D., Taipeti, K. W. T. & Hokianga, H. T. (2014). Learning the ropes: mussel spat ropes improve fish and shrimp passage through culverts. *Journal of Applied Ecology*, 51, 214–223.

DeHaven, J. E., Stouder, D. J., Ratajczak, R., Welch, T. J. & Grossman, G. D. (1992). Reproductive timing in three southern Appalachian stream fishes. *Ecology of Freshwater Fish*, 1, 104–111.

Dudgeon, D. (2000). Conservation of freshwater biodiversity in Oriental Asia: constraints, conflicts, and challenges to science and sustainability. *Limnology*, 1, 237–243.

Dudgeon, D. (2010a). Requiem for a river: extinctions, climate change and the last of the Yangtze. *Aquatic Conservation: Marine and Freshwater Ecosystems*, 20, 127–131.

Dudgeon, D. (2010b). Prospects for sustaining freshwater biodiversity in the 21st century: linking ecosystem structure and function. *Current Opinion in Environmental Sustainability*, 2, 422–430.

Dudgeon, D. (2011). Asian rivers fishes in the Anthropocene: threats and conservation challenges in an era of rapid environmental change. *Journal of Fish Biology*, 79, 1487–1524.

Dudgeon, D., Arthington, A. H., Gessner, M. O., *et al.* (2006). Freshwater biodiversity: importance, threats, status and conservation challenges. *Biological Reviews*, 81, 163–182.

Dunham, J. B. & Rieman, B. E. (1999). Metapopulation structure of bull trout: influences of physical, biotic, and geometrical landscape characteristics. *Ecological Applications*, 9, 642–655.

Dunham, J., Baxter, C., Fausch, K., *et al.* (2008). Evolution, ecology, and conservation of dolly varden, white spotted char and bull trout. *Fisheries*, 33, 537–550.

Fagan, W. F. (2002). Connectivity, fragmentation, and extinction risk in dendritic metapopulations. *Ecology*, 83, 3243–3249.

Fagan, W. F., Unmack, P. J., Burgess, C. & Minckley, W. L. (2002). Rarity, fragmentation, and extinction risk in desert fishes. *Ecology*, 83, 3250–3256.

Fausch, K. D., Torgersen, C. E., Baxter, C. V. & Li, H. W. (2002). Landscapes to riverscapes: bridging the gap between research and conservation of stream fishes. *Bioscience*, 52, 483–498.

Filipe, A. F., Magalhães, M. F. & Collares-Pereira, M. J. (2010). Native and introduced fish species richness in Mediterranean streams: the role of multiple landscape influences. *Diversity and Distributions*, 16, 773–785.

Frimpong, E. A. & Angermeier, P. L. (2010). Trait-based approaches in the analysis of stream fish communities. *American Fisheries Society Symposium*, 73, 109–136.

Genito, D., Gburek, W. J. & Sharpley, A. N. (2002). Response of stream macroinvertebrates to agricultural land cover in a small watershed. *Journal of Freshwater Ecology*, 17, 109–119.

Gido, K. B. & Brown, J. H. (1999). Invasions of North American drainages by alien fish species. *Freshwater Biology*, 42, 387–399.

Grant, G. E. (2012). The geomorphic response of gravel-bed rivers to dams. In *Perspectives and Prospects, in Gravel-Bed Rivers: Processes, Tools, Environments*. Chichester: John Wiley & Sons, Ltd, pp. 165–181.

Gupta, J., Pahl-Wostl, C. & Zondervan, R. (2013). 'Glocal' water governance: a multi-level challenge in the anthropocene. *Current Opinion in Environmental Sustainability*, 5, 573–580.

Hardin, G. (1968). The tragedy of the commons. *Science*, 162, 1243–1248.

Harris, J. H., Bond, N. R., Closs, G. P., *et al.* (2013). Dynamics of populations. In: *Ecology of Australian Freshwater Fishes*. Humphries, P. & Walker, K. (Eds). Collingwood: CSIRO Publishing, pp. 223–244.

Heino, J., Schmera, D. & Erős, T. (2013). A macroecological perspective of trait patterns in stream communities. *Freshwater Biology*, 58, 1539–1555.

Humphries, P. & Lake, P. S. (2000). Fish larvae and the mangement of regulated rivers. *Regulated Rivers: Research & Management*, 16, 421–432.

Humphries, P. & Winemiller, K. O. (2009). Historical impacts on river fauna, shifting baselines, and challenges for restoration. *Bioscience*, 59, 673–684.

Humphries, P., King, A. J. & Koehn, J. D. (1999). Fish, flows and flood plains: links between freshwater fishes and their environment in the Murray–Darling River system, Australia. *Environmental Biology of Fishes*, 56, 129–151.

Kaiser, M. J. & Edwards-Jones, G. (2006). The role of ecolabeling in fisheries management and conservation. *Conservation Biology*, 20, 392–398.

Keith, P. (2003). Biology and ecology of amphidromous Gobiidae of the Indo-Pacific and Caribbean regions. *Journal of Fish Biology*, 63, 831–847.

King, A. J., Humphries, P. & Lake, P. S. (2003). Fish recruitment on floodplains: the roles of patterns of flooding and life history characteristics. *Canadian Journal of Fisheries and Aquatic Sciences*, 60, 773–786.

Lamouroux, N., Poff, N. L. & Angermeier, P. L. (2002). Convergence of stream fish community traits in France and Virginia (USA) along geomorphic and hydraulic gradients. *Ecology*, 83, 1792–1807.

Lapointe, N. W. R., Pendleton, R. M. & Angermeier, P. L. (2012). A comparison of approaches for estimating relative impacts of nonnative fishes. *Environmental Management*, 49, 82–95.

Leprieur, F., Beauchard, O., Blanchet, S., Oberdorff, T. & Brosse, S. (2008). Fish invasions in the world's river systems: when natural processes are blurred by human activities. *PLoS Biology*, 6, e28.

Lord, C., Lorion, J., Dettai, A., *et al.* (2012). From endemism to widespread distribution: phylogeography of three amphidromous *Sicyopterus* species (Teleostei: Gobioidei: Sicydiinae). *Marine Ecology Progress Series*, 455, 269–285.

Lu, J. Y., Schroeder, W. H., Barrie, L. A., *et al.* (2001). Magnification of atmospheric mercury deposition to polar regions in springtime: the link to tropospheric ozone depletion chemistry. *Geophysical Research Letters*, 28, 3219–3222.

Ludwig, D., Hilborn, R. & Walters, C. (1993). Uncertainty, resource exploitation, and conservation: Lessons from history. *Science*, 260, 17, 36.

Ma, J., Hung, H., Tian, C. & Kallenborn, R. (2011). Revolatilization of persistent organic pollutants in the Arctic induced by climate change. *Nature Climate Change*, 1, 255–260.

Mallen-Cooper, M. & Brand, D. A. (2007). Non-salmonids in a salmonid fishway: what do 50 years of data tell us about past and future fish passage? *Fisheries Management and Ecology*, 14, 319–332.

Malmqvist, B. & Rundle, S. (2002). Threats to the running water ecosystems of the world. *Environmental Conservation*, 29, 134–153.

Mant, R. C., Jones, D. L., Reynolds, B., Ormerod, S. J. & Pullin, A. S. (2013). A systematic review of the effectiveness of liming to mitigate the impacts of river acidification on fish and macro-invertebrates. *Environmental Pollution*, 179, 285–293.

Marks, J. C., Haden, G. A., O'Neill, M. & Pace, C. (2010). Effects of flow restoration and exotic species removal on recovery of native fish: lessons from a dam decommissioning. *Restoration Ecology*, 18, 934–943.

Maruyama, A., Yuma, M. & Onoda, Y. (2004). Egg size variation between the fluvial-lacustrine and lacustrine types of a landlocked *Rhinogobius* goby in the Lake Biwa water system. *Ichthyological Research*, 51, 172–175.

Matthews, W. J. (1998). *Patterns in Freshwater Fish Ecology*. New York, NY: Chapman & Hall.

McDowall, R. M. (2006). Crying wolf, crying foul, or crying shame: alien salmonids and a biodiversity crisis in the southern cool-temperate galaxioid fishes? *Reviews in Fish Biology and Fisheries*, 16, 233–422.

Mensing, D. M., Galatowitsch, S. M. & Tester, J. R. (1998). Anthropogenic effects on the biodiversity of riparian wetlands of a northern temperate landscape. *Journal of Environmental Management*, 53, 349–377.

Mims, M. C., Olden, J. D., Shattuck, Z. R. & Poff, N. L. (2010). Life history trait diversity of native freshwater fishes in North America. *Ecology of Freshwater Fish*, 19, 390–400.

Morgan, D. L., Gill, H. S., Maddern, M. G. & Beatty, S. J. (2004). Distribution and impacts of introduced freshwater fishes in Western Australia. *New Zealand Journal of Marine and Freshwater Research*, 38, 511–523.

Morita, K. & Yamamoto, S. (2002). Effects of habitat fragmentation by damming on the persistence of stream-dwelling charr populations. *Conservation Biology*, 16, 1318–1323.

Moyle, P. B. & Mount, J. F. (2007). Homogenous rivers, homogenous faunas. *Proceedings of the National Academy of Sciences*, 104, 5711–5712.

Myers, G. S. (1938). Fresh-water fishes and West Indian zoogeography. Annual Report of the Board of Regents of the Smithsonian Institution 92 (1938, for the year 1937), 339–364.

Naiman, R. J. (2013). Socio-ecological complexity and the restoration of river ecosystems. *Inland Waters*, 3, 391–410.

Naiman, R. J. & Dudgeon, D. (2011). Global alteration of freshwaters: influences on human and environmental well-being. *Ecological Research*, 26, 865–873.

Nakano, S., Kitano, F. & Maekawa, K. (1996). Potential fragmentation and loss of thermal habitats for charrs in the Japanese archipelago due to climatic warming. *Freshwater Biology*, 36, 711–722.

Naylor, R. L., Goldburg, R. J., Primavera, J. H., *et al.* (2000). Effect of aquaculture on world fish supplies. *Nature*, 405, 1017–1024.

Novak, M., Moore, J. W. & Leidy, R. A. (2011). Nestedness patterns and the dual nature of community reassembly in California streams: a multivariate permutation-based approach. *Global Change Biology*, 17, 3714–3723.

Novinger, D. C. & Rahel, F. J. (2003). Isolation management with artificial barriers as a conservation strategy for cutthroat trout in headwater streams. *Conservation Biology*, 17, 772–781.

Oberdorff, T., Tedesco, P. A., Hugueny, B., *et al.* (2011). Global and regional patterns in riverine fish species richness: a review. *International Journal of Ecology*, 2011, Article ID 967631.

O'Grady, J. J., Reed, D. H., Brook, B. W. & Frankham, R. (2004). What are the best correlates of predicted extinction risk? *Biological Conservation*, 118, 513–520.

Olden, J. D. & Naiman, R. J. (2010). Incorporating thermal regimes into environmental flows assessments: modifying dam operations to restore freshwater ecosystem integrity. *Freshwater Biology*, 55, 86–107.

Olden, J. D., Hogan, Z. S. & Zanden, M. (2007). Small fish, big fish, red fish, blue fish: size-biased extinction risk of the world's freshwater and marine fishes. *Global Ecology and Biogeography*, 16, 694–701.

Olden, J. D., Poff, N. L. & Bestgen, K. R. (2008). Trait synergisms and the rarity, extirpation, and extinction risk of desert fishes. *Ecology*, 89, 847–856.

Olden, J. D., Kennard, M. J., Leprieur, F., *et al.* (2010). Conservation biogeography of freshwater fishes: recent progress and future challenges. *Diversity and Distributions*, 16, 496–513.

Ouboter, P. E., Landburg, G. A., Quik, J. H. M., Mol, J. H. A. & van der Lugt, F. (2012). Mercury levels in pristine and gold mining impacted aquatic ecosystems of Suriname, South America. *Ambio*, 41, 873–882.

Palmer, M. A., Bernhardt, E. S., Schlesinger, W. H., *et al.* (2010). Mountaintop mining consequences. *Science*, 327, 148–149.

Pauly, D., Christenson, V., Dalsgaard, J., Froese, R. & Torres, F. Jr. (1998). Fishing down marine food webs. *Science*, 279, 860–863.

Perkin, J. S. & Gido, K. B. (2011). Stream fragmentation thresholds for a reproductive guild of Great Plains fishes. *Fisheries*, 36, 371–383.

Poff, N. L., Allan, J. D. Bain, M. B., *et al.* (1997). The natural flow regime: a new paradigm for riverine conservation and restoration *Bioscience*, 47, 769–784.

Poff, N. L., Allan, J. D., Palmer, M. A., *et al.* (2003). River flows and waters wars: emerging science for environmental decision making. *Frontiers in Ecology and Environment*, 1, 298–306.

Poff, N. L, Olden, J. D., Merritt, D. M. & Pepin, D. M. (2007). Homogenization of regional river dynamics by dams and global biodiversity implications. *Proceedings of the National Academy of Sciences*, 104, 5732–5737.

Pompeu, P. S., Agostinho, A. A. & Pelicice, F. M. (2012). Existing and future challenges: the concept of successful fish passage in South America. *River Research and Applications*, 28, 504–512.

Rahel, F. J. (2007). Biogeographic barriers, connectivity and homogenization of freshwater faunas: it's a small world after all. *Freshwater Biology*, 52, 696–710.

Reid, B. L., Hernandez, K. L., Frangopulos, M., *et al.* (2012). The invasion of the freshwater diatom *Didymosphenia geminata* in Patagonia: prospects, strategies, and implications for biosecurity of invasive microorganisms in continental waters. *Conservation Letters*, 5, 432–440.

Reyes-Gavilán, F. G., Garrido, R., Nicieza, A. G., Toledo, M. M. & Brana, F. (1996). Fish community variation along physical gradients in short streams of northern Spain and the disruptive effect of dams. *Hydrobiologia*, 321, 155–163.

Reynolds, J. D., Webb, T. J. & Hawkins, L. A. (2005). Life history and ecological correlates of extinction risk in European freshwater fishes. *Canadian Journal of Fisheries and Aquatic Sciences*, 62, 854–862.

Richter, B. D., Postel, S., Revenga, C., *et al.* (2010). Lost in development's shadow: the downstream human consequences of dams. *Water Alternatives*, 3, 14–42.

Rieman, B. E. & Dunham, J. B. (2000). Metapopulations and salmonids: a synthesis of life history patterns and empirical observations. *Ecology of Freshwater Fish*, 9, 51–64.

Shellenberger, M. & Nordhaus, T. (2005). The death of environmentalism: Global warming politics in a post-environmental world. www.grist.org/cgi-bin/printthis.pl.

Schindler, D. E., Armstrong, J. B., Bentley, K. T., *et al.* (2013). Riding the crimson tide: mobile terrestrial consumers track phenological variation in spawning of an anadromous fish. *Biology Letters*, 9, Article Number 20130048.

Schlosser, I. J. & Angermeier, P. L. (1995). Spatial variation in demographic processes of lotic fishes: conceptual models, empirical evidence, and implications for conservation. *American Fisheries Society Symposium*, 17, 392–401.

Smith, S. S. & Bermingham, E. (2005). The biogeography of lower Mesoamerican freshwater fishes. *Journal of Biogeography*, 32, 1835–1854.

Snoeks, J. (2000). How well known is the ichthyodiversity of the large East African lakes? *Advances in Ecological Research*, 31, 17–38.

Stone, R. (2007). The last of the leviathans. *Science*, 316, 1684–1688.

Storey, A. E., Yarrao, M., Tenakanai, C., Figa, B. & Lynas, J. (2008). Use of changes in fish assemblages in the Fly River System, Papua New Guinea, to assess effects of the OK Tedi Copper Mine. *Developments in Earth and Environmental Sciences*, 9, 427–462.

Strayer, D. L. (2010). Alien species in fresh waters: ecological effects, interactions with other stressors, and prospects for the future. *Freshwater Biology*, 55 (Suppl. 1), 152–174.

Strayer, D. L. & Dudgeon, D. (2010). Freshwater biodiversity conservation: recent progress and future challenges. *Journal of the North American Benthological Society*, 29, 344–358.

Strayer, D. L., Caraco, N. F., Cole, J. J., Findlay, S. & Pace, M. L. (1999). Transformation of freshwater ecosystems by bivalves. A case study of the zebra mussels in the Hudson River. *Bioscience*, 49, 19–27.

Taillebois, L., Maeda, K., Vigne, S. & Keith, P. (2012). Pelagic duration of three amphidromous Sicydiinae obies (Teleostei: Gobiodei) including widespread and endemic species. *Ecology of Freshwater Fish*, 21, 552–559.

Taylor, B. W., Flecker, A. S. & Hall, R. O. (2006). Loss of a harvested fish species disrupts carbon flow in a diverse tropical river. *Science*, 313, 833–836.

Tisseuil, C., Cornu, J., Beauchard, O., *et al.* (2013). Global diversity patterns and cross-taxa convergence in freshwater systems. *Journal of Animal Ecology*, 82, 365–376.

Tonn, W. M., Magnuson, J. J., Rask, M. & Toivonen, J. (1990). Intercontinental comparison of small-lake fish assemblages: the balance between local and regional processes. *American Naturalist*, 136, 345–375.

Twenge, J. M., Campbell, W. K. & Freeman, E. C. (2012). Generational differences in young adults' life goals, concern for others, and civic orientation, 1966–2009. *Journal of Personality and Social Psychology*, 102, 1045–1062.

Unmack, P. J. (2001a). Biogeography of Australian freshwater fishes. *Journal of Biogeography*, 28, 1053–1089.

Unmack, P. J. (2001b). Fish persistence and fluvial geomorphology in central Australia. *Journal of Arid Environments*, 49, 653–669.

Unmack, P. J. (2013). Biogeography. In: *Ecology of Australian Freshwater Fishes*. Humphries, P. & Walker, K. (Eds). Collingwood: CSIRO Publishing, pp. 25–48.

Van Allen, B. G., Dunham, A. E., Asquith, C. M. & Rudolf, V. H. W. (2012). Life history predicts risk of species decline in a stochastic world. *Proceedings of the Royal Society B: Biological Sciences*, 279, 2691–2697.

Verburg, P., Horrox, J., Chaney, E., *et al.* (2013). Effects of nutrient loading on the trophic state of Lake Brunner. *Marine and Freshwater Research*, 64, 436–446.

Villamagna, A. M., Angermeier, P. L. & Bennett, E. M. (2013). Capacity, pressure, demand, and flow: a conceptual framework for analyzing ecosystem service provision and delivery. *Ecological Complexity*, 15, 114–121.

Vörösmarty, C. J., Pahl-Wostl, C., Bunn, S. E. & Lawford, R. (2013). Global water, the anthropocene and the transformation of a science. *Current Opinion in Environmental Sustainability*, 5, 539–550.

Wagenhoff, A., Townsend, C. R. & Matthaei, C. D. (2012). Macroinvertebrate responses along broad stressor gradients of deposited fine sediment and dissolved nutrients: a stream mesocosm experiment. *Journal of Applied Ecology*, 49, 892–902.

Walsh, C. J., Fletcher, T. D. & Ladson, A. R. (2005). Stream restoration in urban catchments through redesigning stormwater systems: looking to the catchment to save the stream. *Journal of the North American Benthological Society*, 24, 690–705.

Wedderburn, S. D., Hammer, M. P. & Bice, C. M. (2012). Shifts in small-bodied fish assemblages resulting from drought-induced water level recession in terminating lakes of the Murray–Darling Basin, Australia. *Hydrobiologia*, 691, 35–46.

Welcomme, R. L., Cowx, I. G., Coates, D., *et al.* (2010). Inland capture fisheries. *Philosophical Transactions of the Royal Society B*, 365, 2881–2896.

Wells, B. K., Rieman, B. E., Clayton, J. L., Horan, D. L. & Jones, C. M. (2003). Relationships between water, otolith and scale chemistries of westslope cutthroat trout from the Coeur d'Alene River Idaho: the potential application of hard part chemistry to describe movements in freshwater. *Transactions of the American Fisheries Society*, 132, 409–424.

Whitton, B. A., Ellwood, N. T. W. & Kawecka, B. (2009). Biology of the freshwater diatom *Didymosphenia*: a review. *Hydrobiologia*, 630, 1–37.

Wilcock, R. J., Monaghan, R. M., Quinn, J. M., *et al.* (2013). Trends in water quality of five dairy farming streams in response to adoption of best practice and benefits of long term monitoring at the catchment scale. *Marine and Freshwater Research*, 64, 401–412.

Winemiller, K. O. (2005). Life history strategies, population regulation, and implications for fisheries management. *Canadian Journal of Fisheries and Aquatic Science*, 62, 872–885.

Winemiller, K. O. & Rose, K. A. (1992). Patterns of life-history diversification in North American fishes: implications for population regulation. *Canadian Journal of Fisheries and Aquatic Sciences*, 49, 2196–2218.

Winston, M. R., Taylor, C. M. & Pigg, J. (1991). Upstream extirpation of four minnow species due to damming of a prairie stream. *Transactions of the American Fisheries Society*, 120, 98–105.

Witte, F., Seehausen, O., Wanink, J. H., *et al.* (2013). Cichlid species diversity in naturally and anthropogenically turbid habitats of Lake Victoria, East Africa. *Aquatic Sciences*, 75, 169–183.

Woodford, D. J. & McIntosh, A. R. (2010). Evidence of source-sink metapopulations in a vulnerable native galaxiid fish driven by introduced trout. *Ecological Applications*, 20, 967–977.

Woodford, D. J. & McIntosh, A. R. (2011). Location of demographic sources affects the distributions of a vulnerable native fish in invaded river networks. *Freshwater Biology*, 56, 311–324.

Wright, J. (2013). *On a Pathway to Extinction? Parliamentary Commissioner for the Environment.* Wellington, New Zealand.

Climate change effects on freshwater fishes, conservation and management

JANI HEINO, JAAKKO ERKINARO,
ARI HUUSKO AND MISKA LUOTO

3.1 INTRODUCTION

Fresh waters harbour exceptional levels of biodiversity (Chapter 1). However, various anthropogenic environmental forces threaten this splendid richness (Carpenter *et al.*, 1992; Dudgeon *et al.*, 2006; Strayer & Dudgeon, 2010; Chapter 2). These threats include invasive species (Rahel & Olden, 2008; Chapter 8), land-use changes (Allan, 2004), decreased connectivity (Jackson *et al.*, 2001; Chapter 10) and various changes to water quantity and quality (Malmqvist & Rundle, 2002; Chapters 4–6). Furthermore, global climate change has been increasingly recognised as a pressing environmental change affecting many other stressors as well as having direct effects on fish diversity (Tonn, 1990; Comte *et al.*, 2013). Climate change affects fish individuals, populations, species and communities at various spatial and temporal scales (Tonn, 1990; Carpenter *et al.*, 1992; Graham & Harrod, 2009), ranging from effects on the behaviour of fish at a local scale to species distributions at biogeographic scales (Chu *et al.*, 2005; Hickling *et al.*, 2006; Ficke *et al.*, 2007; Hein *et al.*, 2011; Markovic *et al.*, 2012).

Although climates have changed throughout the history of the Earth, the current rates of increases in temperature, changes in rainfall and occurrence of exceptional weather conditions are unprecedented (IPCC, 2007). For example, global annual average air temperatures are, depending on a climate change scenario, projected to increase from 1°C to 5°C by the end of this century, with much among-region variation. Highest changes in temperature have been seen and are likely to be recorded at high latitudes (IPCC, 2007), where the effects of increased water

Conservation of Freshwater Fishes, eds G. P. Closs, M. Krkosek and J. D. Olden. Published by Cambridge University Press. © Cambridge University Press 2016.

temperature on fish are also likely to be most profound (Lehtonen, 1996; Chu *et al.*, 2005; Sharma *et al.*, 2007; Heino *et al.*, 2009). By contrast, closer to the equator, changes in temperature are not likely to be that pronounced, but changes in rainfall, decreases in river discharge and increases in human water withdrawal are likely to increase in the future (Xenopoulos *et al.*, 2005; Thieme *et al.*, 2010). Such changes suggest that there may be more irregular droughts and floods, which affect fish at various levels of organisation from individual physiology and population abundance to community structure and range shifts (Graham & Harrod, 2009; Morrongiello *et al.*, 2011). Predicting the effects of climate change on the abundance and distribution of fish is thus highly challenging and regionally variable, but important for the conservation of fish diversity (Cochrane *et al.*, 2009; Comte *et al.*, 2013).

Two recently emerged fields of research are particularly important for the conservation of freshwater fishes in a changing climate. First, conservation biogeography is a broad subfield of conservation biology, the aim of which is to aid conservation through 'the application of biogeographical principles, theories, and analyses, being those concerned with the distributional dynamics of taxa individually and collectively, to problems concerning the conservation of biodiversity' (Whittaker *et al.*, 2005). In this sense, examining the effects of climate change on fish populations, species and communities at large spatial and temporal scales is situated in this field of research (Olden *et al.*, 2010). At smaller spatial and temporal scales, findings from conservation physiology are also important for advancing our knowledge of climate change impact on fish individuals, populations and species (Wikelski & Cooke, 2006). Conservation physiology can be defined as 'the study of physiological responses of organisms to human alteration of the environment that might cause or contribute to population declines' (Wikelski & Cooke, 2006). As we cannot only concentrate on either the larger-scale views of conservation biogeography or finer-scale approaches of conservation physiology, we should attempt to gain greater understanding by combining these two fields of research in our inquiry into the patterns and processes of climate change effects on freshwater fishes.

The aims of this chapter are to review the direct and indirect effects of climate change on fish individuals, populations, species and communities. We begin by reviewing the main patterns in fish diversity along major ecological gradients, including climatic gradients, as understanding variation in the distribution and abundance of fish along these gradients is a prerequisite for predicting future changes in the face of climate change. We next proceed to examining evidence of climate change effects on fish individuals, populations, species and communities, outline typical modelling approaches and provide predictions for the future of fish distribution based on information provided by modelling techniques. We also consider the applied importance of climate change research in fish conservation and management by addressing questions related to climate change adaptations and conservation efforts.

3.2 RESPONSES OF FRESHWATER FISH TO GEOGRAPHIC, ENVIRONMENTAL AND CLIMATIC GRADIENTS

3.2.1 Freshwater fish diversity along ecological gradients

Biological diversity can be understood at different levels of biological organisation (e.g. genetic diversity vs. species diversity) or based on different ways to characterise fish communities (e.g. taxonomic structure vs. functional structure; Figure 3.1). The responses of fish diversity to ecological gradients can be examined as alpha (e.g. local number of species), beta (e.g. turnover or variation of species composition) and gamma diversity (e.g. regional number of species) (Whittaker, 1960). Components of diversity may respond differently to ecological gradients and, therefore, to climate change. In terms of climate change effects, understanding the responses of fish to latitude, longitude and altitude is of fundamental importance (Figure 3.2). Fish species diversity indeed shows clear lows and highs along these major ecological gradients (Matthews, 1998; Oberdorff *et al.*, 2011). However, the responses of genetic, life history, functional and phylogenetic diversity are less well understood in this regard for multispecies communities (Mims *et al.*, 2010; Strecker *et al.*, 2011), although detailed population genetic and life-history analyses have appeared recently for populations of some iconic, charismatic and commercially valuable species (Crozier *et al.*, 2011; Jonsson & Jonsson, 2011). This paucity of information is due to a lack of adequate resources to study all species in multispecies communities in detail.

Latitude is one of the main proxy variables correlated with fish species diversity (Chapter 2). Gamma diversity typically attains the highest levels in tropical regions (e.g. Amazon, Congo and Mekong basins) and declines sharply towards higher latitudes (Matthews, 1998; Griffiths, 2006; Oberdorff *et al.*, 2011). Local-scale alpha diversity (e.g. a lake or a stream) is also generally higher in the tropics than at high latitudes, although within-region variation in alpha diversity may be high in various regions due to local environmental factors (Griffiths, 1997; Heino,

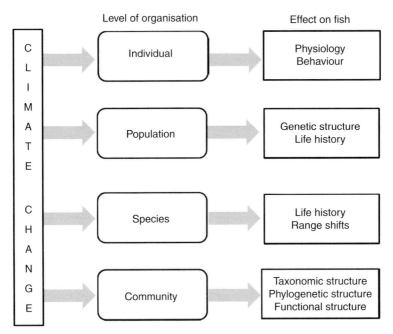

Figure 3.1 Effects of climate change on different levels of biological organisation in freshwater fish.

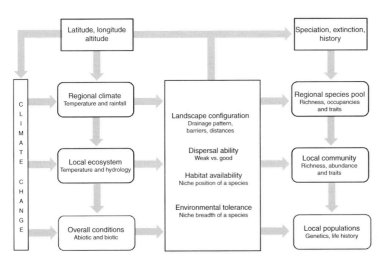

Figure 3.2 A schematic view of factors determining regional species pool (species richness, species occupancy frequencies and biological traits), local community structure (species richness, relative abundances and biological traits) and local population characteristics (genetic structure, life-history characteristics), and how climate change leads to the novel regional species pools and local communities. The reassembly process is particularly affected by drainage patterns and habitat availability for species, and species dispersal ability and environmental tolerance in the face of climate change.

2011). These latitudinal gradients in fish diversity are moulded by various processes and factors, including evolutionary diversification in different regions and historical climate changes, although two of the primary factors are present-day climate and energy (Oberdorff *et al.*, 2011). For example, across European lakes, fish alpha diversity was more strongly related to climatic factors than to local environmental gradients and local anthropogenic stressors (Brucet *et al.*, 2013). This finding suggests that anthropogenic climate change is likely to profoundly affect latitudinal gradients in fish diversity. For example, gamma and alpha diversities are likely to increase at high and mid latitudes in the face of climate change (see Section 3.3). The degree to which fish are able to follow changing climate by latitudinal range shifts is, however, contingent on the degree to which they can disperse to other geographical regions, drainage basins and freshwater ecosystems or adapt to new climates (Poff *et al.*, 2012).

Fish species diversity also differs across longitude. In Europe, the highest diversity in riverine systems can be found in the eastern part near the Ponto-Caspian refugia from which the colonisation of the continent took place after the latest major glaciation (Reyjol *et al.*, 2007). In North America, higher diversity is found in the eastern part of the continent, particularly in the rivers connected to the River Mississippi drainage basin, whereas diversity is lower in the western part of the continent (Matthews, 1998). These patterns have resulted largely from the effects of glaciation and consequent dispersal processes after the ice sheets receded, but may also be related to present-day climatic patterns in terms of rainfall and temperature (Oberdorff *et al.*, 2011).

Fish species diversity typically decreases with increasing altitude (Matthews, 1998). Temperature is one of the key factors influencing fish diversity along altitudinal gradients, although other influential environmental factors, including stream size and habitat heterogeneity, also decrease with increasing altitude. Fish communities at high altitude are typically dominated by cold-water species, while cool- and warm-water species increase in abundance and diversity in downstream reaches at lower altitude.

3.2.2 Characterising freshwater fishes in relation to climate change effects

Temperature is undoubtedly one of the most important niche parameters for ectothermic organisms such as fish. Magnuson *et al.* (1979) divided temperate fish species into cold-, cool- and warm-water guilds, referring to their respective temperature preferences: cold-water species (preferred summer temperature centred upon 11–15°C), cool-water species (21–25°C) and warm-water (27–31°C). Thus, cold-water species are naturally found at high latitudes, higher altitude and smaller streams (e.g. salmonids), whereas cool- and warm-water species are more common in lower latitudes, lower altitude and larger rivers (e.g. cyprinids). These temperature preferences and responses to major ecological gradients are highly important when describing and predicting climate change effects on fish diversity (Daufreshne & Boët, 2007; Buisson et al., 2008; Shuter et al., 2012). Fish species within each of the thermal guilds may show either large (eurythermal) or small (stenothermal) temperature ranges, although typically species are categorised as cold-water stenothermal or generally eurythermal species.

Regarding multiple interacting environmental factors, a distinction should be made between environmental specialists (i.e. stenotopic species) and environmental generalists (i.e. eurytopic species). Specialist and generalist species may be differently sensitive to anthropogenic environmental and climatic changes (Poff et al., 2012; Heino, 2013). For example, specialist species are likely to be driven towards extinction faster than generalist species due to their different environmental niche breadth. Similarly, weak and strong dispersers are able to track differently changed environmental conditions. Species with limited dispersal abilities are likely to be driven to extinction faster than those with better dispersal powers, as the latter are more able to track changes in environmental conditions. Based on this reasoning, specialist species with limited dispersal abilities are under considerable risk in the face of climate change, as they may show limited capacity to find rare habitats suitable for continuing population persistence. Such species may be prevented from extinction through assisted colonisation (Olden et al., 2011).

3.3 CLIMATE CHANGE EFFECTS ON FRESHWATER FISHES

3.3.1 General effects at the individual, population and community levels

Effects of climate change on freshwater fishes at different levels of biological organisation (i.e. individual, population and community) primarily result from physiological changes at the organismal level. Physiological changes at the organismal level are subsequently mirrored in the responses of populations, species and communities to climate warming and hydrological changes. For example, the body size of an individual fish is a basic biological characteristic that scales with many ecological properties, affecting directly energy and water requirements for thermoregulation, energy/mass acquisition and utilisation rates, and life-history characteristics (Gardner *et al.*, 2011). Changes in body size can thus have profound implications for freshwater fish individuals, populations and communities in the face of climate change. Based on long-term surveys, experimental data and published results, Daufresne *et al.* (2009) observed a significant increase in the proportion of small-sized species and young age classes, and a decrease in 'size-at-age' for freshwater fishes and other aquatic animals under global warming. The metabolic rate of fish increases in warm water and, therefore, they need more oxygen. Yet, warm water provides less oxygen and, hence, fish growth is limited, resulting in smaller body sizes. Declining body size of fish and other aquatic animals is a universal response to warming climate along with changes in phenology and range shifts (Daufresne *et al.*, 2009). Although declining body size certainly has physiological consequences for individual fish, one can only imagine the implications of numerous small fish individuals for population- and community-level phenomena.

Fish populations are naturally variable in both space and time (Matthews, 1998), yet this variation is likely to increase with climate change (Jeppesen *et al.*, 2012). For example, among the most well-studied species, migratory salmonid populations are naturally fluctuating and, typically, an interval of seven to nine years has been detected between peak-level abundances for Atlantic salmon (*Salmo salar*) in several long-term monitoring programmes (Niemelä *et al.*, 2005; Jonsson & Jonsson, 2011). There is a growing body of evidence suggesting that global climate patterns affect growth, survival and ultimately abundance of populations of anadromous salmonids in Pacific, Atlantic and Baltic areas (Mantua *et al.*, 1997; Friedland *et al.*, 2009; Huusko & Hyvärinen, 2012). In this regard, less charismatic species and those with limited commercial importance are generally poorly known, and only recently has there been increasing interest in long-term changes in their populations (Jeppesen *et al.*, 2012).

Climate change is likely to change the structure of fish communities in riverine and lacustrine systems. These changes are likely to be seen in the distribution and dominance of cold-water versus cool- and warm-water fishes (Eaton & Scheller, 1996; Mohseni *et al.*, 2003; Rahel & Olden, 2008; Poff *et al.*, 2012). For example, Tisseuil *et al.* (2012a) suggested that climate change will homogenise fish community structure in the Adour–Garonne River system in France, leading to a decrease in spatial species turnover along the upstream–downstream gradient. This was because cool- and warm-water species are likely to expand their distributions, while the few cold-water species present will experience a clear reduction in their distributions across the basin in the near future (Tisseuil *et al.*, 2012b). The expansion of cool- and warm-water species is thus responsible for the homogenisation of fish communities across this temperate river system.

Climate change is also likely to have profound effects on fish distribution and abundance in streams at a larger scale (Eaton & Scheller, 1996; Mohseni *et al.*, 2003). For example, Buisson *et al.* (2008) observed that the few cold-water species (e.g. *Salmo trutta, Cottus gobio*) residing in French streams would experience high reductions in distributions following climate change, while cool- and warm-water species are likely to show increasing distributions. Interestingly, the species with the highest present-day niche breadth and least marginal niche position (e.g. *Salmo trutta, Cottus gobio*) were cold-water inhabitants of headwater streams, i.e. systems which constitute much of stream networks but are also most sensitive to climate change. By contrast, the species that had small present-day niche breadth and relatively high marginality of niche position were warm-water inhabitants of larger streams and rivers, and these species are likely to benefit from warming stream temperatures and to expand their ranges to headwaters previously unsuitable to them with regard to temperature conditions. Buisson *et al.* (2008) noted that such changes are likely to have profound effects on fish community organisation and ecosystem function in streams, and will eventually affect the ecological and economic value of fish communities.

The above findings for temperate streams are, however, opposite to the general intuition that species with small niches should be more sensitive to environmental changes than those with large niches (Swihart *et al.*, 2003; Thuiller *et al.*, 2005; Poff *et al.*, 2012; Botts *et al.*, 2013). These contrasting findings for stream fish are likely to be due to the general characteristics of stream systems, where headwater streams by being more ubiquitous and colder than larger rivers are more sensitive to climate changes, including increases in temperature and variation in flow (Poff *et al.*, 2002; Allan & Castillo, 2007; Heino *et al.*, 2009). Hence, in many drainage basins, it is likely that cold-water species with large niches, non-marginal niche position and broad present-day distributions are more prone to the negative effects of climate change than warm-water species with small niches, marginal niche position and narrower present-day distributions within drainage basins (Buisson *et al.*, 2008). Thus, climate change is likely to have serious effects on the organisation of fish communities at the reach, stream and drainage basin scales if there will be changes in the abundance and dominance of cold-water fish (negative effect) versus cool- and warm-water fishes (positive effect). Comte *et al.* (2013) stress, however, that we still lack adequate knowledge of the effects of climate change on most fish species (e.g. most species that are not iconic or commercially valuable) and fish communities in regions other than Europe and North America. Further studies may indeed show some unexpected responses of fish to climate change, with some cool- and warm-water species showing either increases, reductions or no apparent changes in their distributions in the future (Pascual *et al.*, 2007; Comte *et al.*, 2013).

Climate change is likely to modify not only the taxonomic structure but also the functional structure of stream fish communities (Schleuter *et al.*, 2012). Again, research on temperate stream fish communities has suggested the increase of both taxonomic and functional diversity of fish communities at different parts of drainage basins (i.e. upstream, midstream and downstream). Increases of diversity were predicted to be highest in upstream and midstream reaches, while downstream reaches were assumed to experience lesser changes (Buisson & Grenouillet, 2009). Buisson and Grenouillet (2009) also found that upstream communities would differentiate more (i.e. more taxonomic and functional variation in community composition), while mid- and downstream reaches were expected to become homogenised (i.e. less taxonomic and functional variation in community composition). This homogenisation was predicted to be due to the increase of cool- and warm-water species through increased colonisation, while increased differentiation in upstream reaches was expected to result from local extinctions. Additional research is needed to determine the degree to which similar upstream–downstream patterns can be presumed to occur in other regions.

Climate change, in association with increased human water use and land use intensification, is also likely to modify fish biodiversity in the rivers of tropical and subtropical regions. However, while temperature changes are not likely to be as pronounced as in higher latitudes, fresh waters near the equator are likely to experience highly altered hydrologic conditions with effects on fish abundance, distribution and community structure (Xenopoulos *et al.*, 2005; Thieme *et al.*, 2010; Dudgeon, 2011; Morrongiello *et al.*, 2011). Although future changes in hydrologic conditions will be regionally variable (IPCC, 2007), rivers of arid regions at low latitudes are likely to experience decreased flows (Xenopoulos *et al.*, 2005), which will threaten the often highly endemic fish faunas of these regions (Tedesco *et al.*, 2012). Furthermore, even small changes in temperature at low latitudes may affect fish, as tropical species are considered to be closer to their upper thermal limits than most species at high latitudes (Deutsch *et al.*, 2008; Dudgeon, 2011). Although very precise predictions of the effects of such hydrological and thermal changes on fish diversity are not yet possible, it appears that climate-induced changes will also be visible in subtropical and tropical regions.

Fish communities of lakes at high latitudes will experience dramatic changes in the face of climate change (Heino *et al.*, 2009). For example, shifts towards warm-water cyprinid and cool-water percid dominance in Finnish lakes are predicted, along with decreases in the populations of salmonid and other cold-water species in fish communities (Lehtonen, 1996). This trend means that many cold-water species (e.g. *Salvelinus alpinus*, *Salmo trutta*, *Coregonus albula*) will experience reductions in the amount of suitable habitats as temperatures increase, and such species are likely to go extinct at least in shallow lakes without thermal stratification. By contrast, many cyprinids (e.g. *Rutilus rutilus*, *Alburnus alburnus*) and percids (e.g. *Perca fluviatilis*, *Sander lucioperca*) are likely to expand their ranges and increase in abundance locally in lakes (Lehtonen, 1996). Similar patterns of cyprinid increases and salmonid decreases have also been found across the whole of Europe, as has been evidenced by long-term studies of lake fish communities (Jeppesen *et al.*, 2012).

3.3.2 A special case: migratory salmonid fish, environmental gradients and climate change

Migratory animals are affected by climatic factors (i.e. variability and change) both in their breeding and non-breeding areas, resulting in potential climate change impacts on their life cycles (Robinson *et al.*, 2009; Chapter 11). Anadromous fish are a group of migratory species where the interactions of climate change with multiple environmental factors and fish life-history characteristics have been studied widely. Anadromous fish species have to cope with different habitats and variable environments during their life cycle, across long geographic distances and a wide salinity gradient from fresh water to full sea water (Todd *et al.*, 2010). For example, among the most widely studied species, Atlantic salmon juveniles (i.e. referred to as smolts at this life stage) emigrate from fresh water after having achieved a certain size and physiological stage, and subsequently increase their growth rate in the ocean.

After one or more years at sea they return to their natal rivers to breed. In general, smolt migration occurs in spring or early summer, and the timing of initiation of the downstream migration differs among rivers (Thorstad *et al.*, 2012).

Climate change may alter environmental conditions (e.g. water temperatures and food availability), leading to a mismatch between the environment and the phenology of smolts. A significant advancement of phenological spring events across the Northern Hemisphere has been documented concurrent with earlier arrival of spring (e.g. Burrows *et al.*, 2011). For Atlantic salmon smolts, some long-term studies have suggested that their downstream migration is currently occurring at earlier dates in association with water temperature warming (e.g. Kennedy & Crozier, 2010). In their analyses of spatial and temporal variation in the timing of salmon smolt migration throughout the basin of the North Atlantic over the last five decades, Otero *et al.* (2014) found that migration among rivers was positively associated with both freshwater and sea-surface temperatures, with the initiation of the smolt migration occurring 2.4 days earlier per decade since the 1960s. This shift in phenology appears to match the change in air, river and ocean temperatures, suggesting that Atlantic salmon emigration, throughout the distribution range, is responding to the current global climate change. This response, in turn, might affect salmon growth and survival through a temporal mismatch with the production of marine prey items, causing reduced feeding opportunities. In addition, the earlier migration and its mismatch with oceanic conditions may, through suboptimal ionoregulatory ability coupled with osmotic shock at sea entry, result in altered antipredatory behaviour of the early post-smolts, such as longer acclimatisation period and delay in further migration (Todd *et al.*, 2011; Otero *et al.*, 2014).

Climate change also influences the return migration and adult salmon ascent to rivers. For example, the fish are likely to encounter increased water temperatures and decreased flows associated with global warming (Todd *et al.*, 2011). Such changes will be most severe at the present southern extremes of the distributions of salmon species (e.g. New England, California, France, Iberian Peninsula), and may also exterminate salmonids from those areas and move their southern range boundaries towards north (Reist *et al.*, 2006; Crozier *et al.*, 2008; Todd *et al.*, 2011). These range shifts are likely to result from the fact that migratory salmonids in temperate and cold areas show optima of water temperature at river entry and upstream migrations, and excessively higher temperatures hinder or may even cease the swimming activity. Steady long-term increases in water temperature have been associated with earlier dates for river entry of salmon, both in the Atlantic (e.g. Jonsson & Jonsson, 2011) and the Pacific (e.g. Crozier et al., 2008, 2011) areas. In Fraser River, British Columbia, episodic high river temperatures caused the collapse of aerobic scope and failure in upstream migration of sockeye salmon (*Oncorhynchus nerka*) (Farrell et al., 2008). On the other hand, as the climate moderates and favourable habitats emerge in the Arctic because of global warming, migratory salmonids may expand their range, for example, to currently ice-covered areas (Nielsen et al., 2013).

Salmonid fishes typically spawn in autumn and their egg incubation is exceptionally slow, with hatching typically occurring several months after spawning. Water temperature and thermal sum regulates this development, and the number of degree-days required varies to some extent between salmonid species (Jonsson & Jonsson, 2011). A long incubation period results in high sensitivity to environmental changes, as alterations of flow and temperature conditions may result in a mismatch between embryo emergence and environmental conditions, causing increased mortality (Todd et al., 2011).

In migratory salmonid fishes with complex life histories and utilisation of multiple habitats, climate change may produce conflicting selection pressures in different life stages, which will interact with plastic changes in different ways (Crozier et al., 2008, 2011). The growth of juveniles and smolt production may increase in northern rivers. Along with increasing productivity in fresh waters, northward range expansion might compensate, or even exceed, the predicted reductions of populations in the southern end of their range boundaries (Todd et al., 2011). In addition, migratory species, such as anadromous forms of brown trout (*Salmo trutta*) and Arctic char (*Salvelinus alpinus*), may show a higher proportion of freshwater resident individuals compared to anadromous individuals because of increased productivity of freshwater habitats (Jonsson & Jonsson, 2011). Nevertheless, anadromous fish will show not only immediate phenotypic responses to climate change, but also long-term genetic changes are likely in traits such as smolt age, age at maturity and disease resistance (Jonsson & Jonsson, 2011). However, such predictions are still uncertain because of lack of data on the strength of selection, heritability, and ecological and genetic linkages among many of the traits (Crozier et al., 2008).

3.3.3 Species range shifts

One of the most obvious and anticipated impacts of climate change on freshwater ecosystems is a shift in the distributional range of fish species (Booth *et al.*, 2011; Comte *et al.*, 2013). A central idea in the climate change debate is that species distributions are generally shifted polewards (e.g. Lehtonen, 1996) and towards higher altitude as a result of temperature rises induced by climate change (e.g. Cooney *et al.*, 2005). Abiotic drivers related to climate change, such as temperature and precipitation, can affect directly fish species ranges or indirectly through effects on habitat factors or food resources. These drivers can also interact with other factors such as overfishing, invasive species, habitat removal, disease and pollution to reduce or shift species ranges spatially (Heino *et al.*, 2009; Comte *et al.*, 2013). Furthermore, habitat degradation and connectivity of water bodies may contribute to current range shifts of fish species (Chapters 6 and 10). Different fish species show range shifts relatively independently, and not all fish species exhibit similar range changes at the same time and speed, affecting community reassembly in both the old and new ranges (Chu *et al.*, 2005; Hickling *et al.*, 2006). In freshwater ecosystems, range shifts alter biotic interactions, influencing species abundance, composition, ecosystem processes and services within ecosystems (Jackson & Mandrak, 2002; Wisz *et al.*, 2013). Thus, range shifts may create 'no-analogue' communities where a few key species and population genetics play unprecedented roles, possibly leading to novel ecosystems (Booth *et al.*, 2011).

Surprisingly few studies have attempted to test range shifts in fish species, and even fewer demonstrated clear links to climate change. The study of Hickling *et al.* (2006) remains as one of the most comprehensive systematic studies investigating recent range shifts of freshwater fish. Based on long-term data covering 25 years in Great Britain for 15 stream fish species, they documented mean poleward shifts in northern range margin and altitudinal shifts in optimum by up to 51 km and 33 m, respectively. Also, Daufresne and Boët (2007) analysed trends in the distribution of fish in the Rhone River in France, and observed longitudinal changes in the distribution of warm-water fish, namely chub (*Leuciscus cephalus*) and barbel (*Barbus barbus*), which began to replace dace (*Leuciscus leuciscus*) in more northerly upstream habitats. These shifts were related to changes in water temperatures over the same period, and appeared independent of variations linked to more regional characteristics of river hydrology and temperature patterns. Furthermore, there is also clear evidence that periodic droughts will extirpate local populations at the margins of a species range. For example, Elliott *et al.* (1997) showed a clear impact of drought years on survivorship and recruitment in brown trout populations in northwestern England. Recently, studies have begun to consider the influences of climate change on the ecological characteristics of the flow regimes, as well as run-off per se (Sanborn & Bledsoe, 2006), and assimilation of this information into species distribution models highlights the importance of hydrology in predicting range shifts (Bond *et al.*, 2011; Booth *et al.*, 2011).

Recent research has indicated that the current climatic change has already affected geographical distributions of fish species (Hickling *et al.*, 2006; Booth *et al.*, 2011; Comte *et al.*, 2013). Furthermore, projected climate changes are likely to have an even greater impact on freshwater fish. Developing models that are able to provide reliable predictions of fish species distributions in new areas or other times is a great challenge for contemporary freshwater ecology (Buisson *et al.*, 2008; Comte *et al.*, 2013). There is a long history of research on the effect of climate change on freshwater fish (Elliot *et al.*, 1997), with particular attention being devoted to changes in species distribution (Chu *et al.*, 2005; Hein *et al.*, 2011; Comte *et al.*, 2013). Fish distribution could be strongly affected by climate warming and, as ectothermic organisms, fish have no physiological ability to control their body temperature (Wood & McDonald, 1997). Furthermore, fish cannot disperse across terrestrial areas and are consequently limited to the water systems they currently live in (Buisson *et al.*, 2008; Olden *et al.*, 2010; Chapters 2 and 10). These constraints should be considered in modelling the effects of climate change on freshwater fish distribution.

There are various methodological frameworks available for examining the potential effects of climate change on freshwater fish, ranging from dynamic ecosystem models and spatially explicit mechanistic models for range shifts of single species to physiological-based models and correlative bioclimatic envelope models (Thuiller, 2005; Heikkinen *et al.*, 2006). The following section concentrates on the latter group, i.e. statistical bioclimatic envelope models, which are among the most popular approaches to simulate species–climate change relationships. Such models are a special case of ecological niche models in which the current geographical distribution or abundance of species are related to climatic variables to enable projections of distributions under future climate change scenarios (Guisan & Zimmermann, 2000).

3.3.4 Methodological approaches for modelling freshwater fish distributions

The main phases in bioclimatic envelope modelling contain the following: (i) correlating the known occurrences and/or abundances of the target species to the ecological characteristics of the study area, (ii) calibrating a model that outlines the ecological dimensions of the species niche, and (iii) forecasting the resulting bioclimatic envelope model back onto the geographical space to identify areas with environmental conditions inside or outside the species niche (Guisan & Zimmermann, 2000). Assessments of future species ranges are developed by applying the models based on the climate variables that best describe the current distributions to simulate future distributions under selected climate change scenarios (Heikkinen *et al.*, 2006).

Forecasting potential distributional shifts of freshwater fish in the face of the projected climate change has recently become a popular study topic (Buisson *et al.*, 2008; Comte *et al.*, 2013). Over the last two decades, climate change scenarios have been unceasingly advanced and ecological modelling techniques have been developed further. This progress has increased our ability to forecast how species and communities could respond to climate change (Comte *et al.*, 2013). However, compared with other organisms such as plants, birds or butterflies, there is a clear lack of studies using species distribution models to assess the impacts of the most recent climate change scenarios on freshwater fish.

Predicting the potential impacts of climate change on species distribution has been done by using a wide variety of modelling techniques (Guisan & Thuiller, 2005). Climatic envelope techniques have been used to calculate a fitted, species-specific, minimal rectilinear envelope in a multidimensional climatic space (Guisan & Zimmermann, 2000). The best-known of these techniques are BIOCLIM, HABITAT and DOMAIN. The most commonly used statistical techniques in studying the effects of climate change on fish distributions are the following: (1) generalised linear models (GLMs) are one of the most popular statistical approaches that are used in climate change impact modelling (Heikkinen *et al.*, 2006). This is due to the ability of GLMs to handle nonlinear relationships and different types of statistical distributions characterising spatial data, and to the fact that they are technically closely related to traditional practices used in linear modelling and analysis of variance (ANOVA). (2) Generalised additive models (GAMs) are semiparametric extensions of GLMs (Hastie & Tibshirani, 1990). They are designed to capitalise on the strengths of GLMs without requiring the problematic steps of postulating a response curve shape or specific parametric response function (Wood, 2006). (3) Classification tree analysis (CTA) provides an alternative to regression techniques. CTA is a rule-based method that generates a binary tree through binary recursive partitioning, which is a process that splits a node based on true/false answers about the values of predictors (Venables & Ripley, 2002). (4) Artificial neural networks (ANN) provide an alternative way to generalise linear regression functions (Venables & Ripley, 2002). Neural networks have received considerable attention in building accurate models for prediction when the functional form of the underlying equations is unknown. They have been shown to be universal and highly flexible function approximates for any data (Lek & Guegan, 1999). (5) The boosted regression trees (BRTs) technique is a modern machine-learning method that estimates the relationship between a response variable and its predictors without *a priori* specification of a data model (Elith *et al.*, 2008). This technique combines large numbers of simple tree-based models to form a final model optimised for prediction, using cross-validation for model-building. BRTs can model complex functions and automatically incorporate interactions between predictors. (6) Random forest (RF) is also a tree-based machine-learning method. It is based on an ensemble classifier that consists of many decision trees and outputs the class that is the mode of the classes output by individual trees (Breiman, 2001). For further information about modelling techniques utilised in studies of freshwater fish distributions, see Olden and Jackson (2002).

Currently, BIOMOD is often used in species distribution modelling and climate change impact studies. BIOMOD is a computer platform for ensemble forecasting of species distributions, enabling the treatment of a range of methodological uncertainties in models and the examination of species–environment relationships. BIOMOD can run analysis for each species using several different modelling techniques or different forms and parameterisations of one particular method. Additionally, BIOMOD includes the ability to project species distributions into different environmental conditions and dispersal functions (Thuiller *et al.*, 2009).

3.3.5 Future directions in modelling climate change impacts on freshwater fish

Most of the studies investigating climate change effects on freshwater fish have concentrated on defining suitable thermal conditions for different species under climate change scenarios. The resulting models typically assume that species would be able to persist in a certain habitat or to colonise it without giving attention to the other environmental determinants that could control fish species distribution (Rahel *et al.*, 1996; Sharma *et al.*, 2007). Although climate undoubtedly largely influences fish species distribution, it is well known that many other abiotic factors as well as biotic interactions are also crucial for freshwater fish. Furthermore, in a warming climate, potential extirpation of vulnerable species may occur as a consequence not only of the direct effect of climate change, but also of the invasion of species favoured by climate warming (Jackson & Mandrak, 2002; Spens *et al.*, 2007; Rahel & Olden, 2008).

In future studies, several factors should be combined with modelling climate change effects on freshwater fish. These factors include fragmentation of water bodies, species dispersal limitations, the impacts of rising carbon dioxide and water pollution driven by human interactions, genetic differences in populations in different parts of the range and changing biotic interactions. Such a multitude of potential drivers of freshwater fish distributions suggest that it is necessary to evaluate realistically the use of various methodological approaches in future climate change impact scenarios. These various factors also pose methodological constraints, uncertainties and restrictions on bioclimatic envelope models, and their results should hence be interpreted with caution (Guisan & Thuiller, 2005). Nevertheless, bioclimatic envelope models have certain advantages. They are highly valuable for providing novel insights into potential climate warming effects on fish species when the physiological factors limiting the range of a studied species are relatively poorly known. They also allow the identification of key relationships between species and the factors governing their distributions (Heikkinen *et al.*, 2006). Furthermore, single-species models may offer more accurate and realistic forecasts than those offered by species-assemblage models, not least because species responses to climate change are thought to be mainly individualistic.

3.4 IMPLICATIONS OF CLIMATE CHANGE FOR CONSERVATION AND MANAGEMENT

Climate change will have obvious effects on the conservation of freshwater fish, and both evolutionary adaptations and human-assisted adaptations are likely to be valuable strategies for conservation. Here, we discuss mean trends in climate effects, as well as rare, extreme and unusual climatic events that may be relevant in the context of practical conservation and management efforts.

3.4.1 Climate change adaptations and conservation actions

Climate change affects freshwater fish individuals, populations, species and communities to such a large extent that adaptations to these changes are both obvious and necessary. These include both evolutionary adaptations of organisms and human-assisted means to accommodate organisms to changed conditions (Carter & Kankaanpää, 2003; Heino *et al.*, 2009). On the one hand, autonomous adaptation is related to the responses that take place after climate change impacts have occurred. Such adaptation is often exhibited by individuals, species, communities and ecosystems, which has been suggested by the responses of organisms to past climate changes by shifts in distributions rather than by genetic adaptations to novel environmental conditions (Pöyry & Toivonen, 2005). The most obvious challenges of climate change adaptation for freshwater fish will be related to the amount to which different types of freshwater ecosystems are diminished, the existence of thermal and flow refugia, the abilities of species to disperse to higher latitudes or higher altitude and chances to overcome the effects of increased isolation of freshwater ecosystems due to human intervention (Poff *et al.*, 2002; Heino *et al.*, 2009). Autonomous adaptation is mostly reactive. On the other hand, planned adaptation comprises the conservation and management of fish populations, species and communities in the face of climate change. Planned adaptation includes the establishment of protected areas (Chapter 14), riparian management (Chapter 9), planning of dispersal corridors for various types of species (Chapters 4 and 10), and minimising anthropogenic effects, such as habitat destruction and exotic species introductions, on freshwater ecosystems (Chapter 8). Planned adaptation is both reactive and proactive. In this regard, proactive stands for adaptation to climate change that may be achieved before the predicted impacts, and is the most efficient way for humans to intervene against the negative effects of climate change on freshwater fish.

Managers and scientists will be challenged to understand the importance of autonomous adaptations in the face of climate change. This will entail predicting changes in local selection pressures on populations and their possible adaptive responses (Youngson *et al.*, 2003). For example, success in the conservation of anadromous salmonid fishes with complex life histories and multiplicative impact of selection over their life cycle requires understanding the range of possible evolutionary responses to climate change (Crozier *et al.*, 2008). Assuming that fish populations are genetically structured (e.g. Vähä *et al.*, 2007) and show local adaptation (e.g. Garcia de Leaniz *et al.*, 2007), it is imperative to incorporate genetic aspects into conservation strategies to tackle both the present and future challenges in guaranteeing the persistence of fish populations in a changing climate (Youngson *et al.*, 2003). The task of including genetic aspects in conservation efforts is formidable even for single fish species, and it may be largely impossible as a conservation strategy for multispecies communities.

Planned adaptations incorporate human interference with ecological systems. For example, both riparian conservation (e.g. an increase of forest cover benefits fish through buffering temperature changes on freshwater ecosystems; Chapter 9) and fisheries management (e.g. helping native fish by controlling the abundance of invasive fish in freshwater ecosystems; Chapter 8) are efficient means that may aid the conservation of freshwater fish (Lawrence *et al.*, 2014). Other complicating factors related to fish life histories should be considered as well. For example, although riparian conservation and fisheries management can be used to maximise the production of salmonid juveniles in fresh waters, the main problems with anadromous salmonids have increasingly been recognised at sea because the survivorship there continues to decline (Jonsson & Jonsson, 2011; Todd *et al.*, 2011). There is a clear need for predictive models integrating the various life-history parameters, corresponding environments and their changes (Lassalle *et al.*, 2008). In addition to changes at a larger scale (e.g. changes in annual and seasonal means), high-magnitude and more frequent temperature and flow extremes should be taken into account in the models (Todd *et al.*, 2011).

In the context of climate change adaptation, researchers and managers should also consider the fact that temperature and hydrological changes may differ between different parts of a drainage basin. Ruesch *et al.* (2012) pointed out that the research on effects of climate change on the habitat of freshwater salmonid fishes typically 'focuses on coarse spatial resolutions at which network structure may be irrelevant, and on migration within downstream reaches, where temperature variability is often limited'. Furthermore, they emphasised that temperature and hydrological conditions in headwater reaches are often more susceptible to weather extremes than those in downstream reaches during summer months. Because headwater streams are important as nurseries for juveniles and as oversummering habitats for adults, management and conservation efforts should be directed to these sensitive parts of drainage basins (Ruesch *et al.*, 2012).

3.4.2 Altered environmental conditions in the face of climate change

Climate change may affect fish not only directly, but also indirectly through effects on other environmental factors (Heino *et al.*, 2009; Poff *et al.*, 2012). Effects of climate change on fish through altered habitat conditions have been most intensively studied in the Northern Hemisphere, which are characterised by pronounced seasonality in environmental conditions. In addition to the effects of summer warming and temperature extremes, cold adaptation and altered winter conditions in northern areas have also been the topic of numerous studies (Ficke *et al.*, 2007; Huusko *et al.*, 2007; Linnansaari & Cunjak, 2012). In winter, natural freshwater ecosystems generally exhibit low water temperatures, different forms of ice (e.g. frazil, anchor ice and surface ice), low discharge rates, short periods of daylight and low heat radiation (Prowse, 2001a, 2001b). For fish living in such seasonal environments, the changes in the duration of periods with ice and snow cover are one of the most pronounced impacts of global warming (Huusko *et al.*, 2007; Linnansaari & Cunjak, 2012). As global warming influences fish via ice cover-related processes, even small changes in important factors such as ambient light and temperature may have considerable impacts on the physiological performance, species interactions and habitats of fish (Shuter *et al.*, 2012).

Freshwater fish typically have local intraspecific adaptions to ice cover. Fish are often able to deal with the allostatic load, i.e. chronic stress resulting from activities and experiences of individuals during daily life under the variable ice conditions (McEwen & Wingfield, 2003; Finstad *et al.*, 2004; Wikelski & Cooke, 2006). For example, Finstad and Forseth (2006) showed that young Atlantic salmon from northern populations that experienced extensive natural ice cover lost more energy when kept in ice-free conditions compared to the individuals from southern populations, indicating that northern populations are more susceptible to the disappearance of ice cover. The rate of evolvability in the responses of fish to ice-cover phenology is thus of importance for the conservation of populations that are adapted to long ice-cover periods. Loss of relevant adaptations may lead to allostatic overload resulting from energy requirements that are beyond the capacity of the fish to replace from environmental resources, eventually leading to population declines. Such population declines related to the loss of evolutionary adaptations may be difficult to prevent through conservation and management actions.

Changed habitat conditions in the face of climate change may also affect species interactions (Kuehne *et al.*, 2012). For example, Helland *et al.* (2011) showed that changes in winter conditions, measured as ice-cover phenology, altered interactions between fish species in northern freshwater systems. Based on research on brown trout and Arctic char populations in lakes in Norway, Helland *et al.* (2011) found that the two species differed in their physiological performance under ice. Furthermore, the outcome of competition in natural populations was strongly dependent on the duration of the ice-cover period, with long ice-cover periods favouring Arctic char, while brown trout was a better competitor in more seasonal ice-free conditions. Thus, a decreasing period of ice cover resulted in more brown trout and fewer char in lakes at high altitude and high latitudes. Similarly, Taniguchi and Nagano (2000) showed that differences in the competitive abilities of two riverine char species (*Salvelinus malma* and *S. leucomaenis*) were mediated by temperature and mechanistic linkages among behavioural responses to the abiotic environment, resource use and demographic processes. The outcome of these mechanistic linkages was that *S. leucomaenis* was dominant in warmer water conditions and *S. malma* in colder temperatures. Both these examples suggest that under environmental conditions affected by climate change, the species better adapted to such conditions can extract more resources and thus dominate over the other species. Such climate-mediated shifts in competitive success may occur in a wide range of freshwater fish and other animals as well (Gilman *et al.*, 2010).

Most studies examining the direct and indirect effects of climate change on fish have focused primarily on temperature. Wenger *et al.* (2011) also noted that broad-scale studies have stressed the importance of temperature and, at the same time, have largely ignored other critical drivers, such as flow regime and biotic interactions. Using downscaled outputs from general circulation models coupled with a hydrologic model, they aimed at forecasting the effects of altered flows and increased temperatures on four interacting trout species in the western United States. They found that considering biotic interactions and abiotic variables other than temperature provided a better understanding of species–climate relationships than approaches focusing only on temperature changes (Wenger et al., 2011). An interesting feature of their study was that the four trout species examined (i.e. cutthroat trout, brown trout, rainbow trout, brook trout) differed markedly in their responses to temperature, flood seasonality and the presence of other trout species. Of these four species, brook trout and cutthroat trout occurred in the coldest streams, whereas rainbow trout inhabited warmer locations, and brown trout resided in the warmest locations. Furthermore, the two autumn-spawning species, brook trout and brown trout, exhibited a strong negative relationship with winter high-flow frequency, whereas the spring-spawning cutthroat trout displayed a weak negative relationship, and the spring-spawning rainbow trout had a positive relationship. As a sign of potential biotic interactions, cutthroat trout showed a negative relationship to the occurrence of the other species of trout (Wenger et al., 2011). Hence, this case study clearly underlines the fact that, in order to acquire a comprehensive understanding of climate change effects on fishes, one needs to simultaneously consider temperature, hydrological conditions and various other abiotic and biotic factors.

3.4.3 Effects of extremes in climatic conditions on fish conservation

In addition to mean climatic conditions, more extreme and rarely occurring climatic events may also affect fish in freshwater ecosystems. Extreme hydrological events, such as unusual floods and droughts at unusual times, are projected to become more frequent and more intense with ongoing climate change (IPCC, 2007). High discharges in rivers may have both positive and negative effects on fish. For example, long-term studies of brown trout populations have recognised discharge as one of the most important environmental drivers affecting brown trout recruitment (Lobon-Cervia, 2009). The density of brown trout has been shown to be positively related to high discharge during the spawning period, whereas high flows during fry emergence have frequently been reported to result in weak year-classes (Carline, 2006; Warren et al., 2009). On the other side of the discharge effects, Matthews and Marsh-Matthews (2003) reviewed 50 studies on the impacts of drought on fish and habitat loss, and found that population decline was the most frequent consequence of drought. The occurrences of irregular droughts are likely to be more common in the future, and fish may thus face additional problems related to low-flow conditions, especially in subtropical and tropical regions (Dudgeon, 2011; Morrongiello *et al.*, 2011).

Both high- and low-flow conditions may interfere with practical conservation and management efforts. For example, various river restoration projects have been conducted around the world to enhance habitat conditions for stream-dwelling fishes and to guarantee their population persistence (Roni *et al.*, 2008). However, many recent studies have shown that regional climate variability, especially extreme climatic events, may overwhelm any effects of local habitat management efforts (Straile *et al.*, 2003). Thus, remedial conservation actions should ensure the availability of flood and drought refugia for stream biota. If such refugia are not considered, local restoration of in-stream structures, even when applied across whole stream lengths, may provide limited benefits for fish populations in a rapidly changing world.

3.4.4 Climate change, dispersal and fish conservation

Adequate dispersal is necessary for fish to track changed conditions in the face of climate change (Poff *et al.*, 2012; Heino, 2013). Easy tracking of changed conditions is not self-evident for freshwater fish due to both anthropogenic obstacles (e.g. dams) and natural barriers (e.g. drainage basin boundaries) to dispersal (Chapters 4 and 10). For example, many drainage basins in the Northern Hemisphere are west–east or east–west oriented, meaning that fish have limited capacity to track changes in climate warming by south–north dispersal (Carpenter *et al.*, 1992; Poff *et al.*, 2012). This is largely the situation in Europe, where the River Danube basin drains from west to east, whereas the situation is different in eastern North America, where the Mississippi River drains from north to south. These differences in the orientation of the main drainage basins have had both historical influences and are likely to affect the future of fish diversity in these large geographical regions (Oberdorff *et al.*, 2011). A large share of fish species has thus limited possibilities for dispersal in North America and Europe.

3.4.5 Managed introductions of freshwater fishes

Managed introductions or reintroductions (i.e. intentional movement of a species outside its historical native range or into part of its former native range; Olden *et al.*, 2011) to locations where the probability of the future persistence of a species is high offers a potential conservation means in a changing climate. Concerns have been expressed about the potential risks associated with the introductions of fish, especially the subsequent risks of harmful and unintended interactions with the native fish community in the recipient watershed, questioning the value of introductions (e.g. Ricciardi & Simberloff, 2009). Furthermore, Richardson *et al.* (2009) and Olden *et al.* (2011) have provided recommendations for fish stocking and managed introductions in freshwater environments, stressing the proper multifaceted risk assessment before the actions will take place. Without such prior assessments, managed fish introductions may potentially result in more negative than positive consequences for the native fauna and flora.

In the context of climate change, the most interesting cases of fish stocking experiments are those where a species has been moved outside its historical native range. For example, Salonen and Mutenia (2007) reported the introduction efforts to establish a self-reproducing population of land-locked salmon (*Salmo salar* m. *sebago*) into the Lake Inari watershed in northern Finland. This introduction was carried out by releasing hundreds of thousands of hatchery-reared, fry- to smolt-sized salmon into the lake and in-flowing rivers between 1971 and 2001. The native range of this endangered salmon form is in the Lake Saimaa watershed in southeastern Finland (i.e. 700 km south from Lake Inari). Despite the long-lasting introduction efforts, no viable and self-reproducing population was established by 2006, although the basic environmental characteristics of the two watersheds, except seasonal temperature and photoperiodic conditions, are very similar (Salonen & Mutenia, 2007). This example of salmon indicates that the success of managed introductions of fish species, although thoroughly planned and executed, is not obvious. However, some managed introductions are more successful. For example, Spens *et al.* (2007) found that introduced northern pike populations have persisted in 95% of lakes they studied in central Sweden. It is likely that northern pike is more of a generalist species than the land-locked form of salmon and, hence, its introductions may be expected to be more successful. Furthermore, northern pike in the Swedish example was introduced to lakes within its current distribution range, which also increased the likelihood of the success of the introductions. To improve the success of managed introductions in the face of climate change, a synthesis of previous experimentations of species introductions and stocking could provide a good basis for further endeavours in this field of fish conservation and management (Olden *et al.*, 2011).

3.5 CONCLUSION

Climate change affects freshwater fish both directly and indirectly, ranging from direct effects of warming temperatures and changed hydrologic conditions to indirect effects on fish habitats and biotic interactions. These effects can be seen at different levels of the biological organisation, varying from the genetic structure of populations and life-history changes to community structure and range shifts. Although it is increasingly recognised that freshwater fish will be severely affected by climate change, our knowledge of these effects is severely biased towards populations of a small number of iconic, charismatic and commercially valuable fish species. Similarly, studies on climate change effects on fish communities and range shifts are strongly biased towards temperate regions in Europe and North America. In order to advance our knowledge of climate change effects on freshwater fish, researchers should (i) devote more effort to studying less charismatic and typically commercially less valuable species (Comte *et al.*, 2013), (ii) expand studies on fish community changes to regions outside Europe and North America (Heino, 2011), (iii) compare the effects of climate change on taxonomic, functional and phylogenetic structure of fish communities (Strecker *et al.*, 2011). Only by expanding the focal species, communities and regions will we increase our knowledge on the effects of climate change on freshwater fish diversity, conservation and management.

3.6 DISCUSSION QUESTIONS

1. Freshwater fish can be divided into three broad thermal guilds. How is climate change likely to affect the distributions of species in each guild within a drainage basin, across drainage basins and across the world?
2. Which are the basic differences, mechanisms and environmental drivers that affect fish at high latitudes and near the equator in the face of climate change?
3. What have we learned about the factors affecting the distribution, abundance and life history of anadromous salmonids in the face of climate change?
4. What types of autonomous and planned adaptations may be of relevance in the conservation and management of freshwater fish in the face of climate change?
5. What are the most serious gaps in knowledge of climate change effects on freshwater fish?

3.7 ACKNOWLEDGEMENTS

We thank the editors for inviting us to contribute to this book, and appreciate excellent comments by Julian Olden and two anonymous reviewers on earlier drafts. The Academy of Finland provided financial support that facilitated the writing of this paper.

3.8 REFERENCES

Allan, J. D. (2004). Landscape and riverscapes: the influence of land use on river ecosystems. *Annual Reviews of Ecology, Evolution and Systematics*, 35, 257–284.

Allan, J. D. & Castillo, M. M. (2007). *Stream Ecology*, second edition. New York, NY: Springer.

Bond, N. R., Thomson, J., Reich, P. & Stein, J. A. (2011). Using species distribution models to infer potential climate change-induced range shifts of freshwater fish in south-eastern Australia. *Marine and Freshwater Research*, 62, 1043–1061.

Booth, D. J., Bond, N. & MacReadie, P. (2011). Detecting range shifts among Australian fishes in response to climate change. *Marine and Freshwater Research*, 62, 1027–1042.

Botts, E. A., Erasmus, B. F. N. & Alexander, G. J. (2013). Small range size and narrow niche breadth predict range contractions on South African frogs. *Global Ecology and Biogeography*, 22, 567–576.

Breiman, L. (2001). Random forests. *Machine Learning*, 45, 5–32.

Brucet, S., Pédron, S., Mehner, T., et al. (2013). Fish diversity in European lakes: geographical factors dominate over anthropogenic pressures. *Freshwater Biology*, 58, 1779–1793.

Buisson, L. & Grenouillet, G. (2009). Contrasted impacts of climate change on stream fish assemblages along an environmental gradient. *Diversity and Distributions*, 15, 613–626.

Buisson, L., Thuiller, W., Lek, S., Lim, P. & Grenouillet, G. (2008). Climate change hastens turnover of stream fish assemblages. *Global Change Biology*, 14, 2232–2248.

Burrows, M. T., Schoeman, D. S., Buckley, L. B., et al. (2011). The pace of shifting climate in marine and terrestrial ecosystems. *Science*, 334, 652–655.

Carline, R. F. (2006). Regulation of an unexploited brown trout population in Spruce Creek, Pennsylvania. *Transactions of the American Fisheries Society*, 135, 943–954.

Carpenter, S. R., Fisher, S. G., Grimm, N. B. & Kitchell, J. F. (1992). Global change and freshwater ecosystems. *Annual Review of Ecology and Systematics*, 23, 119–139.

Carter, T. R. & Kankaanpää, S. (2003). A preliminary examination of adaptation to climate change in Finland. *The Finnish Environment*, 640, 1–66.

Chu, C., Mandrak, N. E. & Minns, C. K. (2005). Potential impacts of climate change on the distributions of several common and rare freshwater fish in Canada. *Diversity and Distributions*, 11, 299–310.

Cochrane, K., De Young, C., Soto, D. & Bahri, T. (Eds). (2009). Climate change implications for fisheries and aquaculture: overview of current scientific knowledge. *FAO Fisheries and Aquaculture Technical Paper*, 530, 1–212.

Comte, L., Buisson, L., Daufresne, M. & Grenouillet, G. (2013). Climate-induced changes in the distribution of freshwater fish: observed and predicted trends. *Freshwater Biology*, 58, 625–639.

Cooney, S. J., Covich, A. P., Lukacs, P. M., Harig, A. L. & Fausch, K. D. (2005). Modeling global warming scenarios in greenback cutthroat trout (*Oncorhynchus clarki stomias*) streams: implications for species recovery. *Western North American Naturalist*, 65, 371–381.

Crozier, L. G., Hendry, A. P., Lawson, P. W., et al. (2008). Potential responses to climate change in organisms with complex life histories: evolution and plasticity in Pacific salmon. *Evolutionary Applications*, 1, 252–270.

Crozier, L. G., Scheuerell, M. D. & Zabel, R. W. (2011). Using time series analysis to characterize evolutionary and plastic responses to environmental change: a case study of a shift toward earlier migration date in sockeye salmon. *American Naturalist*, 178, 755–773.

Daufresne, M. & Boët, P. (2007). Climate change impacts on structure and diversity of fish communities in rivers. *Global Change Biology*, 13, 2467–2478.

Daufresne, M., Lengfeller, K. & Sommer, U. (2009). Global warming benefits the small in aquatic ecosystems. *Proceedings of the National Academy of Sciences*, 106, 12788–12793.

Deutsch, C. A., Tewksbury, J. J., Huey, R. B., *et al.* (2008). Impacts of climate warming on terrestrial ectotherms across latitude. *Proceedings of the National Academy of Sciences*, 105, 6668–6672.

Dudgeon, D. (2011). Asian freshwater fishes in the Anthropocene: threats and conservation challenges in an era of rapid environmental change. *Journal of Fish Biology*, 79, 1487–1524.

Dudgeon, D., Arthington, A. H., Gessner, M. O., *et al.* (2006). Freshwater biodiversity: importance, threats, status and conservation challenges. *Biological Reviews*, 81, 163–182.

Eaton, J. G. & Scheller, R. M. (1996). Effects of climate warming on fish thermal habitat in streams of the United States. *Limnology & Oceanography*, 41, 1109–1115.

Elith, J., Leathwick, J. R. & Hastie, T. (2008). Boosted regression trees as a new technique for modelling ecological data. *Journal of Animal Ecology*, 77, 802–813.

Elliott, J. M., Hurley, M. A. & Elliott, J. A. (1997). Variable effects of droughts on the density of a sea-trout *Salmo trutta* population over 30 years. *Journal of Applied Ecology*, 34, 1229–1238.

Farrell, A. P., Hinch, S. G., Cooke, S. J., *et al.* (2008). Pacific salmon in hot water: applying aerobic scope models and biotelemetry to predict the success of spawning migrations. *Physiological and Biochemical Zoology*, 81, 697–709.

Ficke, A. D., Myrick, C. A. & Hansen, L. J. (2007). Potential impacts of global climate change on freshwater fisheries. *Reviews in Fish Biology and Fisheries*, 17, 581–613.

Finstad, A. G. & Forseth, T. (2006). Adaptation to ice-cover conditions in Atlantic salmon, *Salmo salar* L. *Evolutionary Ecology Research*, 8, 1249–1262.

Finstad, A. G., Forseth, T., Nasje, T. F. & Ugedal, O. (2004). The importance of ice-cover for energy turnover in juvenile Atlantic salmon. *Journal of Animal Ecology*, 73, 959–966.

Friedland, K. D., MacLean, J. C., Hansen, L. P., *et al.* (2009). The recruitment of Atlantic salmon in Europe. *ICES Journal of Marine Science*, 66, 289–304.

Garcia de Leaniz, C., Fleming, I. A., Einum, S., *et al.* (2007). A critical review of adaptive genetic variation in Atlantic salmon: implications for conservation. *Biological Reviews*, 82, 173–211.

Gardner, J. L., Peters, A., Kearney, M. R., Joseph, L. & Heinsohn, R. (2011). Declining body size: a third universal response to warming. *Trends in Ecology and Evolution*, 26, 285–291.

Gilman, S. E., Urban, M. C., Tewksbury, J., Gilchrist, G. W. & Holt, R. D. (2010). A framework for community interactions under climate change. *Trends in Ecology and Evolution*, 25, 325–331.

Graham, C. T. & Harrod, C. (2009). Implications of climate change for the fishes of the British Isles. *Journal of Fish Biology*, 74, 1143–1205.

Griffiths, D. (1997). Local and regional species richness in North American lacustrine fish. *Journal of Animal Ecology*, 66, 49–56.

Griffiths, D. (2006). Pattern and process in the ecological biogeography of European freshwater fish. *Journal of Animal Ecology*, 75, 734–751.

Guisan, A. & Thuiller, W. (2005). Predicting species distribution: offering more than simple habitat models. *Ecology Letters*, 8, 993–1009.

Guisan, A. & Zimmermann, N. E. (2000). Predictive habitat distribution models in ecology. *Ecological Modelling*, 135, 147–186.

Hannah, L., Midgley, G. F., Lovejoy, T., *et al.* (2002). Conservation of biodiversity in a changing climate. *Conservation Biology*, 16, 264–268.

Hastie, T. & Tibshirani, R. (1990). *Generalized Additive Models*. London: Chapman and Hall.

Heikkinen, R. K., Luoto, M., Araújo, M. B., *et al.* (2006). Methods and uncertainties in bioclimatic envelope modelling under climate change. *Progress in Physical Geography*, 30, 751–777.

Hein, C. L., Öhlund, G. & Englund, G. (2011). Dispersal through stream networks: modelling climate-driven range expansions of fishes. *Diversity and Distributions*, 17, 641–651.

Heino, J. (2011). A macroecological perspective of diversity patterns in the freshwater realm. *Freshwater Biology*, 56, 1703–1722.

Heino, J. (2013). The importance of metacommunity ecology for environmental assessment research in the freshwater realm. *Biological Reviews*, 88, 166–178.

Heino, J., Virkkala, R. & Toivonen, H. (2009). Climate change and freshwater biodiversity: detected patterns, future trends and adaptations in northern regions. *Biological Reviews*, 84, 39–54.

Helland, I. P., Finstad, A. G., Forseth, T., Hesthagen, T. & Ugedal, O. (2011). Ice-cover effects on competitive interactions between two fish species. *Journal of Animal Ecology*, 80, 539–547.

Hickling, R., Roy, D. B., Hill, J. K., Fox, R. & Thomas, C. D. (2006). The distributions of a wide range of taxonomic groups are expanding polewards. *Global Change Biology*, 12, 450–455.

Huusko, A. & Hyvärinen, P. (2012). Atlantic salmon abundance and size track climate regimes in the Baltic Sea. *Boreal Environment Research*, 17, 139–149.

Huusko, A., Greenberg, L., Stickler, M., *et al.* (2007). Life in the ice lane: the winter ecology of stream salmonids. *River Research and Applications*, 23, 469–491.

IPCC. (2007). *Climate Change 2007: The Physical Science Basis. Contribution of Working Group I to the Fourth Assessment Report of the IPCC.* Cambridge University Press.

Jackson, D. A. & Mandrak, N. E. (2002). Changing fish biodiversity: predicting the loss of cyprinid biodiversity due to global climate change. In *Fisheries in a Changing Climate.* American Fisheries Society Symposium 32. Bethesda, MD: American Fisheries Society, pp. 89–98.

Jackson, D. A., Peres-Neto, P. R. & Olden, J. D. (2001). What controls who is where in freshwater fish communities – the roles of biotic, abiotic, and spatial factors. *Canadian Journal of Fisheries and Aquatic Sciences*, 58, 157–170.

Jeppesen, E., Mehner, T., Winfield, I., *et al.* (2012). Impacts of climate warming on the long-term dynamics of key fish species in 24 European lakes. *Hydrobiologia*, 694, 1–39.

Jonsson, B. & Jonsson, N. (2011). *Ecology of Atlantic Salmon and Brown Trout. Habitat as a Template for Life Histories.* Berlin: Springer.

Kennedy, R. J. & Crozier, W. W. (2010). Evidence of changing migratory patterns of wild Atlantic salmon *Salmo salar* smolts in the River Bush, Northern Ireland, and possible associations with climate change. *Journal of Fish Biology*, 76, 1786–1805.

Kuehne, L. M., Olden, J. D. & Duda, J. J. (2012). Costs of living for juvenile Chinook salmon in an increasingly warming and invaded world. *Canadian Journal of Fisheries and Aquatic Sciences*, 69, 1621–1630.

Lassalle, G., Béguer, M., Beaulaton, L. & Rochard, E. (2008). Diadromous fish conservation plans need to consider global warming issues: an approach using biogeographical models. *Biological Conservation*, 141, 1105–1118.

Lawrence, D. J., Stewart-Koster, B., Olden, J. D., *et al.* (2014). The interactive effects of climate change, riparian management, and a non-native predator on stream-rearing salmon. *Ecological Applications* 24, 895–912.

Lehtonen, H. (1996). Potential effects of global warming on northern European freshwater fish and fisheries. *Fisheries Ecology and Management*, 3, 59–71.

Lek, S. & Guegan, J. (1999). Artificial neural networks as a tool in ecological modelling, an introduction. *Ecological Modelling*, 120, 65–73.

Linnansaari, T. & Cunjak, R. A. (2012). Fish: freshwater ecosystems. In *Temperature Adaptation in a Changing Climate. Nature at Risk.* Wallingford: CAB International, pp. 80–97.

Lobon-Cervia, J. (2009). Why, when and how do fish populations decline, collapse and recover? The example of brown trout (*Salmo trutta*) in Rio Chaballos (northwestern Spain). *Freshwater Biology*, 54, 1149–1162.

Magnuson, J. J., Crowder, L. B. & Medvick, P. A. (1979). Temperature as an ecological resource. *American Zoologist*, 19, 331–343.

Malmqvist, B. & Rundle, S. J. (2002). Threats to the running water ecosystems of the world. *Environmental Conservation*, 29, 134–153.

Mantua, N. J., Hare, S. R., Zhang, Y., Wallace, J. M. & Francis, R. C. (1997). A Pacific interdecadal climate oscillation with impacts on salmon production. *Bulletin of the American Meteorological Society*, 78, 1069–1079.

Markovic, D., Freyhof, J. & Wolter, C. (2012). Where are all the fish: potential of biogeographical maps to project current and future distribution patterns of freshwater species. *PLoS ONE*, 7, e40530.

Matthews, W. J. (1998). *Patterns in Freshwater Fish Ecology.* New York, NY: Chapman and Hall.

Matthews, W. J. & Marsh-Matthews, E. (2003). Effects of drought on fish across axes of space, time and ecological complexity. *Freshwater Biology*, 48, 1232–1253.

McEwen, B. S. & Wingfield, J. C. (2003). The concept of allostasis in biology and biomedicine. *Hormones and Behavior*, 42, 2–15.

Mims, M. C. & Olden, J. D. (2012). Life history theory predicts fish assemblage response to hydrologic regimes. *Ecology*, 93, 35–45.

Mims, M. C., Olden, J. D., Shattuck, Z. R. & Poff, N. L. (2010). Life history trait diversity of native freshwater fishes in North America. *Ecology of Freshwater Fish*, 19, 390–400.

Mohseni, O., Stefan, H. G. & Eaton, J. G. (2003). Global warming and potential changes in fish habitat in U.S. streams. *Climatic Change*, 5, 389–409.

Morrongiello, J. R., Beatty, S. J., Bennett, J. C., *et al.* (2011). Climate change and its implications for Australia's freshwater fish. *Marine and Freshwater Research*, 62, 1082–1098.

Nielsen, J. L., Ruggerone, G. T. & Zimmerman, C. E. (2013). Adaptive strategies and life history charac-teristics in a warming climate: salmon in the Arctic? *Environmental Biology of Fishes*, 96, 1187–1226.

Niemelä, E., Erkinaro, J., Julkunen, M. & Hassinen, E. (2005). Is juvenile salmon abundance related to subsequent and preceding catches? Perspectives from a long-term monitoring programme. *ICES Journal of Marine Science*, 62, 1617–1629.

Oberdorff, T., Tedesco, P. A., Hugueny, B., *et al.* (2011). Global and regional patterns in riverine fish species richness: a review. *International Journal of Ecology*, 2011, Article ID 967631.

Olden, J. D. & Jackson, D. A. (2002). A comparison of statistical models for modelling fish species distributions. *Freshwater Biology*, 47, 1976–1995.

Olden, J. D., Kennard, M. K., Leprieur, F., *et al.* (2010). Conservation biogeography of freshwater fishes: past progress and future directions. *Diversity and Distributions*, 16, 496–513.

Olden, J. D., Kennard, M. J., Lawler, J. J. & Poff, N. L. (2011). Challenges and opportunities for implementing managed relocation of species for freshwater conservation. *Conservation Biology*, 25, 40–47.

Otero, J., L'Abée-Lund, J. H., Castro-Santos, T., *et al.* (2014). Basin-scale phenology and effects of cli-mate variability on global timing of initial seaward migration of Atlantic salmon (*Salmo salar*). *Global Change Biology*, 20, 61–75.

Pascual, M. A., Cussac, V., Dyer, B., *et al.* (2007). Freshwater fishes of Patagonia in the 21st century after a hundred years of human settlement, species introductions, and environmental change. *Aquatic Ecosystem Health & Management*, 10, 212–227.

Poff, N. L., Brinson, M. M. & Day, J.W. (2002). *Aquatic Ecosystems and Global Climate Change. Potential Impacts on Inland Freshwater and Coastal Wetland Ecosystems in the United States*. Arlington, VA: Pew Center on Global Climate Change.

Poff, N. L., Olden, J. D. & Strayer, D. L. (2012). Climate change and freshwater fauna extinction risk. In *Saving A Million Species. Extinction Risk From Climate Change*. Washington, DC: Island Press, pp. 309–336.

Pöyry, J. & Toivonen, H. (2005). Climate change adaptation and biological diversity. *Finnish Environment Institute Mimeographs*, 333, 1–46.

Prowse, T. D. (2001a). River ice ecology. Part A: hydrologic, geomorphic, and water-quality aspects. *Journal of Cold Regions Engineering*, 15, 1–16.

Prowse, T. D. (2001b). River-ice ecology. II: Biological aspects. *Journal of Cold Regions Engineering*, 15, 17–33.

Rahel, F. J. & Olden, J. D. (2008). Assessing the effects of climate change on aquatic invasive species. *Conservation Biology*, 22, 521–533.

Rahel, F. J., Keleher, C. J. & Anderson, J. L. (1996). Potential habitat loss and population fragmentation for cold water fish in the North Platte River drainage of the Rocky Mountains: response to climate warming. *Limnology and Oceanography*, 41, 1116–1123.

Reist, J. D., Wrona, F. J., Prowse, T. D., *et al.* (2006). General effects of climate change on arctic fishes and fish populations. *Ambio*, 35, 370–380.

Reyjol, Y., Hugueny, B., Pont, D., *et al.* (2007). Patterns in species richness and endemism of European freshwater fish. *Global Ecology and Biogeography*, 16, 65–75.

Ricciardi, A. & Simberloff, D. (2009). Assisted colonization is not a viable conservation strategy. *Trends in Ecology and Evolution*, 24, 248–253.

Richardson, D. M., Hellmann, J. J., McLachlan, J., *et al.* (2009). Multidimensional evaluation of man-aged relocation. *Proceedings of the National Academy of Sciences*, 106, 9721–9724.

Robinson, R. A., Crick, H. Q. P., Learmonth, J. A., *et al.* (2009). Travelling through a warming world: climate change and migratory species. *Endangered Species Research*, 7, 87–99.

Roni, P., Hanson, K. & Beechie, T. (2008). Global review of the physical and biological effectiveness of stream habitat rehabilitation techniques. *North American Journal of Fisheries Management*, 28, 856–890.

Ruesch, A. S., Torgersen, C. E., Lawler, J. J., *et al.* (2012). Projected climate-induced habitat loss for salmonids in the John Day River network, Oregon, U.S.A. *Conservation Biology*, 26, 873–882.

Salonen, E. & Mutenia, A. (2007). Alien fish species in northernmost Finland. *Riista- ja kalatalous. Tutkimuksia*, 2, 1–16.

Sanborn, S. C. & Bledsoe, B. P. (2006). Predicting streamflow regime metrics for ungauged streams in Colorado, Washington, and Oregon. *Journal of Hydrology*, 325, 241–261.

Schleuter, D., Daufresne, M., Veslot, J., *et al.* (2012). Geographic isolation and climate govern the functional diversity of native fish communities in European drainage basins. *Global Ecology and Biogeography*, 21, 1083–1095.

Sharma, S., Jackson, D. A., Minns, C. K. & Shuter, B. J. (2007). Will northern fish populations be in hot water because of climate change? *Global Change Biology*, 13, 2052–2064.

Shuter, B. J., Finstad, A. G., Helland, I. P., Zweimuller, I. & Hölker, F. (2012). The role of winter phenology in shaping the ecology of freshwater fish and their sensitivities to climate change. *Aquatic Sciences*, 74, 637–657.

Spens, J., Englund, G. & Lundqvist, H. (2007). Network connectivity and dispersal barriers: using geographical information system (GIS) tools to predict landscape scale distribution of a key predator (*Esox lucius*) among lakes. *Journal of Applied Ecology*, 44, 1127–1137.

Straile, D., Livingstone, D. M., Weyhenmeyer, G. A. & George, G. (2003). The response of freshwater ecosystems to climate variability associated with the North Atlantic Oscillation. In *The North Atlantic Oscillation: Climatic Significance and Environmental Impact*. Geophysical Monograph Series 134. Washington, DC: American Geophysical Union, pp. 263–279.

Strayer, D. L. & Dudgeon, D. (2010). Freshwater biodiversity conservation: recent progress and future challenges. *Journal of the North American Benthological Society*, 29, 344–358.

Strecker, A. L., Olden, J. D., Whittier, J. B. & Paukert, C. P. (2011). Defining conservation priorities for freshwater fishes according to taxonomic, functional, and phylogenetic diversity. *Ecological Applications*, 21, 3002–3013.

Swihart, R. K., Gehring, T. M., Kolozsvary, M. B. & Nupp, T. E. (2003). Responses of "resistant" vertebrates to habitat loss and fragmentation: the importance of niche breadth and range boundaries. *Diversity and Distributions*, 9, 1–18.

Taniguchi, Y. & Nakano, S. (2000). Condition-specific competition: implications for the altitudinal distribution of stream fishes. *Ecology*, 81, 2027–2039.

Tedesco, P. A., Leprieur, F., Hugueny, B., *et al.* (2012). Patterns and processes of global freshwater fish endemism. *Global Ecology and Biogeography*, 21, 977–987.

Thieme, M. L., Lehner, B., Abell, R. & Matthews, J. (2010). Exposure of Africa's freshwater biodiversity to a changing climate. *Conservation Letters*, 3, 324–331.

Thorstad, E. B., Whoriskey, F., Uglem, I., *et al.* (2012). A critical life stage of the Atlantic salmon *Salmo salar*: behaviour and survival during the smolt and initial post-smolt migration. *Journal of Fish Biology*, 81, 500–542.

Thuiller, W., Lavorel, S. & Araujo, M. (2005). Niche properties and geographical extent as predictors of species sensitivity to climate change. *Global Ecology and Biogeography*, 14, 347–357.

Thuiller, W., Lafourcade, B., Engler, R. & Araújo, M. B. (2009). BIOMOD – a platform for ensemble forecasting of species distributions. *Ecography*, 32, 369–373.

Tisseuil, C., Leprieur, F., Grenouillet, G., Vrac, M. & Lek, S. (2012a). Projected impacts of climate change on spatio-temporal patterns of freshwater fish beta diversity: a deconstructing approach. *Global Ecology and Biogeography*, 21, 1213–1222.

Tisseuil, M., Vrac, M., Grenouillet, G., *et al.* (2012b) Strengthening the link between climate, hydrological and species distribution modelling to assess the impacts of climate change on freshwater biodiversity. *Science of the Total Environment*, 424, 193–201.

Todd, C. D., Friedland, K. D., MacLean, J. C., Hazon, N. & Jensen, A. J. (2011). Getting into hot water? Atlantic salmon responses to climate change in freshwater and marine environments. In *Atlantic Salmon Ecology*. Oxford: Blackwell Publishing, pp. 409–443.

Tonn, W. M. (1990). Climate change and fish communities: a conceptual framework. *Transactions of the American Fisheries Society*, 119, 337–352.

Vähä, J.-P., Erkinaro, J., Niemelä, E. & Primmer, C. R. (2007). Life-history and habitat features influence the within–river genetic structure of Atlantic salmon. *Molecular Ecology*, 16, 2638–2654.

Venables, W. N. & Ripley, B. D. (2002). *Modern Applied Statistics with S*. Berlin: Springer-Verlag.

Warren, D. R., Ernst, A. G. & Baldigo, B. P. (2009). Influence of spring floods on year-class strength of fall- and spring-spawning salmonids in Catskill Mountain streams. *Transactions of the American Fisheries Society*, 138, 200–210.

Wenger, S. J., Isaak, D. J., Luce, C. H., *et al.* (2011). Flow regime, temperature, and biotic interactions drive differential declines of trout species under climate change. *Proceedings of the National Academy of Sciences*, 108, 14175–14180.

Whittaker, R. H. (1960) Vegetation of the Siskiyou Mountains, Oregon and California. *Ecological Monographs*, 30, 279–338.

Whittaker, R. J., Araújo, M. B., Jepson, P., *et al.* (2005). Conservation biogeography: assessment and prospect. *Diversity and Distributions*, 11, 3–23.

Wikelski, M. & Cooke, S. (2006). Conservation physiology. *Trends in Ecology and Evolution*, 21, 38–46.

Wisz, M. S., Pottier, J., Kissling, W. D., *et al.* (2013). The role of biotic interactions in shaping distributions and realised assemblages of species: implications for species distribution modelling. *Biological Reviews*, 88, 15–30.

Wood, C. M. & McDonald, G. (1997). *Global Warming – Implications for Freshwater and Marine Fish*. Cambridge University Press.

Wood, S. N. 2006. *Generalized Additive Models: An Introduction with R*. London: Chapman and Hall/ CRC.

Xenopoulos, M. A., Lodge, D. M., Alcamo, J., *et al.* (2005). Scenarios of freshwater fish extinctions from climate change and water withdrawal. *Global Change Biology*, 11, 1557–1564.

Youngson, A. F., Jordan, W. C., Verspoor, E., *et al.* (2003). Management of salmonid fisheries in the British Isles: towards a practical approach based on population genetics. *Fisheries Research*, 62, 193–209.

Challenges and opportunities for fish conservation in dam-impacted waters

JULIAN D. OLDEN

4.1 DAMMED IF YOU DO, DAMNED IF YOU DON'T

Naturally flowing rivers are among the most dynamic ecosystems on Earth, with enormous spatial and temporal complexity. Streamflow defines the physical template of riverine ecosystems (Poff *et al.*, 1997), provides longitudinal and lateral access to foraging, spawning and recruitment habitat (Junk *et al.*, 1989), and acts as an evolutionary selective force and an ecological filter of various survival strategies employed by aquatic and riparian organisms (Townsend & Hildrew, 1994; Jackson *et al.*, 2001; Lytle & Poff, 2004). At the same time, human society requires water for life. Over the millennia, humans have altered streamflow in riverine systems for myriad reasons including harnessing water for drinking, irrigation and recreation, and providing flood control and hydropower (Gleick, 2003). The freshwater footprint of humanity stamps the entire globe, with nearly half of major river systems affected by dams (Vörösmarty *et al.*, 2010; Lehner *et al.*, 2011). The future construction of dams, particularly in economically developing nations, is an inevitable consequence of human population growth and increasing freshwater and electricity needs in a changing climate (Palmer *et al.*, 2008; McDonald *et al.*, 2012).

Despite providing many societal benefits, river regulation by dams has also caused considerable ecological damage and the loss of important ecosystem services valued by society. Dams fragment rivers, creating strings of artificial lakes punctuated by barriers that are often impassable by fish, and they alter physical riverine habitat and water quality in both upstream and downstream directions (Nilsson *et al.*, 2005; Reidy Liermann *et al.*, 2012). In particular, dams are the primary

Conservation of Freshwater Fishes, eds G. P. Closs, M. Krkosek and J. D. Olden. Published by Cambridge University Press. © Cambridge University Press 2016.

driver of hydrologic change throughout the United States (Carlisle *et al.*, 2011), resulting in reduced flow seasonality and variability and generally increasing short-term minimum flows while decreasing short-term maximum peaks (Poff *et al.*, 2007). These changes alter the historical disturbance regime, rendering some biotic adaptations to these regimes obsolete while potentially favouring others. For example, reduced flow variability by dams has been associated with significant losses of native fish species (Meador & Carlisle, 2012) while concurrently creating new niche opportunities above and below dams that are often occupied by non-native fishes (e.g. Olden *et al.*, 2006; Johnson *et al.*, 2008).

Social, political and scientific pressure to recognise rivers as water users has resulted in heightened desires to find compromises in dam management to meet both human and ecosystem water needs (Naiman *et al.*, 2002) through the provisioning of environmental flows (Arthington *et al.*, 2006) and large-scale flood events (Konrad *et al.*, 2011). In fact, trade-offs between economic and environmental benefits and costs of dams have also led to the retirement of many small structures and some large ones (e.g. Bednarek, 2001; Stanley & Doyle, 2003). Achieving compatibility between human and natural ecosystem needs is the ultimate goal of ecologically sustainable water management (Richter *et al.*, 2003), and despite the mounting pressures facing the conservation of freshwater fishes, the prospects for overcoming these challenges are even greater (Olden *et al.*, 2010).

The objectives of this chapter are to review the challenges and opportunities for freshwater fish conservation in response to threats posed by dams and diversions. After taking stock of the dammed world in which we all live, I explore the primary ways that dams influence the environment of freshwater ecosystems and their constituent fish species. Next, I examine current efforts to restore and conserve fish biodiversity in rivers that have been dammed or subjected to flow diversions, including the mimicking of historical or natural flow and thermal regimes below dams, establishing opportunities to circumvent dams via fish passages, and in some cases even restoring and reconnecting rivers by removing dams. Central concepts are introduced and then illustrated with case studies taken from across the world.

4.2 DAMMED WORLD

Rivers were the arteries for the development of ancient civilisations and modern societies, so it comes as no surprise that humans have built dams and impoundments for thousands of years. These dams served various purposes, including flood control, water supply, irrigation, recreation, navigation and the generation of hydropower (WCD, 2000). Records indicate that gravity dams were first built in Jordan around 3000 BC and the earliest known remains of the ancient Sadd el-Kafara earthen dam built by Egyptians are dated to 2600 BC. Around 100 AD, Roman planners introduced the novel concept of large reservoir dams to secure a permanent water supply for urban settlements over the dry season. Their pioneering use of waterproof hydraulic mortar and concrete allowed for the construction of much larger dam structures than had been previously built (Smith, 1971). This was followed by early dam construction in Mesopotamia and the Middle East circa 1280 AD, for example to tame the unpredictable flow regimes of the Tigris and Euphrates rivers. By the seventeenth century, Spanish dam-building was superior to all other civilisations, which they brought to the New World and paved the way for centuries of dam construction.

American rivers were symbols of a burgeoning nation in the eighteenth and nineteenth centuries. They inspired romantic renderings at the hands of artists, and in some cases they were depicted as detailed landscape features with physical and even human qualities (Smith, 1971). However, at times they were regarded as untapped or underutilised resources, raw material waiting to be harnessed, managed and exploited for human benefit. In the early colonial period, dams were constructed on small rivers and tributaries to power gristmills, and for improvement of river navigation by construction of diversion dams and canal systems to enable riverboats to bypass rapids and shoal areas. Later, in the mid-1800s, the rise of the Industrial Revolution ushered in a period of unprecedented dam construction to provide hydromechanical power for larger textile and industrial mills. The history of federal involvement in dam construction goes back at least to this time, when the US Army Corps of Engineers built wing dams to improve navigation on the Ohio River (Billington *et al.*, 2005). The work expanded after the Civil War, when Congress authorised the Corps to build storage dams on the upper Mississippi River. The turn of the nineteenth century witnessed the establishment of the Bureau of Reclamation (then called the 'Reclamation Service'), which with the rise of hydroelectric power led to a great dam construction rush that resulted in much larger dams on nearly all of the nation's major rivers up to the present day.

Dams are now a ubiquitous feature of the global landscape. Over the past half-century, the number and storage volumes of dams and reservoirs have increased markedly, today ranging from small weirs and low run-of-the-river structures with negligible storage, to large reservoirs capable of retaining huge amounts of water and sediment (Figure 4.1). There are an estimated 2.8 million impoundments larger than 0.1 ha (0.001 km²) and 16.7 million when including those larger than 0.01 ha (100 m²) worldwide (Lehner et al., 2011). The total storage volume of all reservoirs amounts to over 8000 km³, a value equivalent to nearly 10% of the water stored in all natural freshwater lakes on Earth (Gleick, 2000), or about one-sixth of the total annual river flow into the oceans (Hanasaki et al., 2006). The end result is that dams trap 25% of the global sediment load before it reaches the oceans (Syvitski et al., 2005). Graf (1999) reported that over 80,000 dams in the continental US are capable of storing a volume of water almost equalling one year's mean run-off. The potential surface water impacts of dams range from more than 3 years' run-off stored in some western mountain and plains reservoirs to as little as a quarter of the annual run-off in reservoirs located in the more mesic regions of northeast and northwest United States.

With a heavy concentration of dams in the northern third of the world and few medium or large rivers left to impound (Nilsson et al., 2005), neotropical regions are now the primary frontier for new dam construction (e.g. Pringle et al., 2000; Anderson et al., 2006). In 1950, Asia had 1541 large dams (more than 15 m high), accounting for about one-third of the global total, but by 1982 that figure had grown to 22,701, representing two-thirds of the global total (Dudgeon, 2000). Today, close to 90 new dams have been proposed for the lower Mekong River Basin with imminent impacts on the world's largest freshwater fishery (Ziv et al., 2012) and about 300 dams are planned in India to meet a projected doubling of electrical demand by the end of this decade (Pandit & Grumbine, 2012). Similarly, due to rising energy demands and abundant untapped potential, hydropower projects are rapidly increasing in regions such as the Andean Amazon (Figure 4.2), where regional governments are prioritising new hydroelectric dams as the centrepiece of long-term energy plans (Finer & Jenkins, 2012).

Although hydropower offers a reliable source of domestically produced electricity to neotropical countries and the chance to diversify away from thermoelectric facilities and the use of fossil fuels, dams may also lead to significant ecological and social impacts (Nogueira et al., 2010). Finer and Jenkins (2012) examined the potential ecological impacts, in terms of river connectivity and forest loss, of the planned

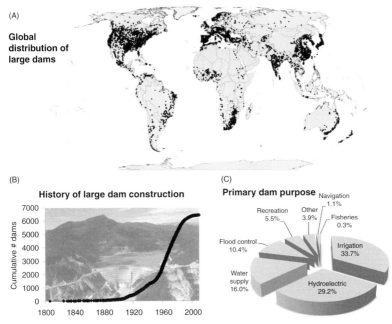

Figure 4.1 (A) Global distribution of large dams; (B) cumulative number of large dam constructions over time; (C) primary purpose of large dams across the globe. Data are sourced from the GRanD database (Lehner *et al.*, 2011). Pictured in panel B is the Inguri Dam, a hydroelectric dam on the Inguri River (Georgia), considered the world's second highest concrete arch dam at a height of 272 m.

proliferation of hydroelectric dams across all Andean tributaries of the Amazon River. Considering data on the full portfolio of existing and planned dams, this study identified over 150 large dams slated for construction in the next 20 years (representing a threefold increase). Sixty percent of the planned dams would cause the first major break in connectivity between protected Andean headwaters and the lowland Amazon, and more than 80% would drive deforestation due to new roads, transmission lines, or inundation. Controversies involving these projects focus primarily on the potential direct impacts on indigenous human settlements (i.e. flooding and displacement), whereas effects on fish and freshwater ecosystems have received little attention (Ferreira *et al.*, 2014). In short, although the dam-building era is largely over in developed nations (although still discussed and occurring), current and future dam construction is rampant in many other developing regions of the world.

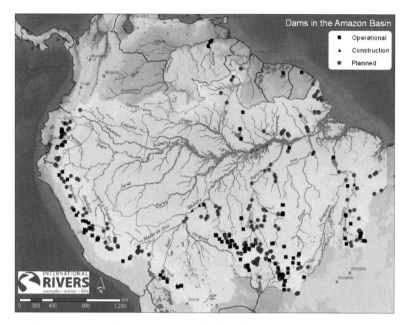

Figure 4.2 Dams that are operational, planned, or under construction in the Amazon River Basin. Map courtesy of International Rivers.

4.3 ENVIRONMENTAL AND ECOLOGICAL IMPACTS OF DAMS

The relative young history of large-dam construction explains why the ecological costs of these structures have only recently captured scientific attention. The maximum potential for the downstream environmental disruptions through reservoir storage has been in place for only a few decades, and in some cases the effects are only now becoming obvious. Despite this, one could argue that the construction and operation of dams has already had greater environmental impacts on the world's rivers than any changes that might reasonably be expected from global climate changes in the near future. Below, I turn our attention to briefly explore the environmental impacts of dams and the associated consequences for freshwater fishes, focusing almost exclusively on downstream patterns and processes and not those manifested in the reservoir created by impoundment. I recommend the following papers for more comprehensive treatments of these topics: Cushman (1985), Ligon *et al.* (1995), Stanford *et al.* (1996), Bunn & Arthington (2002), Helfman (2007), Poff *et al.* (2007), Olden & Naiman (2010).

4.3.1 Flow regime

Streamflow is viewed as a 'maestro' or 'master variable' that shapes many fundamental properties of riverine ecosystems (Poff *et al.*, 1997; Naiman *et al.*, 2008). A preponderance of scientific evidence shows that the streamflow is a major determinant of physical habitat in streams and rivers, which in turn is a primary agent shaping the distribution, abundance and diversity of fishes (Figure 4.3). The geometry of river channels, the arrangement of riffle and pool habitats, and the types and stability of the substrate are all largely defined by the interaction between the flow regime and local geology and landform (Frissell *et al.*, 1986). Long-term variations in flow magnitude, and the timing, frequency and duration of flow events, define the physical habitat template over large spatial scales (e.g. catchments and subcatchments), whereas the short-term history of hydrological events influences habitat availability and connectivity at smaller scales (e.g. within and among river reaches). These factors collectively influence the availability of refuges, food resources, opportunities for movement and migration, and conditions suitable for spawning and recruitment of fish (e.g. Horwitz, 1978; Schlosser, 1982, 1991; Bain *et al.*, 1988; Humphries *et al.*, 1999; Kennard *et al.*, 2007; Mims & Olden, 2012).

Given the broad scientific consensus that the function of freshwater ecosystems is explicitly linked to the river's natural flow regime, it should come as no surprise that the modification of streamflow is perhaps the greatest adverse ecological consequence of dams and reservoirs (Bunn & Arthington, 2002). The goal of many dam operations is to eliminate peak flows, to stabilise low flows, or to impound or divert river flows partially or entirely, to meet human needs. As a consequence, the downstream impacts of dams are manifested across multiple facets of the flow regime, including effects on the seasonal patterning of flows, timing of extreme flows, the frequency, predictability and duration of low- and high-flow events, daily, seasonal and annual flow variability, and rates of change (Ligon *et al.*, 1995; Poff *et al.*, 2007; Rolls *et al.*, 2012). In the US, for example, operations of the dams have modified the magnitude and timing of high and low flows during the twentieth century. Maximum flows have consistently declined below dams, whereas minimum flows have generally increased. Changes in the timing of maximum flows below dams vary geographically, with more eastern, rainfall-dominated regions showing earlier timing of peak flows compared with delayed peak flows in more snowmelt-dominated western

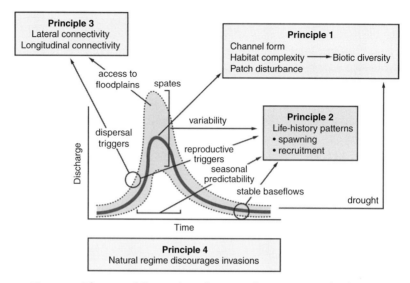

Figure 4.3 The natural flow regime of a river influences aquatic biodiversity via several interrelated mechanisms that operate over different spatial and temporal scales. The relationship between biodiversity and the physical nature of the aquatic habitat is likely to be driven primarily by large events that influence channel form and shape (principle 1). However, droughts and low-flow events are also likely to play a role by limiting overall habitat availability. Many features of the flow regime influence life-history patterns, especially the seasonality and predictability of the overall pattern, but also the timing of particular flow events (principle 2). Some flow events trigger longitudinal dispersal of migratory aquatic organisms and other large events allow access to otherwise disconnected floodplain habitats (principle 3). The native biota have evolved in response to the overall flow regime. Catchment land-use change and associated water resource development inevitably lead to changes in one or more aspects of the flow regime resulting in declines in aquatic biodiversity via these mechanisms. Invasions by introduced or exotic species are more likely to succeed at the expense of native biota if the former are adapted to the modified flow regime (principle 4). Reproduced from Bunn and Arthington (2002).

regions (Poff *et al.*, 2007). Taken together, by modifying fluvial processes and habitat dynamics, dams pose a significant threat to fish biodiversity on local, regional and global scales.

The numerous negative effects of individual dams on freshwater fishes are documented repeatedly in the literature; unfortunately, review-ing the enormity of this topic is well beyond the current chapter. However, Poff and Zimmerman (2010) conducted an extensive review of ecological responses to flow regulation and reported that the biotic integrity of fish assemblages generally decreased with increased flow alteration by dams and diversions. Under modified flows, native fish diversity and abundance showed a 20–100% and 16–95% reduction, respectively. In New Zealand, for example, Jellyman and Harding (2012) found that fish species richness was lower above dams compared to below dams, whereas no differences were present at undammed sites. These differences were attributed to a lower percentage of diadromous species above dams caused by the loss of riverine habitats and restricted upstream movement (discussed in more detail below). In short, the effect of dams on river fish assemblages is a reoccurring theme, regardless of geography.

In another study, Mims and Olden (2013) found that on ecological time scales (i.e. the order of years to decades), dams in the US have effectively changed the functional composition of fish communities that have established over millennia. In agreement with predictions from life-history theory, dam operations that have reduced flow variability and increased flow predictability have led to fish assemblages downstream of dams being characterised by many equilibrium species (a strategy favoured in more stable, predictable environments) and few opportunistic species (a strategy favoured in variable environments) when compared to free-flowing, neighbouring rivers. Furthermore, there is considerable evidence that by altering downstream flow regimes from their natural conditions, fluvial specialists (i.e. those species usually found in flowing waters) are replaced by habitat generalists (e.g. Kinsolving & Bain, 1993; Poff & Allan, 1995; Chapter 12). This has been attributed to the ability of habitat generalists to tolerate changes in flows from lotic to lentic conditions; a condition that has also promoted the establishment, spread and ecological impacts of non-native species (Johnson *et al.*, 2008; discussed in detail below).

4.3.2 Sediment conditions

Flow depletion below dams can fundamentally alter channel hydraulics and lead to changes in channel geomorphology and sediment conditions (Ligon *et al.*, 1995; Brandt, 2000). Depending on the size of the reservoir, large amounts of sediment will be trapped, releasing only a proportion of the former load into the downstream reaches. Typically, retention of sediment by the reservoir results in downstream scouring and bed armouring (i.e. hardening and coarsening of the stream bottom that lacks small-size fractions of sediment), which can diminish habitat and water-quality conditions for fishes (Waters, 1995). Riverine fishes are adapted to particular levels of sediment concentration and deposition;

therefore, just as excess sedimentation has serious consequences (Wood & Armitage, 1997) unnaturally low concentrations also may cause problems (Waters, 1995). This is the case for razorback suckers, *Xyrauchen texanus*, in the Colorado River downstream from Glen Canyon Dam. The endangered razorback sucker experiences higher predation rates by both native and introduced predators under conditions of unnaturally reduced turbidity (Helfman, 2007).

4.3.3 Thermal and chemical regimes

Dams and diversions can impart significant impacts on downstream water quality, including effects such as changes in water temperature, anoxia, dissolved gas supersaturation, heightened levels of hydrogen sulphide and reduced productivity (Petts, 1986; Nilsson & Renöfält, 2008). Below, I review how dam construction has raised long overdue concerns regarding the potential impacts of altered thermal regimes on fishes, but first let's briefly explore dam impacts on other water-quality parameters.

Water-quality problems in the downstream discharge of hydroelectric dams are a common response to the seasonal warming and thermal stratification of impounded waters. Dissolved oxygen concentrations can decline throughout the summer in the relatively stagnant hypolimnion until either anoxic conditions predominate or reoxygenation occurs when the reservoir mixes again. Repeated or prolonged exposure to low oxygen levels is detrimental for fish activity, feeding, growth rates, and other normal biological functions. Conditions brought about by an absence of dissolved oxygen in the hypolimnion may also affect other water-quality parameters. For example, decomposition of organic matter under anaerobic conditions results in the accumulation of toxic hydrogen sulphide and ammonia (Nilsson & Renöfält, 2008). These conditions also increase the solubility of iron, manganese and some heavy metals, and these materials may consequently be mobilised from the reservoir sediments and be released downstream with suspended particulate matter.

Dams have both direct and indirect effects on riverine thermal regimes by affecting the energy budget and the thermal capacity of the channel (Olden & Naiman, 2010). The extent to which dams modify downstream water temperatures largely depends on stratification behaviour of the reservoir, the depth at which water is released from the dam (i.e. above or below the thermocline), and changes in water volume that affect the rate at which water heats and cools in response to heat exchange at the air/surface water and streambed/water interface (Poole & Berman, 2001). Dams releasing cold water from a single deep portal (often associated with hydroelectric generation) below the thermocline have caused the annual thermal maxima of numerous systems to be both reduced and displaced in time; for example, maximum temperatures downstream of Burrendong Dam on the Macquarie River (Australia) are 8–12°C lower and are delayed by 1–3 months compared to natural conditions (Preece, 2004). Similarly, Jackson *et al.* (2007) found that summer water temperatures in the regulated Lyon River were 5–6°C cooler than the adjacent unregulated Lochay River (Scotland), and Angilletta *et al.* (2008) reported a 3°C decrease in summer water temperature after the construction of Hills Creek Dam on the Willamette River (USA). In general, hypolimnetic release below large dams causes water temperatures to be lower in the spring and summer months, higher in the winter months, fluctuate less seasonally, and exhibit delayed timing of maxima compared to natural conditions (Olden & Naiman, 2010).

Empirical evidence suggests that thermal alteration by dams has significant implications for downstream productivity and the reproduction, growth, distribution and assemblage structure of fishes (Haxton & Findlay, 2008). For example, cooling and delayed timing of maximum temperatures in the Namoi River below Keepit Dam (Australia) had significant consequences for the spawning success of several native fish species (Preece & Jones, 2002). Based on the percentage of time in the spawning period that the mean daily water temperature exceeded the temperature threshold for reproduction, spawning opportunities for silver perch (*Bidyanus bidyanus*) and golden perch (*Macquaria ambigua*) were reduced to 25–70% and 44–87% of pre-dam years, respectively. In the Qiantang River Basin (China), the construction of Xinanjiang Dam decreased the annual sum of degree days > 15°C (considered the positive-growth threshold temperature for warmwater fishes) by one-third, resulting in numerous extirpations of warmwater fishes (Zhong & Power, 1996). Similarly, significant declines of native fishes in the Colorado River Basin (USA) have been attributed to reductions in spring–summer temperatures caused by hypolimnetic releases from dams (Clarkson & Childs, 2000).

Dam-induced changes in thermal regimes may also have long-term evolutionary consequences for riverine biota by inducing a mismatch between a species' life-history and other critical environmental conditions. For example, Angilletta *et al.* (2008) hypothesised that warmer temperatures during the autumn and winter below Lost Creek Dam on the Rogue River (United States) may indirectly influence the fitness of Chinook salmon (*Oncorhynchus tshawytscha*) by accelerating the development of embryos, leading to earlier timing of emergence. Shifts to earlier emergence could lead to mortality from high-flow events, elevated predation or insufficient resources. Using an age-based population model, the authors predicted a decrease in mean fitness of Chinook salmon after dam construction. Although the likelihood for these impacts is unknown (i.e. temperature changes may also result in strong compensatory strategies, such as delayed spawning by adults or slowed development by embryos), the potential evolutionary consequences of thermal alteration should not be overlooked.

Many other smaller dams (typically in cooler summer climates) cause elevated spring–summer water temperatures by releasing epilimnetic water from above the thermocline of the reservoir. Lessard and Hayes (2003) examined 10 small dams on Michigan streams (USA) and found that summer temperature below dams increased by an average of 2.7°C. Increased temperatures downstream of overflow dams can eliminate thermal cues vital to the life cycles of some invertebrate prey while concurrently increasing the metabolic rates of fishes and elevating demand for food to support growth and survival.

Despite broad recognition of the importance of flow and thermal regimes for river systems, there are relatively few studies that have explicitly linked dam-induced changes in both flow and thermal factors to the structure of fish assemblages. In a revealing study, Murchie *et al.* (2008) conducted a systematic review of the literature to identify studies that examined the response of fish to modified flow regimes in regulated rivers. Of the 131 studies identified in their review, they found that although almost half of the studies included the collection of water temperature data, the majority of these studies (57%) failed to examine the potential consequences of thermal alteration.

4.3.4 Habitat connectivity and fragmentation

Dams are obstructing rivers worldwide by isolating populations and habitats, creating physical obstructions for migrating fishes, and disrupting interactions between freshwater, terrestrial and coastal systems (Nilsson *et al.*, 2005; Reidy Liermann *et al.*, 2012). Today, the literature is replete with studies documenting significant declines in large river fishes that require long-distance dispersal (see Chapter 12). Most notable are diadromous fishes (e.g. salmonids, eels, lampreys, shad, striped bass), because barriers on large rivers inhibit movement between critical spawning, rearing and adult foraging habitats (e.g. Gehrke *et al.*, 2002; Kiffney *et al.*, 2009; Limburg & Waldman, 2009). The delays in migration time from encountering dams cause energy needed for reproduction to be expended while fish are pooling above or below the dam. For example, the American shad reabsorbs its gonads when returning to the ocean if it is delayed, without releasing eggs or sperm. Potamodromous fishes (e.g. sturgeon, suckers) that move considerable distances within freshwater are similarly affected by fragmentation. By contrast, species that display little dispersal prior to spawning are less likely to show severe responses to dam fragmentation (e.g. Gehrke *et al.*, 2002; Fukushima *et al.*, 2007). In summary, the impacts of connectivity alteration on riverine fish assemblages are dependent on the migration characteristics, life history and habitat requirements of the constituent species.

Dams on large rivers are impassable to fishes at all life stages in the upstream direction unless specifically designed to allow passage. Downstream movement at some or all life stages (i.e. drifting eggs and larvae, juveniles, adults) is more likely, albeit with increased mortality that is dependent on the outflow structure of the dam (Baumgartner *et al.*, 2006; Agostinho *et al.*, 2007). For example, migration through hydropower dam turbines causes significant mortality, especially of adult fish. There are many examples of the extirpation of anadromous Pacific salmon stocks that are unable to reach their spawning grounds after the construction of impassable dams. In the Columbia River Basin (USA), an extensive network of hydroelectric dams, constructed between 1939 and 1975, has blocked access to or inundated most riverine habitats suitable for spawning by autumn Chinook salmon (*Oncorhynchus tshawytscha*). More than 75% of the original 2500 km of mainstem spawning and rearing habitats has been eliminated (Schindler *et al.*, 2003). For resident or potamodromous fish, competition for spawning sites and food can increase as dams disconnect, isolate, and reduce the number and size of habitats (Dudley & Platania, 2007). By partially or totally restricting movement between populations, distinctly different fish assemblages are often formed in reaches separated by barriers. In New Zealand, Joy and Death (2001) found that fish species richness in Taranaki Ring Plain streams was significantly reduced above dams due to their influence on movements of diadromous species. In short, dams may decrease the long-term persistence of fish (e.g. Morita & Yamamoto, 2002) and result in upstream extirpation (e.g. Winston *et al.*, 1991).

Available data suggest that fragmentation associated with dams has interrupted gene flow between subpopulations of riverine fish species, leading to the loss of genetic diversity and over- or underrepresentation of genotypes (e.g. Yamamoto *et al.*, 2004; Roberts *et al.*, 2013). For example, in the Rio Grande River (USA), dam fragmentation now restricts the endangered Rio Grande silvery minnow (*Hybognathus amarus*) to 5% of its original range – represented by only 280 km of river (Bestgen & Platania, 1991). Genetic diversity is low and homozygous individuals are overrepresented because dams degrade spawning habitat and prevent movement of adults, drifting eggs and recruiting juveniles (Alò & Turner, 2005).

Somewhat counter-intuitively, this isolation may prove beneficial for native biota by blocking the upstream spread of invasive species (Rahel, 2013; discussed later in this chapter). In contrast to the impacts of large dams, small barriers such as low-head weirs are less likely to result in reach-scale species extinctions as some level of connectivity is often maintained (McLaughlin *et al.*, 2006). Furthermore, smaller (and often semipermeable) barriers such as road crossings or water withdrawals that dry sections of a river network also have consequences for freshwater fishes (see Chapter 10).

By dampening high flows and reducing the frequency, extent and duration of floodplain inundation, large dams can also reduce lateral connectivity in floodplain rivers. This is often compounded by other structural modifications such as channelisation and levee banks. The elimination of smaller channel-forming flows also modify patterns of channel migration, lowering habitat diversification on the floodplain. These changes can have significant implications for freshwater fishes (Bunn & Arthington, 2002). For example, prior to river regulation, many species of fish in Australian rivers were thought to have used inundated floodplain wetlands of lowland rivers for breeding and juvenile habitat. River regulation to prevent flooding has seriously affected such recruitment, and there is an obvious trend of reduced native species abundance with increasingly regulated catchments (Gehrke & Harris, 2001). The effects of dams on floodplain-dependent fish species are widespread across the world, including many neotropical rivers (e.g. Agostinho *et al.*, 2004).

Lastly, estuaries are often impacted by upstream dams. These effects include drying of wetlands, increased estuarine and nearshore salinity, reduced sediment inputs to coastal regions, and reduced inputs of nutrients to estuaries. One telling example is the San Francisco Bay/ Sacramento–San Joaquin estuary (California, USA), where reduced freshwater inputs due to dams (and other land-use activities) are impacting the federally threatened delta smelt, *Hypomesus transpacificus* (Helfman, 2007).

4.3.5 Creation of reservoirs

An expanding network of reservoirs has dramatically increased the number, area and spatial distribution of standing waters across the globe. In fact, vast regions where lakes were formerly rare are now home to hundreds of reservoirs, while other regions historically dominated by natural lakes now also include numerous reservoirs. For example, the construction of dams in Oklahoma, USA, a region historically characterised by natural oxbow and playa lakes, has led to an over 200-fold increase in area of standing water caused by dams inundating upstream reaches (Havel *et al.*, 2005). Perhaps not surprisingly, reservoirs support very different fish assemblages dominated by lentic species compared to lotic species formerly occupying the flowing river.

Mounting research suggests that reservoirs may function as invasion 'hubs' for freshwater invaders because they are frequently larger and more accessible to humans, younger in age and higher in disturbance relative to natural lakes; factors that facilitate greater establishment and spread across the landscape. Consistent with this hypothesis, Johnson *et al.* (2008) found that reservoirs were up to eight times more likely than natural lakes to have established populations for five widespread aquatic invaders (including one fish species). Species introductions in reservoirs can also spread to connected streams (Gido *et al.*, 2004), where the effects on fish community composition weaken with increasing distance from the reservoir (Falke & Gido, 2006).

Whereas the upstream effects of dams are still relatively poorly understood, studies have reported changes in fish assemblage structure associated with stream bank destabilisation (Kruk & Penczak, 2003), increased richness of fish macrohabitat generalists (Herbert & Gelwick, 2003), decreased juvenile fish survival (Ponton & Copp, 1997) and decreased native fish diversity (Reyes-Gavilan *et al.*, 1996) in streams above reservoirs. Species with drifting larvae rely on large reaches of free-flowing river habitat and are negatively affected when their larvae or eggs drift into a reservoir and are either consumed by predators or settle to the substrate. This may ultimately lead to fish species extirpations upstream of reservoirs (Winston *et al.*, 1991; Matthews & Marsh-Matthews, 2007).

4.3.6 Serial discontinuity caused by dams

It is well recognised that dams can cause rivers to experience breaks or discontinuities in the longitudinal trends of biotic and abiotic variables; this has been formalised as the serial discontinuity concept (Ward & Stanford, 1983). Discontinuities in the river can reset the continuum, creating a longitudinal shift or departure from a reference condition of a given variable defined as 'discontinuity distance' (also known as recovery distance or zone of influence). The recovery of each variable depends on dam size, position along the river, and tributary inputs among others (Ward & Stanford, 1983). After receiving criticism for its simplistic 'linear' view, the serial discontinuity concept was expanded to include lateral and vertical connectivity (Ward & Stanford, 1995). Although the serial discontinuity concept does not make explicit predictions for fish assemblages, it does describe the dynamics of energy and materials that have direct relevance to fish. Interestingly, this concept has been widely cited, but its hypotheses have rarely been tested (Ellis & Jones, 2013). In one example, however, Reyjol *et al.* (2001) showed that in unregulated sections of a river in France, temperature determined the transition from the salmonid-dominated region to the cyprinid-dominated region, but that this changed with regulation and the presence of dams.

4.4 OPPORTUNITIES FOR FISH CONSERVATION

Considerable research in freshwater ecology is now devoted to understanding and mitigating the ecological impacts of dams. The widespread and intensive nature of dams suggests that they also offer means of mitigating some hydrologic impacts, if not through retirement of dams, through changes in their structure or operating rules. Here, we will examine current efforts to restore and conserve fish biodiversity in rivers that have been dammed or subjected to flow diversions, including the decommissioning and removal of dams, mimicking of historical or natural flow regimes below dams and establishing opportunities to circumvent dams via fish passages. Many other opportunities exist, but these approaches represent the most significant for fish conservation.

4.4.1 Dam removal

For a long time, no one except environmental advocates talked much about removing dams, which were seen as necessary for water storage, flood control, power generation and recreation. When author Edward Abbey famously wrote *The Monkey Wrench Gang* in 1975, he was roundly condemned for portraying eco-terrorists plotting to blow up Glen Canyon Dam on the Colorado River. However, the ageing of the world's dams, coupled with increasing awareness of their environmental costs, has brought dam decommissioning and removal to the attention of the scientific community, management agencies and the general public (WCD, 2000). In the United States – where hydropower dams must often be relicensed every 30–50 years and many are economically inefficient or at high risk of structural failure – the rate of dam removal now exceeds the rate of construction. In the past two decades alone, about 600 dams have purposely fallen, and many more are slated for removal, especially those smaller in size (Doyle & Havlick, 2009). Internationally, Denmark has removed several dams from its highly impacted river systems and has succeeded in improving habitat and fish passage on many of its rivers as a result (Iversen *et al.*, 1993). Dams have also been (or plan to be) removed in Australia, Canada, Japan and France. Future dam removals will be a result of relicensing requirements, heightened environmental literacy of their impacts, and because many of these structures are older than their life expectancies, structurally compromised and prone to fail under current or future flooding regimes (Bednarek, 2001).

Dam removal may eliminate several of the problems associated with their downstream environmental effects and issues associated with fish migration or movement discussed previously in this chapter. In the simplest sense, once a dam is removed, so are the downstream impacts on water quantity (i.e. stream discharge), water quality (e.g. temperature, dissolved oxygen) and sediment transport. However, dam removal can result in decades of accumulated material being released downstream in a rapid and catastrophic fashion. Fine sediment is mobilised from the slow-moving reservoir and redistributed, exposing gravel, cobble and boulders within formerly impounded areas. In very small impoundments, or those with limited accumulation, sediments can be flushed out relatively fast. In contrast, the development of new channels in larger reservoirs is a more prolonged and dynamic process that may sustain downstream sediment export for months or even years (Stanley & Doyle, 2003).

The most striking consequence of dam removal is the conversion of slow-moving tail-waters and reservoir habitat into stretches of flowing water. Replacement of reservoir fish assemblages by more typical riverine species can occur relatively quickly after the dam is removed. For example, fishes adapted to slow-moving water and fine sediments gave way to riverine fishes within a year of removal of a Wisconsin dam (Kanehl *et al.*, 1997). Local anglers and conservationists were delighted to witness declines in non-native common carp (*Cyprinus carpio*) and increases in smallmouth bass (*Micropterus dolomieu*) and darters (*Etheostoma* and *Percina* spp). Fish recovery reflected both recolonisation of individuals that had previously resided upstream or downstream from the dam and successful reproduction within this newly created habitat. However, there are also instances where dam removal did not automatically trigger recolonisation and changes in fish assemblages. Stanley *et al.* (2007) examined fish community composition two years prior to and two years after the removal of a pair of low-head dams from Boulder Creek (Wisconsin, USA) to determine if the removal of these potential barriers affected the resident population of native brook trout (*Salvelinus fontinalis*). Despite the presence of source populations in downstream reaches and other adjacent streams, Boulder Creek received no new colonists of trout in the two years following dam removal. In short, the current evidence of fish responses to dam removal is often anecdotal, and only recently have quantitative investigations been conducted.

In the case where a dam has no fish passage structure (see Section 4.4.3 for more details), removal eliminates mortality due to the inability to pass around the dam, and allows organisms to inhabit previously impounded areas. For example, removal of small dams in Denmark resulted in salmonids and other fish being able to reach optimum spawning grounds and enhance their chances of survival (Iversen *et al.*, 1993). Similarly, within years of taking apart the Edwards Dam on the Kennebec River (Maine, USA), biologists observed the return of scores of striped bass, alewives, American shad, Atlantic salmon, sturgeon and numerous birds including ospreys, kingfishers, cormorants and bald eagles that feed on fish. In another example, a success story of dam removal has emerged from the Loire River Basin (France). Atlantic salmon were abundant in the nineteenth century – tens of thousands would migrate annually – but by 1997, only a couple of hundred were observed making the trip. Despite the incorporation of fish ladders and elevators, the eight dams along the Loire River and its major tributaries – as well as their turbines and pumps – had decimated the salmon population. Non-governmental organisations, including the European Rivers Network, led a campaign to bring the salmon back, and in response the French government decommissioned four of the dams from 1998 to 2005. Within a few months of each dam removal, five species of fish, Atlantic salmon and shad among them, began to re-establish their historical migratory pathways. Similar success stories of dam removal for migratory fishes are evident (see Chapter 11); here, it is worth highlighting the world's largest removal effort involving the Elwha and Glines Canyon dams on the Elwha River (USA) (Box 4.1).

Box 4.1. Dam removal on the Elwha River: the largest in United States history

The Elwha River arises from headwaters in the Olympic Mountains and flows northward across Washington State (USA), emptying into the Strait of Juan de Fuca. In the early twentieth century, Thomas Aldwell – a politician, banker and entrepreneur – saw an opportunity in the river's rapid waters and narrow gorges. He decided to harness the river's energy and secured enough funding to establish the Olympic Power Company and support the construction of the Elwha Dam (1910) and later the Glines Canyon Dam (1927). The dams fuelled economic development, but terminated the spawning migration of fish upriver because they lacked fish passage capabilities. The Elwha River was historically one of the most productive salmon rivers in the Pacific Northwest, but today all five salmon species and steelhead are locally extirpated or persist in critically low numbers. The dams also changed the landscape by holding back not just water but also sediment. The trapped sediment could no longer replenish the riverbanks and coastline downstream.

Although it took decades of legal wrangling, the world's largest dam removal project began in September 2011 on the Elwha River (Figures 4.4, 4.5 and 4.6). The Elwha and Glines Canyon Dams were slowly dismantled, reopening more than 145 km of pristine salmon habitat in the Olympic National Park that has been blocked to anadromous fish passage for over a century. The demolition of the dams has released more than 18 million cubic metres of silt and other debris downstream and scientists continue to study its effects on fish and other wildlife.

It is still too early to assess the longer-term ecological recovery of the Elwha River; however, it is clear that it has altered the landscape for large-scale dam removals in the United States. By 2020, 70% of the 84,000 dams in the US will be more than 50 years old, and it is estimated that many will require repair, removal, or replacement. If successful, the Elwha Dam removals may pave the way for future decommissioning of large dams across the United States and globally.

Figure 4.4 Elwha Dam under initial deconstruction in September 2011.

Figure 4.5 The remains of the Elwha Dam in February 2012.

Figure 4.6 Elwha Dam after demolition in May 2013.

For dams with fish passage capabilities, dam removal eliminates death or injury to fishes caused by passage mechanisms, such as turbine entrainment and fish ladder mortality. Furthermore, it also eliminates delays occurring at crowded upstream passage devices and downstream delays from swimming through the slow-moving reservoir. Fish passage structures cannot typically accommodate large numbers of fish at the same time; therefore, dam removal can speed fish movement and increase the odds of successful reproduction (e.g. Wik, 1995). Analyses of fish passage versus dam removal for the Enloe Dam on the Similameen River (USA), for example, suggested that added fish passage would not successfully accommodate the large number of migrating fish attempting to pass (Winter, 1990).

Today, ecological restoration is a billion dollar per annum industry in the US alone (Bernhardt et al., 2005), and a growing body of research has focused on the effectiveness of dam removal as a method of restoration (Hart et al., 2002). However, this has not been without many controversies. The actual process of dam removal can be thought of as a significant disturbance, with the various economic, ecological and social implications being very situation-specific (Stanley & Doyle, 2003). For example, dam removal can increase property value, improve local fisheries, water quality, dissolved oxygen levels and habitat connectivity, but at the expense of reduced wetland distribution, mortality from intense short-term sedimentation, increased potential for invasive species spread, or decreased groundwater discharge (Hart et al., 2002). Other risks involve nutrient and sediment contamination to downstream reaches, potential connectivity created between wild and farm-raised fisheries, and some public opposition due to the sentimental and historic value some place on these structures (Doyle et al., 2005).

One emerging controversy is worth discussing. Previously cited examples demonstrate that dam removal can increase habitat availability and restore migratory and movement routes for fishes; however, these barriers can also isolate and protect native species from aggressive downstream invasive species (see Chapter 8). This challenge is especially acute in inland western North America, where remnant populations of native trout (e.g. cutthroat trout, bull trout) have been invaded by non-native brook, brown and rainbow trout while simultaneously encountering countless impassable small and large dams, diversions and road culverts. Biologists charged with conserving native salmonids often disagree over the relative merits of building new barriers (or converting temporary barriers, such as impassable culverts, into permanent structures) to limit upstream invasions versus removing barriers to allow recolonisation, demographic support, or movement to complementary habitats to enhance population persistence (Fausch *et al.*, 2009). This invasion–isolation dilemma clearly illustrates that dam removal is not a panacea for fish conservation. Managers face a clear trade-off because the barriers designed to protect native populations from invasions (i.e. purposeful small dam construction for conservation) may also hasten their extinction by creating small populations isolated in habitat fragments (e.g. Novinger & Rahel, 2003). This dilemma will become more prevalent globally as barriers continue to be removed in some regions, but are constructed in other regions to limit the spread of non-native fishes. A good example of linking dam removal activities with non-native fish concerns is found on Fossil Creek (USA) (Box 4.2).

Box 4.2. Dam removal and non-native fish species control

Few streams in the arid American Southwest have discharge as large and steady as Fossil Creek, Arizona. Despite its remoteness, the stream was seen in the early twentieth century as a good place to produce electricity for growing copper, gold and silver mining in the region. In 1909, the creek was diverted to generate electricity in the Irving Power Plant. In 1916, the Childs Power Plant was built further downstream, and the creek was left dry except for storm run-off. Its water was shunted through flumes and penstocks, captive until it emptied into the Verde River just upstream from the creek's natural confluence.

In 1999, Arizona Public Service decided to voluntarily shut down operations on both power plants and decommission the diversion dam. This provided an opportunity to compare the responses of native fish to flow restoration and non-native fish removal, using a before–after control-impact design. From 1909 to 2005, the majority of flow was diverted from the river, at the diversion dam, through a flume to the two power plants. In 2004, non-native fish were removed from 16 km upstream where a fish barrier was constructed. Flow was restored in June 2005 when the diversion dam was decommissioned. The control reach above the dam always experienced full flows and was not invaded by non-native fish. Restoration treatments include: flow restoration alone where non-natives had not been present, flow restoration and non-native fish removal, and flow restoration where non-natives remain (see Figure 4.7).

Marks *et al.* (2010) found that removal of non-native fish dramatically increased native fish abundance (see Figure 4.8). Flow restoration also increased native fish abundance, but the effect was smaller than that from removing non-natives. Flow restoration had no effect where non-native fish remained, although it may have had other benefits to the ecosystem. The cost to restore flow ($12 million) was considerably higher than that to eradicate non-native fish ($1.1 million), suggesting that dam removal projects should also consider the additional investment required to eradicate non-native species.

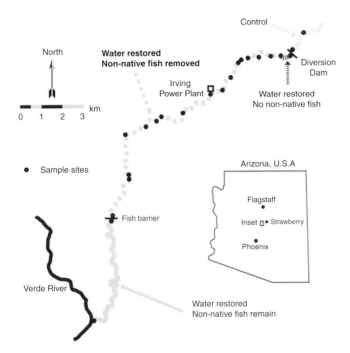

Figure 4.7 Map of Fossil Creek depicting the flow and non-native species removal treatments.

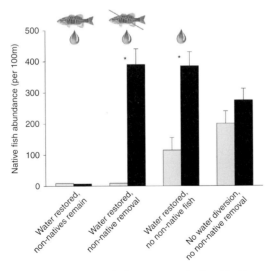

Figure 4.8 Native fish densities before (grey) and after (black) restoration across treatments (ordered from left to right as progressing upstream: see Figure 4.7). Significant responses to restoration are denoted by an asterisk. Map and data for graph from Marks *et al.* (2010).

Although abandoned or obsolete dams often provide an opportunity for complete removal and full restoration of the riverine conditions, most dams will provide opportunities for river restoration either through environmental flows or installation of fish passage facilities. Both of these management strategies are discussed below.

4.4.2 Dam releases to mimic natural flow regimes

Dam removal is neither a panacea nor Pandora for river restoration and the conservation of freshwater fishes. Dams were built for a purpose; they provide valuable socioeconomic goods and services, including hydropower, flood control and recreation (Figure 4.1), and dam removal will not be warranted or even possible in many instances. Therefore, considerable interest remains in learning how to balance river restoration efforts with the continued use of dams through various operational changes, including increased minimum flows, periodic releases of high flows and mimicking flow seasonality (Poff *et al.*, 2003).

Methods designed to quantify minimum 'in-stream flows' to sustain fish species first appeared in the US in the late 1940s. These flow-management schemes range in complexity from simple discharge limits thought to maintain a desired ecological outcome (Tennant, 1976) to the instream flow incremental methodology that explicitly accounts for the microhabitat requirements of individual fish species and life stages (Bovee, 1982). The physical habitat simulation model (one tool in the instream flow incremental methodology) is perhaps the most widely adopted by natural resource agencies (in its original or some modified form) despite its required hydraulic and geomorphological surveys and uncertainties in defining habitat preferences for species of interest (Castleberry *et al.*, 1996). Examples are much too numerous to describe here, but in one instance, O'Brien (1987) defined the minimum stream flow required to maintain existing habitat, food supplies and spawning potential of the endangered Colorado River pikeminnow (*Ptychocheilus lucius*) in terms of four flow characteristics. To develop this minimum hydrograph, the approach combined the results of a two-year field study, a physical model, laboratory simulation of flows over cobble substrate and a mathematical sediment transport model.

In a surge of developments over the past decade or so, scientists now recognise that arbitrary 'minimum' flows are inadequate – the structure and function of a riverine ecosystem and many adaptations of its biota are dictated by patterns of temporal variation in river flows (see previous discussion). There is now general agreement among scientists and many managers that to protect freshwater biodiversity, including fishes, and maintain the essential goods and services provided by rivers, we need to mimic components of natural flow variability, taking into consideration the magnitude, frequency, timing, duration, rate of change and predictability of flow events (e.g. floods and droughts) (Arthington *et al.*, 2006). This has spawned the field of scientific research termed environmental flow assessment (Tharme, 2003). In simple terms, such an assessment addresses how much and which specific temporal characteristics of the original flow regime of a river should continue to flow down it and onto its floodplains in order to maintain specified features of the riverine ecosystem (Tharme, 2003). I refer the reader to the book by Angela Arthington (2012) for an excellent treatment of the science and implementation of environmental flow assessments across the globe.

Hydroelectric dams are often operated in a peaking mode, releasing large volumes of water during the day when the value of power is greatest, and smaller volumes at night, when economic return is low. The environmental consequence of peaking operation is that downstream fishes may experience rapid increases in flow when turbines are placed into operation and prolonged reduction of water level and velocity during the off-peak periods (Cushman, 1985). In 1981, a summer minimum flow requirement (from 3 to 142 m^3/s) was established below the Conowingo Dam on the Susquehanna River (USA) to protect fishes from becoming stranded during the off-peak periods. Fish condition (weight at length) and growth of three abundant downstream species (white perch, yellow perch, channel catfish) was significantly greater after the minimal flow was established. These increases were driven by decreased stranding and increased fish prey consumption in response to enhanced benthic invertebrate production (Weisberg & Burton, 1993).

Resource managers have manipulated streamflow below dams with the goal of benefiting fishes with certain life histories. In response to a continuous-flow release below a diversion dam in the Bridge River (Canada), adult salmon quickly recolonised and spawned in a 4-km streambed that had been without flow for 37 years (Decker *et al.*, 2008). In the Tallapoosa River (USA), Travnichek *et al.* (1995) demonstrated increased species richness and abundance of fluvial-specialist fishes relative to habitat generalists in response to increased minimum flows below Thurlow Dam. The relative abundance of species preferring fast-flowing and/or deep microhabitats increased from two- to fourfold after minimum flow increase on the Rhône River below the Pierre-Bénite dam (Lamouroux *et al.*, 2006). In response to managed high flows in the Murray River (Australia), King *et al.* (2010) observed benefits to native fishes. Different taxa responded through distinct mechanisms: some fishes increased spawning, whereas others had high larval survivorship. In the San Juan River (USA), native fish densities increased in response to elevated releases from Navajo Dam in the spring to mimic snowmelt run-off; however, the non-native fish density increased in response to the lowering of summer flows, which also reflected unregulated flow patterns (Propst & Gido, 2004). Increases in non-native fish abundance were also observed after high-flow pulses in the Colorado River below Glen Canyon Dam that were aimed at flushing sediment and creating important habitat for endangered native fishes (Korman *et al.*, 2011). In many instances, achieving management objectives of environmental flows depends on flow management over the life cycle of targeted taxa. In the Skagit River (USA), successful salmon reproduction depended on flows to cue spawning that were maintained throughout the incubation period (Connor & Pflug, 2004).

In summary, changing societal values have compelled the modification of dam operations and water diversions to mitigate physical and biological impacts on aquatic systems (Konrad *et al.*, 2011). Scientists have increasingly advocated for an experimental framework to evaluate and develop operations that provide ecological benefits, to create a more rational basis for water-management decisions, and to advance broader scientific knowledge (Souchon *et al.*, 2008; Konrad *et al.*, 2011). Indeed, flow experiments have been used globally to evaluate the effects of alternative dam operations on rivers, floodplains and estuaries, and increasingly are more commonplace in the lexicon of river management (Olden *et al.*, 2014; see Figure 4.9 for examples). However, such management actions cost money and potentially forego other benefits that humans derive from dams. For example, the four large controlled floods from Glen Canyon Dam to the Colorado River (1996–2012) and the purposeful spills at four Columbia River dams (McNary, John Day, Dalles, Bonneville) for fish passage (2005) in the US were estimated to cost $12 and $57–81 million (USD) in foregone or replaced power revenue, respectively. Similarly, multiyear flow releases from Friant Dam (San Joaquin River, USA) to benefit Chinook salmon cost water users approximately $8 million annually in environmental fees, and the Low Summer Steady Flow experiment downstream of Glen Canyon

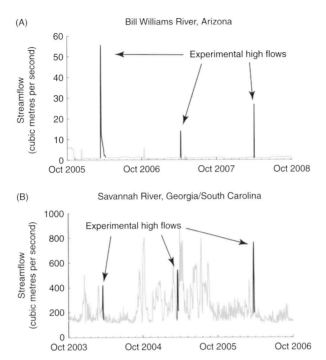

Figure 4.9 Experimental high flows have been delivered to a number of rivers globally (Olden *et al.*, 2014) including (A) the Bill Williams River (Arizona, USA) and (B) the Savannah River (Georgia and South Carolina, USA).

Dam was estimated to result in about $25 million in replacement energy costs.

4.4.3 Dam releases to mimic natural thermal regimes

The ecological significance of water temperature for freshwater fishes is widely acknowledged, yet the mitigation of thermal impacts below dams has received little attention compared to flow when environmental flow assessments are conducted (Olden & Naiman, 2010). As discussed previously, hypolimnetic releases of cold water below dams provide unique and highly desirable fishing opportunities for trout or salmon in geographic regions that could otherwise not support recreational fisheries (e.g. Krause *et al.*, 2005). Although these dam operations may benefit a few (often non-native) species, it also often comes at the detriment of entire warm-water assemblages of native fishes and other aquatic organisms. Below, current technologies for directly modifying water temperature released from dams to benefit the conservation of fresh-water fishes are explored.

Current management options for mitigating thermal impacts from dams fall into two general categories: exploit the temperature stratification of the reservoir by selective withdrawal of water of the desired temperature, or artificially break up the stratification prior to discharging water from the dam (Sherman, 2000). Selective offtake using a multi-level intake structure is the most common and effective means of controlling the water temperature of dam releases. A selective withdrawal system (also called a temperature control device) can extract water from selected depths of a thermally stratified reservoir to produce a release with desired characteristics (Olden & Naiman, 2010). This technology provides the flexibility to increase water temperatures by preferentially selecting warm epilimnetic water from the surface, or decrease water temperature by drawing cold hypolimnetic water from below the thermocline (Figure 4.10A). For example, Flaming Gorge Dam on the Green River (USA) was installed with a multilevel intake structure in 1978 with the goal of increasing summer water temperatures for native species in the trout-dominated tail-waters. After the installation of the intake structure, average stream temperatures during the summer months of May to August doubled from a pre-installation temperature of 5.7°C to 11.0°C; a temperature approaching the pre-dam summer temperatures that averaged 17.2°C (Figure 4.10B). With another management goal in mind, Shasta Dam on the Sacramento River (California, USA) was retrofitted with a multilevel intake structure in 1997 to improve downstream temperatures for endangered cold-water salmonids. Dam management is focused on releasing warmer surface waters in the winter/spring and colder deep waters in the summer/autumn. These two examples illustrate the flexibility of multilevel intake structures for controlling downstream temperatures for a defined ecological goal.

Most selective withdrawal intake structures were built during initial reservoir construction (ironically, most were originally designed to support tail-water trout fisheries) and the capital costs of refitting dams with these structures may be prohibitive. For example, the installation cost for several large dams in Australia was estimated to range between 3 and 30 million AUD (Sherman, 2000), and the actual cost for retrofitting Shasta Dam was 80 million USD. Estimated installation costs for modifying eight intake portals of Glen Canyon Dam (Colorado River, USA) is 15 million USD. Despite the costs of installing multilevel intakes, investigations suggest that these structures provide the most flexible means

(A)

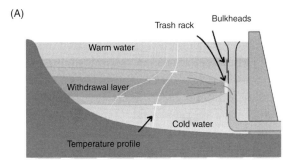

(B)

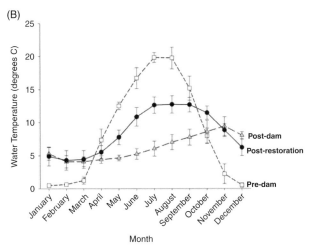

Figure 4.10 (A) Current management options for mitigating thermal impacts from dams include selective withdrawal using multilevel intake structures. Figure modified from Olden and Naiman (2010). (B) Selective discharge below Flaming Gorge Dam using a multilevel intake structure has markedly decreased the degree of thermal alteration in the Green River (USA). Comparisons of monthly water temperature during pre-dam (1958–1962, squares), post-dam (1963–1977, triangles) and post-thermal restoration years (1978–2013, circles) shows significant increases in spring–summer temperatures toward unregulated conditions.

of modifying downstream water temperatures, even at low to medium release volumes, and may represent the best opportunities for ecological restoration. For example, Sherman *et al.* (2007) predicted that the installation of a multilevel intake structure in Hume Dam, Australia, would increase discharge temperatures by 4–6°C during the spring–early summer post-spawning period for Murray cod (*Maccullochella peelii*). These temperature increases were forecasted to increase minimum female population abundance in the Murray River by 30–300% depending on the assumed spawning behaviour.

Evidence suggests that defining environmental flows with respect to either flow or temperature in isolation is highly unlikely to support fish conservation; both must be considered in tandem (Rolls *et al.*, 2013). One example supporting this notion is the study of King *et al.* (1998), who examined patterns of temperature and discharge resulting from an experimental dam release, and their relative importance in triggering successful spawning of a threatened large cyprinid endemic, the Clanwilliam yellowfish *Barbus capensis*, downstream of Clanwilliam Dam on the Olifants River (South Africa). A critical flow component for many rivers in the winter-rainfall region of South Africa, including the Olifants River, is the small pulses of higher flow that occur in the dry season (called freshes). Experimental freshes released from Clanwilliam Dam in the early 1990s were strongly correlated with the suspected spawning success of yellowfish; however, in subsequent years, freshes delivered during the species' breeding season failed to induce spawning. King *et al.* (1998) found that differences in the temperatures of the water release were the key factor related to spawning success. Specifically, warm epilimnetic freshes (19–21°C) triggered fish spawning behaviour and the movement of individuals onto spawning beds, whereas cool hypolimnetic baseflows (16–18°C) released immediately after the experimental freshes caused fish to abort spawning activities. Moreover, the occurrence of dead and deformed young suggested that the cold water has a detrimental effect on the offspring of those individuals who spawned during the warm-water events. This research showed that freshes released from Clanwilliam Dam at the appropriate time should be able to induce spawning and support early life stages of yellowfish only if the required water temperatures are maintained for an extended period after spawning to provide for the development of the embryos and larvae.

Similar examples exist in the lower Mississippi River (USA), where research has shown that the growth and abundance of juvenile fishes are only linked to floodplain inundation when water temperatures are greater than a particular threshold. Schramm and Eggleton (2006) reported that the growth of catfishes (Ictaluridae spp.) was significantly related to the extent of floodplain inundation only when water temperature exceeded 15°C, a threshold temperature for active feeding and growth by catfishes. Under the current hydrographic conditions in the lower Mississippi River, the authors report that the duration of floodplain inundation when water temperature exceeds the threshold is insufficient for floodplain-foraging catfishes to achieve a detectable energetic benefit. These results are consistent with the 'thermal coupling' hypothesis offered by Junk et al. (1989), whereby the concordance of both hydrologic and thermal cycles is required for maximum ecological benefit.

Today, ecologists are seeking ways to define environmental flows that provide the critical flow and thermal requirements of resident fishes, while continuing to provide the goods and services expected by human society. Will there be conflicts between flow and thermal targets in environmental flow prescriptions? Yes, quite simply because stream temperature co-varies with flow. For example, Krause et al. (2005) found that proposed modifications to a regulated discharge regime below Philpott Dam on the Smith River (USA) designed to improve the physical habitat for brown trout (Salmo trutta) would concurrently reduce the frequency of optimal growth temperatures. This study illustrated that the best management option for improving flow regimes (e.g. for providing critical fish habitat) was not necessarily the best for improving thermal regimes (e.g. for providing optimal growth conditions). Consequently, different aspects of the temperature regime for any stream require careful analysis before the prescription of an environmental flow for fish conservation.

4.4.4 Fish passage structures to restore longitudinal connectivity

The long-term persistence of migratory and resident fish species that would otherwise move to and from different habitats within the river system may diminish substantially, if not completely, due to the effects of dams that act as barriers. As dam construction expanded and many fisheries declined in parts of America and Europe during the seventeenth century, conflicts between mill-industrialists and fisheries ensued (Smith, 1971). In response to collapsing fisheries, the design of fish passage facilities began. Early fish passages in France consisted of steep, constructed channels 'roughed' with bundles of tree branches to dissipate energy and provide passage for some fish species over low-head dams (Smith, 1971). In the late eighteenth century, various laws were enacted to require dams to include 'fish sluices', or open gaps in dams for boats and fish in the United States (Watson, 1996). During the same century, Marshall McDonald of the US Commission of Fish and Fisheries helped develop designs for pool and weir fish ladders for the passage of shad in rivers along the Atlantic coast (McDonald, 1887). McDonald's designs provided a foundation for the further development of effective fish passage designs in France, eventually resulting in an effective fish ladder around 1910. Fish passage research waned after the early 1900s, but was rejuvenated in the 1970s by researchers and engineers.

Recent advances in fish passage (or fishway) technology are resulting in the development of practical and effective passage designs for most dams that aim to pass important diadromous fish species. Fish passage facilities that are appropriately designed and operated may help mitigate the impact of barriers on fish by providing passage to and from habitats for spawning, rearing, feeding, growth to maturity, dispersion, migration and seasonal use of habitat. Upstream and downstream fish passage facilities also mitigate entrainment and mortality at hydropower turbines and impingement and entrainment at dead-end water intakes and diversions. It is true that various types of fish passages (Figure 4.11) can partially connect stretches of rivers fragmented by dams; however, aspects of their efficacy are still very obscure and many may have unexpected impacts on fish populations.

Fish passage efficiency data for the large Atlantic Coast rivers of the USA have shown that, despite using modern engineering approaches, fish passages have been largely ineffective. Historically, Atlantic drainages supported runs of about a dozen diadromous fishes, including American shad, *Alosa sapidissima*. Today, of American shad that passed the first dam, on average 4%, 16% and 32% passed the second dam in the Connecticut, Merrimack and Susquehanna Rivers, respectively (Brown *et al.*, 2013). System-wide efficiencies for these three rivers in past years have been < 3%. Low passage efficiency and delays accrued at fishways impose significant energetic costs that reduce American shad condition and therefore iteroparity (repeat spawning). The phenomenon of reduced repeat-spawning and consequent decreased mean size and age of adult American shad has occurred in the Connecticut River, resulting in lower population fecundity and probably lower annual recruitment survival (Castro-Santos & Letcher, 2010) in addition to the lost production from obstructed access to spawning grounds. Similar poor upstream movement efficiency has been noted for fish passages in Europe (Larinier & Travade, 2002), South America (Agostinho *et al.*, 2007; Oldani & Baigún, 2002) and Australia (Mallen-Cooper & Brand, 2007). Here, we highlight the history and challenges associated with the use of fish passage facilities in neotropical rivers of Brazil.

Figure 4.11 The four main types of fishways. (A) Pool and weir (John Day Dam, Columbia River, USA); (B) baffle fishway (River Dart, England); (C) rock-ramp fishway (Brewarrina Weir, Barwon River, Australia); and (D) fish elevator (Paradise Dam, Burnett River, Australia). Photos are licensed under the Creative Commons (A, B), New South Wales Government (C) and ABC Brisbane (D).

Authorities are increasingly concerned about the consequences of dam blockage on fish migration routes, and they have taken to building fish passages, specifically fish ladders (Figure 4.11) to facilitate the movement of fish through dams in Brazil (Agostinho *et al.*, 2004). Today, many Brazilian dams are equipped with ladders and other such facilities, copying strategies used to manage salmon populations in North America and Europe, and there is continued pressure from the public and government agencies to install new fish passages. The first fish pass constructed in Brazil was the Igarapava fish ladder, on the Pardo River, built in 1911. More than a dozen fish ladders were constructed in other small rivers before the 1960s, as well as fish lifts (Porto Primavera and Funil dams) and bypass channels (Itaipu dam). After the construction of these structures, evaluation was rarely conducted until recent decades. Even now, most of the existing studies are restricted to the efficiency of the ladder allowing fish to go upstream, and their effectiveness for fish conservation has not been evaluated.

An impressive body of literature suggests that fish ladders often fail to preserve migratory patterns in several key ways and may be damaging neotropical migratory fishes (Agostinho *et al.*, 2002, 2007). The weaknesses of fish passages include the unidirectional movement of individuals (upriver), the strong species and numerical selectivity, the high predation pressure throughout the ladder and the impoverishment effect (genetic and demographic) on stocks from below the dam. Despite these effects, there is little doubt that large migratory fish can ascend fish ladders and other passages, depending on the adequacy of the design. However, the main restriction is related to the downstream movements of eggs, larvae and adults through the reservoir. Yet another common difficulty is the installation of ladders in areas that do not warrant such action; this occurs when spawning grounds or other critical habitats are absent from upstream stretches. This problem often occurs when some Brazilian states mandate the construction of ladders by law. For these reasons, some fish ladders may develop into ecological traps for migratory fish species by promoting the transference of populations from high-quality (riverine) to low-quality (reservoir) environments that lack the conditions required for fish recruitment (Pelicice & Agostinho, 2008).

To date, much of the current knowledge on fish passage design and success comes from research conducted on high-priority species, particularly salmonids (Bunt *et al.*, 2012). In some cases, the transfer of salmonid fish passage designs to non-salmonid waters has resulted in poor success (Mallen-Cooper & Brand, 2007), with changes in design often proving to be a more suitable alternative for some species. Roscoe and Hinch (2010) reported that only 4% of published North American fishway studies examined the entire fish community, compared with the broader taxonomic scopes of fishway studies conducted in Europe (38%), South America and Australia (94% combined). Despite this broader scope in other regions, fishways reported to successfully pass entire fish communities remain uncommon (Mallen-Cooper & Brand, 2007), with failure often attributed to the diversity in behaviours, morphology, physiological capacity and swimming ability of the different species using these structures.

Despite many unanswered questions regarding the efficiency of fish passage facilities to mitigate the fragmenting effect of dams, many continue to claim that fish ladders are the ecological solution to future dam construction (Pompeu *et al.*, 2012). For example, the start of construction of the Xayaburi Dam on the Mekong River (Laos) – the world's largest inland fishery – is expected to block the migration route of between 23 and 100 fish species. The dam's proponents argue that a state-of-the-art fish ladder will allow migrating fish to safely pass through the dam, while many scientists claim that this will not be the case. For the migratory fishes of the Mekong River, the nearly 200 mainstem and tributary dams that are installed, under construction, or planned will undoubtedly have profound ecological effects. Solutions beyond fish passage facilities must be considered.

4.5 CONCLUSION

The chapter's opening statement 'Dammed if you do, damned if you don't' reflects the clear dilemma associated with dams. Water is a valuable commodity as well as a destructive force. Human society requires water for life while seeking protection from floods and droughts. As a consequence, the majority of medium and large rivers of the world have been heavily modified to enable water managers to control flows to meet human needs (e.g. industry, agriculture, development of historical floodplains) while dampening or eliminating their original flow variability upon which the persistence of freshwater fishes depends. Here, we have explored the challenges and opportunities for freshwater fish conservation in response to threats posed by dams and diversions. Current efforts to restore and conserve fish biodiversity in regulated rivers include mimicking the natural flow and thermal regimes below dams, installing fish passage facilities to allow fish to circumvent dams, and in some cases even restoring and reconnecting rivers by removing dams. However, these remedies are often costly or difficult to implement, and in some cases may not be possible or warranted.

Our ability to meet future conservation challenges will benefit from what I call the Janus approach. In Roman mythology, Janus was the god of gates, doors, doorways, beginnings and endings, and is most often depicted as having two faces or heads, facing in opposite directions; one viewing what is behind and the other looking towards what lies ahead. In modern culture, his most prominent namesakes are the month of January, which begins the New Year, and the janitor who is a caretaker of doors and halls. To me, Janus reminds us that we must always be aware of where we have been by using knowledge from the past, while at the same time being conscious of the future in which humans will continue to depend on rivers for both water and energy security as well as the numerous goods and services provided by healthy fish communities.

Achieving a delicate balance between the practice of backward reflection and forward-looking is critical for ecologists to advance the conservation of freshwater fishes in the third millennium.

4.6 DISCUSSION QUESTIONS

1. What benefits do dams provide humans?
2. What are the major ecological impacts of dams?
3. Describe three primary conservation strategies for freshwater fishes in response to dams.
4. Explain how environmental flows have been used to conserve fishes in dam-affected waters.
5. Describe the strengths and weaknesses of fish passage facilities.

4.7 ACKNOWLEDGEMENTS

I thank Keith Gido (Kansas State University) and Paul Humphries (Charles Sturt University) for thoughtful comments that improved the chapter, and Zachary Hurwitz (International Rivers) for providing Figure 4.2. Support was provided by an H. Mason Keeler Endowed Professorship from the School of Aquatic and Fishery Sciences (University of Washington) and a Short Term Visiting Research Fellowship from the Australian Rivers Institute (Griffith University).

4.8 REFERENCES

Agostinho, A. A., Gomes L. C., Fernandez, D. R. & Suzuki, H. I. (2002). Efficiency of fish ladders for neotropical ichthyofauna. *River Research and Applications*, 18, 299–306.

Agostinho, A. A., Gomes, L. C., Verissimo, S. & Okada, E. K. (2004). Flood regime, dam regulation and fish in the Upper Paraná River: effects on assemblage attributes, reproduction and recruitment. *Reviews in Fish Biology and Fisheries*, 14, 11–19.

Agostinho, C. S., Pereira, C. R., de Oliveira, R. J., Freitas, I. S. & Marques, E. E. (2007). Movements through a fish ladder: temporal patterns and motivations to move upstream. *Neotropical Ichthyology*, 5, 161–167.

Alò, D. & Turner, T. F. (2005). Effects of habitat fragmentation on effective population size in the endangered Rio Grande silvery minnow. *Conservation Biology*, 19, 1138–1148.

Anderson, E. P., Freeman, M. C. & Pringle, C. M. (2006). Ecological consequences of hydropower development in Central America: impacts of small dams and water diversion on neotropical stream fish assemblages. *River Research and Applications*, 22, 397–411.

Angilletta, M. J., Steel, E. A., Bartz, K. K., *et al.* (2008). Big dams and salmon evolution: changes in thermal regimes and their potential evolutionary consequences. *Evolutionary Applications* 1, 286–299.

Arthington, A. H. (2012). *Environmental Flows: Saving Rivers in the Third Millennium*. Berkeley, CA: University of California Press.

Arthington, A. H., Bunn, S. E., Poff, N. L. & Naiman, R. J. (2006). The challenge of providing environmental flow rules to sustain river ecosystems. *Ecological Applications*, 16, 1311–1318.

Bain, M. B., Finn, J. T. & and Booke, H. E. (1988). Streamflow regulation and fish community structure. *Ecology*, 69, 382–392.

Baumgartner, L. J., Reynoldson, N. & Gilligan, D. M. (2006). Mortality of larval Murray cod (*Maccullochella peelii peelii*) and golden perch (*Macquaria ambigua*) associated with passage through two types of low-head weirs. *Marine and Freshwater Research*, 57, 187–191.

Bednarek, A. T. (2001). Undamming rivers: a review of the ecological impacts of dam removal. *Environmental Management*, 27, 803–814.

Bernhardt, E. S., Palmer, M. A., Allan, J. D., *et al.* (2005). Synthesizing U.S. river restoration efforts. *Science*, 308, 636–637.

Bestgen, K. R. & Platania, S. P. (1991). Status and conservation of the Rio Grande silvery minnow, *Hybognathus amarus*. *Southwestern Naturalist*, 36, 225–232.

Billington, D. P., Jackson, D. C. & Melosi, M. V. (2005). *The History of Large Federal Dams: Planning, Design, and Construction in the Era of Big Dams*. Denver, CO: US Department of the Interior.

Bovee, K. D. (1982). A guide to stream habitat analysis using the Instream Flow Incremental Methodology. Instream Flow Information Paper 12. U.S.D.I. Fish and Wildlife Service, Office of Biological Services. FWS/OBS-82/26. 248 pp.

Brandt, S. A. (2000). Classification of geomorphological effects downstream of dams. *Catena*, 40, 375–401.

Brown, J. J., Limburg, K. E., Waldman, J. R., *et al.* (2013). Fish and hydropower on the US Atlantic coast: failed fisheries policies from half-way technologies. *Conservation Letters*, 6, 280–286.

Bunn, S. E. & Arthington, A. H. (2002). Basic principles and ecological consequences of altered flow regimes for aquatic biodiversity. *Environmental Management*, 30, 492–507.

Bunt, C. M., Castro-Santos, T. & Haro, A. (2012). Performance of fish passage structures at upstream barriers to migration. *River Research and Applications*, 28, 457–478.

Carlisle, D. M., Wolock, D. M. & Meador, M. R. (2011). Alteration of streamflow magnitudes and potential ecological consequences: a multiregional assessment. *Frontiers in Ecology and the Environment*, 9, 264–270.

Castleberry, D. T., Cech, J. J. Jr., Erman, D. C., *et al.* (1996). Uncertainty and instream flow standards. *Fisheries*, 20, 21–22.

Castro-Santos, T. & Letcher, B. H. (2010). Modeling migratory energetics of Connecticut River American shad (*Alosa sapidissima*): implications for the conservation of an iteroparous anadromous fish. *Canadian Journal of Fisheries and Aquatic Sciences*, 67, 806–830.

Clarkson, R. W. & Childs, M. R. (2000). Temperature effects of hypolimnial-release dams on early life stages of Colorado River Basin big-river fishes. *Copeia*, 402–412.

Connor, E. J. & Pflug, D. E. (2004). Changes in the distribution and density of pink, chum, and Chinook salmon spawning in the upper Skagit River in response to flow management measures. *North American Journal of Fisheries Management*, 24, 835–852.

Cushman, R. M. (1985). Review of ecological effects of rapidly varying flows downstream of hydroelectric facilities. *North American Journal of Fisheries Management*, 5, 330–339.

Decker, A. S., Bradford, M. J. & Higgins, P. S. (2008). Rate of biotic colonization following flow restoration below a diversion dam in the Bridge River, British Columbia. *River Research and Applications*, 24, 876–883.

Doyle, M. W. & Havlick, D. G. (2009). Infrastructure and the environment. *Annual Review of Environment and Resources*, 34, 349–373.

Doyle, M. W., Harbor, J. M. & Stanley, E. H. (2005). Toward policies and decision-making for dam removal. *Environmental Management*, 31, 453–465.

Dudgeon, D. (2000). Large-scale hydrological changes in tropical Asia: prospects for riverine biodiversity. *Bioscience*, 50, 793–806.

Dudley, R. K. & Platania, S.P. (2007). Flow regulation and fragmentation imperil pelagic-spawning riverine fishes. *Ecological Applications*, 17, 2074–2086.

Ellis, L. E. & Jones, N. E. (2013). Longitudinal trends in regulated rivers: a review and synthesis within the context of the serial discontinuity concept. *Environmental Reviews*, 21, 136–148.

Falke, J. A. & Gido, K. B. (2006). Spatial effects of reservoirs on fish assemblages in Great Plains streams in Kansas, USA. *River Research and Applications*, 22, 55–68.

Fausch, K. D., Rieman, B. E., Dunham, J. B., Young, M. K. & Peterson, D. P. (2009). Invasion versus isolation: trade-offs in managing native salmonids with barriers to upstream movement. *Conservation Biology*, 23, 859–870.

Ferreira, J., Aragão, L. E. O. C., Barlow, J., *et al.* (2014). Brazil's environmental leadership at risk. *Science*, 346, 706–707.

Finer, M. & Jenkins, C. N. (2012). Proliferation of hydroelectric dams in the Andean Amazon and implications for Andes–Amazon connectivity. *Plos ONE* 7(4), e35126.

Frissell, C. A., Liss, W. J., Warren, C. E. & Hurley, M. C. (1986). A hierarchical framework for stream habitat classification: viewing streams in a watershed context. *Environmental Management*, 10, 199–214.

Fukushima, M., Kameyama, S., Kaneko, M., Nakao, K. & Steel, E. A. (2007). Modelling the effects of dams on freshwater fish distributions in Hokkaido, Japan. *Freshwater Biology*, 52, 1511–1524.

Gehrke, P. C. & Harris, J. H. (2001). Regional-scale effects of flow regulation on lowland riverine fish communities in New South Wales, Australia. *River Research and Applications*, 17, 369–391.

Gehrke, P. C., Gilligan, D. M. & Barwick, M. (2002). Changes in fish communities of the Shoalhaven River 20 years after construction of Tallowa Dam, Australia. *River Research and Applications*, 18, 265–286.

Gido, K. B., Schaefer, J. F. & Pigg, J. (2004). Patterns of fish invasions in the Great Plains of North America. *Biological Conservation*, 118, 121–131.

Gleick, P. H. (2000). The changing water paradigm – a look at twenty-first century water resources development. *Water International*, 25, 127–138.

Gleick, P. H. (2003). Water use. *Annual Review of Environmental Resources*, 28, 275–314.

Graf, W. (1999). Dam nation: a geographic census of American dams and their large-scale hydrologic impacts. *Water Resources Research*, 35, 1305–1311.

Hanasaki, N., Kanae, S. & Oki, T. (2006). A reservoir operation scheme for global river routing models. *Journal of Hydrology*, 327, 22–41.

Hart, D. D., Johnson, T. E., Bushaw-Newton, K. L., *et al.* (2002). Dam removal: challenges and opportunities for ecological research and river restoration. *Bioscience*, 52, 669–681.

Havel, J. E., Lee, C. E. & Vander Zanden, M. J. (2005). Do reservoirs facilitate invasions into landscapes? *Bioscience*, 55, 518–525.

Haxton, T. J. & Findlay, C. S. (2008). Meta-analysis of the impacts of water management on aquatic communities. *Canadian Journal of Fisheries and Aquatic Sciences*, 65, 437–447.

Helfman, G. S. (2007). *Fish Conservation*. Washington, DC: Island Press.

Herbert, M. E. & Gelwick, F. P. (2003). Spatial variation of headwater fish assemblages explained by hydrologic variability and upstream effects of impoundment. *Copeia*, 2003, 273–284.

Horwitz, R. J. (1978). Temporal variability patterns and the distributional patterns of stream fishes. *Ecological Monographs*, 48, 307–321.

Humphries, P., King, A. J. & Koehn, J. D. (1999). Fish, flows and floodplains: Links between freshwater fishes and their environment in the Murray–Darling River system, Australia. *Environmental Biology of Fishes*, 56, 129–151.

Iversen, T. M., Kronvang, B., Madsen, B. L., Markmann, P. & Nielsen, M. B. (1993). Re-establishment of Danish streams: restoration and maintenance measures. *Aquatic Conservation: Marine and Freshwater Ecosystems*, 3, 73–92.

Jackson, D. A., Peres-Neto, P. R. & Olden, J. D. (2001). What controls who is where in freshwater fish communities: the roles of biotic, abiotic and spatial factors? *Canadian Journal of Fisheries and Aquatic Science*, 58, 157–170.

Jackson, H. M., Gibbins, C. N. & Soulsby, C. (2007). Role of discharge and temperature variation in determining invertebrate community structure in a regulated river. *River Research and Applications*, 23, 651–669.

Jellyman, P. G. & Harding, J. S. (2012). The role of dams in altering freshwater fish communities in New Zealand. *New Zealand Journal of Marine and Freshwater Research*, 46, 475–489.

Johnson, P. T. J., Olden, J. D. & vander Zanden, M. J. (2008). Dam invaders: impoundments facilitate biological invasions into freshwaters. *Frontiers in Ecology and the Environment*, 6, 359–365.

Joy, M. K. & Death, R. G. (2001). Control of fish and crayfish community structure in Taranaki, New Zealand: dams, diadromy or habitat structure. *Freshwater Biology*, 46, 417–429.

Junk, W. J., Bayley, P. B. & Sparks, R. E. (1989). The flood pulse concept in river-floodplain systems. In: *Proceedings of the International Large River Symposium* Dodge, D. P. (Ed.) Canadian Special Publications in Fisheries and Aquatic Sciences. Ottawa: Canadian Government Publishing Centre/Canada Communication Group, pp. 110–27.

Kanehl, P. D., Lyons, J. & Nelson, J. E. (1997). Changes in the habitat and fish community of the Milwaukee River, Wisconsin, following removal of the Woolen Mills Dam. *North American Journal of Fisheries Management*, 17, 387–400.

Kennard, M. J., Olden, J. D., Arthington, A. H., Pusey, B. J. & Poff, N. L. (2007). Multiscale effects of flow regime and habitat and their interaction on fish assemblage structure in eastern Australia. *Canadian Journal of Fisheries and Aquatic Sciences*, 64, 1346–1359.

Kiffney, P. M., Pess, G. R., Anderson, J. H., *et al.* (2009). Changes in fish communities following recolonization of the Cedar River, WA, USA by Pacific salmon after 103 years of local extirpation. *River Research and Applications*, 25, 438–452.

King, A. J., Ward, K. A., O'Connor, P., *et al.* (2010). Adaptive management of an environmental water event to enhance native fish spawning and recruitment. *Freshwater Biology*, 55, 17–31.

King, J., Cambray, J. A. & Impson, N. D. (1998). Linked effects of dam-released floods and water temperature on spawning of the Clanwilliam yellowfish *Barbus capensis*. *Hydrobiologia*, 384, 245–265.

Kinsolving, A. D. & Bain, M. B. (1993). Fish assemblage recovery along a riverine disturbance gradient. *Ecological Applications*, 3, 531–544.

Konrad, C. P., Olden, J. D., Lytle, D. A., *et al.* (2011). Large-scale flow experiments for managing river systems. *Bioscience*, 61, 948–959.

Korman, J., Kaplinski, M. & Melis, T. S. (2011). Effects of fluctuating flows and a controlled flood on incubation success and early survival rates and growth of age-0 rainbow trout in a large regulated river. *Transactions of the American Fisheries Society*, 140, 487–505.

Krause, C. W., Newcomb, T. J. & Orth, D. J. (2005). Thermal habitat assessment of alternative flow scenarios in a tailwater fishery. *River Research and Applications*, 21, 581–593.

Kruk, A. & Penczak, T. (2003). Impoundment impact on populations of facultative riverine fish. *International Journal of Limnology*, 39, 197–210.

Lamouroux, N., Olivier, J.-M., Capra, H., *et al.* (2006). Fish community changes after minimum flow increase: testing quantitative predictions in the Rhone River at Pierre-Benite, France. *Freshwater Biology*, 51, 1730–1743.

Larinier, M. & Travade, F. (2002). Design of fishways for shad. *Bulletin Français de Pêche et Pisciculture*, 364 (Suppl.), 135–146.

Lehner, B., Liermann, C. R., Revenga, C., *et al.* (2011). High-resolution mapping of the world's reservoirs and dams for sustainable river-flow management. *Frontiers in Ecology and the Environment*, 9, 494–502.

Lessard, J. L. & Hayes, D. B. (2003). Effects of elevated water temperature on fish and macroinverebrate communities below small dams. *River Research and Applications*, 19, 721–732.

Ligon, F. K., Dietrich, W. E. & Trush, W. J. (1995). Downstream ecological effects of dams. *Bioscience*, 45, 183–192.

Limburg, K. E. & Waldman, J. R. (2009). Dramatic declines in north Atlantic diadromous fishes. *Bioscience*, 59, 955–965.

Lytle, D. A. & Poff, N. L. (2004). Adaptation to natural flow regimes. *Trends in Ecology & Evolution*, 19, 94–100.

Mallen-Cooper, M. & Brand, D. A. (2007). Non-salmonids in a salmonid fishway: what do 50 years of data tell us about past and future fish passage? *Fisheries Management and Ecology*, 14, 319–332.

Marks, J. C., Haden, G. A., O'Neill, M. & Pace, C. (2010). Effects of flow restoration and exotic species removal on recovery of native fish: lessons from a dam decommissioning. *Restoration Ecology*, 18, 934–943.

Matthews, W. J. & Marsh-Mathews, E. (2007). Extirpation of red shiner in direct tributaries of lake texoma (Oklahoma–Texas): a cautionary case history from a fragmented river-reservoir system. *Transactions of the American Fisheries Society*, 136, 1041–1062.

McDonald, M. (1887). The River Fisheries of the Atlantic States: The rivers of eastern Florida, Georgia, and South Carolina. In: *Report of the U.S. Commission of Fish & Fisheries*. Mis. Doc. 124, pp. 613–625.

McDonald, R. I., Olden, J. D., Opperman, J. J., *et al.* (2012). Energy, water and fish: biodiversity impacts of energy-sector water demand in the United States depend on efficiency and policy measures. *PLoS ONE* 7(11), e50219.

McLaughlin, R. L., Porto, L., Noakes, D. L. G., *et al.* (2006). Effects of low-head barriers on stream fishes: taxonomic affiliations and morphological correlates of sensitive species. *Canadian Journal of Fisheries and Aquatic Sciences*, 63, 766–779.

Meador, M. R. & Carlisle, D. M. (2012). Relations between altered streamflow variability and fish assemblages in eastern USA streams. *River Research and Applications*, 28, 1359–1368.

Mims, M. C. & Olden, J. D. (2012). Life history theory predicts fish assemblage response to hydrologic regimes. *Ecology*, 93, 35–45.

Mims, M. C. & Olden, J. D. (2013). Fish assemblages respond to altered flow regimes via ecological filtering of life history strategies. *Freshwater Biology*, 58, 50–62.

Morita, K. & Yamamoto, S. (2002). Effects of habitat fragmentation by damming on the persistence of stream-dwelling charr populations. *Conservation Biology*, 16, 1318–1323.

Murchie, K. J., Hair, K. P. E., Pullen, C. E., *et al.* (2008). Fish response to modified flow regimes in regulated rivers: research methods, effects and opportunities. *River Research and Applications*, 24, 197–217.

Naiman, R. J., Bunn, S. E., Nilsson, C., *et al.* (2002). Legitimizing fluvial ecosystems as users of water: an overview. *Environmental Management*, 30, 455–467.

Naiman, R. J., Latterell, J. J., Pettit, N. E. & Olden, J. D. (2008). Flow variability and the biophysical vitality of river systems. *Comptes Rendus Geoscience*, 340, 629–643.

Nilsson, C. & Renöfält, B. M. (2008). Linking flow regime and water quality in rivers: a challenge to adaptive catchment management. *Ecology and Society*, 13(2), 18.

Nilsson, C., Reidy, C. A., Dynesius, M. & Revenga, C. (2005). Fragmentation and flow regulation of the world's large river systems. *Science*, 308, 405–408.

Nogueira, C., Buckup, P. A., Menezes, N. A., *et al.* (2010). Restricted-range fishes and the conservation of Brazilian freshwaters. *PLoS ONE* 5(6), e11390.

Novinger, D. C. & Rahel, F. J. (2003). Isolation management with artificial barriers as a conservation strategy for cutthroat trout in headwater streams. *Conservation Biology*, 17, 772–781.

O'Brien, J. S. (1987). A case study of minimum streamflow for fishery habitat in the Yampa River. In: *Sediment Transport in Gravel-bed Rivers*. Thorne, C. R., Bathurst, J. C. & Hey, R. D. (Eds). Chichester: John Wiley & Sons, pp. 921–946.

Oldani, N. O. & Baigun, C. R. M. (2002). Performance of a fishway system in a major South American dam on the Parana River (Argentina–Paraguay). *River Research and Applications*, 18, 171–183.

Olden, J. D. & Naiman, R. J. (2010). Incorporating thermal regimes into environmental flows assessments: modifying dam operations to restore freshwater ecosystem integrity. *Freshwater Biology*, 55, 86–107.

Olden, J. D., Poff, N. L. & Bestgen, K. R. (2006). Life-history strategies predict fish invasions and extirpations in the Colorado River Basin. *Ecological Monographs*, 76, 25–40.

Olden, J. D., Kennard, M. J., Leprieur, F., *et al.* (2010). Conservation biogeography of freshwater fishes: recent progress and future challenges. *Diversity and Distributions*, 16, 496–513.

Olden, J. D., Konrad, C. P., Melis, T. S., *et al.* (2014). Are large-scale flow experiments informing the science and management of freshwater ecosystems? *Frontiers in Ecology and the Environment*, 12, 176–185.

Palmer, M. A., Reidy, C., Nilsson, C., *et al.* (2008). Climate change and world's river basins: anticipating management options. *Frontiers in Ecology and the Environment*, 6, 81–89.

Pandit, M. K. & Grumbine, R. E. (2012). Potential effects of ongoing and proposed hydropower development on terrestrial biological diversity in the Indian Himalaya. *Conservation Biology*, 26, 1061–1071.

Pelicice, F. M. & Agostinho, A. A. (2008). Fish-passage facilities as ecological traps in large neotropical rivers. *Conservation Biology*, 22, 180–188.

Petts, G. E. (1986). Water quality characteristics of regulated rivers. *Progress in Physical Geography*, 10, 492–516.

Poff, N. L. & Allan, J. D. (1995). Functional organization of stream fish assemblages in relation to hydrological variability. *Ecology*, 76, 606–627.

Poff, N. L. & Zimmerman, J. K. H. (2010). Ecological responses to altered flow regimes: a literature review to inform the science and management of environmental flows. *Freshwater Biology*, 55, 194–205.

Poff, N. L., Allan, J. D., Bain, M. B., *et al.* (1997). The natural flow regime: a paradigm for river conservation and restoration. *Bioscience*, 47, 769–784.

Poff, N. L., Allan, J. D., Palmer, M. A., *et al.* (2003). River flows and water wars? Emerging science for environmental decision-making. *Frontiers in Ecology and the Environment*, 1, 298–306.

Poff, N. L., Olden, J. D., Merritt, D. M. & Pepin, D. M. (2007). Homogenization of regional river dynamics by dams and global biodiversity implications. *Proceedings of the National Academy of Sciences of the United States of America*, 104, 5732–5737.

Pompeu, P. S., Agostinho, A. A. & Pelicice, F. M. (2012). Existing and future challenges: the concept of successful fish passage in South America. *River Research and Applications*, 28, 504–512.

Ponton, D. & Copp, G. H. (1997). Early dry-season community structure and habitat use of young fish in tributaries of the River Sinnamary (French Guiana, South America) before and after hydrodam operation. *Environmental Biology of Fishes*, 50, 235–256.

Poole, G. C. & Berman, C. H. (2001). An ecological perspective on in-stream temperature: natural heat dynamics and mechanisms of human-caused thermal degradation. *Environmental Management*, 27, 787–802.

Preece, R. M. (2004). *Cold Water Pollution Below Dams in New South Wales: A Desktop Assessment*. Sydney, NSW: NSW Department of Infrastructure, Planning and Natural Resources.

Preece, R. M. & Jones, H. A. (2002). The effect of Keepit Dam on the temperature regime of the Namoi River, Australia. *River Research and Applications*, 18, 397–414.

Pringle, C. M., Freeman, M. C. & Freeman, B. J. (2000). Regional effects of hydrologic alterations on riverine macrobiota in the new world: tropical–temperate comparisons. *Bioscience*, 50, 807–823.

Propst, D. L. & Gido, K. B. (2004). Responses of native and nonnative fishes to natural flow regime mimicry in the San Juan River. *Transactions of the American Fisheries Society*, 133, 922–931.

Rahel, F. J. (2013). Intentional fragmentation as a management strategy in aquatic systems. *Bioscience*, 63, 362–372.

Reidy Liermann, C., Nilsson, C., Robertson, J. & Ng, R. Y. (2012). Implications of dam obstruction for global freshwater fish diversity. *Bioscience*, 62, 539–548.

Reyes-Gavilan, F. G., Garrido, R., Nicieza, A. G. & Brana, F. (1996). Fish community variation along physical gradients in short streams of northern Spain and the disruptive effect of dams. *Hydrobiologia*, 321, 155–163.

Reyjol, Y., Lim, P., Dauba, F., Baran, P. & Belaud, A. (2001). Role of temperature and flow regulation on the Salmoniform–Cypriniform transition. *Archiv fur Hydrobiologie*, 152, 567–582.

Richter, B. D., Matthews, R. A., Harrison, D. L. & Wigington, R. (2003). Ecologically sustainable water management: managing river flows for river integrity. *Ecological Applications*, 13, 206–224.

Roberts, J. H., Angermeier, P. L. & Hallerman, E. M. (2013). Distance, dams and drift: what structures populations of an endangered, benthic stream fish? *Freshwater Biology*, 58, 2050–2064.

Rolls, R. J., Leigh, C. & Sheldon, F. (2012). Mechanistic effects of low-flow hydrology on riverine eco-systems: ecological principles and consequences of alteration. *Freshwater Science*, 31, 1163–1186.

Rolls, R. J., Growns, I. O., Khan, T. A., *et al.* (2013). Fish recruitment in rivers with modified discharge depends on the interacting effects of flow and thermal regimes. *Freshwater Biology*, 58, 1804–1819.

Roscoe, D. W. & Hinch, S. G. (2010). Effectiveness monitoring of fish passage facilities: historical trends, geographic patterns and future directions. *Fish and Fisheries*, 11, 12–33.

Schindler, D. E., Scheuerell, M. D., Moore, J. W., *et al.* (2003). Pacific salmon and the ecology of coastal ecosystems. *Frontiers in Ecology and the Environment*, 1, 31–7.

Schlosser, I. J. (1982). Trophic structure, reproductive success, and growth rates of fishes in natural and modified streams. *Canadian Journal of Fisheries and Aquatic Sciences*, 39, 968–978.

Schlosser, I. J. (1991). Stream fish ecology: a landscape perspective. *Bioscience*, 41, 704–712.

Schramm, H. L. & Eggleton, M. A. (2006). Applicability of the flood-pulse concept in a temperate floodplain river ecosystem: thermal and temporal components. *River Research and Applications*, 22, 543–553.

Sherman, B. (2000). *Scoping Options for Nitigating Cold Water Discharges from Dams*. CSIRO Land and Water.

Sherman, B., Todd, C. R., Koehn, J. D. & Ryan, T. (2007). Modelling the impact and potential mitigation of cold water pollution on Murray cod populations downstream of Hume Dam, Australia. *River Research and Applications*, 23, 377–389.

Smith, N. A. F. (1971). *A History of Dams*. London: Citadel Press

Souchon, Y., Sabaton, C., Deibel, R., *et al.* (2008). Detecting biological responses to flow management: missed opportunities and future directions. *River Research and Applications*, 24, 506–518.

Stanford, J. A., Ward, J. V., Liss, W. J., *et al.* (1996). A general protocol for restoration of regulated rivers. *Regulated Rivers: Research & Management*, 12, 391–414.

Stanley, E. H. & Doyle, M. W. (2003). Trading off: the ecological removal effects of dam removal. *Frontiers in Ecology and the Environment*, 1, 15–22.

Stanley, E. H., Catalano, M. J., Mercado-Silva, N. & Orr, C. H. (2007). Effects of dam removal on brook trout in a Wisconsin stream. *River Research and Applications*, 23, 792–798.

Syvitski, J. P. M., Vorosmarty, C. J., Kettner, A. J. & Green, P. (2005). Impact of humans on the flux of terrestrial sediment to the global coastal ocean. *Science*, 308, 376–380.

Tennant, D. L. (1976). Instream flow regimens for fish, wildlife, recreation and related environmental resources. *Fisheries*, 1, 6–10.

Tharme, R. E. (2003). A global perspective on environmental flow assessment: emerging trends in the development and application of environmental flow methodologies for rivers. *River Research and Applications*, 19, 397–441.

Townsend, C. R. & Hildrew, A. G. (1994). Species traits in relation to a habitat templet for river systems. *Freshwater Biology*, 31, 265–275.

Travnichek, V. H., Bain, M. B. & Maceina, M. J. (1995). Recovery of a warmwater fish assemblage after initiation of a minimum-flow release downstream of a hydroelectric dam. *Transactions of the American Fisheries Society*, 124, 836–844.

Vörösmarty, C. J., McIntyre, P. B., Gessner, M. O., *et al.* (2010). Global threats to human water security and river biodiversity. *Nature*, 467, 555–561.

Ward, J. V. & Stanford, J. A. (1983). The serial discontinuity concept of lotic ecosystems. In *Dynamics of Lotic Ecosystems*. Fontaine, T. D. & Bartell, S. M. (Eds). Ann Arbor, MI: Ann Arbor Scientific Publishers, pp. 29–42.

Ward, J. V. & Stanford, J. A. (1995). The serial discontinuity concept: extending the model to floodplain rivers. *Regulated Rivers – Research & Management*, 10, 159–168.

Waters, T. F. (1995). *Sediment in Streams – Sources, Biological Effects, and Control*. Bethesda, MD: American Fisheries Society Monograph.

Watson, H. L. (1996). The common rights of mankind: subsistence, shad, and commerce in the early Republican South. *Journal of American History*, 83, 13–43.

[WCD] World Commission on Dams. (2000). Dams and development. A new frame for decision-making. In: *Report of the World Commission on Dams*, London: WCD.

Weisberg, S. B. & Burton, W. H. (1993). Enhancement of fish feeding and growth after an increase in minimum flow below the Conowingo Dam. *North American Journal of Fisheries Management*, 13, 103–109.

Wik, S. J. (1995). Reservoir drawdown: case study in flow changes to potentially improve fisheries. *Journal of Energy Engineering*, 121, 89–96.

Winston, M. R., Taylor, C. M. & Pigg, J. (1991). Upstream extirpation of four minnow species due to damming of a prairie stream. *Transactions of the American Fisheries Society*, 120, 98–105.

Winter, B. D. (1990). A brief review of dam removal efforts in Washington, Oregon, Idaho, and California. In: *US Department of Commerce, NOAA Tech. Memo. NMFS F/NWR-28*, p. 13.

Wood, P. J. & Armitage, P. D. (1997). Biological effects of fine sediment in the lotic environment. *Environmental Management*, 21, 203–217.

Yamamoto, S., Morita, K., Koizumi, I. & Maekawa, K. (2004). Genetic differentiation of white-spotted charr (*Salvelinus leucomaenis*) populations after habitat fragmentation: dpatial–temporal changes in gene frequencies. *Conservation Genetics*, 5, 529–538.

Zhong, Y. G. & Power, G. (1996). Environmental impacts of hydroelectric projects on fish resources in China. *Regulated Rivers – Research & Management*, 12, 81–98.

Ziv, G., Baran, E., Nam, S., Rodríguez-Iturbe, I. & Levin, S. A. (2012). Trading-off fish biodiversity, food security, and hydropower in the Mekong River Basin. *Proceedings of the National Academy of Sciences*, 109, 5609–5614.

Chemical pollution

NATHANIEL L. SCHOLZ
AND JENIFER K. MCINTYRE

5.1 INTRODUCTION

Chemical forms of water pollution are a major cause of freshwater habitat degradation worldwide. There are many sources of toxic contaminants, and these reflect past and present human activities and land uses. Toxics can have adverse health impacts on all components of aquatic ecosystems, including threatened fish species and the biological communities they rely on, particularly for food. Toxics can also interact in complex ways with other non-chemical habitat stressors such as water temperature, disease vectors and non-native species (Chapter 2). Therefore, chemical pollution poses important challenges for the conservation of freshwater fish and their habitats. The pollution problem scales roughly in proportion to the global human population, and is therefore expected to grow in significance throughout many parts of the world in the first half of the twenty-first century.

This chapter provides an introduction to freshwater pollution science, with an emphasis on current and emerging threats to vulnerable fish populations. There are now more than 80,000 individual chemicals in societal use, derived from commercial product manufacturing, drug development, pest control practices and many other processes that underpin modern economies. A large fraction of these chemicals eventually ends up in aquatic habitats via direct discharges, land-based run-off and atmospheric deposition. An overview of water quality threats on a chemical-by-chemical basis is impracticable. Rather, we will focus on central themes in freshwater ecotoxicology and common challenges for the conservation and recovery of threatened fish. Additional important

Conservation of Freshwater Fishes, eds G. P. Closs, M. Krkosek and J. D. Olden. Published by Cambridge University Press. © Cambridge University Press 2016.

categories of water pollution are beyond the scope of this chapter. For more information on these other topic areas, the reader is referred to reviews on nutrients and sediments (Bouwman et al., 2013), acid rain (Schindler, 1981), microbial pathogens (Ferguson et al., 2003) and natural toxins produced by biological organisms (e.g. microcystins from cyanobacteria; Landsberg, 2002).

Several core concepts in aquatic ecotoxicology are provided in Table 5.1. These terms are generally used to draw distinctions between contaminants that: (1) are historical use (legacy) vs. modern use, (2) are released from focal vs. diffuse sources, (3) are short-lived vs. persistent in the environment, (4) are metabolised and eliminated by organisms vs. passed through food webs to higher trophic levels, (5) impact fish directly vs. indirectly via loss of prey, (6) are acutely lethal (i.e. cause fish kills) vs. causing more nuanced sublethal toxicity, and (7) have effects on fish at the individual vs. the population scale.

There are many freshwater taxa in global decline. This notably includes amphibians, which are also susceptible to chemical contaminants, many of which interact with other stressors such as pathogens to cause mortality, reproductive failure and other losses to individual fitness (e.g. Rohr et al., 2008). The life histories of these and other aquatic species are generally limited in geographic range. By contrast, many fish species are highly mobile. Some endangered Pacific salmon, for example, traverse hundreds of kilometres of freshwater habitat in western North America on their migration paths to and from the ocean (Quinn, 2005). This degree of mobility makes the aggregate impacts of pollution at larger watershed and basin scales a particularly important consideration for fish.

Watershed-scale inputs of toxic chemicals to a representative river network are illustrated in Figure 5.1. These inputs are usually tied to specific human activities that correspond to different categories of land use (Allan, 2004). Common examples include hardrock mining, industrial discharges, run-off from agricultural lands, discharges from municipal wastewater treatment plants, legacy pollution from historically contaminated sites, urban run-off and accidental spills from natural and man-made disasters. These myriad sources of contaminant inputs influence habitat quality and ecological resiliency across scales, from localised toxic 'hot spots' at the river reach scale to large ecosystems such as the Great Lakes in North America.

Depending on life history, habitat use patterns, migration and lifetime dispersal, vulnerable fish species may be exposed to complex

Table 5.1 *Definitions of key terms used in Chapter 5.*

Term	Definition
Biota	Living organisms, e.g. plants, animals, bacteria
Pollution/contaminant	A chemical not naturally found in the environment or found at concentrations not normally present
Point source	Originating from identifiable, unique sources such as effluent from a sewage treatment plant or an industrial discharge pipe
Non-point source	Originating from dispersed sources such as agricultural lands, residential areas, public spaces, vehicles
Persistent contaminant	A chemical that degrades very slowly in the environment, e.g. PCBs
Persistent bioaccumulative toxicant (PBT)	Any contaminant taken up by biota faster than it is metabolised and discharged
Persistent organic pollutant (POP)	Organic contaminants that are persistent, toxic, undergo long-range transport and bioaccumulate, e.g. organochlorines such as PCBs and dioxins, organometalloids such as methylmercury or tributyltin
Legacy contaminant	Persistent contaminants from past industrial, commercial and/or agricultural practices, e.g. PCBs, DDTs
Half-life	Time required for a contaminant to decrease to half its initial concentration, e.g. the half-life of methylmercury in aquatic organisms is approximately 72 days
Trophic transfer	Contaminant transfer to higher trophic levels through dietary intake, e.g. transfer of PCBs to killer whales from salmon prey, usually resulting in bioaccumulation and biomagnification
Bioaccumulation	Accumulation of a contaminant in an individual organism due to faster uptake (ingestion with food or absorption from water) than elimination (metabolism and excretion)
Biomagnification	Accumulation of a contaminant at higher trophic levels in a food web due to transfer from prey to predator
Environmental fate	The destiny of a contaminant after release into the environment, e.g. partitioning into air, soil, water
Chemical metabolism/ biotransformation	Process whereby contaminants in an organism are chemically altered to facilitate excretion
Direct toxicity	Caused by a contaminant acting at a physiological site on or in an organism
Indirect toxicity	Caused by a change in the physical, chemical or biological environment, e.g. food web-mediated effects

Figure 5.1 Common sources of chemical pollution to aquatic habitats. Myriad human land-use activities are represented at the watershed scale, reflecting pollution inputs from point and non-point sources. Depending on life history and movement patterns, fish species may be exposed to one or more of these typical inputs.

mixtures of chemicals from one or more of the above sources. There are two primary pathways for toxicant uptake. The first is via the gills, wherein chemicals dissolved in the surrounding water cross into the bloodstream and are then transported to the brain, the gonads, the liver and other target organs. The second pathway is via ingestion of contaminated prey or, for some species, polluted sediments. Uptake across the skin can also be important, particularly for benthic fish that burrow or otherwise live in close proximity to toxic sediments. Elimination of toxics from the body primarily occurs via the kidney or liver. Highly water-soluble contaminants (e.g. elemental metals) are filtered out of the blood by the kidneys and excreted in urine. Less water-soluble contaminants (e.g. persistent organic pollutants, discussed below) are extracted from the blood by the liver, metabolised (or biotransformed) enzymatically to increase their solubility, and then excreted via the bile in faeces.

5.2 A LEGACY OF TOXICS IN FRESHWATER HABITATS

Persistent organic pollutants, including many that were a legacy of industrial activities during the mid-twentieth century, remain a conservation challenge to this day. These chemicals were widely used in society until it was discovered they were toxic to humans and wildlife, after which many were banned. They include the early organochlorine insecticides, the most widely known of which is dichlorodiphenyltrichloroethane, or DDT. The discovery of DDT's insecticidal properties led to a Nobel Prize in 1948, and use rates worldwide soared in the decades that followed, initially as part of public health campaigns to control mosquito-borne illnesses such as malaria, typhus and dengue fever. Concurrently, DDT and related organochlorines (e.g. chlordane, aldrin, dieldrin, endrin and mirex) saw increasing use as agricultural insecticides. However, growing concerns over environmental impacts culminated in the publication of Rachel Carson's landmark book *Silent Spring* in 1962, which led to a large public outcry and an eventual ban on DDT for agricultural use in the US in 1972. In 2001, the use of DDT in agriculture was banned worldwide by the Stockholm Convention on Persistent Organic Pollutants.

The persistent organic pollutant category of toxics also includes the polychlorinated biphenyls, or PCBs. These chemicals were incorporated widely into industrial coolant fluids for transformers, capacitors and other electrical equipment, with manufacturing peaking in the 1960s. They were eventually found to cause cancer and other adverse effects in humans, birds, fish and other wildlife, and were banned in the US in 1979. In the years since the US ban and similar restrictions in many other countries, environmental surveillance has shown that PCBs remain measurable in the tissues of fish around the world, with levels particularly elevated in freshwater systems near sites of PCB manufacture or disposal. The effects of PCBs on fish health are myriad. In addition to carcinogenesis, for example, they disrupt immune system function, thereby rendering exposed fish more susceptible to disease.

The persistent pollutants are so-named because they resist degradation in the environment, by processes that are either biotic (e.g. microbial breakdown) or abiotic (e.g. photolysis in response to sunlight). They often accumulate in sediments, and can be resuspended and redistributed in aquatic habitats by burrowing aquatic animals (bioturbation), severe storms and scour, and by dredging and similar forms of human disturbance. As a consequence of their persistence, levels in aquatic systems have been slow to decline, even over decadal time scales. In addition to dispersal in water and through eroded soils, persistent organic pollutants are spread globally by atmospheric transport. Consequently, these contaminants are present in fish habitats that are otherwise pristine.

At high northern latitudes, for example, persistent pollutants accumulated for many years in Arctic ice because of the cold condensation effect (Grannas *et al.*, 2013). With Arctic ice now receding as a consequence of ongoing climate change, these contaminants are increasingly remobilised into the air, water, and biota (Ma *et al.*, 2011).

Organochlorines and various other persistent organic pollutants are also bioaccumulative, a term for toxicants with physico-chemical properties that make them hydrophobic (repelled by water) and lipophilic (attracted to fat in living organisms). Bioaccumulative contaminants tend to biomagnify in aquatic food webs; that is, they become proportionately more concentrated at higher trophic levels (Table 5.1). As a consequence, they represent a greater exposure risk for animals that feed at or near the top of aquatic food webs, as indicated in Figure 5.2. The biomagnification of DDT (or its primary metabolite, DDE) caused the precipitous decline of fish-eating (piscivorous) bird populations in the 1960s and 1970s. Lipophilic contaminants are maternally transferred in yolk. Accumulation of DDTs in eggs resulted in eggshell thinning and nest failure among breeding pairs of raptors. In the US, these declines led to the listing of bald eagles and brown pelicans under the Endangered Species Act. The present-day recovery of both species is credited primarily to the ban on DDT and its subsequent decline in aquatic food webs. Despite a ban on agricultural applications by the Stockholm Convention, the use of DDT is still allowed for mosquito control in many equatorial countries where malaria and other vector-borne diseases remain a serious public health problem. For freshwater fish, persistent and bioaccumulative pollutants will continue to pose a health risk for many years to come, particularly among top predators.

Legacy pollution offers insights into how toxics can influence fish population dynamics at the scale of large ecosystems. A case example is the decline and extinction of lake trout (*Salvelinus namaycush*) in Lake Ontario in the mid-twentieth century, and in particular the role of the persistent and maternally transferred dioxin 2,3,7,8-tetra-chlorodibenzo-*p*-dioxin (TCDD). Lake trout are an extensively studied and economically important species. Great Lakes populations collapsed in the 1940s due to the twin pressures of overfishing and parasitism by the invasive sea lamprey. However, despite sea lamprey control and a ban on commercial harvest, fisheries managers were unable to rebuild lake trout populations in subsequent decades, a time during which there were significant inputs of dioxins into the lower Great Lakes from pulp mills and other industrial sources. Annual re-stocking efforts

Figure 5.2 Persistent organic pollutants are transported to aquatic ecosystems from industrial discharges, atmospheric deposition, surface run-off and the erosion or resuspension of historically contaminated sediments. Persistent chemicals resist environmental degradation and accumulate in living organisms. Biomagnification occurs as they are transferred up food webs, thereby concentrating at higher trophic levels. This is illustrated here as the movement of contaminants (circles) in a representative freshwater food web, from chironomid pupae to stoneflies, and to juvenile salmonids that are ultimately eaten by a top fish predator. Exposure risks for persistent pollutants are therefore greater for fish that feed at higher trophic levels.

with millions of juvenile trout yielded adult fish but no sustainable natural recruitment. Several decades later, it was discovered that lake trout embryos are very sensitive to the toxic effects of TCDD at very low exposure concentrations – i.e. in the low parts per trillion (Walker *et al.*, 1991). TCDD causes a range of developmental defects, including heart abnormalities and craniofacial deformities, and these sublethal effects ultimately lead to mortality at the fry stage. Lake trout are long-lived top predators, and bioaccumulated TCDD is maternally transferred from reproductively mature females to embryos (Walker *et al.*, 1994).

Environmental levels of TCDD from the 1940s through the 1980s, as predicted from sediment cores and estimated sediment–biota accumulation factors, were sufficient to explain the sustained reproductive failure and population-scale extirpation of the species (Cook *et al.*, 2003). As dioxin levels have fallen in more recent years, natural recruitment among lake trout populations has gradually improved.

Research on legacy pollution has also revealed how toxics can act as a selective pressure to drive the local adaptation of fish populations. A major hotspot for PCB contamination in the US is the Hudson River in New York State, following decades of discharges in the years before PCBs were banned in the late 1970s. As noted above, PCBs are toxic to fish, and this toxicity is primarily mediated by the aryl hydrocarbon receptor (AhR). Levels of PCBs (and related dioxins and furans) in the livers of Atlantic tomcod (*Microgadus tomcod*) collected from the Hudson River are among the highest for fish from anywhere in the world (Fernandez *et al.*, 2004). This population of tomcod has evolved a heritable resistance to the toxic effects of PCBs, as evidenced by a lack of early life-stage defects and corresponding induction of the protective metabolic enzyme cytochrome P4501A, the latter a classic biomarker for PCB exposure in fish and other animals (Yuan *et al.*, 2006). This resistance was recently attributed to sequence variation in the gene that codes for AhR2 (Wirgin *et al.*, 2011), the more functionally active of the two forms of AhR in fish. The resulting receptor is much less effective at binding PCBs and triggering the cascade of events that lead to toxicity from PCB exposure. The gene deletion was not present in tomcod from nearby rivers that were relatively cleaner, suggesting the Hudson River population has undergone rapid local evolution (Wirgin *et al.*, 2011) in response to a chemical habitat stress.

Persistent bioaccumulative contaminants continue to enter freshwater fish habitats. Present-day sources include the worldwide use of polybrominated diphenyl ethers (PBDEs) as flame retardants in textiles, electronics and other manufacturing applications (de Wit, 2002). Certain PBDEs are now listed under the Stockholm Convention, which will eliminate their production, use and trade by participating parties. Mercury, emitted primarily from modern coal-fired electrical power plants, is a widespread aquatic contaminant. Methylation by bacteria converts mercury into a highly toxic form that is readily bioaccumulated in aquatic biota. Biomagnification of methylmercury has raised human health concerns and corresponding fish consumption advisories, particularly for indigenous peoples and others with relatively higher rates of consumption of fish (Mozaffarian & Rimm, 2006). Lastly, legacy contaminants periodically become a new challenge for fish conservation when they are remobilised by modern human activities. Examples include the dredging of contaminated sediments (common near waterfronts with a history of heavy industry) and the removal of dams that releases sediments contaminated with DDT and other toxics (see Chapter 4).

5.3 EXPANDING FROM POINT SOURCES TO NON-POINT SOURCES OF POLLUTION

'Point source' and 'non-point source' are phrases in water quality science that distinguish between discharges from a discrete location and more spatially diffuse sources. Non-point sources of pollution are primarily in the form of land-based run-off, but can also include atmospheric inputs directly to water bodies. The line between point and non-point sources is indistinct at times; for example, many stormwater conveyance systems collect contaminants in surface run-off from urban landscapes (diffuse sources) and discharge these to a river or a lake at a discrete location, usually a pipe. Nevertheless, the distinction is meaningful from the standpoint of our evolving understanding of toxic exposures and corresponding consequences for the health of threatened fish species and their habitats.

Throughout most of the world, regulations to protect water quality were weak or non-existent until the latter half of the twentieth century. Lakes and rivers served as receiving waters, or conduits for dumping toxic waste. Hence the common phrase, 'the solution to pollution is dilution'. In developed countries, the Industrial Revolution had a major impact on freshwater and sediment quality. In the US, rivers became so polluted that several actually caught fire (Hartig, 2010). A notable example is the burning of the Cuyahoga River in Cleveland in 1969, an event so egregious it helped rally public support for the eventual passage of the Clean Water Act in 1972. Over the four decades that followed, the Clean Water Act proved effective at reducing end-of-pipe discharges to freshwater habitats. In the restored Cuyahoga River, for example, fish communities have transformed from a few pollution-tolerant species, such as the gizzard shad (*Dorosoma cepedianum*), to more diverse and abundant assemblage dominated by pollution-sensitive species (e.g. spotfin shiner, *Notropis spilopterus*) (Rahel, 2010). The gradual reversal of the industrial river syndrome as a positive response to point-source controls has played out similarly in waterways around the world, clearly demonstrating that pollution reduction can ensure and promote fresh-water fish biodiversity.

While end-of-pipe discharges remain an important source of tox-ics, particularly in developing countries that do not have a regulatory framework equivalent to the Clean Water Act in the US, the challenge for the twenty-first century is increasingly non-point source pollution. Chemical releases from point sources generally involve a single polluter, are relatively easy to profile using conventional analytical chemistry and are amenable to traditional single-species toxicity bioassays using model fish and invertebrate species. By contrast, non-point pollution at the watershed scale may involve many sources. Water quality monitoring over larger spatial scales and longer timelines can be labour-intensive and expensive. A watershed-scale perspective also necessitates a consideration of multiple native species interacting in dynamic communities. The ecological impacts of non-point source pollution must therefore be evaluated against a backdrop of non-chemical stressors. Lastly, fish kills are generally rare, so impacts on fish health are usually assessed in terms of sublethal effects at the suborganismal level – e.g. adverse changes in predator avoidance, disease resistance, reproductive success and migration. For these reasons, many of the traditional tools in aquatic toxicology are a mismatch for the scientific and resource management questions that are specific to non-point source pollution. This includes, for example, the 96-hour median lethal concentration (LC_{50}; the concentration killing 50% of a test population of model fish) for individual chemicals, a cornerstone metric for traditional risk assessment.

The sections that follow profile two broad and present-day categories of pollution threats to freshwater habitats, both non-point sources. The first is pesticide-containing run-off (or irrigation return flows) from agricultural lands, and the second is urban stormwater run-off. As noted earlier, this chapter is focused on toxic chemical contaminants. There are other major water quality categories beyond this scope that also fall under the heading of non-point source pollution, most notably nutrients and sediments.

5.4 AGRICULTURAL RUN-OFF

Agrochemicals have been used to kill or otherwise control undesired biological organisms for millennia – i.e. for as long as there have been crops to protect from pests. Ancient texts specify the use of sulphur and toxic plants for insect control, and in the Middle Ages arsenic and mercury were commonly applied. Natural products were subsequently developed as pesticides, most notably pyrethrum (pyrethrins) from chrysanthemums and nicotine sulphate from the leaves of tobacco plants. In the mid-twentieth century, arsenic-based and natural products gave way to synthetic pesticides, including the aforementioned DDT and related organochlorines. Many of the pesticides in modern use are synthetic derivatives of earlier natural products, including the pyrethroid and neonicotinoid insecticides that are chemically related to pyrethrin and nicotine, respectively. As an aside, agricultural uses of neonicotinoids (e.g. imidacloprid) have received considerable attention in recent years for their implication in the recent 'colony collapse disorder' among honeybees in North America and Europe. Because of the high risk to pollinators, and therefore the agricultural economy, they were banned by the European Union in 2013.

The term 'pesticide' encompasses a large diversity of chemicals that target specific taxa – i.e. insecticides, herbicides, algaecides, fungicides, rodenticides, avicides, molluscicides, nematicides, miticides and piscicides (e.g. rotenone, also a natural plant derivative). There are hundreds of different pesticides currently used worldwide, applied in more than a thousand different product formulations. These uses go far beyond agriculture, and include commercial and residential pest control, right-of-way maintenance, disease vector control, aquatic vegetation removal, forestry management and invasive species control. From the standpoint of threatened fish, the use of pesticides (primarily herbicides) to kill non-native species via in-water or riparian zone applications creates a trade-off between the known habitat benefits of chemical control and the often poorly understood potential for toxicity to fish and other so-called non-target species (Stehr et al., 2009).

This section will focus on agricultural pesticides, and current-use insecticides in particular, as these are generally more toxic to freshwater fish and their prey. The ongoing effort to recover threatened and endangered Pacific salmon populations in western North America serves as a case example of pesticide risks to imperilled fish species. Pacific salmonids are highly migratory, and include distinct population segments of Chinook (*Oncorhynchus tshawytscha*), coho (*O. kisutch*) and steelhead (*O. mykiss*), several of which are currently listed for protection under the United States Endangered Species Act (for species status and range, see National Marine Fisheries Service listings at www.nmfs.noaa.gov/pr/species/esa/fish.htm). Salmonids with stream-type life histories spend a year or longer foraging in freshwater habitats as juveniles before migrating to the ocean (Quinn, 2005). These rearing habitats span several large western United States river basins with extensive agricultural land uses, and complex mixtures of current-use pesticides have been detected in surface waters for many years (for a synthesis of the most comprehensive pesticide monitoring in United States surface waters, see the United States Geological Survey's National Water Quality Assessment Program at http://water.usgs.gov/nawqa/).

Exposures to current-use insecticides belonging to organophosphate (e.g. diazinon, malathion, chlorpyrifos) and carbamate (e.g. carbaryl, carbofuran) classes cause a variety of direct sublethal effects in salmon. These chemicals are toxic to the nervous system, and thus primarily disrupt normal behaviours, including feeding (Sandahl et al., 2005), predator avoidance (Scholz et al., 2000) and migration (Scholz et al., 2000). For juveniles, the depression of feeding behaviour and the corresponding reduction in ration, juvenile growth and size at ocean migration has been linked (via modelling) to an increase in size-dependent individual mortality, thereby reducing population-scale growth and abundance over time (Baldwin et al., 2009). Moreover, certain combinations of these insecticides interact in mixtures to cause synergistic (greater-than-additive) neurotoxicity in juvenile salmon (Laetz et al., 2009) at very low, environmentally relevant concentrations (Laetz et al., 2013). Juvenile salmon exposed to another common pesticide, copper sulphate, lose their ability to smell. As a consequence, they become unresponsive to olfactory predator cues and are captured at higher rates in encounters with larger fish predators (McIntyre et al., 2012).

By design, insecticides are highly toxic to terrestrial and aquatic insects, and to date there have been no chemicals synthesised that are simultaneously toxic to targeted insect pests and non-toxic to the non-target arthropod taxa that juvenile salmon and many other freshwater fishes rely on for food (reviewed by Macneale et al., 2010; Scholz et al., 2012). Pesticides are a significant cause of macroinvertebrate biodiversity decline in freshwater ecosystems on a global scale (e.g. Europe and Australia; Beketov et al., 2013) and are a priority concern for the integrity of food webs in the large Columbia River Basin that supports many threatened Pacific salmon and steelhead stocks (Naiman et al., 2012). Although less well studied, pesticides also affect freshwater biodiversity via their impact on microorganisms (algae, bacteria and protozoans) at the base of the food web (DeLorenzo et al., 2001). Reducing agrochemical inputs to river systems can be expected to promote biodiversity, improve the resiliency of food webs for fish, and enhance the likelihood that more conventional physical habitat restoration activities are successful (Naiman *et al.*, 2012).

5.5 URBAN RUN-OFF

By most measures, the biological integrity of urban watersheds is poor. This 'urban stream syndrome' has been widely documented (reviewed by Walsh *et al.*, 2005) and is attributable to physical, biological and chemical forms of habitat degradation. The science in recent decades has focused to a large extent on hydrologic and geomorphic changes at the catchment and drainage scales, including altered flows, sedimentation and loss of streambed substrates, channel incision and loss of side channel habitat, and non-chemical changes in water quality (e.g. increasing stream temperatures). In some cases, stream headwaters are simply buried (Elmore & Kaushal, 2008). Recurring biological impacts include a loss of functioning riparian streamside habitat and the invasion of non-native species. Toxics have received comparatively less attention, in part because urban stormwater run-off is exceptionally complicated in terms of chemical composition. Also, in cities with ageing infrastructure, severe storms deliver stormwater mixed with wastewater in combined sewer overflows.

The multitude of pressures on urban stream habitats presents significant challenges for stream restoration efforts (Bernhardt & Palmer, 2007). Moreover, non-urban watersheds (forested, grassland, desert, etc.) that presently support viable fish populations are widely vulnerable to the impacts of future urbanisation, in tandem with projected climate change (e.g. Nelson *et al.*, 2008). In cities worldwide, urban run-off is consistently toxic to fish and stream invertebrates (e.g. McIntyre *et al.*, 2014, 2015). Depending on how, when and where within a watershed the stormwater is collected, the toxicity is variously attributable to metals from motor vehicle brake pads; petroleum hydrocarbons (polycyclic aromatic hydrocarbons, or PAHs) from vehicle emissions of oil, grease and exhaust; de-icing salts; residential and commercial pesticide use; metals from building materials (e.g. zinc and copper from metal galvanised roofs and treated wood); and a large diversity of toxics that are deposited atmospherically in urban environments, including mercury, PBDEs and plasticisers (phthalates).

Chemical pollution in urban stormwater has the potential to undermine fish habitat restoration efforts that focus exclusively on physical and biological processes. Toxicological studies to date have mostly relied on model test organisms to bioassay collected stormwater, but *in situ* habitat surveys have also shown how run-off can turn restored urban streams into unintended ecological traps. For example, in the Pacific Northwest of the USA, it was discovered from post-project monitoring of restored urban streams that adult coho salmon spawners were entering the newly accessible habitats and dying before spawning. Coho return to freshwater during the autumn months when rainfall elevates surface flows in streams. In urban areas, these surface flows are mostly run-off from streets, car parks and other impervious surfaces. Affected adult spawners show a consistent suite of symptoms that include surface swimming and gaping, followed by a loss of orientation, a loss of equilibrium, and eventually death within 1–2 hours. Across a decade of autumn habitat surveys, pre-spawn mortality rates have been high – as much as 90% of the entire annual run (Scholz *et al.*, 2011). The severity of the recurring fish kills varies among watersheds with more or less urban land cover, and can be predicted from the relative proportion of impervious surfaces (Feist *et al.*, 2011). The documented mortality rates are too high to sustain local population abundance, thus urban streams are sink habitats for larger coho salmon metapopulations that are interconnected through adult straying (Spromberg & Scholz, 2011). As in many areas of the world, human population growth continues to drive urban development across the lowland Pacific Northwest. The challenge ahead is to protect the remaining coho population segments that are currently healthy from the future impacts of toxic urban run-off.

5.6 CONTAMINANTS OF EMERGING CONCERN

'Contaminants of emerging concern', or CECs, is a catchall phrase for pharmaceuticals, personal care products and other chemicals that are increasingly detected worldwide in water, sediments and biota. They originate from both point and non-point sources. Wastewater effluent is a typical point source. Modern municipal sewage treatment systems process wastewater from households, industrial facilities and sometimes stormwater, with the aim of removing contaminants. The steps usually involve primary treatment (settling and surface skimming), secondary treatment (microbial digestion of biodegradable organic compounds) and tertiary treatment (nutrient removal and disinfection to remove pathogenic organisms). Notably, conventional treatment plants are not designed to remove CECs. Consequently, prescription medications, antibiotics, perfumes, nanoparticles and any other chemicals that go down a household drain are potential CECs. In rural areas, these same chemicals can enter streams and lakes from more diffuse sources, for example through the hyporheic inflow of groundwater contaminated by leaky septic systems. Concentrated animal feeding operations (CAFOs) for cattle, pigs and poultry are also important sources of CECs in run-off, particularly where antibiotics and growth hormones are used to maximise agricultural yield. Notably, CECs are not always new or emerging chemicals; on occasion, new scientific information raises concerns over familiar chemicals – e.g. the toxicity of imidacloprid to honeybees and the ensuing European ban, as mentioned earlier.

Many CECs are hormones or hormone mimics that cause endocrine disruption in fish, particularly oestrogens, androgens and their chemical antagonists. Much of the initial work on wastewater-driven impacts grew out of studies in the UK. Upon discovering hermaphroditic fish in sewage treatment lagoons, researchers subsequently showed that effluent contained oestrogenic compounds that caused an upregulation of vitellogenin in the plasma of caged rainbow trout (Purdom *et al.*, 1994). Vitellogenin, an egg yolk protein normally produced only by females, has since proven to be a very sensitive bioindicator of oestrogenic endocrine disruption in male fish from freshwater and marine habitats around the world. Field assessments in the UK (Jobling *et al.*, 1998) demonstrated high frequencies of intersexuality (feminised male fish) in wild populations of the riverine roach (*Rutilus rutilus*). Although most wild intersex roach are able to breed, their reproductive success is impaired in proportion to the extent of their feminisation (Harris *et al.*, 2011). Although there are many oestrogenic CECs, much of the research over the past two decades has focused on the potent oestrogen ethynylestradiol, derived from birth control pills. In a landmark study, Canadian scientists working at the Experimental Lakes Area in northwestern Ontario conducted an ecosystem assessment of endocrine disruption by releasing low levels of ethynylestradiol into an otherwise pristine experimental lake over the course of three years. The chronic exposure to ethynylestradiol feminised resident male fathead minnows (*Pimephales promelas*), as evidenced by vitellogenin induction and intersex gonadal development, and in subsequent years the minnow population in the lake collapsed (Kidd *et al.*, 2007). More broadly, the theme of endocrine disruption in fish and other species has grown to be one of the most active research areas in aquatic ecotoxicology (Sumpter, 2005).

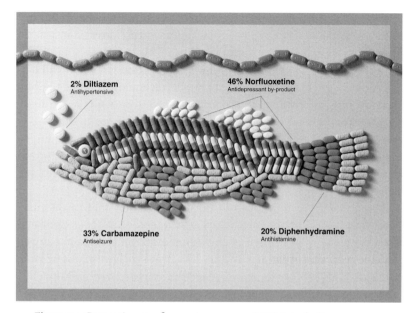

Figure 5.3 Contaminants of emerging concern (CECs), including pharmaceuticals and personal care products, are commonly detected in the tissues of fish from lakes and rivers that receive treated municipal wastewater. Above are the relative amounts of four different drugs measured in fish from a representative waterway near Chicago, USA, shown proportionally in the number of pills. *Source*: Oliveri Uberti/National Geographic Creative.

In addition to endocrine disruption, CECs can have many other types of effects on fish. For example, fish from streams with significant inputs of wastewater effluent have measurable levels of pharmaceuticals in their tissues (Figure 5.3). These include selective serotonin reuptake inhibitors (SSRIs) such as fluoxetine, a common antidepressant drug more familiarly known as Prozac (e.g. Brooks *et al.*, 2005). SSRIs in freshwater habitats selectively accumulate in the fish brain, as demonstrated recently for native white suckers (*Catastomus commersoni*) from central North America (Schultz *et al.*, 2010). Upon reaching the brain, these drugs are pharmacologically active. In adult fathead minnows, for example, sertraline (trade name: Zoloft) binds to the serotonin reuptake transporter and inhibits shelter-seeking behaviour (Valenti *et al.*, 2012). In a twist with potential implications for human health, a gene set analysis for fathead minnows exposed to fluoxetine and other unmetabolised psychoactive pharmaceuticals revealed an autism-like pattern of gene expression in the brain (Thomas & Klaper, 2012). Examples of neuroactive CECs altering the physiology and behaviour of fish in effluent-impacted habitats are likely to grow in the years ahead.

5.7 RESOURCE EXTRACTION AND TRANSPORT

Throughout the world there have been relatively massive and usually unintended inputs of toxics to freshwater ecosystems. These can be caused by natural disasters, such as earthquakes, floods and hurricanes. In 2006, for example, the winds and floodwaters from Hurricane Katrina in the USA caused hundreds of onshore releases of petroleum and other hazardous substances to Lake Pontchartrain and the Mississippi River in the vicinity of New Orleans (Santella et al., 2010). Other disasters, including those caused by warfare, are human-caused. Often these involve industrial accidents such as the fire that consumed a pesticide manufacturing facility near Basel, Switzerland in 1986. More than 1300 metric tons of various pesticides, solvents and intermediate chemicals were stored in the facility, and large volumes were washed into the Rhine River as the blaze was battled (Capel et al., 1998). The results were catastrophic, with massive die-offs of eels, salmonids and other aquatic life for hundreds of kilometres downriver through Germany and the Netherlands. Natural disasters and the ensuing human response can sometimes intermix to cause fish kills. This includes, for example, the use of toxic fire retardants and foams to suppress wildfires across the globe (e.g. Adams & Simmons, 1999).

Large-scale pollution inputs that are a consequence of natural resource extraction and transport pose considerable past, present and future threats to the conservation of freshwater fishes. Certain forms of extraction (e.g. in-river gravel mining) have overtly negative impacts on fish habitats. A prominent example is the worldwide increase in surface mining. In the Appalachian region of the eastern US, for example, so-called 'mountaintop removal' is used to access buried seams of coal. The overlying deciduous forest and topsoil are removed and explosives are used to break up the underlying rock. The debris buries nearby stream headwaters ('valley fill'), thereby destroying habitat for endemic fishes (Palmer et al., 2010). Furthermore, the loss of vegetation and topsoil reduces rainwater infiltration, leading to substantial increases in storm run-off and downstream flooding. Water flowing out of valley fills is chemically degraded in many ways, including increased levels of sulphuric acid, metals and selenium (Palmer et al., 2013). A bioaccumulative contaminant, selenium is widely known to be toxic to wildlife.

In fish, exposures cause reproductive failure and developmental deformities (Hamilton, 2004).

Other types of mining can also have ecosystem-scale impacts on freshwater systems. For example, the Amazon has experienced multiple gold rushes over the centuries. In recent decades, hundreds of thousands of wildcat prospectors have used primitive tools to extract gold from small mines scattered over vast distances. In a process known as amalgamation, mercury is used to extract fine gold particles from alluvial placer deposits. As consequence of this practice, thousands of metric tons of mercury have been released into the Amazon (Malm, 1998). Methylated mercury is bioaccumulative, neurotoxic and generally considered one of the world's most hazardous chemicals. The biomagnification of mercury through biodiverse aquatic communities (Uryu et al., 2001) has yielded a complex mosaic of risks to top predators and the humans that rely on fish for food (e.g. Sampaio Da Silva, 2005). Notably, the mercury problem is not unique to the Amazon. There has been a recent boom in artisanal gold mining in impoverished areas throughout the world. Although countries such as Indonesia have banned the use of mercury in mining because of the environmental risks, the populace still face widespread food safety issues associated with the consumption of freshwater fish (e.g. Castilhos et al., 2006).

Large-scale mining for metals presents a wide array of pollution threats to fish and their habitats. The companies conducting large operations generally moved away from mercury amalgamation in the twentieth century in favour of cyanide leaching and other alternative practices. Nonetheless, hardrock mining in the USA has left a legacy of toxic habitat destruction. Sources of toxics include waste piles (tailings), metal-contaminated dust, processing chemicals (e.g. sodium cyanide, sulphuric acid) and acid mine drainage (Woody et al., 2010). Acidic conditions leach metals into the dissolved phase, and thus acid drainage is one of the biggest resource extraction threats to freshwater systems. Individual mines across all the western USA have caused fish kills or otherwise impacted habitats (reduced fish distribution, reduced invertebrate diversity and abundance, etc.; reviewed by Woody et al., 2010).

Beyond the USA, accidents related to large-scale mining have produced some of the most high-profile ecotoxicological disasters in modern history. In 1996, for example, a holding dam at the Los Frailes mine in Aznalcóllar, Andalusia (southern Spain) failed. Millions of cubic metres of mine tailings were discharged. Acidic, metals-polluted waters killed nearly all aquatic life in the nearby River Agrio. The toxic waste then flowed downstream unchecked into the Doñana National Park, a UNESCO World Heritage Site (Grimalt *et al.*, 1999). A few years later, in 2000, a dike holding back millions of litres of cyanide-laced water failed in the Romanian gold mining town of Baia Mare. The ensuing toxic plume entered the Tisza River and devastated in-river aquatic life across Hungary en route to the Danube River delta, hundreds of kilometres distant on the shores of the Black Sea (Koenig, 2000). Fish species endemic to the Tisza were nearly eradicated, and the disaster had a significant impact on the ~ 60 fish taxa downstream, a third of which were in a protected status. The question of whether past disasters are a prologue for the future is currently under debate in Alaska, where a huge open-pit copper, gold and molybdenum mining project (known familiarly as Pebble) is currently proposed in the headwaters of Bristol Bay. The pristine headwater river systems in the vicinity of the proposed mining support the largest salmon run in the world, and the world's largest commercial sockeye (*O. nerka*) fishery. Similar issues are at play in the transboundary region of southeast Alaska, USA, and northeastern British Columbia, Canada.

The extraction and transport of petroleum products, particularly crude oil, is also a worldwide threat to freshwater habitats. Crude oils from different geological sources all contain PAHs that cause various adverse health effects in fish, including cancer (Myers *et al.*, 2003) and immunotoxicity (Reynaud & Deschaux, 2006). Moreover, research in the aftermath of the 1989 *Exxon Valdez* oil spill in Prince William Sound, Alaska, has shown that marine and freshwater fish embryos are exceptionally sensitive to the cardiotoxic effects of PAHs. Transient low-level exposures cause heart failure during embryogenesis (Incardona *et al.*, 2011), and fish that survive have cardiovascular abnormalities that reduce cardiac output and swimming performance later in life (Hicken *et al.*, 2011).

Oil exploration, extraction and transport have polluted many freshwater systems in recent decades. Seepage from production pits and failing infrastructure has created a major environmental problem in the Ecuadorian Amazon (e.g. Wernersson, 2004). The controversial development of oil sands in northern Alberta, Canada, has polluted the Athabasca River with PAHs and a variety of metals (e.g. Kelly *et al.*, 2010). The heavy crude oil from the Athabasca tar sands has proven especially difficult to clean up, as evidenced by the expense and complexity of restoration efforts in the Kalamazoo River following the 2010 pipeline disaster in Michigan, the largest on-land oil spill in US history. Pipeline infrastructure is also a major issue in Nigeria, where thousands of crude and refined oil spills have occurred in the Niger River delta since the 1970s. Warfare, illegal dumping, pipeline vandalism and petroleum theft have been causal factors. The ecological effects on the river delta ecosystem have not been extensively studied, but are likely to be massive in scale. Overall, accidental spills related to the global transport of petroleum via pipeline, train, ship and trucking will continue to be an important category of chemical pollution in fish habitats.

5.8 POLLUTION SOLUTIONS, BEYOND DILUTION

Pollution of freshwater ecosystems can be reduced through a variety of approaches that range in scale from changing the behaviour of individual people to passing regulatory legislation at the national (e.g. United States Clean Water Act) and international (e.g. Stockholm Convention) scales. As described above, most non-point pollution is a consequence of individual activities. At this scale, environmental education has proven to be an effective alternative to regulatory action. For example, the Global Mercury Project, a campaign by the United Nations, targeted more than 4000 artisanal gold miners in Brazil for education on the environmental and human health risks of mercury, as well as alternative methods to produce more gold using less mercury. In response, the per-prospector use of mercury dropped by nearly 90% (Sousa & Veiga, 2009). In rural Scotland, streams were extensively degraded in the 1980s and 1990s because farmers were treating sheep with organophosphate insecticides to control ectoparasites (Virtue & Clayton, 1997). Water-quality conditions in fish habitats improved considerably when education campaigns reformed farming practices, including the use of less-toxic, alternative sheep dips.

Regulatory policies can promote pollution source control without necessarily changing the behaviour of individual people. In the USA, for example, the state of Washington has taken several proactive measures in recent years to reduce pollution loading to aquatic habitats. These include a ban on dishwasher detergents for residential use that contain more than 0.5% phosphorus (2010), a ban on certain PBDEs in consumer products in favour of safer alternatives (2009), a ban on the use of coal tar-derived (and thus PAH-enriched) sealants for car parking lots and other impervious surfaces (2011), and 'better brakes' legislation to phase out copper, zinc, cadmium and other metals from motor vehicles (e.g. brake pads) in favour of non-toxic ceramic alternatives (2010). These regulatory actions have generally enjoyed broad public support, and have not required significant change in individual lifestyle.

Green stormwater infrastructure, or low-impact development, encompasses a suite of strategies and technologies to capture and, where possible, detoxify pollutants in stormwater run-off before they reach rivers, lakes and streams. Green infrastructure is primarily used in urban watersheds in situations where source control is not possible. The approaches fall into the categories of green roofs, permeable pavements and bioretention (Dietz, 2007; Ahiablame *et al.*, 2012). All use soil to filter particulate-bound and dissolved-phase contaminants. For metals, removal success using green infrastructure can be quite high (e.g. up to 99% for zinc; Davis *et al.*, 2009). Moreover, field assessments of bioretention effectiveness in dense urban areas report large removals of PAHs (87%; DiBlasi *et al.*, 2009) and motor oil (up to 96%; Chapman & Horner, 2010). Retained or sorbed organic contaminants such as PAHs are often biodegraded in soils through microbial processes (LeFevre *et al.*, 2012). Notably, green infrastructure may not be as effective for some CECs. For example, in a study with five pharmaceuticals, carbamazepine and gemfibrozil were found to be relatively mobile in soils and more resistant to degradation (Yu *et al.*, 2013).

Lastly, green chemistry describes a set of guidelines for conducting chemical reactions in industrial processes that minimise harm to the environment, including aquatic habitats. These principles promote reusable catalysts, safe solvents (i.e. water), energy efficiency, and the use of chemicals that break down into non-toxic substances (Ahuja & Hristovski, 2013). Green chemistry is proving increasingly useful for minimising industrial forms of water pollution, as well as restoring water and sediment quality in aquatic systems that have been impacted by legacy contamination. As an example, ferrate(VI) is a powerful oxidant that undergoes a reaction and conversion into a non-toxic byproduct, Fe(III) (Sharma, 2002). Ferrate(VI) can be successfully substituted for weaker and more toxic oxidants such as permanganate, ozone and hypochlorite to break down anionic surfactants (Eng *et al.*, 2006), endocrine-disrupting toxicants (Jiang *et al.*, 2005) and pharmaceuticals (Sharma & Mishra, 2006; Sharma *et al.*, 2006), among other contaminants.

5.9 CHEMICAL POLLUTION AND FRESHWATER FISH CONSERVATION: A LOOK FORWARD

This chapter has provided a broad overview of toxic chemical threats to freshwater fishes and their habitats. While pollution continues to be an important driver for species extinction and biodiversity loss world-wide, positive lessons have been learned. Aquatic communities generally recover when toxic inputs are identified and reduced, as evidenced by the ongoing revitalisation of urban rivers in the US that served as dumping grounds during the mid-twentieth century. Rivers devastated by accidental disasters return to a state of diversity and productivity after the pollution is abated by natural decline, or removed by dredging and other restoration actions. Across Europe, Atlantic salmon and other fish species that were driven to local extinction by pollution are recolonising and thriving in cleaner rivers. Examples include the River Purtse in northern Estonia (Kesler *et al.*, 2011), the River Mersey in northwest England (Ikediashi *et al.*, 2012) and the Seine River in France (Perrier *et al.*, 2010). Improvements in water quality have therefore proven to be an effective strategy for conserving and recovering freshwater fishes, in terms of expanding geographical range, increasing abundance and increasing diversity.

The above notwithstanding, many complex challenges are looming for the next generation of conservationists. Water-quality conditions are generally getting worse, not better, in the developing world. The toxic degradation of fish habitats is severe in many parts of China (Qu & Fan, 2010), India (e.g. Lokhande *et al.*, 2011) and other countries where industrialisation is currently lifting the populace out of poverty at the expense of the environment. Also, ecotoxicology has not kept pace with chemical engineering, or the rate at which new chemicals are identified, synthesised and brought into production for societal use. As a consequence, an awareness of the ecological impacts is most often delayed, typically coming years after a novel contaminant is first detected in aquatic systems. The upward global trends in environmental PBDEs in the late 1990s are an example of this. Many novel chemicals and compounds are on the horizon, including the use of RNA interference (RNAi)-based pesticides (small RNAs) to silence targeted genes in pest organisms (Lundgren & Duan, 2013). To shift the burden of proof to industry – i.e. to demonstrate that a chemical is safe as a precondition for approving commercial use, the European Union promulgated the Registration, Evaluation, Authorisation and Restriction of Chemicals (REACH) regulatory initiative in 2006. However, the adoption of the so-called 'precautionary principle' in regulatory decision-making remains controversial in Europe and beyond. Lastly, toxics are likely to interact increasingly with other large-scale forcing pressures in freshwater habitats – particularly those that reduce water quantity (e.g. less rainfall and snowpack due to climate change, instream flow diversions, etc.). In effect, less dilution will make it more difficult to find solutions to pollution.

To meet these challenges, conservation biologists focusing on freshwater fishes and chemical pollution will need a diverse toolbox of knowledge and skills. This toolbox necessarily spans several disciplines that have traditionally been somewhat distinct. For example, conservation biologists, fisheries biologists and ecotoxicologists typically train in separate academic departments. They join scientific societies and attend conferences that often have minimal substantive overlap. In the USA, for example, these include the Society for Conservation Biology, the American Fisheries Society and the Society of Environmental Toxicology and Chemistry. This separation leads in turn to the partitioning of new scientific information in discipline-specific journals. For example, the journal *Conservation Biology* rarely features pollution studies. Conversely, advances at the leading edge of theory and practice in conservation biology are rarely published in *Aquatic Toxicology*. On the path forward to freshwater fish conservation, modern scientific institutions have become barriers to collaboration, innovation and the effective exchange of new ideas and technologies. In the practice of conservation, this makes it very difficult to assess the side-by-side importance of pollution and other habitat factors (water quantity, invasive species, temperature, etc.) as limiting factors for the recovery of imperilled fish species. A different approach is needed, beginning with a more diverse and inclusive curriculum for undergraduate and graduate training programmes.

Technology is expected to play an outsized role in pollution science in the near term. Methods for analytical chemistry are improving constantly, allowing for the detection of more and more chemicals at ever-lower concentrations in water, sediments and fish tissues. This conveys an increasingly accurate picture of chemical habitat quality, and has been the impetus for the recent expansion of new research on the environmental occurrence and toxicity of pharmaceuticals, nanomaterials and other CECs. A key challenge ahead is the development of real-time observing systems for toxics, to circumvent the need to bring discrete samples back to a laboratory for analyses that may take days or weeks. These observing systems are in place for other water-quality parameters (e.g. temperature, conductivity) and, increasingly, for pathogenic biological organisms. *In situ* observing systems for toxics will have numerous applications, from early warning to integrated profiling of complex non-point source pollution. For example, a monoclonal antibody-based biosensor was recently used to monitor the real-time resuspension of

PAHs during a habitat restoration (dredging) operation, to minimise risks to fish (Spier *et al.*, 2011).

Whereas advances in analytical chemistry are improving our understanding of chemical exposure in aquatic systems, parallel advances in the biological sciences are rapidly expanding our understanding of toxicological response – i.e. the impacts of pollutants on aquatic organisms, populations and communities. Traditionally, the practice of aquatic toxicology has been chemical-centric (Scholz *et al.*, 2012). Under this paradigm, toxicity is determined on a chemical-by-chemical basis, typically using a few biological endpoints that all living things have in common (growth, mortality and reproduction). Standardised toxicity tests (e.g. the 96-hour LC_{50}) are conducted using a few model species that are particularly amenable to laboratory assays, such as rainbow trout and fathead minnows. This approach has been useful for measuring the relative hazard, or toxic potential, of individual chemicals and environmental mixtures. However, it has also proven too simplistic to capture the full spectrum of biological responses in aquatic organisms, particularly in habitats where low (sublethal) levels of chemical pollution are the norm. Also, conservation management priorities almost always involve fish species that are not model test organisms in aquatic toxicology. As a consequence, cross-species extrapolation has been a constant source of scientific uncertainty for decades.

Advances in so-called 'omics' technologies are changing this scientific landscape rapidly. It is becoming increasingly practical to monitor the toxicological responses of aquatic species at the scale of the expressed gene (transcriptomics), the translated protein (proteomics) and the integrated physiology pathway (metabolomics). Moreover, improvements in direct sequencing methodologies (Mehinto *et al.*, 2012) have made it possible to assess the global transcriptional response of non-model fish species to individual chemicals (laboratory assays) or toxics in combination with other habitat variables (field assessments). The entire genomes of more than 65 aquatic species have now been sequenced, and this number is expanding monthly (Garcia-Reyero & Perkins, 2011). The future of pollution science in aquatic conservation biology will be transformational and increasingly aligned with biomedical research. The decoding of interacting networks of genes and corresponding biochemical cascades will greatly expand our understanding of 'sublethal effects' on non-model fish species. As we move forward, the longstanding scientific goals remain the same – i.e. to identify and minimise the impacts of chemical contaminants on aquatic ecosystems. The challenge ahead will be to mine large data sets via bioinformatics. New and meaningful information about fish health will still need to be scaled, to forecast impacts on wild populations and changes in the dynamics of aquatic communities. An emerging framework for this is the increasing use of 'adverse outcome pathways' (AOPs), which sequentially link a molecular initiating event to endpoints at the individual or population level in ways that are relevant for ecological risk assessment (e.g. growth, survival, reproduction) (Ankley *et al.*, 2010).

In closing, chemical pollution science continues to be a dynamic and evolving facet of freshwater conservation biology. Habitat pressures are likely to expand in many areas of the world in response to ongoing human population growth and industrialisation (Bakker, 2012). New technologies are revealing new threats to endangered fish species, and new strategies are being implemented to reduce toxic run-off and restore degraded habitats. Many fish populations are resilient, and begin a positive recovery trajectory when pollution sources are controlled. Collaboration is essential, across the disciplines of environmental chemistry, molecular biology, physiology, animal behaviour, experimental ecology, fisheries biology, mathematical modelling, geographic information systems, computational informatics, social science, natural resource policy, conservation management and environmental law. Future conservationists should prepare to study genes, ecosystems and everything in between.

5.10 DISCUSSION QUESTIONS

1. Which sets of laws in your country are intended to protect threatened freshwater fishes from the harmful effects of chemical pollution? How much protection do they afford?
2. What is the concept of a sentinel species, and how are sentinels used in freshwater ecotoxicology?
3. In the larger picture of habitat pollution as it relates to species conservation, what is the role of the environmental chemist? How is this changing?
4. How might the life history of a particular threatened fish species influence its vulnerability to chemical contaminants?
5. Designing studies to address chemical mixtures has been an enormous challenge in ecotoxicology for decades. How might emerging technologies change this?

5.11 REFERENCES

Adams, R. & Simmons, D. (1999). Ecological effects of fire fighting foams and retardants: a summary. *Australian Forestry*, 62, 307–314.

Ahiablame, L. M., Engel, B. A. & Chaubey, I. (2012). Effectiveness of low impact development practices: literature review and suggestions for future research. *Water Air and Soil Pollution*, 223, 4253–4273.

Ahuja, S. & Hristovski, K. (2013). *Novel solutions to water pollution.* ACS Symposium Series. Washington, DC: ACS Publications.

Allan, J. D. (2004). Landscapes and riverscapes: the influence of land use on stream ecosystems. *Annual Review of Ecology, Evolution, and Systematics*, 35, 257–284.

Ankley, G. T., Bennett, R. S., Erickson, R. J., *et al.* (2010). Adverse outcome pathways: a conceptual framework to support ecotoxicology research and risk assessment. *Environmental Toxicology and Chemistry*, 29, 730–741.

Bakker, K. (2012). Water security: research challenges and opportunities. *Science*, 337, 914–915.

Baldwin, D. H., Spromberg, J. A., Collier, T. K. & Scholz, N. L. (2009). A fish of many scales: extrapolating sublethal pesticide exposures to the productivity of wild salmon populations. *Ecological Applications*, 19, 2004–2015.

Beketov, M. A., Kefford, B. J., Shäfer, R. B. & Liess, M. (2013). Pesticides reduce regional biodiversity of stream invertebrates. *Proceedings of the National Academy of Sciences*, 110, 11039–11043.

Bernhardt, E. S. & Palmer, M. A. (2007). Restoring streams in an urbanizing world. *Freshwater Biology*, 52, 738–751.

Bouwman, A. F., Bierkens, M. F. P., Griffioe J., *et al.* (2013). Nutrient dynamics, transfer and retention along the aquatic continuum from land to ocean: towards integration of ecological and biogeochemical models. *Biogeosciences*, 10, 1–22.

Brooks, B. W., Chambliss, C. K., Stanley, J. K., *et al.* (2005). Determination of select antidepressants in fish from an effluent-dominated stream. *Environmental Toxicology and Chemistry*, 24, 464–469.

Capel, P. D., Giger, W., Reichert, P. & Wanner, O. (1988). Accidental input of pesticides into the Rhine River. *Environmental Science and Technology*, 22, 992–997.

Castilhos, Z. C., Rodrigues-Filho, S., Rodrigues, A. P. C., *et al.* (2006). Mercury contamination in fish from gold mining areas in Indonesia and human health risk assessment. *Science of the Total Environment*, 368, 320–325.

Chapman, C. & Horner, R. R. (2010). Performance assessment of a street-drainage bioretention system. *Water Environment Research*, 82, 109.

Cook, P. M., Robbins, J. A., Endicott, D. D., *et al.* (2003). Effects of aryl hydrocarbon receptor-mediated early life stage toxicity on lake trout populations in Lake Ontario during the twentieth century. *Environmental Science & Technology*, 37, 3864–3877.

Davis, A. P., Hunt, W. F., Traver, R. G. & Clar, M. (2009). Bioretention technology: overview of current practice and future needs. *Journal of Environmental Engineering – ASCE*, 135, 109–117.

DeLorenzo, M. E., Scott, G. I. & Ross, P. E. (2001). Toxicity of pesticides to aquatic microorganisms: a review. *Environmental Toxicology and Chemistry*, 20, 84–98.

de Wit, C. A. (2002). An overview of brominated flame retardants in the environment. *Chemosphere*, 46, 583–624.

DiBlasi, C. J., Li, H., Davis, A. P. & Ghosh, U. (2009). Removal and fate of polycyclic aromatic hydrocarbon pollutants in an urban stormwater bioretention facility. *Environmental Science and Technology*, 43, 494–502.

Dietz, M. E. (2007). Low impact development practices: a review of current research and recommendations for future directions. *Water Air and Soil Pollution*, 186, 351–363.

Elmore, A. J. & Kaushal, S. S. (2008). Disappearing headwaters: patterns of stream burial due to urbanization. *Frontiers in Ecology and the Environment*, 6, 308–312.

Eng, Y. Y., Sharma, V. K. & Ray, A. K. (2006). Ferrate(VI): green chemistry oxidant for degradation of cationic surfactant. *Chemosphere*, 63, 1785–1790.

Feist, B. E., Buhle, E. R., Arnold, P., Davis, J. W. & Scholz, N. L. (2011). Landscape ecotoxicology of salmon spawner mortality in urban streams. *PLoS ONE*, 6, e23424.

Ferguson, C., de Roda Husman, A., Altavilla, N., Deere, D. & Ashbolt, N. (2003). Fate and transport of surface water pathogens in watersheds. *Critical Reviews in Environmental Science and Technology*, 33, 299–361.

Fernandez, M. P., Ikonomou, M. G., Courtenay, S. C. & Wirgin, I. I. (2004). Spatial variation in hepatic levels and patterns of PCBs and PCDD/Fs among young-of-the-year and adult Atlantic tomcod (*Microgadus tomcod*) in the Hudson River estuary. *Environmental Science & Technology*, 38, 976–983.

Garcia-Reyero, N. & Perkins, E. J. (2011). Systems biology: leading the revolution in ecotoxicology. *Environmental Toxicology and Chemistry*, 30, 265–273.

Grannas, A. M., Bogdal, C., Hageman, K. J., *et al.* (2013). The role of the global cryosphere in the fate of organic contaminants. *Atmospheric Chemistry and Physics*, 13, 3271–3305.

Grimalt, J. O., Ferrer, M. & Macpherson, E. (1999). The mine tailing accident in Aznalcollar. *Science of the Total Environment*, 242, 3–11.

Hamilton, S. J. (2004). Review of selenium toxicity in the aquatic food chain. *Science of the Total Environment*, 326, 1–31.

Harris, C. A., Hamilton, P. B., Runnalls, T. J., *et al.* (2011). The consequence of feminization in breeding groups of wild fish. *Environmental Health Perspectives*, 119, 306–311.

Hartig, J. H. (2010). *Burning Rivers: Revival of Four Urban-Industrial Rivers that Caught on Fire*. Hickley: Multi-Science Publishing Co.

Hicken, C. L., Linbo, T. L., Baldwin, D. W., *et al.* (2011). Sublethal exposure to crude oil during embryonic development alters cardiac morphology and reduces aerobic capacity in adult fish. *Proceedings of the National Academy of Sciences*, 108, 7086–7090.

Ikediashi, C., Billington, S. & Stevens, J. R. (2012). The origins of Atlantic salmon (*Salmo salar* L.) recolonizing the River Mersey in northwest England. *Ecology and Evolution*, 2, 2537–2548.

Incardona, J. P., Collier, T. K. & Scholz, N. L. (2011). Oil spills and fish health: exposing the heart of the matter. *Journal of Exposure Science and Environmental Epidemiology*, 21, 3–4.

Jiang, J. Q., Yin, Q., Zhou, J. L. & Pearce, P. (2005). Occurrence and treatment trials of endocrine disrupting chemicals (EDCs) in wastewaters. *Chemosphere*, 61, 544–550.

Jobling, S., Nolan, M., Tyler, C. R., Brighty, G. & Sumpter, J. P. (1998). Widespread sexual disruption in wild fish. *Environmental Science and Technology*, 32, 2498–2506.

Kelly, E. N., Schindler, D. W., Hodson, P. V., *et al.* (2010). Oil sands development contributes elements toxic at low concentrations to the Athabasca River and its tributaries. *Proceedings of the National Academy of Sciences*, 107, 16178–16183.

Kesler, M., Kangur, M. & Vetemaa, M. (2011). Natural re-establishment of Atlantic salmon reproduction and the fish community in the previously heavily polluted River Purtse, Baltic Sea. *Ecology of Freshwater Fish*, 20, 472–474.

Kidd, K. A., Blanchfield, P. J., Mills, K. H., *et al.* (2007). Collapse of a fish population after exposure to a synthetic estrogen. *Proceedings of the National Academy of Sciences*, 104, 8897–8901.

Koenig, R. (2000). Wildlife deaths are a grim wake-up call in Eastern Europe. *Science*, 287, 1737–1738.

Laetz, C. A., Baldwin, D. H., Collier, T. K., *et al.* (2009). The synergistic toxicity of pesticide mixtures: implications for ecological risk assessment and the conservation of threatened Pacific salmon. *Environmental Health Perspectives*, 117, 348–353.

Laetz, C. A., Baldwin, D. H., Hebert, V. R., Stark, J. D. & Scholz, N. L. (2013). The interactive neurobehavioral toxicity of diazinon, malathion, and ethoprop to juvenile coho salmon. *Environmental Science and Technology*, 47, 2925–2931.

Landsberg, J. H. (2002). The effects of harmful algal blooms on aquatic organisms. *Reviews in Fisheries Science*, 10, 113–390.

LeFevre, G. H., Hozalski, R. M. & Novak, P. J. (2012). The role of biodegradation in limiting the accumulation of petroleum hydrocarbons in raingarden soils. *Water Research*, 46, 6753–6762.

Lokhande, R. S., Singare, P. U. & Pimple, D. S. (2011). Pollution in water of Kasardi River flowing along Taloja industrial area of Mumbai, India. *World Environment*, 1, 6–13.

Lundgen, J. G. & Duan, J. J. (2013). RNAi-based insecticidal crops: potential effects on nontarget species. *BioScience*, 63, 657–665.

Ma, J., Hung, H., Tian, C. & Kallenborn, R. (2011). Revolatilization of persistent organic pollutants in the Arctic induced by climate change. *Nature Climate Change*, 1, 255–260.

Macneale, K. H., Kiffney, P. M. & Scholz, N. L. (2010). Pesticides, aquatic food webs, and the conservation of Pacific salmonids. *Frontiers in Ecology and the Environment*, 9, 475–482.

Malm, O. (1998). Gold mining as a source of mercury exposure in the Brazilian Amazon. *Environmental Research*, 77, 73–78.

McIntyre, J. K., Baldwin, D. H., Beauchamp, D. A. & Scholz, N. L. (2012). Low-level copper exposures increase the visibility and vulnerability of juvenile coho salmon to cutthroat trout predators. *Ecological Applications*, 22, 1460–1471.

McIntyre, J. K., Davis, J. W., Incardona, J. P., *et al.* (2014). Zebrafish and clean water technology: assessing the protective effects of bioinfiltration as a treatment for toxic urban runoff. *Science of the Total Environment*, 500, 173–180.

McIntyre, J. K., Davis, J. W., Hinman, C., *et al.* (2015). Soil bioretention protects juvenile salmon and their prey from the toxic impacts of urban stormwater runoff. *Chemosphere*, 132, 213–219.

Mehinto, A. C., Martyniuk, C. J., Spade, D. J. & Denslow, N. D. (2012). Applications for next-generation sequencing in fish ecotoxicogenomics. *Frontiers in Genetics*, 3, Article 62.

Mozaffarian, D. & Rimm, E. B. (2006). Fish intake, contaminants, and human health – evaluating the risks and the benefits. *Journal of the American Medical Association*, 296, 1885–1899.

Myers, M. S., Johnson, L. L. & Collier, T. K. (2003). Establishing the causal relationship between polycyclic aromatic hydrocarbon (PAH) exposure and hepatic neoplasms and neoplasia-related liver lesions in English sole (*Pleuronectes vetulus*). *Human and Ecological Risk Assessment*, 9, 67–94.

Naiman, R. J., Alldredge, J. R., Beauchamp, D. A., *et al.* (2012). Developing a broader scientific foundation for river restoration: Columbia River food webs. *Proceedings of the National Academy of Sciences*, 109, 21201–21207.

Nelson, K. C., Palmer, M. A., Pizzuto, J. E., *et al.* (2008). Forecasting the combined effects of urbanization and climate change on stream ecosystems: from impacts to management options. *Journal of Applied Ecology*, 46, 154–163.

Palmer, M. A., Bernhardt, E. S., Schlesinger, W. H., *et al.* (2010). Mountaintop mining consequences. *Science*, 327, 148–149.

Palmer, M. E., Keller, W. & Yan, N. D. (2013). Gauging recovery of zooplankton from historical acid and metal contamination: the influence of temporal changes in restoration targets. *Journal of Applied Ecology*, 50, 107–118.

Perrier, C., Evanno, G., Belliard, J., Guyomard, R. & Baglinière, J. L. (2010). Natural recolonization of the Seine River by Atlantic salmon (*Salmo salar*) of multiple origins. *Canadian Journal of Fisheries and Aquatic Sciences*, 67, 1–4.

Purdom, C. E., Hardiman, P. A., Bye, V. J., *et al.* (1994). Estrogenic effects of effluents from sewage treatment works. *Chemistry and Ecology*, 8, 275–285.

Qu, J. & Fan, M. (2010). The current state of water quality and and technology development for water pollution control in China. *Critical Reviews in Environmental Science and Technology*, 40, 519–560.

Quinn, T. P. (2005). *The Behavior and Ecology of Pacific Salmon and Trout.* Bethesda, MD: American Fisheries Society and University of Washington Press.

Rahel, F. J. (2010). Homogenization, differentiation, and the widespread alteration of fish faunas. *American Fisheries Society Symposium*, 73, 311–326.

Reynaud, S. & Deschaux, P. (2006). The effects of polycyclic aromatic hydrocarbons on the immune system of fish: a review. *Aquatic Toxicology*, 77, 229–238.

Rohr, J. R., Schotthoefer, A. M., Raffel, T. R., *et al.* (2008). Agrochemicals increase trematode infections in a declining amphibian species. *Nature*, 455, 1235–1239.

Sampaio Da Silva, D., Lucotte, M., Roulet, M., *et al.* (2005). Trophic structure and bioaccumulation of mercury in fish of three natural lakes of the Brazilian Amazon. *Water, Air, and Soil Pollution*, 165, 77–94.

Sandahl, J. F., Baldwin, D. H., Jenkins, J. J. & Scholz, N. L. (2005). Comparative thresholds for acetylcholinesterase inhibition and behavioral impairment in coho salmon exposed to chlorpyrifos. *Environmental Toxicology and Chemistry*, 24, 136–145.

Santella, N., Steinberg, L. J. & Sengul, H. (2010). Petroleum and hazardous material releases from industrial facilities associated with Hurricane Katrina. *Risk Analysis*, 30, 635–649.

Schindler, D. W. (1981). Effects of acid rain on freshwater ecosystems. *Science*, 239, 149–157.

Scholz, N. L., Truelove, N., French, B., *et al.* (2000). Diazinon disrupts antipredator and homing behaviors in chinook salmon (*Oncorhynchus tshawytscha*). *Canadian Journal of Fisheries and Aquatic Sciences*, 57, 1911–1918.

Scholz, N. L., Myers, M. S., McCarthy, S. G., *et al.* (2011). Recurrent die-offs of adult coho salmon returning to spawn in Puget Sound lowland urban streams. *PLoS ONE*, 6, e28013.

Scholz, N. L., Fleishman, E., Brooks, M. L., *et al.* (2012). A perspective on modern pesticides, pelagic fish declines, and unknown ecological resiliency in highly managed ecosystems. *BioScience*, 62, 428–434.

Schultz, M. M, Furlong, E. T., Kolpin, D. W., *et al.* (2010). Antidepressant pharmaceuticals in two US effluent-impacted streams: occurrence and fate in water and sediment, and selective uptake in fish neural tissue. *Environmental Science and Technology*, 44, 1918–1925.

Sharma, V. K. (2002). Potassium ferrate(VI): an environmentally friendly oxidant. *Advances in Environmental Research*, 6, 143–156.

Sharma, V. K. & Mishra, S. K. (2006). Ferrate(VI) oxidation of ibuprofen: a kinetic study. *Environmental Chemistry Letters*, 3, 182–185.

Sharma, V. K., Mishra, S. K. & Ray, A. K. (2006). Kinetic assessment of the potassium ferrate(VI) oxidation of antibacterial drug sulfamethoxazole. *Chemosphere*, 62, 128–134.

Sousa, R. N. & Veiga, M. M. (2009). Using performance indicators to evaluate an environmental education program in artisanal gold mining communities in the Brazilian Amazon. *Ambio*, 38, 40–46.

Spier, C. R., Vadas, G. G., Kaattari, S. L. & Unger, M. A. (2011). Near real-time, on-site, quantitative analysis of PAHs in the aqueous environment using an antibody-based biosensor. *Environmental Toxicology and Chemistry*, 30, 1557–1563.

Spromberg, J. A. & Scholz, N. L. (2011). Estimating the decline of wild coho salmon populations due to premature spawner mortality in urbanizing watersheds of the Pacific Northwest. *Integrated Environmental Assessment and Management*, 4, 648–656.

Stehr, C. M., Linbo, T. L., Scholz, N. L. & Incardona, J. P. (2009). Evaluating effects of forestry herbicides on fish development using zebrafish rapid phenotypic screens. *North American Journal of Fisheries Management*, 29, 975–984.

Sumpter, J. P. (2005). Endocrine disrupters in the aquatic environment: An overview. *Acta Hydrochimica et Hydrobiologica*, 33, 9–16.

Thomas, M. A. & Klaper, R. C. (2012). Psychoactive pharmaceuticals induce gene expression profiles associated with human idiopathic autism. *PLoS ONE*, 7, e32917.

Uryu, Y., Malm, O., Thornton, I., Payne, I. & Cleary, D. (2001). Mercury contamination of fish and its implications for other wildlife of the Tapajós Basin, Brazilian Amazon. *Conservation Biology*, 15, 438–446.

Valenti, T. W., Gould, G. G., Berninger, J. P., *et al.* (2012). Human therapeutic plasma levels of the selective serotonin reuptake inhibitor (SSRI) sertraline decrease serotonin reuptake transporter binding and shelter-seeking behavior in adult male fathead minnows. *Environmental Science & Technology*, 46, 2427–2435.

Virtue, W. A. & Clayton, J. W. (1997). Sheep dip chemicals and water pollution. *Science of the Total Environment*, 194, 207–217.

Walker, M. K., Spitsbergen, J. M., Olson, J. R. & Peterson, R. E. (1991). 2,3,7,8-Tetrachlorodibenzo-*para*-dioxin (TCDD) toxicity during early life stage development of lake trout (*Salvelinus namaycush*). *Canadian Journal of Fisheries and Aquatic Sciences*, 48, 875–883.

Walker, M. K., Cook, P. M., Batterman, A. R., *et al.* (1994). Translocation of 2,3,7,8-tetrachlorodibenzo-*p*-dioxin from adult female lake trout (*Salvelinus namaycush*) to oocytes – effects on early life stage development and sac fry survival. *Canadian Journal of Fisheries and Aquatic Sciences*, 51, 1410–1419.

Walsh, C. J., Roy, A. H., Feminella, J. W., *et al.* (2005). The urban stream syndrome: current knowledge and the search for a cure. *Journal of the North American Benthological Society*, 24, 706–723.

Wernersson, A. S. (2004). Aquatic ecotoxicity due to oil pollution in the Ecuadorian Amazon. *Aquatic Ecosystem Health and Management*, 7, 127–136.

Wirgin, I., Roy, N. K., Loftus, M., *et al.* (2011). Mechanistic basis of resistance to PCBs in Atlantic tomcod from the Hudson River. *Science*, 331, 1322–1325.

Woody, C. A., Hughes, R. M., Wagner, E. J., *et al.* (2010). The Mining Law of 1872: change is overdue. *Fisheries*, 35, 321–331.

Yu, Y., Liu, Y. & Wu, L. S. (2013). Sorption and degradation of pharmaceuticals and personal care products (PPCPs) in soils. *Environmental Science and Pollution Research*, 20, 4261–4267.

Yuan, Z. P., Courtenay, S., Chambers, R. C. & Wirgin, I. (2006). Evidence of spatially extensive resistance to PCBs in an anadromous fish of the Hudson River. *Environmental Health Perspectives*, 114, 77–84.

Multiple-stressor effects on freshwater fish: a review and meta-analysis

CHRISTOPH D. MATTHAEI AND KATHARINA LANGE

6.1 INTRODUCTION

A stressor can be defined as a variable that, as a result of human activity, exceeds its normal range of variation and affects individual species or communities (modified after Townsend *et al.*, 2008). Most present-day ecosystems are exposed to multiple stressors acting simultaneously (Vinebrooke *et al.*, 2004; Crain *et al.*, 2008; Dudgeon, 2010) or sequentially (Christensen *et al.*, 2006). Therefore, multiple stressors research is highly relevant for both fundamental and applied science, and when trying to deal with complex global problems. For example, multiple-stressor effects are believed to be responsible for the ongoing global declines of honeybees (reviews by Potts *et al.*, 2010; Aebi & Neumann, 2011), amphibians (reviews by Sodhi *et al.*, 2008; Mann *et al.*, 2009; Hof *et al.*, 2011), coral reefs (reviews by Harvey *et al.*, 2013; McLeod *et al.*, 2013) and freshwater biodiversity (reviews by Allan, 2004; Dudgeon, 2010; Ormerod *et al.*, 2010; Vörösmarty *et al.*, 2010; Woodward *et al.*, 2010). Multiple-stressors effects are also a central concern when trying to understand and predict the all-pervasive impacts of global climate change (reviews by Lindenmayer *et al.*, 2010; Woodward *et al.*, 2010; Chmura *et al.*, 2011; Hof *et al.*, 2011; Harvey *et al.*, 2013; McLeod *et al.*, 2013).

The development of a general theory of multiple stressors has started relatively recently (Hay *et al.*, 1994; Hay, 1996; Folt *et al.*, 1999; Swanson, 2004; Vinebrooke *et al.*, 2004) and is still ongoing (e.g. Crain *et al.*, 2008; Downes, 2010; Statzner & Bêche, 2010; Boone *et al.*, 2011; Verberk *et al.*, 2013). The combined action of multiple stressors can produce either simple or complex responses. According to Folt *et al.*

Conservation of Freshwater Fishes, eds G. P. Closs, M. Krkosek and J. D. Olden. Published by Cambridge University Press. © Cambridge University Press 2016.

(1999), 'simple' additive responses are patterns where the effect of all stressors combined is equal to the sum of the effects of each individual stressor. By contrast, 'complex' outcomes can be synergistic or antagonistic, when the combined effect is either larger or smaller than one would expect based on the individual effects of each stressor.

From the viewpoint of a resource manager or conservation ecologist, complex outcomes of interacting multiple stressors are particularly important because they can lead to 'ecological surprises' (Paine et al., 1998; Lindenmayer et al., 2010). For example, due to synergistic interactions between stressors, the actual threat to a given endangered ecosystem may be far more severe than expected based on the combined knowledge from all single-stressor studies conducted to help protect this ecosystem. During the past decade, the number of multiple-stressor studies has grown rapidly and complex interactions between anthropogenic stressors have been identified in a wide range of ecosystems. Nevertheless, the specific processes that cause these complex interactions are only partly understood (see e.g. Crain et al., 2008; Darling & Côté, 2008; Townsend et al., 2008; Wagenhoff et al., 2012). We will return to this knowledge gap later on.

6.2 MULTIPLE-STRESSORS RESEARCH ON FRESHWATER FISH

The existing research on this topic has been summarised in at least 12 reviews (Morgan et al., 2001; Walsh et al., 2005; Couillard et al., 2008a, 2008b; Wenger et al., 2009; Fausch et al., 2010; Olden et al., 2010; Johnson et al., 2012; Stasko et al., 2012; Vanhoudt et al., 2012; Hughes et al., 2014; Lapointe et al., 2014). Seven of these reviews focus primarily on fish, whereas the others also include other freshwater biota such as benthic invertebrates or algae.

6.2.1 Main stressor origins and key stressors

Most of the stressors affecting freshwater fish are closely linked to the four main threats these fish are exposed to (Dudgeon et al., 2006; Chapter 2): water pollution, habitat degradation/destruction (including altered flow regimes), overexploitation, and invasion by non-native species. Global climate change is also likely to play an important role (Olden et al., 2010), both as a stressor in its own right and via interactions with other stressors (Brook et al., 2008; Stasko et al., 2012). Within these five broad categories, stressor origins at the landscape/catchment scale include agriculture (Allan, 2004), urbanisation/industry and large-scale disturbances such as mining, oil exploration or logging (Couillard *et al.*, 2008a, 2008b; Fausch *et al.*, 2010; Chapter 5), dams/diversions (Poff *et al.*, 1997; Fausch *et al.*, 2010) and acidification (Morgan *et al.*, 2001). The potential of these human activities to give rise to multiple stressors impacting freshwaters is particularly high if they affect the riparian zones of freshwaters (Chapter 9).

The resulting key stressors affecting freshwater fish in their aquatic environment include nutrient enrichment/eutrophication (Allan, 2004), high levels of suspended or deposited fine sediment (Kemp *et al.*, 2011; Lange *et al.*, 2014), chemical pollutants (toxic contaminants; Couillard *et al.*, 2008a, 2008b; Chapter 5), increased water temperature, reduced pH and low oxygen levels (Morgan *et al.*, 2001; Johnson *et al.*, 2012). Most of these stressors have predominantly negative effects on fish individuals and communities. At the level of individual fish, stressors can affect fish survival, growth, condition, physiology, behaviour and reproduction rates. Stressors can act either directly (e.g. toxins or elevated water temperature; Lapointe *et al.*, 2011) or indirectly, through biological pathways (e.g. via food availability and predator-prey or non-consumptive interactions among individuals of the same or different species; Pilati *et al.*, 2009). Such indirect effects on individuals have consequences for fish populations because they affect size and genetic diversity of the population. The sensitivity of fish to stressors also depends on their life stage, and fish eggs and/or larval stages are usually more sensitive to stressors compared to adults (McKim, 1977; Power, 1997).

6.2.2 Urban stream syndrome

The reviews by Walsh *et al.* (2005), Wenger *et al.* (2009) and Hughes *et al.* (2014) deal with a particular multiple-stressor scenario, the 'urban stream syndrome'. In urban streams, aquatic communities are usually exposed to a number of stressors that can interact with each other but also co-vary because they originate from the same source (Walsh *et al.*, 2005). Thus, most stressors originate in a few major sources, primarily stormwater run-off delivered to streams by efficient urban drainage systems. Stressors from other sources, such as combined or sanitary sewer overflows, wastewater treatment plant effluent and legacy pollutants (long-lived pollutants from earlier land uses), often add further impacts. Symptoms of this syndrome include a flashier stream hydrograph, homogenised and incised channel morphology, raised concentrations of nutrients and contaminants, and deficits in dissolved oxygen concentrations (Wenger *et al.*, 2009; Hughes *et al.*, 2014). The cumulative effect of these stressors on fish communities is generally a loss or reduced abundance of sensitive species, resulting in a less-diverse assemblage dominated by disturbance-tolerant species (Walsh *et al.*, 2005).

6.2.3 Rehabilitation projects

Given that most freshwater systems are affected by multiple stressors, especially those draining agricultural or urban catchments, it is encouraging that rehabilitation/restoration projects have recently become common in many parts of the world, in particular for streams and rivers in North America and Europe. The ultimate goal of these projects is to regain the historical ecological status by restoring natural flow regimes, in-stream habitat connectivity, intact riparian vegetation and high water quality (Hughes *et al.*, 2014). Additional goals are to make waters safe for body contact as prescribed by the Clean Water Act in the United States (U.S.C. 33 § 1251) or to meet the target of 'good ecological status' specified in the Water Framework Directive (WFD) of the European Union (European Commission, 2000). Especially the latter has become a powerful legislative tool driving restoration projects, because almost all rivers in Central Europe require restoration to achieve the WFD ecological quality target by 2027 as they are affected by non-point source pollution and/or alterations to hydromorphology and habitat structure (Haase *et al.*, 2013).

The success or failure of river restoration projects has been widely debated (see reviews by Palmer *et al.*, 2010; Hughes *et al.*, 2014). One of the challenges in this debate is that it remains uncommon to evaluate success with a sound monitoring programme or a standardised before-and-after sampling programme (Bernhardt *et al.*, 2005; Jähnig *et al.*, 2011). Most existing restoration projects have been conducted at the site/reach scale (10–100 m) or river segment scale (one to a few kilometres). While many of these projects were successful in improving stream hydromorphology, positive effects on aquatic biota are much rarer (Palmer *et al.*, 2010; Bernhardt & Palmer, 2011). This general statement also holds true for freshwater fish which often show no, or only minor, improvements in terms of species richness and/or the prevalence of pollution-sensitive species (see e.g. Jähnig *et al.*, 2011; Haase *et al.*, 2013; Stoll *et al.*, 2013). Possible reasons for these weak responses include that (i) stream segments are still affected by water pollution or other stressors acting at the catchment scale, (ii) long periods of degradation cause several sensitive organisms to become extinct in entire catchments, thus source populations for recolonisation are missing, and (iii) long-distance dispersal (> 5 km) of recolonising fish may be rarer than previously thought (Haase et al., 2013; Stoll et al., 2013).

In recent years, restoration ecologists have become increasingly aware that it is more effective to rehabilitate at the catchment scale than at the reach/segment scale (Palmer et al., 2010; Bernhardt & Palmer, 2011), with particular focus on restoring natural flow regimes (Poff et al., 1997). Therefore, the priorities for future river restoration projects should be to protect existing upstream high-quality catchments and to re-establish natural ecosystem processes and connectivity in altered upstream river segments, before attempting to restore sites lower down in the catchment (Hughes et al., 2014). In heavily impacted urban catchments, some scientists also recommend managing streams as 'designed ecosystems' and aiming to optimise their services to urban human populations (Grimm et al., 2008), instead of trying to achieve the impossible by attempting to restore them to a natural state.

6.2.4 Ecosystem-based fisheries management

Because overexploitation is one of the four main threats for freshwater fish (Dudgeon et al., 2006; Chapter 2), there is an urgent need to improve the sustainability of freshwater fisheries. In this context, ecosystem-based fisheries management (EBFM) could become a valuable tool. EBFM is a relatively recent concept (reviewed by Pikitch et al., 2004; Espinoza-Tenorio et al., 2012; Fogarty & Rose, 2013; Lapointe et al., 2014) that has been promoted by international resource management agreements for marine ecosystems (Law of the Sea Convention, Code of Conduct for Responsible Fisheries). One of the motivations behind developing this concept has been an increasing awareness that global fisheries are exposed to multiple stressors (Minns, 2013; Lapointe et al., 2014). By reversing the order of management priorities (i.e. starting with the ecosystem rather than the target species), EBFM represents a holistic approach that emphasises an understanding of the reciprocal and complex interactions between humans and marine resources (Pikitch et al., 2004). The overall objective of EBFM is to sustain healthy marine ecosystems and the fisheries they support, and its specific goals are to avoid degradation of ecosystems, minimise the risk of irreversible changes to natural assemblages of species and ecosystem processes, obtain long-term socioeconomic benefits without compromising ecosystems, and generate enough knowledge of ecosystem processes to understand the likely consequences of human actions (Pikitch *et al.*, 2004).

Most of the conceptual development and progress in the implementation of EBFM has occurred in marine ecosystems. However, in recent years scientists are increasingly advocating its application to freshwater fisheries (Lapointe *et al.*, 2014), arguing that EBFM should be particularly relevant to freshwater ecosystems given the complexity of these systems and the diversity of their users (Beard *et al.*, 2011). For some of the large lakes of the world, encouraging progress towards implementing EBFM has been made, for example in Lake Titicaca in South America (Vera Cartas *et al.*, 2013), Lake Victoria in East Africa (Everson *et al.*, 2013) and especially in the Laurentian Great Lakes in North America where the state of fishery resources is seen as the best overall indicator of ecosystem health (Minns, 2013). The latter study identified a whole-system view, active adaptive management, acceptance of science-based evidence and shared goals with common objectives as essential features of effective EBFM (Minns, 2013).

In their review, Lapointe *et al.* (2014) identified three important challenges when trying to apply EBFM to freshwater fisheries: (i) the magnitude and direction of cumulative stressor effects are difficult to predict because of complex interactions between these multiple stressors and potentially long lag times, (ii) a disconnect exists between the concept of EBFM and its application because the concept is seen as difficult to understand, and (iii) the current regulatory system is still based on dissecting the ecosystem to focus management on distinct, 'simple' aspects. In an ecological modelling study, Drake and Mandrak (2014) highlighted a fourth challenge: even though reducing the probability of by-catch (the incidental capture of non-target species) is a core goal of EBFM, reliable by-catch data for freshwater fisheries are scarce in the scientific literature.

In a rare study in rivers and adjoining marine ecosystems, Levi *et al.* (2012) focused on a fifth challenge for successful EBFM: the need for a clear conceptual and quantitative framework that allows assessing how different harvest options can modify benefits to ecosystems and humans. They focused on Pacific salmon fisheries, which are economically valuable but intercept much of the annual pulse of nutrient subsidies this salmon species provides to terrestrial and aquatic food webs, and used grizzly bears as surrogates for 'salmon ecosystem function' (because bears are vectors of salmon nutrients and bear densities are strongly linked to salmon abundance). Combining salmon biomass and stock-recruitment data with stable isotope analysis, the authors assessed potential trade-offs between fishery yields and bear densities for six Pacific salmon stocks in Alaska and British Columbia. For coastal stocks, they found that both bear densities and fishery yields would increase substantially if ecosystem allocations of salmon increased from currently applied lower to upper goals and beyond. In contrast, in interior stocks in a British Columbia river where biomass from other salmon species is low, acute trade-offs between bear densities and fishery yields occurred, implying that increasing salmon allocations to the river ecosystem would benefit threatened bear populations at the cost of reduced long-term fishery yields. To resolve this conflict, the authors proposed an EBFM goal that values fisheries and bears (and thus the ecosystem) equally, but they also generated trade-off curves providing a stock-specific accounting of the expected loss to fishers and gain to bears as more salmon are allocated to the river ecosystem. Perhaps a similar approach, modified to suit multiple scenarios, could provide a generalisable method to resolve conflicts over shared resources in other ecosystems.

6.2.5 A key knowledge gap: interactive multiple-stressor effects on freshwater fish

Scientists have just begun to gain an understanding of the mechanisms underlying the effects of multiple stressors acting directly on individual organisms. Such stressors are usually of physico-chemical origin and tend to generate antagonisms if they act through the same physiological pathways within an organism and have strong individual effects (Folt et al., 1999). Antagonisms may be the result of situations where organisms have developed a tolerance to one stressor and are therefore more tolerant to another stressor acting through the same mechanism (Christensen et al., 2006). If stressors act through different mechanisms and stressor intensities are relatively low, however, synergistic outcomes seem to be more common. Then synergisms occur when stressed individuals have a lower resistance to additional stressors (Darling & Côté, 2008).

Stressors acting on individuals have consequences at the population level because they affect population size and genetic diversity. Selection for certain traits (e.g. resistance to a particular stressor) may make populations more vulnerable to the impacts of additional stressors acting through different pathways. Such ecological trade-offs at the population level can therefore lead to synergistic interactions of stressors acting simultaneously or sequentially (Christensen *et al.*, 2006; Darling & Côté, 2008). Conversely, if stressors act through similar mechanisms, certain physiological adaptations could be beneficial against exposure to some stressors and thus lead to antagonistic responses (Christensen *et al.*, 2006). Consequently, knowledge of stressor identities and mechanisms of action is crucial for determining their combined outcomes (Vinebrooke *et al.*, 2004; Crain *et al.*, 2008).

Local populations of freshwater fish or other organisms are at risk of extirpation if their population size is reduced due to direct stressor impacts (e.g. overexploitation) and then exposed to other stressors acting at the population level (e.g. invasive species or inbreeding depression) that can be disconnected from the original cause of decline (Brook *et al.*, 2008). In such situations, the synergistic effects of these stressors can push vulnerable species (e.g. with small geographic distribution, low reproduction rate, low dispersal ability and high specialisation) towards extinction of the entire species (Brook *et al.*, 2008). For instance, long-lived and slow-reproducing species may be especially vulnerable to stressors reducing their population size when their response to another stressor, such as global warming, depends on resilience via genetic diversity (Darling & Côté, 2008).

Against this conceptual background, it is important to note that very few studies included in the existing reviews of multiple-stressor research on freshwater fish investigated interactive effects of stressors (see a few exceptions each in Morgan *et al.*, 2001; Couillard *et al.*, 2008a, 2008b; Fausch *et al.*, 2010; Vanhoudt *et al.*, 2012). Therefore, interactive stressor effects on freshwater fish remain poorly understood, a key knowledge gap identified in 9 of the 12 reviews cited in our Introduction. Moreover, none of these reviews comprised a systematic meta-analysis that allowed classifying the types of interactions between stressors (see Brook *et al.*, 2008). Consequently, the two aims of our remaining chapter were (1) to determine the extent of the first of these knowledge gaps by reviewing the existing multiple-stressor experiments involving freshwater fish, and (2) to address the second knowledge gap by conducting a meta-analysis combined with an interaction classification on these experiments. For the latter, we used an analysis similar to those performed by Crain *et al.* (2008) for marine ecosystems and by Darling and Côté (2008) for animal mortality in freshwater, marine and terrestrial communities. Because manipulative experiments are arguably the most powerful tool for studying inter-actions between multiple stressors (see Crain *et al.*, 2008; Townsend *et al.*, 2008), we focused on experiments (rather than surveys or mod-elling studies) involving at least two stressors and their interactions.

The remaining text is structured in four sections. The first summa-rises the methods of our qualitative and quantitative literature reviews. The second provides a tabular and narrative overview of the studies included in the qualitative review. The third presents and discusses the findings of the meta-analysis and interaction classification. The final section summarises existing knowledge gaps and makes specific recom-mendations for future research.

6.3 MULTIPLE-STRESSOR EXPERIMENTS ON FRESHWATER FISH – REVIEWING METHODS

6.3.1 Qualitative review

We reviewed experiments on fish in running and standing freshwaters; estuarine habitats were beyond the scope of this work. Studies inves-tigating stressor effects in full-factorial, controlled experiments with at least two stressors were obtained from the ISI Web of Knowledge (1960–2013). We used 11 stressors (adapted from Crain *et al.*, 2008: nutrients, fine sediment, water temperature, pH, flow reduction, radiation, density, toxins/toxic, parasites, hypoxia and invasive species) as search keywords (together with closely related terms for each stressor, e.g. 'anox*', 'hypox*' and 'low oxygen' for 'hypoxia'), in conjunction with the additional keywords fish, experiment, mesocosm and factorial. This initial search yielded 7825 articles, but only 36 of these were suitable for inclusion in the qualitative review.

6.3.2 Quantitative review (meta-analysis)

Twenty-four studies fulfilled the six criteria we had selected for inclu-sion in the meta-analysis: (i) use true freshwater or pre-smolting juven-ile anadromous fish (which reside in freshwaters) as focal species, (ii) experimentally manipulate at least two anthropogenic stressors in a full-factorial design with true replicates (no pseudoreplication as defined by Hurlbert, 1984), (iii) report effects of all treatments (e.g. control and effects of stressor A, stressor B and stressors A and B combined), (iv) measure either survival, growth or physiological responses to the focal stressors or quantify behavioural responses, (v) report sample sizes for each treatment group, and (vi) report sample variance or some measure from which variance can be calculated (standard error or standard deviation). From these articles we extracted means, sample size and variance for control and treatment groups. For uniformity, all variances were converted to standard deviations.

As focal response variables, we chose fish biomass, growth, blood physiology (cortisol levels, number of lymphocytes), feeding rate, oxygen consumption and survival rate. To maintain independence of all data points, we extracted only the 'most inclusive' response variable from studies reporting several response types (see Crain *et al.*, 2008). Responses of studies with time-series data were based on the final time point. We extracted 31 sets of effect sizes for individual and interactive effects from the 24 articles (some studies included up to 4 independent experiments).

For each experiment, the interaction strength between paired stressors was computed according to the methods for factorial meta-analysis described in Hedges and Olkin (1985), Gurevitch *et al.* (2000) and Crain *et al.* (2008), using Hedge's d to calculate individual stressor effects. An interaction between two stressors was classified as synergistic if the cumulative effect of both exceeded the added effects of each stressor acting in isolation, as antagonistic if the cumulative effect was smaller than expected (Folt *et al.*, 1999), and as additive if the 95% confidence intervals of the effect size included zero. In cases where individual effects were both negative, a negative interaction effect size indicated synergisms and a positive one antagonisms, and vice versa where both were positive (see Crain *et al.*, 2008). In cases where individual effects were opposing, we compared the direction of the interactive effect to that of the individual effect with the higher absolute value (in contrast with the procedure for such cases in Crain *et al.*, 2008). We then inverted the interactive effect sizes in all cases where both individual effects were positive or in cases with opposing effects where the positive effect had the higher absolute value. This allowed direct comparisons with stressor pairs where both stressors had negative or opposing individual effects where the negative effect had the higher absolute value. Effect sizes were categorised after Cohen (1988): not biologically relevant ($d < 0.2$), small ($d < 0.5$), medium ($d < 0.8$) and large ($d > 0.8$) and interpreted as statistically significant if their 95% confidence intervals did not include zero (Nakagawa & Cuthill, 2007).

All analyses were computed in R (version 3.0.2, R Development Core Team, 2013) on untransformed data. A linear mixed-effects model was used to calculate the overall effect size in a weighted random-effects

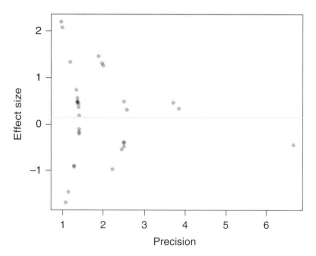

Figure 6.1 Funnel plots of effect sizes (Hedge's *d*) versus sample precision (1/ standard error). For more details, see text.

meta-analysis (Pinheiro *et al.*, 2014). Study ID was the only random factor in this model; there were insufficient observations for ecosystem (lentic vs. lotic) or experiment type (laboratory vs. mesocosm). Two outliers were removed by visual inspection of a funnel plot before model fitting (Figure 6.1). In this plot, decreasing effect sizes with increasing sample size indicated that there was no publication bias, and Spearman-rank correlation revealed no relationship between effect size and sample size ($P = 0.66$).

6.4 QUALITATIVE LITERATURE REVIEW: MAIN FINDINGS AND DISCUSSION

Of the 36 studies included in the review (Table 6.1), 64% (23) focused on real or simulated stream/river habitats and 36% (13) on real or simulated lake/pond habitats. Several experiments were conducted on fish species found in both. Thirty-five studies manipulated two stressors, and a single experiment (Qiang *et al.*, 2012) involved three stressors studied simultaneously (Table 6.1).

We were struck by 4 attributes of the 36 articles. First, this is a small number, and less than 0.5% of the articles scanned. By comparison, Crain *et al.* (2008) found 202 studies that manipulated at least two stressors in marine or coastal ecosystems. Consequently, multiple-stressor experiments on freshwater fish are generally quite rare. Second, 92% of the studies (33 of 36) focused on single fish species (or, in other words, on the population level). By contrast, only 8% (3 studies) were conducted at the community level (i.e. in simulated stream or lake environments containing not just the focal (predatory) fish species, but also other species including representatives of lower trophic levels). Third, a clear majority (80%) was conducted in the laboratory, plus 20% in outdoor mesocosms, whereas field experiments were lacking completely. Fourth, we rated the realism of all experiments (as low, medium or high), and almost all laboratory experiments were assessed as having a low realism. Consequently, the limited existing knowledge regarding multiple-stressors effects on fish is mainly based on single-species laboratory experiments conducted under highly artificial conditions.

6.4.1 Interactions between stressors and related mechanisms

Interactions between the focal stressors were not examined with appropriate statistical tests (i.e. ANOVAs or GLMs with interaction terms) in at least 10 of the 36 papers, and observed interactions between focal stressors (or a lack thereof) were discussed in less than half the papers (44%). Attempts to explain causal mechanisms behind any observed or expected synergistic or antagonistic interactions were often brief and sometimes highly speculative or lacking completely. Moreover, several of the studies that discussed possible causal mechanisms behind 'apparent' interactions between stressors had included no interaction term in their analysis, raising concerns that non-significant patterns were interpreted. Overall, these results indicate that the existing knowledge gap regarding interaction mechanisms between multiple stressors identified in other fields of ecology (see e.g. Crain et al., 2008; and Darling & Côté, 2008) has to be extended to studies on freshwater fish.

6.4.2 Stressor origins and stressors studied

Stressors investigated in the 36 papers were linked to a number of human activities. In order of declining frequency, the origins of the focal stressors included aquaculture, agriculture, climate change, urbanisation/industry, landscape disturbances (mining, logging), eutrophication, pollution, acidification, habitat fragmentation and invasive species. Apart from the dominant role of aquaculture, this list agrees well with the stressor origins specified in existing reviews of multiple-stressor research on freshwater fish (see Section 6.1 Introduction).

Table 6.1 *Overview of the 36 factorial experiments on freshwater fish involving at least two stressors and their interactions. (Please see related text for further explanations and comments on some of the stressors included here.)*

#	Stressors	Stressor origin	Fish species	Fish response variables	Interactions: tested for and possible mode discussed?	Study design	Response level category	Type of experiment and overall realism of study	Habitat	Reference
1	SED × NUT	Agriculture	*Dorosoma cepedianum*	Growth rate and body size variables	Tested but mode not discussed	2 × 2	1 fish species (larvae)	Mesocosm (5000-l outdoor tanks) Medium–high	Lentic	Gonzalez et al. (2010), Ohio, USA
2	SED × NUT	Agriculture	*Dorosoma cepedianum, Lepomis macrochirus*	Growth rate, body size variables and survival	Tested but mode not discussed	2 × 2	2 fish species (larvae, adults)	Mesocosm (22.5 × 15 × 2.5-m outdoor ponds) High	Lentic	Pilati et al. (2009), Ohio, USA
3	TOX × OXY	Agriculture	*Cyprinus carpio*	Survival and physiological characteristics	Not tested; mode not discussed	2 × 2	1 fish species (embryos)	Laboratory (15-ml glass vials) Low	Lentic	Palikova et al. (2007), Czech Republic
4	NUT × DEN	Aquaculture	*Micropterus salmoides*	Survival, viral load and body size variables	Not tested; mode not discussed	4 × 2	1 fish species (juveniles)	Laboratory (38 or 76-l aquaria) Low	Lentic	Inendino et al. (2005), Illinois, USA
5	WQ × DEN	Aquaculture	*Oncorhynchus mykiss*	Survival, growth rate and physiological variables	Tested but mode not discussed	3 × 2	1 fish species (juveniles)	Mesocosm (1-m³ flow-through outdoor tanks) Low–medium	Lotic	Person-Le Ruyet et al. (2008), France

	Interaction	Stressor	Species	Response	Mode tested	Design	Subjects	Setting	Habitat	Reference
6	OXY × TOX	Agriculture	*Retropinna retropinna, Gobiomorphus cotidianus, Galaxias maculatus*	Behavioural response	Not tested; mode not discussed	5 ×2	3 fish species (juveniles)	Laboratory (135-l 'fluviarium') Low–medium	Lotic	Richardson et al., (2001), New Zealand
7	OXY × TOX	Urbanisation	*Oncorhynchus mykiss*	Survival	Tested and mode discussed	4 × 3	1 fish species (juveniles)	Laboratory (55-l tank) Low	Lotic	Magaud et al. (1997), France
8	SED × TEMP	Human land use (logging and mining)	*Oncorhynchus tshawytscha*	Growth rate and physiological characteristics	Not tested; mode not discussed	2 × 2	1 fish species (juveniles)	Laboratory (140-l flow-through tanks) Low	Lotic	Shrimpton et al. 2007), British Columbia, Canada
9	TOX × PARA	Agriculture	*Salvelinus alpinus*	Growth rate, body size and physiological variables	Not tested; mode not discussed	2 × 2	1 fish species (juveniles)	Laboratory (300-l tanks) Low	Lentic	Blanar et al. (2005), Quebec, Canada
10	TEMP × NUT	Climate change and Agriculture	*Gasterosteus aculeatus*	Biomass and body size variables	Not tested but mode discussed	2 × 3	1 fish species (adults)	Mesocosms (314-l outdoor tanks) Medium–high	Lentic	Moran et al. (2010), UK
11	TEMP × TOX	Climate change	*Oncorhynchus mykiss*	Growth rate, body size and physiological variables	Tested but mode not discussed	2 × 2	1 fish species (juveniles)	Laboratory (270-l flow-through tanks) Low	Lentic	Linton et al. (1999), Ontario, Canada

Table 6.1 (cont)

#	Stressors	Stressor origin	Fish species	Fish response variables	Interactions: tested for and possible mode discussed?	Study design	Response level category	Type of experiment and overall realism of study	Habitat	Reference
12	TEMP × TOX TEMP × pH	Climate change and 'Pollution'	Oncorhynchus mykiss	Food consumption, growth rate, body size and physiological variables	Not tested; mode not discussed	2 × 2 (2 exper.)	1 fish species (juveniles)	Laboratory (270-l flow-through tanks) Low	Lentic	Morgan et al. (1998), Ontario, Canada
13	OXY × pH	Macrophyte beds (Agriculture)	Fundulus diaphanus, Lepomis macrochirus, Morone saxatilis	Behavioural response	Not tested; mode not discussed	3 treatm. involving 2 factors (no control)	3 fish species (juvenile M. saxatilis; others adults)	Laboratory (2.4 × 1.2-m flow-through channels) Low	Lotic	Serafy & Harrell (1993), Maryland, USA
14	TEMP × TOX	Climate change and Human land use	Pimephales promelas	Physiological variables	Not tested but mode discussed	3 × 2	1 fish species (adults)	Laboratory (45-l aquaria) Low	Lentic	Lapointe et al. (2011), Quebec, Canada
15	TOX × pH	Eutrophication	Pimephales promelas	Survival	Not tested but mode discussed	7 × 3	1 fish species (larvae)	Laboratory (100-ml beakers) Low–medium	Lentic	Valenti et al. (2010), Texas, USA

	Interaction	Stressor origin	Species	Response variable	Tested/mode	Design	Organisms	Setting	Lotic/Lentic	Reference
16	TEMP × TOX	Climate change and Human land use	*Oncorhynchus kisutch*	Physiological variables and behavioural responses	Tested; observed only additive effects and discussed these	2 × 2	1 fish species (juveniles)	Laboratory (40-l tanks) Low	Lotic	Bowen et al. (2006), California, USA
17	pH × TOX	Urbanisation	*Pimephales promelas*	Survival, growth and feeding rates	Not tested but mode discussed	3×6	1 fish species (larvae)	Laboratory (600-ml beakers) Low	Lentic	Valenti et al. (2009), Texas, USA
18	HARD × TOX	No stressor origin given	*Pimephales promelas*	Survival	Not tested; mode not discussed	2 × 10	1 fish species (larvae)	Laboratory (700-ml chambers) Low	Lentic	Erickson et al. (1997), Minnesota, USA
19	UV × TEMP	Climate change	*Salmo salar*	Body size and physiological variables	Tested; observed only additive effects (not discussed)	2 × 2	1 fish species (juveniles)	Mesocosm (4500-l outdoor tanks) Low–medium	Lentic	Jokinen et al. (2011), Norway
20	UV × TOX	Climate change and Industry	*Esox lucius*	Survival, physiological variables and behavioural response	Tested but mode not discussed	4 × 6	1 fish species (larvae)	Laboratory (1-l beakers) Low	Lentic	Häkkinen et al. (2004), Finland
21	TEMP × DEN × SALIN	Aquaculture	*Oreochromis niloticus*	Growth rate	Tested and mode discussed	3 × 3 × 3	1 fish species (juveniles)	Laboratory (50-l buckets) Low	Lentic	Qiang et al. (2012), China
22	CURR × DEN	Aquaculture	*Oncorhynchus mykiss*	Physiological variables	Tested but mode not discussed	2 × 2	1 fish species (juveniles)	Laboratory (600-l circular tanks) Low	Lotic	Skov et al. (2011), Denmark

Table 6.1 *(cont)*

#	Stressors	Stressor origin	Fish species	Fish response variables	Interactions: tested for and possible mode discussed?	Study design	Response level category	Type of experiment and overall realism of study	Habitat	Reference
23	TEMP × INV	Human land use and Species introduction	*Oncorhynchus mykiss*	Growth rate	Tested; observed only additive effects and discussed these	2 × 3	1 fish species (juveniles)	Mesocosms (10-m laboratory streams) Low–medium	Lotic	Reese & Harvey (2002), California, USA
24	TEMP × pH	No stressor origin given	*Oncorhynchus mykiss*	Survival and physiological characteristics	Tested and mode discussed	2 × 2	1 fish species (adults)	Mesocosms (800-l tanks) Low–medium	Lotic	Wagner *et al.* (1997), Utah, USA
25	CURR × DEN	Aquaculture	*Oncorhynchus tshawytscha*	Survival and physiological variables	Tested but mode not discussed	3 × 3	1 fish species (juveniles)	Mesocosms (23 × 2.4-m outdoor raceways) Low–medium	Lotic	Banks (1994), Washington, USA
26	SED × INBR	Human land use and Aquaculture	*Plecoglossus altivelis*	Physiological characteristics	Tested; observed only additive effects and discussed these	2 × 3	1 fish species (adults)	Mesocosm (95-l outdoor tanks) Low–medium	Lotic	Awata *et al.* (2011), Japan

	Driver	Context	Species	Response measured	Tested	Design	Biological units	Setting	Water	Reference, location
27	TEMP × INV	Climate change and Species introduction	*Iothichthys phlegethontis*	Survival and behavioural responses	Tested; observed only additive effects and discussed these	2 × 4	1 fish species (juveniles)	Laboratory (20-l aquaria) Low	Lentic	Priddis et al. (2009), Utah, USA
28	TEMP × DIS	Habitat fragmentation and Species introduction	*Salvelinus confluentus*	Survival	Tested; observed only additive effects and discussed these	2 × 2	1 fish species (juveniles)	Laboratory (35- and 100-l tanks) Low	Lentic	Jones et al. (2007), Montana, USA
29	TEMP × DEN	Aquaculture	*Oncorhynchus mykiss*	Survival and physiological characteristics	Tested and mode discussed	2 × 2	1 fish species (juveniles)	Laboratory (15-l aquaria) Low	Lentic	Suomalainen et al. (2005), Finland
30	TEMP × UV	Summer conditions	*Salmo trutta*	Physiological characteristics	Tested but mode not discussed	2 × 2	1 fish species (adult)	Laboratory (100-l aquaria) Low	Lotic	Lahnsteiner et al. (2011), Austria
31	TEMP × TOX	Human land use and Seasonality	*Thymallus thymallus*	Survival and physiological characteristics	Tested and mode discussed	2 × 2	1 fish species (juvenile)	Laboratory (60-l aquaria) Low	Lentic	Peuranen et al. (2003), Finland
32	TEMP × TOX	Agriculture	*Colossoma macropomum*	Physiological characteristics	Tested and mode discussed	2 × 3	1 fish species (juveniles)	Laboratory (40-l aquaria) Low	Lentic	Salazar-Lugo et al. (2009), Venezuela
33	TEMP × DIS	Climate change	*Oncorhynchus mykiss*	Survival, parasite load and physiological characteristics	Not tested but mode discussed	3 × 2	1 fish species (juveniles)	Laboratory (6.5-l swim chamber for trials; fish tested individually) Low	Lotic	Kocan et al. (2009), Washington, USA

Table 6.1 (cont)

#	Stressors	Stressor origin	Fish species	Fish response variables	Interactions: tested for and possible mode discussed?	Study design	Response level category	Type of experiment and overall realism of study	Habitat	Reference
34	pH × TOX	Acidification and Winter temperatures	*Micropterus salmoides*	Survival, body size variables and physiological characteristics	Not tested; mode not discussed	2 × 2	1 fish species (juveniles)	Laboratory (14-l aquaria) Low	Lentic	Leino and McCormick (1993), Minnesota, USA
35	TOX × TOX	Mining	*Salmo salar*	Survival and physiological characteristics	Tested; observed only additive effects (not discussed)	2 × 2	1 fish species (juveniles)	Laboratory (small aquaria) Low	Lentic	Heier et al. (2013), Norway
36	TOX × UV	Mining and Exposure to radiation	*Salmo salar*	Physiological characteristics	Tested; observed only additive effects (not discussed)	2 × 2	1 fish species (juveniles)	Laboratory (small aquaria) Low	Lentic	Salbu et al. (2008), Norway

Focal stressors are abbreviated as follows: CURR, current velocity; DEN, density; DIS, disease; HARD, water hardness; INBR, inbreeding; INV, invasive species; NUT, nutrients; OXY, oxygen; PARA, parasite; SALIN, water salinity; SED, fine sediment; TEMP, water temperature; TOX, toxin (including ammonium/ammonia); UV, ultraviolet radiation; WQ, water quality.

The conspicuous prominence of aquaculture as a stressor origin in the 36 studies is most likely a consequence of their predominantly single-species, laboratory nature. In some studies, stressor origins were not even specified, probably due to the same reason. Further, several studies included at least one experimental factor which was not a stressor according to the definition given above (because its levels were not linked clearly to human activities), but rather a natural factor (e.g. water hardness, which is driven by regional/local geology, in Erickson *et al.*, 1997; or a non-invasive parasite in Kocan *et al.*, 2009). Therefore, these studies were retained in Table 6.1 but omitted from our meta-analysis.

Ordered by declining importance, the stressors satisfying our definition investigated in the 36 papers included water temperature, toxins (e.g. heavy metals, pesticides, ammonia), high density of the focal fish species, nutrient enrichment, pH (mostly low), high levels of suspended or deposited fine sediment, UV-B radiation, current velocity (unusually slow), invasive species, (low) oxygen levels, and degree of inbreeding of hatchery fish. Several of these stressors (e.g. high fish density, current velocity, inbreeding; see Table 6.1 for details) reflect that many of the studies simulated an aquaculture setting rather than a real freshwater environment. In the real world, these particular stressors are likely to be much less important for freshwater fish than those originating in key global processes such as agricultural intensification, urbanisation and climate change (see Introduction). In this context the most commonly manipulated factor (18 of 36 studies), water temperature (which was increased experimentally by a median of 5°C across these studies), can be seen as the exception. This is because, while temperature was often manipulated in simulated aquaculture settings, it is also predicted to increase in real freshwater systems due to global climate change. Moreover, land-use intensification (via removal of shading riparian vegetation) can also result in higher water temperatures in agricultural streams (see e.g. Allan, 2004; Rutherford *et al.*, 2004). For stream algae and invertebrates, a recent experiment by Piggott *et al.* (2012) has shown that even a fairly minor water temperature increase of 1.4°C can synergistically interact with other stressors such as sedimentation, resulting in a stronger negative impact of fine sediment on algal and invertebrate taxon richness at raised temperature.

Overall, we can summarise this section as follows: while stressor origins and focal stressors in the 36 studies spanned most of the stressors known to affect freshwater fish communities in the real world, the relative research effort focusing on each stressor and some of the specific stressors chosen reflect an inflated focus on aquaculture scenarios. By comparison, several key stressors known to have pervasive effects on freshwater fish worldwide when acting as single stressors, for example in-stream sedimentation (see reviews by Waters, 1995; Wood & Armitage, 1997; Kemp *et al.*, 2011) are decidedly underrepresented in the existing multiple-stressor experiments on freshwater fish.

6.4.3 Focal fish species, life stages and response variables

Due to the predominance of single-species studies and because certain species were the subjects of several studies each, the 36 articles investigated multiple-stressor effects on just 26 different fish species (in 14 families and 8 orders, including 9 salmonid species, 3 cyprinid species and 6 species in the order Perciformes). This number is extremely small when compared to the global biodiversity of freshwater fish. For example, in North America alone there are an estimated 1213 extant species, and globally 15,750 species have been described (Chapter 1). Moreover, the majority of the 26 focal fish species are commercially and/or recreationally important for humans, and none of these species can be considered as nationally rare and potentially endangered. Consequently, we know very little about the interactive effects of multiple stressors on the vast majority of extant freshwater fish species, including all rare and/or endangered fish species (see also Chapter 1).

Only 3 of the 36 studies (Pilati *et al.*, 2009; Gonzalez *et al.*, 2010; Moran *et al.*, 2010) conducted experiments at the community level, in mesocosms (simulated ponds) involving one or two fish species as the top predators in fairly realistic aquatic food webs. All other studies included no other organisms apart from the single focal fish species in their experimental units. Therefore, our lack of knowledge makes it largely impossible to evaluate how multiple-stressor effects mediate trophic interactions involving fish, or how such effects might scale up to affect fish communities (and thus ultimately fish conservation goals) in the real world. This knowledge gap also prevents us from developing meaningful quantitative ecological models aimed at predicting multiple-stressor effects on freshwater fish communities.

Regarding fish life-cycle stages, 27 of the 36 studies focused on fish larvae or juveniles, plus one on embryos, compared to 5 studies that used only adults and 3 studies using both larvae and adults. The early ontogenetic stages of fish are widely regarded to be more vulnerable than adults to environmental stressors, presumably due to their relatively large body surface area (compared to their volume) and the sensitivity of developmental processes (McKim, 1977; Von Westernhagen, 1988; Power, 1997; Luckenbach et al., 2001). Consequently, these early life stages are likely to be the 'bottlenecks' when it comes to surviving multiple-stressor effects (see also Chapter 2). Therefore, the focus of the existing experimental multiple-stressor literature on early fish life stages is one of its few strengths, because otherwise stressor effects might have been underestimated. Conversely, it is another limitation that at least 15 of the 36 studies used hatchery fish instead of juveniles caught in the wild. Because hatchery fish, especially highly inbred domesticated strains, can respond differently to stressors from wild-bred fish (see e.g. Awata et al., 2011), this fact further reduces the applicability of the existing knowledge to real freshwater ecosystems.

Given the many limitations discussed so far, we were surprised by the wide range of fish response variables examined in the 36 studies. These spanned a broad range of responses and often involved labour-intensive and/or expensive sample analysis techniques. The following list gives examples in the main categories of metrics, but is by no means comprehensive: (i) survival, growth and body condition (assessed via mass and length metrics); (ii) behavioural (e.g. swimming performance, feeding rate, aggression); (iii) physiological/biochemical (e.g. oxygen consumption, nitrogen retention, gill chloride cells, plasma cortisol, muscle lipids, muscle protein including turnover, liver glycogen, depot fat, blood cell metrics such as hematocrit, lymphocytes and erythrocytes, DNA transcription and enzymatic responses in fish muscles/gills, heat stress/shock protein induction/expression, immune status); and (iv) pathological (e.g. viral load, parasite load, damage to fins or gills, skin conditions).

The contrast between this wealth of studied fish response variables and the narrowness of the existing knowledge base on multiple-stressor effects on freshwater fish brings us to the final conclusion of this section: there is a mismatch between the research effort dedicated to quantifying numerous response variables for individual fish and the many serious limitations/knowledge gaps (e.g. number of multiple-stressor experiments and fish species studied, lack of community-level studies or field studies, realism of experiments, interactions between stressors and mechanisms behind these, stressor origins and key stressors studied). Reasons for this mismatch may include that water-quality standards in most countries have traditionally been based on single-stressor studies conducted on a few 'model' species (see e.g. the review by McCullough *et al.*, 2009 for water temperature and stream fish). Moreover, suites of 'biomarkers' from these model fish species are widely regarded as promising tools for environmental risk assessment, especially in the field of environmental toxicology (see e.g. the review by Van der Oost *et al.*, 2003).

6.5 QUANTITATIVE REVIEW: META-ANALYTIC FINDINGS AND DISCUSSION

Overall, 77% of the 31 data sets (henceforth called 'experiments') included in our meta-analysis produced additive stressor effects. The mean inter-active effect size across all experiments was very small (d = 0.15) and non-significant (95% CIs: −0.15 to 0.45; Figure 6.2), indicating an over-all additive effect of the paired stressors. When summarised using effect size categories, individual interactive effect sizes were not ecologically relevant (d < 0.2) in 9.7% (3 of 31) of experiments, but most were either small (45.1%, 14), medium (9.7%, 3) or large (35.5%, 11). Five experiments yielded significant antagonistic and two synergistic interactions (Figure 6.2; Table 6.2). Six of these occurred at large effect sizes (d > 0.8), whereas the remaining interaction occurred at d = 0.45 in combination with a relatively large number of replicates (n = 6) in that study (Priddis *et al.*, 2009).

At first glance, these results might seem reassuring for the conservation of freshwater fish, because additive effects (which are easy to predict) of the paired investigated stressors dominated and synergisms (where combined stressor effects are worse than expected) were rarer than antagonisms. Thus, we found little evidence that stressors affected survival, physiology or behaviour of individual fish in a predominantly synergistic manner, despite the fact that synergisms were the most common interaction among multiple stressors in related meta-analyses (Brook *et al.*, 2008; Crain *et al.*, 2008; Darling & Côté, 2008). However, because of the many limitations of the data included in our meta-analysis (see earlier discussion), extrapolating these findings to the real world has to be done with considerable caution, due to these limitations and three further reasons.

First, most of the 31 experiments had only 2–3 replicates of each treatment combination. This resulted in low statistical power and only interactions with large effect sizes were detected as significant, except for a single experiment with a medium effect size but six replicates. Nevertheless, in terms of effect size alone, almost half the experiments

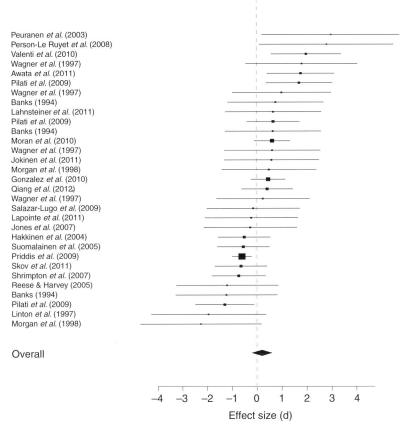

Figure 6.2 Forest plot showing the effect sizes (Hedge's *d*) of the interactive effect for each experiment included in the meta-analysis and the overall average effect size (meta-analytic mean), with 95% confidence intervals (95% CIs) for all effect size estimates. Negative effect size estimates indicate synergisms and positive antagonisms. Estimates with CIs that included zero were regarded as statistically non-significant and indicate additive effects. The size of the symbol for each study indicates the precision of each effect size estimate based on the inverse squared standard error (for more details see text). The number of replicates of each treatment combination is given in Table 6.2.

(45%) yielded medium or large interactive effects, which are clearly ecologically relevant based on Cohen's (1988) effect size categories. In spite of this, only 23% of interactions were classified as non-additive. Consequently, the frequency of antagonisms or synergisms would probably have been higher if the experiments included had used more replicates. In Crain *et al.*'s (2008) meta-analysis for marine and

Table 6.2 Overview of the 24 studies (comprising 31 experiments) included in the meta-analysis. All stressors listed here were linked clearly to human activities. Presented are focal fish species, focal stressors (see Table 6.1 for abbreviations), ecosystem type (lentic/lotic), experiment type (lab: laboratory, meso: mesocosm), fish response variables, number of replicates (N), individual effect of each stressor (stressor A or B compared to control), interactive effect with 95% confidence intervals, and interaction type (add: additive, syn: synergistic, ant: antagonistic).

Study Reference	Species	Stressor A	Stressor B	Habitat B	Type	Response variable	N	Individual effect stressor A	Individual effect stressor B	Interaction effect	Interaction lower 95% CI	Interaction upper 95% CI	Interaction type
01 Awata et al. (2011)	*Plecoglossus altivelis*	SED	INBR	lotic	lab	Cortisol	3	1.90	−1.19	−1.31	−2.30	−0.32	ant
02 Banks (1994)	*Oncorhynchus tshawytscha*	DEN	CURR	lotic	meso	Survival	2	−1.40	−2.34	0.48	−0.95	1.90	add
02 Banks (1994)	*Oncorhynchus tshawytscha*	DEN	CURR	lotic	meso	Survival	2	−3.57	−2.33	0.56	−0.88	2.00	add
02 Banks (1994)	*Oncorhynchus tshawytscha*	DEN	CURR	lotic	meso	Survival	2	1.44	−0.58	0.92	−0.61	2.44	add
03 Gonzalez et al. (2010)	*Dorosoma cepedianum*	NUT	SED	lentic	meso	Biomass	4	19.40	5.69	−0.34	−0.85	0.17	add
04 Häkkinen et al. (2004)	*Esox lucius*	UV	TOX	lentic	lab	Survival	3	−0.57	−0.62	−0.38	−1.16	0.39	add
05 Jokinen et al. (2011)	*Salmo salar*	TEMP	UV	lentic	meso	Weight	2	5.19	−2.38	−0.44	−1.86	0.98	add
06 Jones et al. (2007)	*Salvelinus confluentus*	TEMP	DIS	lentic	lab	Survival	2	0.20	−0.61	−0.20	−1.59	1.19	add
07 Lahnsteiner et al. (2011)	*Salmo trutta*	TEMP	UV	lentic	lab	Lymphocytes	2	−4.41	−1.12	0.49	−0.93	1.92	add

	Reference	Species					Response	n						
08	Lapointe et al. (2011)	*Pimephales promelas*	TEMP	TOX	lentic	lab	Weight	2	-2.74	-0.73	-0.17	-1.56	1.22	add
09	Linton et al. (1997)	*Oncorhynchus mykiss*	TEMP	TOX	lentic	lab	Weight	2	-0.13	1.59	-1.46	-3.18	0.26	add
10	Moran et al. (2010)	*Gasterosteus aculeatus*	TEMP	NUT	lentic	meso	Biomass	4	-1.18	-2.46	0.46	-0.06	0.98	add
11	Morgan et al. (1998)	*Oncorhynchus mykiss*	TEMP	TOX	lentic	lab	Weight	2	2.54	0.40	-0.36	-1.77	1.05	add
11	Morgan et al. (1998)	*Oncorhynchus mykiss*	TEMP	pH	lentic	lab	Weight	2	3.43	-0.13	1.69	-0.12	3.51	add
12	Person–Le Ruyet et al. (2008)	*Oncorhynchus mykiss*	WQ	DEN	lentic	meso	Weight	2	-6.48	-4.72	2.08	0.08	4.08	ant
13	Peuranen et al. (2003)	*Thymallus thymallus*	TEMP	TOX	lentic	lab	Oxygen	2	6.87	-1.82	-2.21	-4.28	-0.15	ant
14	Pilati et al. (2009)	*Dorosoma cepedianum*	NUT	SED	lentic	meso	Biomass	3	1.70	5.91	-1.26	-2.23	-0.28	ant
14	Pilati et al. (2009)	*Dorosoma cepedianum*	NUT	SED	lentic	meso	Biomass	3	1.37	2.37	-0.49	-1.29	0.30	add
14	Pilati et al. (2009)	*Lepomis macrochirus*	NUT	SED	lentic	meso	Biomass	3	5.76	-0.23	0.97	0.08	1.86	syn
15	Priddis et al. (2009)	*Iotichthys phlegethontis*	TEMP	INV	lentic	lab	Feeding	6	1.00	1.37	0.45	0.15	0.75	syn
16	Qiang et al. (2012)	*Oreochromis niloticus*	TEMP	DEN	lentic	lab	Growth	3	3.69	-0.57	-0.31	-1.08	0.46	add
17	Reese & Harvey (2005)	*Oncorhynchus mykiss*	TEMP	INV	lotic	lab	Growth	2	0.03	0.24	-0.90	-2.42	0.62	add

Table 6.2 (cont.)

	Study Reference	Species	Stressor A	Stressor B	Habitat B	Type	Response variable	N	Individual effect stressor A	Individual effect stressor B	Interaction effect	Interaction lower 95% CI	Interaction upper 95% CI	Interaction type
18	Salazar-Lugo et al. (2009)	Colossoma macropomum	TEMP	TOX	lentic	lab	Lymphocytes	2	0.10	−0.64	−0.11	−1.50	1.27	add
19	Shrimpton et al. (2007)	Oncorhynchus tshawytscha	TEMP	SED	lentic	meso	Weight	3	0.55	−0.96	−0.55	−1.35	0.25	add
20	Skov et al. (2011)	Oncorhynchus mykiss	DEN	CURR	lentic	lab	Weight	3	1.69	−0.57	0.48	−0.31	1.27	add
21	Suomalainen et al. (2005)	Oncorhynchus mykiss	TEMP	DEN	lotic	lab	Survival	3	−0.84	−0.45	−0.41	−1.19	0.37	add
23	Valenti et al. (2010)	Pimephales promelas	pH	TOX	lentic	lab	Survival	3	1.31	−5.76	1.47	0.42	2.51	ant
23	Wagner et al. (1997)	Oncorhynchus mykiss	TEMP	pH	lotic	lab	Cortisol	2	−0.13	0.35	−0.18	−1.57	1.21	add
24	Wagner et al. (1997)	Oncorhynchus mykiss	TEMP	pH	lotic	lab	Cortisol	2	1.15	2.46	−1.34	−3.01	0.33	add
24	Wagner et al. (1997)	Oncorhynchus mykiss	TEMP	pH	lotic	lab	Cortisol	2	−0.47	2.26	−0.46	−1.88	0.96	add
24	Wagner et al. (1997)	Oncorhynchus mykiss	TEMP	pH	lotic	lab	Cortisol	2	−0.13	1.80	−0.74	−2.22	0.73	add

coastal ecosystems, 76.5% of all interactions between paired stressors had medium or large effect sizes and significant complex interactions dominated (38% of all experiments yielded antagonisms and 36% synergisms), in contrast to ours where additive effects prevailed. One might interpret these differences as evidence that marine and coastal systems are affected by stronger stressors than freshwaters, and that complex interactions between stressors are more prevalent in the former systems. However, because statistical power depends on sample size, effect size and variance between experimental units (Quinn & Keough, 2002), the differences between the two meta-analyses could be caused by all three of these factors. (Note that Crain *et al.* (2008) provide no sample sizes for the studies included in their meta-analysis.)

Second, it is difficult to find statistical evidence for synergisms when using an additive model (as done by Crain *et al.* (2008) and in our meta-analysis) if both stressors have large individual effects. In such cases, the combined stressor effect is likely to be antagonistic (Folt *et al.*, 1999; Fausch *et al.*, 2010). This may have reduced the number of synergisms we detected given that 36% of all individual interactive effects were large.

Third, antagonistic interactions between multiple stressors (which were reasonably common in our analysis) cannot be regarded as the 'best-case scenario' for resource managers and fish conservationists. Such interactions can represent considerable management challenges because, if an ecosystem is degraded in spite of the presence of antagonisms, all or most co-acting stressors may need to be substantially reduced to achieve a recovery (see e.g. Crain *et al.*, 2008; Brown *et al.*, 2013).

6.6 FUTURE RESEARCH NEEDS

We have identified eight key areas urgently requiring future research efforts aimed at helping resource managers and conservationists understand and successfully alleviate multiple-stressor effects on freshwater fish.

(1) Study many different fish species. We know very little about the interactive effects of multiple anthropogenic stressors on the vast majority of extant freshwater fish species. Consequently, future studies should focus more on species that are not commercially and/or recreationally important, and especially (as far as ethically, legally and logistically feasible) on rare and/or endangered species. Funding entities should be made aware of the pressing need for this type of research.

(2) Design more realistic and powerful manipulative experiments. Current knowledge regarding multiple-stressor effects on freshwater fish is mainly based on single-species laboratory experiments (often involving domesticated hatchery fish) conducted under highly artificial conditions, typically with low replication of treatments. These limitations prevent evaluating how multiple-stressor effects mediate trophic or non-consumptive interactions (e.g. competition, mutualisms) involving fish, or how such effects might scale up to affect fish communities in the real world. The numerous interactions between fish and other species in real freshwater communities could easily modify the net effects of multiple stressors on individual fish species. Moreover, low replication may lead to underestimating the frequency of complex interactions between focal stressors. Consequently, resource managers leading restoration efforts in freshwater systems currently have to make difficult decisions, for example regarding which fish habitat attributes to improve with limited funding, based on what is clearly insufficient information.

To help reduce these problems, future research should focus on mesocosm experiments involving simulated stream/lake communities comprising several different trophic levels, wild-caught fish and better replication. Examples of such highly realistic experimental setups include the pond mesocosms used by Moran *et al.* (2010) or the circular stream mesocosms used by Wagenhoff *et al.* (2012). Besides factorial designs, researchers should also employ replicated regression designs (Cottingham *et al.*, 2005) spanning wide stressor gradients, as such designs are required for detecting potential threshold responses (see e.g. Wagenhoff *et al.*, 2012; and Wagenhoff *et al.*, 2013 for stream algae and invertebrates). Finally, manipulative field experiments in real streams/rivers and ponds/lakes are needed as well. Such experiments are the 'ultimate challenge' for future multiple stressors research on freshwater fish, because they will have to be conducted and replicated at relatively large spatial scales (e.g. at least at the reach scale in running waters) and/or in multiple systems.

(3) Manipulate three or more stressors simultaneously. Many present-day ecosystems are subject to more than two stressors acting together, and it has been postulated that interactions between three or more stressors are likely to become more commonly synergistic (Crain *et al.*, 2008). Consequently, studies manipulating at least three stressors simultaneously should be performed to complement the currently dominating two-stressor experiments.

Mesocosms are probably best suited for such studies due to the large number of experimental units required, or possibly longitudinal streamside channels (see e.g. Matthaei *et al.*, 2010; and Piggott *et al.*, 2012 for two three-stressor experiments involving stream algae and invertebrates).

(4) Conduct 'natural experiments' using carefully designed surveys. Although they do not allow determining true causes and effects, field surveys designed to span wide ranges of two or three focal stressors as evenly as possible can yield many useful insights that can complement experimental research on multiple stressors (see Townsend *et al.*, 2008). In such surveys, stressors should be treated as continuous predictor variables and their interactions included in the statistical analysis. Apart from migratory fish species with large home ranges, which are particularly challenging to study (see Chapter 1), surveys can be conducted at the spatial scales required to investigate fish population dynamics in real freshwater ecosystems. For instance, Lange *et al.* (2014) studied fish communities in 36 stream reaches covering wide gradients of catchment farming intensity and water abstraction intensity in a dryland river catchment. They found that presence of exotic brown trout was best described by an additive multiple-stressor model consisting of a unimodal response to farming intensity and a negative response to abstraction intensity, whereas presence of native upland bullies was unrelated to either predictor. Similarly, long-term temporal data sets monitoring environmental changes related to human activities combined with biological measurements can also reveal interactive effects of multiple stressors, for example in a 23-year study performed by Christensen *et al.* (2006) on plankton communities in a boreal lake.

(5) Choose more pervasive focal stressors. Stressors should be selected primarily based on their pervasiveness and relevance in the real world. Thus future multiple-stressor experiments on freshwater fish should involve stressors originating in the five key global processes threatening these organisms, namely water pollution, habitat degradation/destruction (including altered flow regimes), overexploitation, invasion by non-native species and climate change. Our review revealed an inflated focus on stressors relevant mainly in aquaculture scenarios, and most of the stressors studied were abiotic factors that could be easily manipulated in the laboratory (e.g. water temperature, dissolved toxins or pH).

(6) Understand the mechanisms behind stressor interactions. To direct conservation and restoration efforts efficiently and successfully, it is crucial to gain a comprehensive understanding of how, and through which mechanisms (physical, chemical and/or ecological), multiple stressors affect freshwater fish populations, communities and ecosystems. Therefore, future studies need to (i) develop hypotheses aimed at predicting not just stressor main effects but also their interactions, (ii) examine interactions with appropriate statistical methods, and (iii) discuss all existing interactions (or a lack thereof) in the light of the corresponding a-priori hypotheses. When doing so, the potential underlying ecological mechanisms described by Vinebrooke *et al.* (2004), Christensen *et al.* (2006) and Darling and Côté (2008) may help predict the outcomes of certain multiple-stressor scenarios. Besides studying the physiology of individual fish, focusing on biological traits of fish (see Olden *et al.*, 2008, 2010; Heino *et al.*, 2013; Chapter 2) may also provide new insights in this context.

(7) Improve predictive models of freshwater fish population dynamics. Existing models use fish preference curves for a number of environmental factors including several known stressors (PHABSIM, Stalnaker *et al.*, 1995; AQUATOX, Park *et al.*, 2008; population dynamics models, Yen *et al.*, 2013), but they do not include interactions among stressors. In the longer term, this shortcoming should be addressed (see Brook *et al.*, 2008). However, due to the many knowledge gaps revealed in our review, considerable research effort will be required before sufficient knowledge on the interactions between key stressors will have accumulated to allow the development of such improved models.

(8) Assess the potential role of genetic diversity. A certain level of genetic diversity may be required to ensure resistance and/or resilience of animal or plant populations to multiple stressors, but studies testing this hypothesis are generally rare (Roger *et al.*, 2012), not just for freshwater fish. In one of the few such studies, Ehlers *et al.* (2008) found some evidence that high genetic diversity of coastal eelgrass populations may increase their resilience to global warming. Thus, our final recommendation is to include aspects of genetic diversity (both as an experimental factor and as a response variable) in future multiple-stressor studies on fish. The ongoing rapid advances in molecular biology are likely to provide valuable tools for complementing the existing ecological methods.

6.7 DISCUSSION QUESTIONS

1. What are the key stressors affecting freshwater fish, and where do they originate?
2. What is the urban stream syndrome, and how does it affect freshwater fish?
3. Why do river restoration projects often fail to improve fish community diversity?
4. Why is it particularly important to examine multiple stressor effects on the embryo, larval and juvenile life stages of freshwater fish?
5. Is it currently possible to predict large-scale and/or long-term effects of interacting multiple stressors on freshwater fish communities?

6.8 REFERENCES

Aebi, A. & Neumann, P. (2011). Endosymbionts and honey bee colony losses? *Trends in Ecology and Evolution*, 26, 494–494.

Allan, J. D. (2004). Landscapes and riverscapes: the influence of land use on stream ecosystems. *Annual Review of Ecology, Evolution and Systematics*, 35, 257–284.

Awata, S., Tsuruta, T., Yada, T. & Iguchi, K. (2011). Effects of suspended sediment on cortisol levels in wild and cultured strains of ayu *Plecoglossus altivelis*. *Aquaculture*, 314, 115–121.

Banks, J. L. (1994). Raceway density and water-flows as factors affecting spring chinhook salmon (*Oncorhynchus tshawytscha*) during release and after release. *Aquaculture*, 119, 201–217.

Beard, T. D., Arlinghaus, R., Cooke, S. J., *et al.* (2011). Ecosystem approach to inland fisheries: research needs and implementation strategies. *Biology Letters*, 7, 481–483.

Bernhardt, E. S. & Palmer, M. A. (2011). River restoration: the fuzzy logic of repairing reaches to reverse catchment scale degradation. *Ecological Applications*, 21, 1926–1931.

Bernhardt, E. S., Palmer, M. A., Allan, J. D., *et al.* (2005). Synthesizing U.S. river restoration efforts. *Science*, 308, 636–637.

Blanar, C. A., Curtis, M. A. & Chan, H. M. (2005). Growth, nutritional composition, and hematology of Arctic charr (*Salvelinus alpinus*) exposed to toxaphene and tapeworm (*Diphyllobothrium dendriticum*) larvae. *Archives of Environmental Contamination and Toxicology*, 48, 397–404.

Boone, E., Ye, K. & Smith, E. (2011). Assessing environmental stressors via Bayesian model averaging in the presence of missing data. *Environmetrics*, 22, 13–22.

Bowen, L., Werner, I. & Johnson, M. L. (2006). Physiological and behavioral effects of zinc and temperature on coho salmon (*Oncorhynchus kisutch*). *Hydrobiologia*, 559, 161–168.

Brook, B. W., Sodhi, N. S. & Bradshaw, C. J. (2008). Synergies among extinction drivers under global change. *Trends in Ecology & Evolution*, 23, 453–460.

Brown, C. J., Saunders, M. I., Possingham, H. P. & Richardson, A. J. (2013). Managing for interactions between local and global stressors of ecosystems. *PLoS ONE*, 8, e65765.

Chmura, D. J., Anderson, P. D., Howe, G. T., *et al.* (2011). Forest responses to climate change in the northwestern United States: ecophysiological foundations for adaptive management. *Forest Ecology and Management*, 261, 1121–1142.

Christensen, M. R., Graham, M. D., Vinebrooke, R. D., *et al.* (2006). Multiple anthropogenic stressors cause ecological surprises in boreal lakes. *Global Change Biology*, 12, 2316–2322.

Cohen, J. (1988). *Statistical Power Analysis for the Behavioral Sciences*. Seond edition. Hillsdale, NJ: Lawrence Erlbaum Associates.

Cottingham, K. L., Lennon, J. T. & Brown, B. L. (2005). Knowing when to draw the line: designing more informative ecological experiments. *Frontiers in Ecology and the Environment*, 3, 145–152.

Couillard, C. M., Courtenay, S. C. & Macdonald, R. W. (2008a). Chemical–environment interactions affecting the risk of impacts on aquatic organisms: a review with a Canadian perspective – interactions affecting vulnerability. *Environmental Reviews*, 16, 19–44.

Couillard, C. M., Macdonald, R. W., Courtenay, S. C. & Palace, V. P. (2008b). Chemical–environment interactions affecting the risk of impacts on aquatic organisms: a review with a Canadian perspective – interactions affecting exposure. *Environmental Reviews*, 16, 1–17.

Crain, C. M., Kroeker, K. & Halpern, B. S. (2008). Interactive and cumulative effects of multiple human stressors in marine systems. *Ecology Letters*, 11, 1304–1315.

Darling, E. S. & Côté, I. M. (2008). Quantifying the evidence for ecological synergies. *Ecology Letters*, 11, 1278–1286.

Downes, B. J. (2010). Back to the future: little used tools and principles of scientific inference can help disentangle effects of multiple stressors on freshwater ecosystems. *Freshwater Biology*, 55(S1), 60–79.

Drake, D. A. R. & Mandrak, N. E. (2014). Harvest models and stock co-occurrence: probabilistic methods for estimating bycatch. *Fish and Fisheries*, 15, 23–42.

Dudgeon, D. (2010). Prospects for sustaining freshwater biodiversity in the 21st century: linking ecosystem structure and function. *Current Opinion in Environmental Sustainability*, 2, 422–430.

Dudgeon, D., Arthington, A. H., Gessner, M. O., *et al.* (2006). Freshwater biodiversity: importance, threats, status and conservation challenges. *Biological Reviews*, 81, 163–182.

Ehlers, A., Worm, B. & Reusch, T. B. (2008). Importance of genetic diversity in eelgrass *Zostera marina* for its resilience to global warming. *Marine Ecology Progress Series*, 355, 1–7.

Erickson, R. J., Kleiner, C. F., Fiandt, J. T. & Highland, T. L. (1997). Effect of acclimation period on the relationship of acute copper toxicity to water hardness for fathead minnows. *Environmental Toxicology and Chemistry*, 16, 813–815.

Espinoza-Tenorio, A., Wolff, M., Taylor, M. H. & Espejel, I. (2012). What model suits ecosystem-based fisheries management? A plea for a structured modeling process. *Reviews in Fish Biology and Fisheries*, 22, 81–94.

European Commission. (2000). *The EU Water Framework Directive – Integrated River Basin Management for Europe*. Directive 2000/60/EC.

Everson, I., Taabu-Munyaho, A. & Kayanda, R. (2013). Acoustic estimates of commercial fish species in Lake Victoria: moving towards ecosystem-based fisheries management. *Fisheries Research*, 139, 65–75.

Fausch, K. D., Baxter, C. V. & Murakami, M. (2010). Multiple stressors in north temperate streams: lessons from linked foreststream ecosystems in northern Japan. *Freshwater Biology*, 55, 120–134.

Fogarty, M. J. & Rose, K. (2013). The art of ecosystem-based fishery management. *Canadian Journal of Fisheries and Aquatic Sciences*, 71, 479–490.

Folt, C. L., Chen, C. Y., Moore, M. V. & Burnaford, J. (1999). Synergism and antagonism among multiple stressors. *Limnology and Oceanography*, 44, 864–877.

Gonzalez, M. J., Knoll, L. B. & Vanni, M. J. (2010). Differential effects of elevated nutrient and sediment inputs on survival, growth and biomass of a common larval fish species (*Dorosoma cepedianum*). *Freshwater Biology*, 55, 654–669.

Grimm, N. B., Faeth, S. H., Golubiewski, N. E., *et al.* (2008). Global change and the ecology of cities. *Science*, 319, 756–760.

Gurevitch, J., Morrison, J. A. & Hedges, L. V. (2000). The interaction between competition and predation: a meta-analysis of field experiments. *American Naturalist*, 155, 435–453.

Haase, P., Hering, D., Jaehnig, S. C., Lorenz, A. W. & Sundermann, A. (2013). The impact of hydromorphological river restoration on river ecological status: a comparison of fish, benthic invertebrates, and macrophytes. *Hydrobiologia*, 704, 475–488.

Häkkinen, J., Vehniainen, E. & Oikari, A. (2004). High sensitivity of northern pike larvae to UV-B but no UV-photoinduced toxicity of retene. *Aquatic Toxicology*, 66, 393–404.

Harvey, B. P., Gwynn-Jones, D. & Moore, P. J. (2013). Meta-analysis reveals complex marine biological responses to the interactive effects of ocean acidification and warming. *Ecology and Evolution*, 3, 1016–1030.

Hay, M. E. (1996). Defensive synergisms? Reply to Pennings. *Ecology*, 77, 1950–1952.

Hay, M. E., Kappel, Q. E. & Fenical, W. (1994). Synergisms in plant defenses against herbivores: interactions of chemistry, calcification, and plant quality. *Ecology*, 75, 1714–1726.

Hedges, L. V. & Olkin, I. (1985). *Statistical Methods for Meta-analysis*. New York, NY: Academic Press.

Heier, L. S., Teien, H. C., Oughton, D., *et al.* (2013). Sublethal effects in Atlantic salmon (*Salmo salar*) exposed to mixtures of copper, aluminium and gamma radiation. *Journal of Environmental Radioactivity*, 121, 33–42.

Heino, J., Schmera, D. & Erős, T. (2013). A macroecological perspective of trait patterns in stream communities. *Freshwater Biology*, 58, 1539–1555.

Hof, C., Araújo, M. B., Jetz, W. & Rahbek, C. (2011). Additive threats from pathogens, climate and land-use change for global amphibian diversity. *Nature*, 480, 516–519.

Hughes, R. M., Dunham, S., Maas-Hebner, K. G., *et al.* (2014). A review of urban water body challenges and approaches: (1) rehabilitation and remediation. *Fisheries*, 39, 18–29.

Hurlbert, S. H. (1984). Pseudoreplication and the design of ecological field experiments. *Ecological Monographs*, 54, 187–211.

Inendino, K. R., Grant, E. C., Philipp, D. P. & Goldberg, T. L. (2005). Effects of factors related to water quality and population density on the sensitivity of juvenile largemouth bass to mortality induced by viral infection. *Journal of Aquatic Animal Health*, 17, 304–314.

Jähnig, S., Lorenz, A. W., Hering, D., *et al.* (2011). River restoration success: a question of perception. *Ecological Applications*, 21, 2007–2015.

Johnson, J. E., Patterson, D. A., Martins, E. G., Cooke, S. J. & Hinch, S. G. (2012). Quantitative methods for analysing cumulative effects on fish migration success: a review. *Journal of Fish Biology*, 81, 600–631.

Jokinen, I. E., Salo, H. M., Markkula, E., *et al.* (2011). Additive effects of enhanced ambient ultraviolet B radiation and increased temperature on immune function, growth and physiological condition of juvenile (parr) Atlantic Salmon, *Salmo salar*. *Fish & Shellfish Immunology*, 30, 102–108.

Jones, D. T., Moffitt, C. M. & Peters, K. K. (2007). Temperature-mediated differences in bacterial kidney disease expression and survival in *Renibacterium salmoninarum*-challenged bull trout and other salmonids. *North American Journal of Fisheries Management*, 27, 695–706.

Kemp, P., Sear, D., Collins, A., Naden, P. & Jones, I. (2011). The impacts of fine sediment on riverine fish. *Hydrological Processes*, 25, 1800–1821.

Kocan, R., Hershberger, P., Sanders, G. & Winton, J. (2009). Effects of temperature on disease progression and swimming stamina in *Ichthyophonus*-infected rainbow trout, *Oncorhynchus mykiss* (Walbaum). *Journal of Fish Diseases*, 32, 835–843.

Lahnsteiner, F., Haunschmid, R. & Mansour, N. (2011). Possible reasons for late summer brown trout (*Salmo trutta* Linnaeus 1758) mortality in Austrian prealpine river systems. *Journal of Applied Ichthyology*, 27, 83–93.

Lange, K., Townsend, C. R., Gabrielsson, R., Chanut, P. C. M. & Matthaei, C. D. (2014). Responses of stream fish populations to farming intensity and water abstraction in an agricultural catchment. *Freshwater Biology*, 59, 286–299.

Lapointe, D., Pierron, F. & Couture, P. (2011). Individual and combined effects of heat stress and aqueous or dietary copper exposure in fathead minnows (*Pimephales promelas*). *Aquatic Toxicology*, 104, 80–85.

Lapointe, N. W. R., Cooke, S. J., Imhof, J. G., *et al.* (2014). Principles for ensuring healthy and productive freshwater ecosystems that support sustainable fisheries. *Environmental Reviews*, 22, 110–134.

Leino, R. L. & McCormick, J. H. (1993). Responses of juvenile largemouth bass to different pH and aliminium levels at overwintering temperatures – effects on gill morphology, electrolyte balance, scale calcium, liver glycogen, and depot fat. *Canadian Journal of Zoology*, 71, 531–543.

Levi, T., Darimont, C. T., MacDuffee, M., *et al.* (2012). Using grizzly bears to assess harvest-ecosystem tradeoffs in salmon fisheries. *PLoS Biology*, 10, e1001303.

Lindenmayer, D. B., Likens, G. E., Krebs, C. J. & Hobbs, R. J. (2010). Improved probability of detection of ecological 'surprises'. *Proceedings of the National Academy of Sciences*, 107, 21957–21962.

Linton, T. K., Reid, S. D. & Wood, C. M. (1999). Effects of a restricted ration on the growth and energetics of juvenile rainbow trout exposed to a summer of simulated warming and sublethal ammonia. *Transactions of the American Fisheries Society*, 128, 758–763.

Luckenbach, T., Kilian, M., Triebskorn, R. & Oberemm, A. (2001). Fish early life stage tests as a tool to assess embryotoxic potentials in small streams. *Journal of Aquatic Ecosystem Stress and Recovery*, 8, 355–370.

Magaud, H., Migeon, B., Morfin, P., Garric, J. & Vindimian, E. (1997). Modelling fish mortality due to urban storm run-off: interacting effects of hypoxia and un-ionized ammonia. *Water Research*, 31, 211–218.

Mann, R. M., Hyne, R. V., Choung, C. B. & Wilson, S. P. (2009). Amphibians and agricultural chemicals: review of the risks in a complex environment. *Environmental Pollution*, 157, 2903–2927.

Matthaei, C. D., Piggott, J. J. & Townsend, C. R. (2010). Multiple stressors in agricultural streams: interactions among sediment addition, nutrient enrichment and water abstraction. *Journal of Applied Ecology*, 47, 639–649.

McCullough, D. A., Bartholow, J. M., Jager, H. I., *et al.* (2009). Research in thermal biology: burning questions for coldwater stream fishes. *Reviews in Fisheries Science*, 17, 90–115.

McKim, J. M. (1977). Evaluation of tests with early life stages of fish for predicting long-term toxicity. *Journal of the Fisheries Research Board of Canada*, 34, 1148–1154.

McLeod, E., Anthony, K. R. N., Andersson, A., *et al.* (2013). Preparing to manage coral reefs for ocean acidification: lessons from coral bleaching. *Frontiers in Ecology and the Environment*, 11, 20–27.

Minns, C. K. (2013). The science of ecosystem-based management on a global scale: the Laurentian Great Lakes, Lake Ontario, and the Bay of Quinte as a nested case study. *Aquatic Ecosystem Health and Management*, 16, 229–239.

Moran, R., Harvey, I., Moss, B., *et al.* (2010). Influence of simulated climate change and eutrophication on three-spined stickleback populations: a large scale mesocosm experiment. *Freshwater Biology*, 55(2), 315–325.

Morgan, I. J., D'Cruz, L. M., Dockray, J. J., *et al.* (1998). The effects of elevated winter temperature and sub-lethal pollutants (low pH, elevated ammonia) on protein turnover in the gill and liver of rainbow trout (*Oncorhynchus mykiss*). *Fish Physiology and Biochemistry*, 19, 377–389.

Morgan, I. J., McDonald, D. G. & Wood, C. M. (2001). The cost of living for freshwater fish in a warmer, more polluted world. *Global Change Biology*, 7, 345–355.

Nakagawa, S. & Cuthill, I. C. (2007). Effect size, confidence interval and statistical significance: a practical guide for biologists. *Biological Reviews*, 82, 591–605.

Olden, J. D., Poff, N. L. & Bestgen, K. R. (2008). Trait synergisms and the rarity, extirpation, and extinction risk of desert fishes. *Ecology*, 89, 847–856.

Olden, J. D., Kennard, M. J., Leprieur, F., *et al.* (2010). Conservation biogeography of freshwater fishes: recent progress and future challenges. *Diversity and Distributions*, 16, 496–513.

Ormerod, S., Dobson, M., Hildrew, A. & Townsend, C. (2010). Multiple stressors in freshwater ecosystems. *Freshwater Biology*, 55(S1), 1–4.

Paine, R. T., Tegner, M. J. & Johnson, E. A. (1998). Compounded perturbations yield ecological surprises. *Ecosystems*, 1, 535–545.

Palikova, M., Krejci, R., Hilscherova, K., *et al.* (2007). Effects of different oxygen saturation on activity of complex biomass and aqueous crude extract of cyanobacteria during embryonal development in carp (*Cyprinus carpio* L.). *Acta Veterinaria Brno*, 76, 291–299.

Palmer, M. A., Menninger, H. L. & Bernhardt, E. (2010). River restoration, habitat heterogeneity and biodiversity: a failure of theory or practice? *Freshwater Biology*, 55(S1), 205–222.

Park, R. A., Clough, J. S. & Wellman, M. C. (2008). AQUATOX: modeling environmental fate and ecological effects in aquatic ecosystems. *Ecological Modelling*, 213, 1–15.

Person-Le Ruyet, J., Labbe, L., Le Bayon, N., *et al.* (2008). Combined effects of water quality and stocking density on welfare and growth of rainbow trout (*Oncorhynchus mykiss*). *Aquatic Living Resources*, 21, 185–195.

Peuranen, S., Keinanen, M., Tigerstedt, C. & Vuorinen, P. J. (2003). Effects of temperature on the recovery of juvenile grayling (*Thymallus thymallus*) from exposure to Al+Fe. *Aquatic Toxicology*, 65, 73–84.

Piggott, J. J., Lange, K., Townsend, C. R. & Matthaei, C. D. (2012). Multiple stressors in agricultural streams: a mesocosm study of interactions among raised water temperature, sediment addition and nutrient enrichment. *PLoS ONE*, 7, e49873.

Pikitch, E. K., Santora, C., Babcock, E. A., *et al.* (2004). Ecosystem-based fishery management. *Science*, 305, 346–347.

Pilati, A., Vanni, M. J., Gonzalez, M. J. & Gaulke, A. K. (2009). Effects of agricultural subsidies of nutrients and detritus on fish and plankton of shallow-reservoir ecosystems. *Ecological Applications*, 19, 942–960.

Pinheiro, J., Bates, D., DebRoy, S. & Sarkar, D. (2014). *nlme: Linear and Non-linear Mixed-effects Models*. Vienna.

Poff, N. L., Allan, J. D, Bain, M. B., *et al.* (1997). The natural flow regime. *Bioscience*, 47, 769–784.

Potts, S. G., Biesmeijer, J. C., Kremen, C., *et al.* (2010). Global pollinator declines: trends, impacts and drivers. *Trends in Ecology and Evolution*, 25, 345–353.

Power, M. (1997). Assessing the effects of environmental stressors on fish populations. *Aquatic Toxicology*, 39, 151–169.

Priddis, E., Rader, R., Belk, M., Schaalje, B. & Merkley, S. (2009). Can separation along the temperature niche axis promote coexistence between native and invasive species? *Diversity and Distributions*, 15, 682–691.

Qiang, J., Xu, P., Wang, H., Li, R. & Wang, H. (2012). Combined effect of temperature, salinity and density on the growth and feed utilization of Nile tilapia juveniles (*Oreochromis niloticus*). *Aquaculture Research*, 43, 1344–1356.

Quinn, G. P. & Keough, M. J. (2002). *Experimental Design and Data Analysis for Biologists*. New York, NT: Cambridge University Press.

R Development Core Team. (2013). *R: A Language and Environment for Statistical Computing*. Vienna: R Foundation for Statistical Computing.

Reese, C. D. & Harvey, B. C. (2002). Temperature-dependent interactions between juvenile steelhead and Sacramento pikeminnow in laboratory streams. *Transactions of the American Fisheries Society*, 131, 599–606.

Richardson, J., Williams, E. K. & Hickey, C. W. (2001). Avoidance behaviour of freshwater fish and shrimp exposed to ammonia and low dissolved oxygen separately and in combination. *New Zealand Journal of Marine and Freshwater Research*, 35, 625–633.

Roger, F., Godhe, A. & Gamfeldt, L. (2012). Genetic diversity and ecosystem functioning in the face of multiple stressors. *PLoS ONE*, 7, e45007.

Rutherford, J. C., Marsh, N. A., Davies, P. M. & Bunn, S. E. (2004). Effects of patchy shade on stream water temperature: how quickly do small streams heat and cool? *Marine and Freshwater Research*, 55, 737–748.

Salazar-Lugo, R., Estrella, A., Oliveros, A., *et al.* (2009). Paraquat and temperature affect nonspecific immune response of *Colossoma macropomum*. *Environmental Toxicology and Pharmacology*, 27, 321–326.

Salbu, B., Denbeigh, J., Smith, R. W., *et al.* (2008). Environmentally relevant mixed exposures to radiation and heavy metals induce measurable stress responses in Atlantic salmon. *Environmental Science & Technology*, 42, 3441–3446.

Serafy, J. E. & Harrell, R. M. (1993). Behavioral-response of fishes to increasing pH and dissolved-oxygen – field and laboratory observations. *Freshwater Biology*, 30, 53–61.

Shrimpton, J. M., Zydlewski, J. D. & Heath, J. W. (2007). Effect of daily oscillation in temperature and increased suspended sediment on growth and smolting in juvenile chinook salmon, *Oncorhynchus tshawytscha*. *Aquaculture*, 273, 269–276.

Skov, P. V., Larsen, B. K., Frisk, M. & Jokumsen, A. (2011). Effects of rearing density and water current on the respiratory physiology and haematology in rainbow trout, *Oncorhynchus mykiss* at high temperature. *Aquaculture*, 319, 446–452.

Sodhi, N. S., Bickford, D., Diesmos, A. C., *et al.* (2008). Measuring the meltdown: drivers of global amphibian extinction and decline. *PLoS ONE*, 3, e1636.

Stalnaker, C. B., Lamb, B. L., Henriksen, J., Boveee, K. & Bartholow, J. (1995). *The Instream Flow Incremental Methodology: A Primer for IFIM*. Washington, DC: USGS.

Stasko, A. D., Gunn, J. M. & Johnston, T. A. (2012). Role of ambient light in structuring north-temperate fish communities: potential effects of increasing dissolved organic carbon concentration with a changing climate. *Environmental Reviews*, 20, 173–190.

Statzner, B. & Bêche, L. A. (2010). Can biological invertebrate traits resolve effects of multiple stressors on running water ecosystems? *Freshwater Biology*, 55, 80–119.

Stoll, S., Sundermann, A., Lorenz, A. W., Kail, J. & Haase, P. (2013). Small and impoverished regional species pools constrain colonisation of restored river reaches by fishes. *Freshwater Biology*, 58, 664–674.

Suomalainen, L. R., Tiirola, M. A. & Valtonen, E. T. (2005). Influence of rearing conditions on *Flavobacterium columnare* infection of rainbow trout, *Oncorhynchus mykiss* (Walbaum). *Journal of Fish Diseases*, 28, 271–277.

Swanson, S. M. (2004). *Multiple Stressors: Literature Review and Gap Analysis (Water Environment Research Foundation Report 00-ECO-2B)*. London: International Water Association Publishing.

Townsend, C. R., Uhlmann, S. S. & Matthaei, C. D. (2008). Individual and combined responses of stream ecosystems to multiple stressors. *Journal of Applied Ecology*, 45, 1810–1819.

Valenti, T. W., Perez-Hurtado, P., Chambliss, C. K. & Brooks, B. W. (2009). Aquatic toxicity of sertraline to *Pimephales promelas* at environmentally relevant surface water pH. *Environmental Toxicology and Chemistry*, 28, 2685–2694.

Valenti, T. W., James, S. V., Lahousse, M. J., *et al.* (2010). A mechanistic explanation for pH-dependent ambient aquatic toxicity of *Prymnesium parvum carter*. *Toxicon*, 55, 990–998.

Van der Oost, R., Beyer, J. & Vermeulen, N. P. (2003). Fish bioaccumulation and biomarkers in environmental risk assessment: a review. *Environmental Toxicology and Pharmacology*, 13, 57–149.

Vanhoudt, N., Vandenhove, H., Real, A., Bradshaw, C. & Stark, K. (2012). A review of multiple stressor studies that include ionising radiation. *Environmental Pollution*, 168, 177–192.

Vera Cartas, J., Pucheu, K. & Torres Beristain, B. (2013). Contributions towards an ecosystem based management of Lake Titicaca. *Aquatic Ecosystem Health & Management*, 16, 240–247.

Verberk, W. C. E. P., van Noordwijk, C. G. E. & Hildrew, A. G. (2013). Delivering on a promise: integrating species traits to transform descriptive community ecology into a predictive science. *Freshwater Science*, 32, 531–547.

Vinebrooke, R. D., Cottingham, K. L., Norberg, J., *et al.* (2004). Impacts of multiple stressors on biodiversity and ecosystem functioning: the role of species co-tolerance. *Oikos*, 104, 451–457.

Von Westernhagen, H. (1988). Four sublethal effects of pollutants on fish eggs and larvae. In *Fish Physiology*. London: Academic Press, pp. 253–346.

Vörösmarty, C. J., McIntyre, P. B., Gessner, M. O., *et al.* (2010). Global threats to human water security and river biodiversity. *Nature*, 467, 555–561.

Wagenhoff, A., Lange, K., Townsend, C. R. & Matthaei, C. D. (2013). Patterns of benthic algae and cyanobacteria along twin-stressor gradients of nutrients and fine sediment: a stream mesocosm experiment. *Freshwater Biology*, 58, 1849–1863.

Wagenhoff, A., Townsend, C. R. & Matthaei, C. D. (2012). Macroinvertebrate responses along broad stressor gradients of deposited fine sediment and dissolved nutrients: a stream mesocosm experiment. *Journal of Applied Ecology*, 49, 892–902.

Wagner, E. J., Bosakowski, T. & Intelmann, S. (1997). Combined effects of temperature and high pH on mortality and the stress response of rainbow trout after stocking. *Transactions of the American Fisheries Society*, 126, 985–998.

Walsh, C. J., Fletcher, T. D. & Ladson, A. R. (2005). The urban stream syndrome: current knowledge and the search for a cure. *Journal of the North American Benthological Society*, 24, 706–723.

Waters, T. F. (1995). *Sediment in Streams: Sources, Biological Effects, and Control*. Bethesda, MD: American Fisheries Society.

Wenger, S. J., Roy, A. H., Jackson, C. R., *et al.* (2009). Twenty-six key research questions in urban stream ecology: an assessment of the state of the science. *Journal of the North American Benthological Society*, 28, 1080–1098.

Wood, P. J. & Armitage, P. D. (1997). Biological effects of fine sediment in the lotic environment. *Environmental Management*, 21, 203–217.

Woodward, G., Perkins, D. M. & Brown, L. E. (2010). Climate change and freshwater ecosystems: impacts across multiple levels of organization. *Philosophical Transactions of the Royal Society B: Biological Sciences*, 365, 2093–2106.

Yen, J. D. L., Bond, N. R., Shenton, W., Spring, D. A. & Mac Nally, R. (2013). Identifying effective water-management strategies in variable climates using population dynamics models. *Journal of Applied Ecology*, 50, 691–701.

Infectious disease and the conservation of freshwater fish

MARTIN KRKOSEK AND ROBERT POULIN

7.1 INTRODUCTION

Infectious diseases may be an important component of the conservation of freshwater fish. Rates of infectious diseases in freshwater fish are increasing (Johnson & Paull, 2011), and are likely a consequence of the multiple anthropogenic effects that are making freshwater systems the most degraded ecosystems on Earth (Carpenter *et al.*, 2011). More broadly, emerging diseases across taxa from all habitats tend to be related to environmental changes such as habitat fragmentation, species translocations/invasions, altered food webs, climate change or pollution (Daszak *et al.*, 2000; Dobson & Foufopoulos, 2001; Kelly *et al.*, 2009b). Similar processes may be at play in freshwater systems; however, these systems have received comparatively less attention than their terrestrial and marine counterparts (Johnson & Paull, 2011).

However, the role of infectious disease in species endangerment and extinction is complex and debatable (Lafferty & Gerber, 2002), and so it is not immediately clear if increasing rates of disease imply that disease is a threat to the conservation of freshwater fish. The primary reason for this is that transmission efficiency of many pathogens is linked to the density of hosts, and so when hosts become rare, it is expected that diseases will fade out and therefore may not necessarily cause extinction of their host (Grenfell & Dobson, 1995; Hudson *et al.*, 2001). Indeed, a parasite that eliminates its host population also reduces its own fitness to zero, and so it is not clear if diseases are a direct threat to species persistence.

Conservation of Freshwater Fishes, eds G. P. Closs, M. Krkosek and J. D. Olden. Published by Cambridge University Press. © Cambridge University Press 2016.

Furthermore, fishes have a common life-history characteristic of relatively high fecundity (egg output) but relatively low survival from egg through to reproductive maturity. Such high mortality within the life cycle is often associated with non-disease related factors such as predation. This leads to potentially complex dynamics (Hatcher *et al.*, 2012) where different mortality processes may interact in compensatory or synergistic ways. For example, if most juvenile fish will die anyway due to predation, does an increase in infection level correspond to an increase in overall mortality? If predators selectively remove infected prey, this may actually counteract disease mortality and lead to healthier populations or alternatively by making prey easier to capture may increase overall mortality (Packer *et al.*, 2003; Krkošek *et al.*, 2011a). Such complex ecological dynamics entangle predator–prey, competition, and host–parasite relationships and can thus lead to multiple mortality processes that may exacerbate or dampen the effects of disease on host populations.

Nonetheless, there are many mechanisms by which such density-dependent transmission may not limit the threat that pathogens can pose as agents of species endangerment and extinction (Lafferty & Gerber, 2002; de Castro & Bolker, 2005; Pedersen *et al.*, 2007). For example, a generalist parasite exploiting several host species can maintain high transmission rates even as the abundance of one of its hosts decreases, and can therefore contribute to the further decline of that host's population. Furthermore, there are case studies where disease features prominently in the decline of some freshwater fish species, or populations, as well as conservation and restoration efforts. Examples include migratory and reproductive failure associated with a genomic syndrome of viral infection in declining populations of sockeye salmon (Miller *et al.*, 2011), the role of the intracellular parasite, *Sphaerothecum destruens*, in the decline and extinction of the European endangered cyprinid *Leucaspius delineatus* and threats to other European freshwater fish taxa (Andreou *et al.*, 2012), and the role of trematode infections in developmental abnormalities in threatened New Zealand freshwater fish (Kelly *et al.*, 2010b).

When considering how infectious diseases might affect the conservation of a host population or species, it is important to keep in mind that agents of infectious disease can be divided into two major classes. Collectively, the infectious agents are referred to as parasites, although this may be misleading because the term 'parasites' is often thought to be specific to metazoans such as intestinal worms or lice. However, 'parasites' has a broader meaning that accommodates metazoan, bacterial, fungal and viral agents of infectious disease. The two broad classes are macroparasites and microparasites. Macroparasite typically refers to metazoan parasites, which have not just a larger body size, but also the characteristic of a life cycle in which they must pass through the environment and possibly intermediate hosts to complete their life cycle. In contrast, microparasites can replicate freely within an individual host and these typically refer to protists and viral or bacterial agents of disease. The distinction is important, because the life histories are very different and the models used to understand their epidemiology are also very different. Indeed, the virulence of macroparasites is intensity-dependent and their distribution among individual hosts therefore matters for their population dynamics and impact on the host, whereas the dynamics and impact of microparasites are best understood by treating the host population as distinct subsets based on whether they are infected, susceptible or recovered. However, some parasites do not easily fit into this framework – *Gyrodactylus salaris* is a metazoan ectoparasite that does not have an obligate free-living stage in its life cycle but instead replicates directly on the surface of the host. Overall, though, a common feature to both classes of parasite is that simple models of single host and single parasite dynamics typically involve a host density threshold, below which diseases are expected to fade out.

The mechanisms by which infectious diseases may contribute to conservation threats of endangered species typically involve more complexity than the simple single-host–single-pathogen case where the pathogen is expected to fade out before the host reaches extinction (de Castro & Bolker, 2005). These can include mechanisms of aggregation of the host that are independent of population size, frequency-dependent transmission, alternative host populations that can serve as a reservoir of infection, alternate contributory factors to disease such as contaminants that affect immune function, water abstraction and climate change that can affect the thermal environments of the fish and pathogen, the spread of exotic pathogens, pathogens that mediate other ecological interactions such as competition or predation, genetic effects of host populations at small sizes, and the role of infectious disease in making populations more susceptible to other contributing factors of extinction risk. In this review, we focus on the theoretical and empirical evidence of these potential processes by which infectious disease can threaten the conservation of freshwater fish.

7.2 FREQUENCY-DEPENDENT TRANSMISSION

For single-host–single-parasite dynamics, there are mechanisms by which the density dependence of transmission may not have a strong influence on eliminating disease-induced extinction risk. One obvious factor is the social organisation of a host population (Altizer *et al.*, 2003), which for fish includes common behaviours of shoaling and schooling, or aggregation in cold-water refugia, that lead to non-homogeneous mixing of the host population that forms a basis for density-dependent transmission (Richards *et al.*, 2010). Such aggregating of individuals can cause the density of a host population to remain relatively unchanged while the overall population is declining through a range contraction. In this situation, it is easy to see that changes in overall population size may have little influence on local density-dependent transmission dynamics and thereby allow disease to be a major component of population decline and extinction. Frequency-dependent transmission is thought to charac-terise the myxozoan, *Ceratomyxa shasta*, infecting juvenile Chinook sal-mon perhaps via host aggregation (Ray & Bartholomew, 2013).

There are also examples of infectious diseases that do not involve density-dependent transmission (de Castro & Bolker, 2005; Smith *et al.*, 2009). Such examples are typically frequency-dependent and apply to vector- or sexually transmitted infectious diseases (Thrall *et al.*, 1993, 2000). In these situations, the active mate-searching behaviour of hosts or host-finding behaviour of vectors can maintain higher contact rates between hosts and pathogens than would otherwise occur under nor-mal assumptions of well-mixed populations with passive host behaviour that gives rise to density-dependent transmission dynamics. Sexually transmitted diseases in freshwater fish are not common probably due to typical external fertilisation; however, *Gyrodactylus* monogeneans are a likely candidate to be transmitted during mating in guppies (which have internal fertilisation) (Harris, 1988; Houde & Torio, 1992). For vector-borne diseases, it is possible that macroparasites (ectoparasites) can act as vectors of microparasitic (viral and bacterial) diseases of fish (Cusack & Cone, 1986). For example, *Argulus* spp. are known to act as a vector in the transmission dynamics of infectious dropsy (spring viraemia) caused by *Rhabidovirus* in cyprinids (Ahne, 1985) and it is thought that salmon lice (*Lepeophtheirus salmonis*) may act as a vector of viral and bacterial agents of disease (Nese & Enger, 1993; Barker *et al.*, 2009; Petterson *et al.*, 2009).

7.3 MULTI-HOST DYNAMICS

Perhaps the most obvious route by which infectious disease can cause species endangerment and extinction is the presence of a secondary reservoir host population that maintains high levels of infection in the threatened host population even at very low population size. Many parasites are not actually specific to a single host species, and such multi-host population dynamics of disease may be common and can lead to dynamics that can threaten a particular host species. Alternatively, a reservoir host population can emerge due to the arrival of a non-indigenous species that shares parasites with native species. Such transmission of disease between native and non-native species can also lead to multi-host population dynamics that can affect either host species. Finally, a third common mechanism by which a reservoir host may threaten freshwater fish is the growth of domesticated populations and aquaculture or hatchery systems that are sympatric with wild fish. We review each of these three mechanisms of infectious disease dynamics in the presence of the reservoir host separately.

7.3.1 Natural multi-host dynamics

Parasites by definition have at least one competent host. That is, there is at least one host species in which the parasite population can spread. However, it is probably very common that parasites can infect multiple host species, to varying degrees of effectiveness (Holt *et al.*, 2003). Situations where there are at least two host species whose populations can sustain the infection lead to multi-host dynamics (Dobson, 2004). The key dynamic for conservation in such systems is termed apparent competition, where two host species appear to have negative effects on each other, but which actually occur indirectly through transmission of a shared parasite (Hudson & Greenman, 1998). More generally, apparent competition refers to the situation in which two species interact indirectly in a negative way due to a shared natural entity, which could also be a predator (Holt, 1977). Clear examples of apparent competition in freshwater fish are rare, although it is thought to have happened in some systems, but where the shared natural enemy is a predator (Persson, 1997). Empirical evidence for parasite-mediated apparent competition in non-fish vertebrates is known (Schmitz & Nudds, 1994), but we are unaware of any examples from fish.

7.3.2 Invasive host species

Interactions between infectious disease and invasive species may have dynamics that affect conservation of freshwater fish. An obvious way that biological invasion can cause disease emergence is when invasive species transmit novel parasites to native species (Daszak *et al.*, 2000; Dunn, 2009). For example, the only freshwater crayfish native to the British Isles, *Austropotamobius pallipes*, has suffered a major range contraction likely due to mortality from the oomycete *Aphanomyces astaci* introduced in the 1970s with the invasive American crayfish *Pacifastacus leniusculus* (Holdich & Reeve, 1991; Kozubikova *et al.*, 2009). Similarly, populations of the native European eel, *Anguilla anguilla*, have declined markedly following the introduction to Europe of the eel-specific parasitic nematode *Anguillicola crassus* with its original host from East Asia (Taraschewski, 2006), although overfishing and other causes have doubtless also played large roles. In Australia, the highly endemic freshwater fishes are threatened by a range of anthropogenic processes, and it has become clear that parasitic diseases, particularly those introduced with exotic fish species, may also pose a threat to native fishes (Thompson *et al.*, 2010). In North America, the introduction of the non-native freshwater snail *Melanoides tuberculata* has facilitated the spread of infection of the non-native trematode *Centrocestus formosanus* to native hosts, some of which are of conservation concern (Rader *et al.*, 2003; Mitchell *et al.*, 2005; McDonald *et al.*, 2006).

Infectious diseases, or rather their absence, may also play a role in the success of invasive species due to a tendency for parasites being lost during the invasion process because of random sampling processes or an inability for parasites to complete their life cycle or transmit efficiently in the new habitats of the introduced host (Torchin *et al.*, 2003). For example, release from parasites in their native range may help explain the successful *Neogobius* spp. invasions of European freshwaters (Ondrackova *et al.*, 2012) as well as the North American Great Lakes (Gendron *et al.*, 2012). Introductions of novel parasites along with their non-indigenous hosts may not be as common as one might think, however, because empirical evidence indicates that during introduction non-indigenous species tend to lose most of the parasites they had in their region of origin (Torchin *et al.*, 2003). Such processes appear to have played a role in the invasive spread of the cichlid, *Oreochromis niloticus*, in Central America (Roche *et al.*, 2010). In these scenarios, the role of infectious disease in conservation of freshwater fish is indirect: the process of releasing invasive species from parasites during the colonisation processes may enhance the fitness of invasive species in their introduced range and their subsequent impact on native species (Torchin *et al.*, 2003).

Poulin *et al.* (2011) reviewed the ways in which introduced species may alter the dynamics of endemic diseases in freshwater systems. Introduced species may act as competent hosts for endemic parasites in which infection is amplified and then 'spilled back' to native hosts. Examples include introduced salmonids and the acanthocephalan *Acanthocephalus tumescens* in Lake Moreno, Argentina (Rauque *et al.*, 2003) and potentially invasive cichlids on *Cyprinodon* species from Laguna Chichancanab, Mexico (Strĕcker, 2006). In North America, the introduction of shad (*Alosa sapidissima*) to the Pacific coast has led to the amplification and spread of the endemic parasites *Ichthyophonus* spp. (Hershberger *et al.*, 2010). In contrast, if introduced species are not suitable hosts for endemic parasites but become infected anyway, they may act as sinks for parasites and thus dilute disease risk for native hosts. Examples of such dilution processes include introduced salmonids in New Zealand acting as incompetent hosts for trematodes of native galaxiids (Kelly *et al.*, 2009a). Such effects can also occur by introduced species causing direct mortality of infective stages by feeding on them or causing transmission failure via physical interference (Thieltges *et al.*, 2008). Introduced species may also alter disease incidence and severity in native hosts through trait-mediated indirect effects that, for example, change the exposure or susceptibility of native hosts to infection by causing alterations in their behaviour or immunocompetence (Poulin *et al.*, 2011).

7.3.3 Domesticated host species

Multi-host disease dynamics can also threaten conservation of freshwater fish due to transmission dynamics between wild and domesticated populations (Johansen *et al.*, 2011; Kurath & Winton, 2011). For freshwater fish, such dynamics may occur between hatchery populations that are used for artificial propagation to augment fisheries, stock aquaculture, or ironically, to conserve wild populations. Diseases of bacterial, viral, fungal and metazoan origin are a common problem for freshwater fish hatchery environments, ranging from salmon, trout and char (Johnsen & Jensen, 1994; Jorgensen *et al.*, 2011; Wagner *et al.*, 2012) to catfish, tilapia and seabass (Akoll *et al.*, 2012; Smiley *et al.*, 2012). Movement of hatchery fish can also introduce new diseases, such as the case of furunculosis and salmonids in Norway, where the recovery of Norwegian rivers involved the use of rotenone to clear river systems of wild host populations for the parasite (Johnsen & Jensen, 1994).

Domesticated fishes that are held in aquaculture production systems can also create transmission dynamics between wild and farmed fish (Krkošek et al., 2006; Costello, 2009; Walker and Winton, 2010; Frazer et al., 2012). One consequence of these types of dynamics is that because parasite dynamics in the aquaculture environment are primarily controlled by aquaculture management rather than population dynamics of a free-living host, this can create dynamics in nearby wild host populations that include the erosion of protective dilution effects caused by host aggregation (Poulin & Fitzgerald, 1989), which then leads to Allee effects (Krkosek et al., 2013a), a key dynamic that can accelerate extinction risk (Stephens & Sutherland, 1999). Probably the flagship case study for understanding spill-over and spill-back dynamics of parasites between wild and farmed fish are salmon and their ectoparasitic copepods known as sea lice (Costello, 2009), where mortality in wild fish due to the parasite has been quantified (Krkošek et al., 2011b, 2013b). However, situations where wild and farmed fish share diseases are considered potential challenges to conservation (Diana, 2009).

A third dimension by which the domesticated fish populations may change disease dynamics in wild freshwater fish is due to evolutionary changes in pathogen traits that are selected for in aquaculture environments. As in agriculture and human-health fields, concern surrounds the potential changes in pathogen susceptibility to chemical or antibiotic treatments for controlling disease, traits that can be under intense selection for resistance (Bonhoeffer et al., 1997; Lenormand & Raymond, 1998; Lipsitch et al., 2000; Despres et al., 2007). Empirical evidence suggests that disease control in aquaculture can become problematic due to the evolution of resistance of parasites to disease control treatment (Westcott et al., 2010; Jones et al., 2012). Another process of concern is a potential rise in virulence of the pathogen (Pulkkinen et al., 2010), which determines the rate at which the pathogen induces morbidity and mortality in the host (Gandon et al., 2001; Strange & Scott, 2005; Day & Gandon, 2007). The basic assumption of virulence evolution is that there is a trade-off between transmission and pathogenicity (Day & Gandon, 2007). That is, the higher the rate of exploitation of the host by the pathogen, the more pathogen particles can be produced but at the expense of a reduced life span of the host. A consequence of this trade-off is that when the host population becomes more abundant or dense, the efficiency of transmission increases, sometimes allowing higher pathogen virulence to evolve because the pathogen can rely less heavily on host survival to propagate. Furthermore, applications of drugs or vaccines that are only partially effective or partially applied can also select for higher levels of virulence (Gandon *et al.*, 2001). Although the above predictions are based more on theory than empirical evidence, the necessary ingredients for the evolution of virulence exist in aquaculture systems (Pulkkinen *et al.*, 2010) as do examples of highly virulent outbreaks (Asche *et al.*, 2009).

7.4 POLLUTION AND DISEASE

Infectious diseases may emerge and threaten freshwater fish due to changes in the immunocompetence, physiology, or development of individual fish due to contamination of the environment (Poulin, 1992; Luebke *et al.*, 1997). Various environmental contaminants are known to modulate the immune system (Hill, 1997), and this may play a role in the impact of existing pathogens of freshwater fish. Examples include the effects of contaminants on parasite communities from roach (*Rutilus rutilus*) and perch (*Perca fluviatilis*) in Europe (Valtonen *et al.*, 1997; Kortet *et al.*, 2002), the effects of anti-inflammatory drug residues on immune function of brown trout in Europe (Hoeger *et al.*, 2005), the effects of cadmium and PCBs on immune response of eels (*Anguilla anguilla*) to metazoan parasites (Sures & Knopf, 2004) and the effects of nickel pollution on the immune function of Japanese killifish (*Oryzias latipes*) (Prophete *et al.*, 2006). Contaminants of freshwaters may also interact with parasites to impact freshwater fish, not through immune function, but rather development. For example, synergistic effects of trematode parasitism and glyphosate (a common herbicide) contamination can produce developmental abnormalities in galaxiids in New Zealand that likely have consequences for survival of affected fish (Kelly *et al.*, 2010a). However, it must be pointed out that depending on the type of contamination and the parasite taxon involved, pollutants can also have the opposite effects; that is, they can have direct negative impacts on parasite transmission stages or intermediate hosts, and thus lead to reduced infection of fish (Poulin, 1992; Lafferty, 1997).

7.5 WARMING AND DISEASE

The dynamics of many infectious diseases are likely dependent on temperature (Harvell *et al.*, 2002; Karvonen *et al.*, 2010). For example, developmental, survival and reproductive rates of aquatic trematode parasites are known to be sensitive to temperature (Poulin, 2006). Climate change will likely influence the rate of viral spread and also host immune response to infection, affect host susceptibility via thermal stress, and alter the distribution of vectors and hosts (Walker & Winton, 2010). For macroparasites, the bird tapeworm *Shistocephalus solidus* grows faster in warmer waters and alters the behaviour of its inter-mediate fish host – the stickleback – to seek warmer waters that in turn favours fish growth (Macnab & Barber, 2012). Warming waters may also allow pre-existing subclinical infections of the endoparasite myxozoan *Tetracapsuloides bryosalmonae* to emerge into epidemics of proliferative kidney disease in salmonid populations, or spread into new habitats, by altering spore production and the abundance of primary bryozoan hosts (Okamura *et al.*, 2011). Such effects of climate on diseases of freshwater fish may be compounded by water abstraction as well as modification of riparian areas that may magnify temperature changes (Meyer *et al.*, 1999; Caissie, 2006).

However, while it is clear that climate change may alter disease dynamics, it is not clear if this will increase disease rates overall, or rather cause geographic range shifts in the distribution of disease (Lafferty, 2009). For example, some diseases are linked to cold waters rather than warm, such as outbreaks of coldwater bacterial disease of wild ayu (*Plecoglossus altivelis*) and pale chub (*Zacco platypus*) in Japan (Iida & Mizokami, 1996) as well as bacterial coldwater diseases of rainbow trout (*Oncorhynchus mykiss*) reared in freshwaters in Europe (Dalsgaard & Madsen, 2000). Furthermore, while temperature may accelerate the developmental rates of parasites, this may also involve a reduction in the time period in which infectious stages can infect a host, thereby poten-tially offsetting increased reproductive rates. Nonetheless, some fresh-water fish appear well adapted to local thermal conditions (Eliason *et al.*, 2011), and temperature is associated with disease response in some pop-ulations (Foott *et al.*, 2004). Overall, it may not be possible to predict the general directional consequences of climate change on disease threats to wild fish due to the complexity of thermal adaptation, thermal stress and climate change.

7.6 POPULATION REDUCTION

Infectious diseases may be unlikely to be an ultimate cause of extinc-tion in simple single-host–single-parasite systems due to density depend-ence of transmission causing infection to fade out at low host abundance

(Grenfell & Dobson, 1995; Hudson *et al.*, 2001). However, diseases may nonetheless contribute to population declines and/or crashes which then make host populations more susceptible to other mechanisms of extinction, such as demographic stochasticity or Allee effects (de Castro & Bolker, 2005). Disease-induced/associated population declines are known from many systems, including vertebrates (de Castro & Bolker, 2005), but empirical evidence from freshwater fishes may have been overlooked. Examples from salmonids include declines in sockeye salmon (*O. nerka*) populations that may be associated with gene expression consistent with a viral infection (Miller *et al.*, 2011), declines of pink (*O. gorbuscha*), coho (*O. kisutch*) and Atlantic (*Salmo salar*) salmon populations associated with ectoparasitic sea lice (Figure 7.1) (Connors *et al.*, 2010; Krkošek & Hilborn, 2011; Krkošek *et al.*, 2013b), declines of Atlantic salmon and sea trout (*S. trutta*) populations associated with furunculosis (Johnsen & Jensen, 1994), increased mortality in Chinook salmon associated with the myxozoan parasite *Ceratomyxa shasta* (Fujiwara *et al.*, 2011) and population decline of rainbow trout (*O. mykiss*) following introduction of whirling disease (Miller & Vincent, 2008). Other examples from freshwaters are the decline of crayfish in Europe associated with crayfish plague (Holdich & Reeve, 1991; Taugbol *et al.*, 1993; Edgerton *et al.*, 2004; Kozubikova *et al.*, 2009), and the role of Asian tapeworm (*Bothriocephalus acheilognathi*) in conservation of the endangered bonytail chub (*Gila elegans*) in the Grand Canyon, USA (Hansen *et al.*, 2006).

Many of the above examples are correlative or may not fully account for the potential of compensatory mechanisms to offset parasite-associated mortality, largely due to a lack of ecosystem-scale experimental data. However, perhaps the clearest evidence comes from a meta-analysis of replicated manipulative field experiments on the influence of parasitism by crustaceans on the survival of Atlantic salmon in Europe (Figure 7.1) (Krkošek *et al.*, 2013b). The data included 24 trials in which tagged smolts (totalling 283,347 fish; 1996–2008) were released as paired control and parasiticide-treated groups into 10 areas of Ireland and Norway. All experimental fish were infection-free when released into freshwater, and a proportion of each group was recovered as adult recruits returning to coastal waters one or more years later. Treatment had a significant positive effect on survival to recruitment, with an overall effect size (odds ratio) of 1.29 that corresponds to an estimated loss of 39% of adult salmon recruitment. The results of that meta-analysis provide ecosystem-scale evidence that parasites can have large impacts on fish recruitment, fisheries and conservation (Krkošek *et al.*, 2013b).

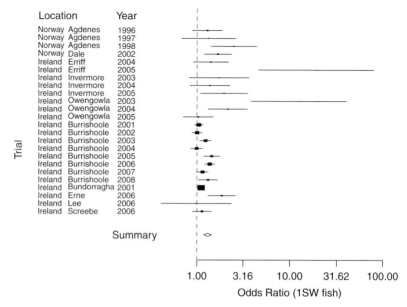

Figure 7.1 Forest plot displaying a random-effects meta-analysis of the effect of parasiticide treatment on the likelihood of a one sea winter (1SW) adult salmon returning (Krkosek et al., 2013b). Horizontal lines represent the 95% confidence intervals of the effect size in each trial, and the relative sizes of solid squares reflect the percentage weighting (based on standard errors of effect sizes) of each trial in the meta-analysis (range 0.42–8.12%). The open diamond shows the overall meta-analytic effect across all studies, with its width corresponding to the 95% confidence interval. Results are given by trial, identified by Location (country and river of release of smolts) and Year (year when smolts were released). See Krkosek et al. (2013b) for details and data sources.

7.7 GENETIC DIVERSITY AND DISEASE

Disease risks for the conservation of freshwater fish may be exacerbated by the loss of genetic diversity in fish populations that have declined for various other reasons or may simply have small population sizes. Such effects are known for non-fish species, where population bottlenecks can reduce genetic variability and immune function, increasing disease risk. There are several examples from non-fish species where genetic bottlenecks have contributed to extinction risk via disease, including cheetahs (*Acinonyx jubatus*) and feline infectious peritonitis (Obrien *et al.*, 1985), pocket gopher (*Thomomys bottae*) (Sanjayan *et al.*, 1996), and wolves (*Canis lupis*) at Isle Royale and canine parvovirus (Peterson *et al.*, 1998). For freshwater fish, there is less direct evidence of extinction risk, but examples relevant to conservation include effects of inbreeding in guppies (*Poecilia reticulata*) on mortality associated with gyrodactylid parasites (Van Oosterhout *et al.*, 2007). For rainbow trout, standing genetic variation is considered to have been an important component of a rapid adaptation of a local rainbow trout population to the introduction of a novel pathogen, *Myxobolus cerebralis*, the causative agent of whirling disease (Miller & Vincent, 2008). There are unique aspects of habitat structure and dispersal of freshwater fish that may influence genetic risks. For example, the fidelity to their natal rivers likely causes populations of Atlantic salmon to exhibit genetic structuring and variability that is adaptive (Garcia de Leaniz *et al.*, 2007), and small river systems that support salmon populations of low effective population size (Nikolic *et al.*, 2009) may be more vulnerable. Furthermore, low diversity of major histocompatibility (MH) genes in Atlantic salmon is associated with disease mortality, and maintaining MH genetic variation is a conservation recommendation (de Eyto *et al.*, 2007).

7.8 PREDATOR–PREY DYNAMICS

Mortality from infectious diseases may be exacerbated or compensated by other mortality processes, and for freshwater fish, predation may be a key component. A common characteristic of freshwater fish life histories is their relatively high fecundity (output of eggs and larvae) but relatively low survival from egg to reproductive maturity, typically due to high predation rates. Understanding how infectious diseases may impact the conservation of freshwater fish therefore entails understanding how infectious disease fits into predator–prey dynamics. Multiple outcomes are possible (Hatcher *et al.*, 2012), and interactions are likely if infection has the probable effect of increasing the risk of predation in prey (Mesa *et al.*, 1998; Mages & Dill, 2010; Krkošek *et al.*, 2011a). For example, trophically transmitted helminth parasites using fish as intermediate hosts can induce subtle changes in fish behaviour that cause disproportionate increases in their rates of predation by the parasites' definitive hosts (Lafferty & Morris, 1996; Seppala *et al.*, 2004; Poulin, 2010) resulting in additive mortality for the fish population. On the other hand, selective predation on infected individuals may reduce infection levels in a prey population, increasing the overall health and viability in the prey population (Packer *et al.*, 2003). Such effects may be density-dependent; when parasite influences are incorporated into a type II functional response in a predator–prey model, compensatory dynamics occur at high prey abundance where predators reduce infection levels without changing overall mortality rates, but at low prey abundance parasitism and predation act synergistically, increasing mortality rates (Krkošek *et al.*, 2011a). Situations where both a parasite and a predator are shared by two host prey populations may lead to exacerbated mortality in one prey population but release from predation by parasites in the other prey population (Peacock *et al.*, 2014), underscoring the complexity of disease dynamics that are possible in ecological networks (Hatcher *et al.*, 2012).

7.9 GEOGRAPHIC AND TAXONOMIC BIASES AND TRENDS

The literature on infectious diseases and conservation of freshwater fish has a clear bias towards developed countries in temperate latitudes and species of importance to aquaculture. This may not be surprising, as the recent rapid growth of aquaculture has been accompanied by the emergence of infectious diseases and literature associated with veterinary management (Meyer, 1991; Murray & Peeler, 2005; Walker & Winton, 2010). For wild freshwater fish, the literature on infectious diseases in conservation seems to be largely focused on salmonids, potentially due to the growth of knowledge from diseases in salmon aquaculture systems (Johnson *et al.*, 2004; Costello, 2006), but also likely due to the overall large research programmes on these fishes motivated by their ecological and fishery importance (Groot & Margolis, 1991; Groot *et al.*, 1995; Quinn, 2005). Even for wild salmon, knowledge of how infectious diseases affect population dynamics has been considered sparse (Bakke & Harris, 1998), but there has been considerable growth in the area in recent years (Downing *et al.*, 2002; Foott *et al.*, 2004; Miller & Vincent, 2008; Costello, 2009; Fujiwara *et al.*, 2011; Miller *et al.*, 2011). More generally, the literature on infectious diseases in freshwater fish has been rapidly growing in recent years, but of the most commonly reported parasites of freshwater fish, four of the top five are viral and myxozoan parasites of salmonids, and the fifth is a viral parasite of carp that is primarily of concern for aquaculture (Johnson & Paull, 2011).

Data on infectious diseases and their role in the conservation of tropical freshwater fishes are very sparse. That is not to say, however, that infectious diseases are not an important component of aquatic ecosystems in these areas. Recent meta-analyses indicate that, for aquaculture systems, outbreaks of infectious diseases develop at faster rates and with higher mortality in the tropics than in temperate regions, potentially due to climatic variables, but other confounding factors, such as

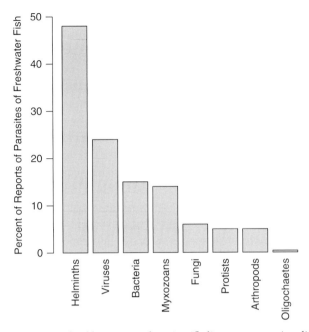

Figure 7.2 Percent of publications in the scientific literature reporting diseases according to the type of parasites involved (Johnson & Paull, 2011).

animal husbandry and veterinary knowledge and practice, may play a role (Leung & Bates, 2013).

In terms of taxonomic biases and trends regarding the parasites themselves, infectious diseases of freshwater fish depart from other freshwater taxa (Johnson & Paull, 2011). In particular, reports of helminth macroparasites tend to be less frequent in freshwater fish than in amphibians, molluscs, reptiles and mammals, whereas myxozoan parasites are reported much more frequently from freshwater fish than other host taxa. Overall, about half of the reports of parasites of freshwater fish concern helminths and about a quarter concern viruses; there are much fewer reports on other parasite taxa (Figure 7.2) (Johnson & Paull, 2011).

7.10 CONCLUSION

It is clear that the role of infectious diseases in the conservation of freshwater fish has received very little attention relative to diseases in terrestrial and marine environments (Harvell *et al.*, 1999; Daszak *et al.*, 2000; McCallum *et al.*, 2004) as well as relative to other factors that may affect the conservation of freshwater fish reviewed in this book. We are unaware of a single case study where an infectious disease has been the cause of extinction of a freshwater fish species; however, there are many cases where infectious diseases may play a role in the decline, conservation, or recovery of freshwater fish (Edgerton *et al.*, 2004; Miller *et al.*, 2011; Andreou *et al.*, 2012; Krkosek *et al.*, 2013b).

We have identified a suite of mechanisms by which infectious diseases may impact the conservation of freshwater fish. Many of the mechanisms are not unique to freshwater or even aquatic habitats, but are instead general principles by which disease may contribute to endangerment or extinction of host species (de Castro & Bolker, 2005). Primary among these are the dynamics of parasite transmission in multi-host systems where spill-over and spill-back dynamics may occur between invasive and native host species, between domestic and wild species, or simply between native host species. These situations can lead to apparent competition, or simply spill-back dynamics that allow parasite levels to remain high even when a focal host population of conservation concern declines to low abundance. There is ample theoretical and empirical evidence for these types of dynamics in domestic–wild host interactions of salmon (Costello, 2006, 2009). Such dynamics between invasive and native species are poorly studied (Kelly *et al.*, 2009b), but include examples where the invasive host species acts as a sink rather than a source in disease transmission (Paterson *et al.*, 2011). We are unaware of any examples of disease-mediated apparent competition between native freshwater fish species, a noted knowledge gap (Tompkins *et al.*, 2011). Of course, the introduction of highly pathogenic parasites to naïve host populations is a component of such multi-host dynamics that has had impacts on conservation of freshwater fauna (Edgerton *et al.*, 2004; Andreou *et al.*, 2012).

In addition to the above known mechanisms, there are some unique characteristics of freshwater fish that may elevate their susceptibility to infectious diseases. Primary among these is the interaction between pollution, immune function and disease. Due to the high levels of pollution in freshwater systems, freshwater fish are likely more chronically exposed to higher concentrations and more diverse pollutants than taxa in other habitats (see Chapter 5). As such, the relationship between environmental contaminants, immune function, infectious disease and host population ecology are a key set of linkages for understanding how infectious disease may affect the conservation of freshwater fish. This area is quite understudied, although examples of effects of contaminants on immune function in freshwater fish exist (Luebke *et al.*, 1997). Comprehensive analyses that scale from toxicology, immunology and population ecology are lacking and should be a research priority.

Another unique feature of freshwater fish that may be important in the role of infectious diseases and conservation is the typical spatial meta-population structure of species across landscapes (see Chapter 16). Such population structuring implies that local adaptation and more limited dispersal may increase many synergies of infectious diseases with population endangerment, such as population reduction and genetic diversity. However, it also implies that diseases may be somewhat slower to spread across landscapes and thereby endanger an entire meta-population as a whole. Examples of these types of dynamics are evident from salmonids in Norway, where local extirpations were actually induced to eradicate the monogenean ectoparasite *Gyrodactylus* (Johnsen & Jensen, 1991).

Overall, infectious diseases may be an important but critically understudied component of the conservation of freshwater fish (Johnson & Paull, 2011). Areas of knowledge strengths are largely limited to temperate fishes of importance to fisheries and aquaculture (i.e. salmonids), although even here research is quite limited and in its infancy. There is nearly a complete dearth of knowledge on how infectious diseases influence the conservation of tropical freshwater fishes. In addition to taxonomic and geographic limitations, perhaps the greatest research challenge is understanding how multiple factors that interact with infectious diseases scale from molecular to ecological processes. This includes compensatory and synergistic dynamics of multiple mortality processes that likely influence how disease dynamics affect the conservation of freshwater fish.

7.11 DISCUSSION QUESTIONS

1. What is a host-density threshold in epidemiology? What types of disease systems may not have such a threshold?
2. Are infectious diseases a primary or secondary factor that may endanger freshwater fish species?
3. Does pollution exacerbate or ameliorate the spread of infectious disease in freshwater fish?
4. Will global climate change increase disease rates in freshwater fish? Why or why not?

5. How does the metapopulation structure of freshwater fish species affect disease spread and extinction risk?

7.12 REFERENCES

Ahne, W.(1985). *Argulus foliaceus* L. and *Piscicola geometra* L. as mechanical vecctors of spring viremia of carp virus (SVCV). *Journal of Fish Diseases*, 8, 241–242.

Akoll, P., Konecny, R. W., Mwanja, W., *et al.* (2012). Parasite fauna of farmed Nile tilapia (*Oreochromis niloticus*) and African catfish (*Clarias gariepinus*) in Uganda. *Parasitology Research*, 110, 315–323.

Altizer, S., Nunn, C. L., Thrall, P. H., *et al.* (2003). Social organisation and parasite risk in mammals: Integrating theory and empirical studies. *Annual Review of Ecology Evolution and Systematics*, 34, 517–547.

Andreou, D., Arkush, K. D., Guegan, J. F. & Gozlan, R. E. (2012). Introduced pathogens and native freshwater biodiversity: a case study of *Sphaerothecum destruens*. *PLoS ONE*, 7, e36998

Asche, F., Hansen, H., Tveteras, R. & Tveteras, S. (2009). The salmon disease crisis in Chile. *Marine Resource Economics*, 24, 405–411.

Bakke, T. A. & Harris, P. D. (1998). Diseases and parasites in wild Atlantic salmon (*Salmo salar*) populations. *Canadian Journal of Fisheries and Aquatic Sciences*, 55(S1), 247–266.

Barker, D. E., Braden, L. M., Coombs, M. P. & Boyce, B. (2009). Preliminary studies on the isolation of bacteria from sea lice, *Lepeophtheirus salmonis*, infecting farmed salmon in British Columbia, Canada. *Parasitology Research*, 105, 1173–1177.

Bonhoeffer, S., Lipsitch, M. & Levin, B. R. (1997). Evaluating treatment protocols to prevent antibiotic resistance. *Proceedings of the National Academy of Sciences of the United States of America*, 94, 12106–12111.

Caissie, D. (2006). The thermal regime of rivers: a review. *Freshwater Biology*, 51, 1389–1406.

Carpenter, S. R., Stanley, E. H. & Vander Zanden, M. J. (2011). State of the world's freshwater ecosystems: physical, chemical, and biological changes. *Annual Review of Environment and Resources*, 36, 75–99.

Connors, B., Krkošek, M., Ford, J. S. & Dill, L. M. (2010). Coho salmon productivity in relation to salmon lice from infected prey and salmon farms. *Journal of Applied Ecology*, 47, 1372–1377.

Costello, M. J. (2006). Ecology of sea lice parasitic on farmed and wild fish. *Trends in Parasitology*, 22, 475–483.

Costello, M. J. (2009). How sea lice from salmon farms may cause wild salmonid declines in Europe and North America and be a threat to fishes elsewhere. *Proceedings of the Royal Society B*, 276, 3385–3394.

Cusack, R. & Cone, D. K. (1986). A review of parasites as vectors of viral and bacterial diseases of fish. *Journal of Fish Diseases*, 9, 169–171.

Dalsgaard, I. & Madsen, L. (2000). Bacterial pathogens in rainbow trout, *Oncorhynchus mykiss* (Walbaum), reared at Danish freshwater farms. *Journal of Fish Diseases*, 23, 199–209.

Daszak, P., Cunningham, A. A. & Hyatt, A. D. (2000). Emerging infectious diseases of wildlife – threats to biodiversity and human health. *Science*, 287, 443–449.

Day, T. & Gandon, S. (2007). Applying population-genetic models in theoretical evolutionary epidemiology. *Ecology Letters*, 10, 876–888.

de Castro, F. & Bolker, B. (2005). Mechanisms of disease-induced extinction. *Ecology Letters*, 8, 117–126.

de Eyto, E., McGinnity, P., Consuegra, S., *et al.* (2007). Natural selection acts on Atlantic salmon major histocompatibility (MH) variability in the wild. *Proceedings of the Royal Society B*, 274, 861–869.

Despres, L., David, J. P. & Gallet, C. (2007). The evolutionary ecology of insect resistance to plant chemicals. *Trends in Ecology and Evolution*, 22, 298–307.

Diana, J. S. (2009). Aquaculture production and biodiversity conservation. *Bioscience*, 59, 27–38.

Dobson, A. (2004). Population dynamics of pathogens with multiple host species. *American Naturalist*, 164, S64–S78.

Dobson, A. & Foufopoulos, J. (2001). Emerging infectious pathogens of wildlife. *Philosophical Transactions of the Royal Society B*, 356, 1001–1012.

Downing, D. C., McMahon, T. E., Kerans, B. L. & Vincent, E. R. (2002). Relation of spawning and rearing life history of rainbow trout and susceptibility to *Myxobolus cerebralis* infection in the Madison River, Montana. *Journal of Aquatic Animal Health*, 14, 191–203.

Dunn, A. M. (2009). Parasites and biological invasions. *Advances in Parasitology*, 68, 161–184.

Edgerton, B. F., Henttonen, P., Jussila, J., *et al.* (2004). Understanding the causes of disease in European freshwater crayfish. *Conservation Biology*, 18, 1466 1474.

Eliason, E. J., Clark, T. D., Hague, M. J., *et al.* (2011). Differences in thermal tolerance among sockeye salmon populations. *Science*, 332, 109–112.

Foott, J. S., Harmon, R. & Stone, R. (2004). Effect of water temperature on non-specific immune function and ceratomyxosis in juvenile chinook salmon and steelhead from the Klamath River. *California Fish and Game*, 90, 71–84.

Frazer, L. N., Morton, A. & Krkošek, M. (2012). Critical thresholds in sea lice epidemics: evidence, sensitivity, and subcritical estimation. *Proceedings of the Royal Society B*, 279, 1950–1958.

Fujiwara, M., Mohr, M. S., Greenberg, A., Foott, J. S. & Bartholomew, J. L. (2011). Effects of ceratomyxosis on population dynamics of Klamath fall-run Chinook salmon. *Transactions of the American Fisheries Society*, 140, 1380–1391.

Gandon, S., Mackinnon, M. J., Nee, S. & Read, A. F. (2001). Imperfect vaccines and the evolution of pathogen virulence. *Nature*, 414, 751–756.

Garcia de Leaniz, C. G., Fleming, I. A., Einum, S., *et al.* (2007). A critical review of adaptive genetic variation in Atlantic salmon: implications for conservation. *Biological Reviews*, 82, 173–211.

Gendron, A. D., Marcogliese, D. J. & Thomas, M. (2012). Invasive species are less parasitised than native competitors, but for how long? The case of the round goby in the Great Lakes–St. Lawrence Basin. *Biological Invasions*, 14, 367–384.

Grenfell, B. T. & Dobson, A. P. (1995). *Ecology of Infectious Diseases in Natural Populations*. Cambridge University Press.

Groot, C. & Margolis, L. (1991). *Pacific Salmon Life Histories*. Vancouver: UBC Press.

Groot, C., Margolis, L. & Clarke, W. C. (1995). *Physiological Ecology of Pacific Salmon*. Vancouver: UBC Press.

Hansen, S. P., Choudhury, A., Heisey, D. M., *et al.* (2006). Experimental infection of the endangered bonytail chub (*Gila elegans*) with the Asian fish tapeworm (*Bothriocephalus acheilognathi*): impacts on survival, growth, and condition. *Canadian Journal of Zoology*, 84, 1383–1394.

Harris, P. D. (1988). Changes in the site specificity of *Gyrodactylus turnbulli* Harris, 1986 (Monogenea) during infections of individual guppies (*Poecilia reticulata* Peters, 1859). *Canadian Journal of Zoology*, 66, 2854–2857.

Harvell, C. D., Kim, K., Burkholder, J. M., *et al.* (1999). Review: Marine ecology – emerging marine diseases – climate links and anthropogenic factors. *Science*, 285, 1505–1510.

Harvell, C. D., Mitchell, C. E., Ward, J. R., *et al.* (2002). Ecology – climate warming and disease risks for terrestrial and marine biota. *Science*, 296, 2158–2162.

Hatcher, M. J., Dick, J. T. A. & Dunn, A. M. (2012). Diverse effects of parasites in ecosystems: linking interdependent processes. *Frontiers in Ecology and the Environment*, 10, 186–194.

Hershberger, P. K., van der Leeuw, B. K., Gregg, J. L., *et al.* (2010). Amplification and transport of an endemic fish disease by an introduced species. *Biological Invasions*, 12, 3665–3675.

Hill, M. (1997). *Understanding Environmental Pollution*. Cambridge University Press.

Hoeger, B., Kollner, B., Dietrich, D. R. & Hitzfeld, B. (2005). Water-borne diclofenac affects kidney and gill integrity and selected immune parameters in brown trout (*Salmo trutta f. fario*). *Aquatic Toxicology*, 75, 53–64.

Holdich, D. M. & Reeve, I. D. (1991). Distribution of freshwater crayfish in the British Isles, with particular reference to crayfish plague, alien introductions, and water quality. *Aquatic Conservation – Marine and Freshwater Ecosystems*, 1, 139–158.

Holt, R. D. (1977). Predation, apparent competition, and structure of prey communities. *Theoretical Population Biology*, 12, 197–229.

Holt, R. D., Dobson, A. P., Begon, M., Bower, R. G. & Schauber, E. M. (2003). Parasite establishment in host communities. *Ecology Letters*, 6, 837–842.

Houde, A. E. & Torio, A. J. (1992). Effect of parasitic infection on male color pattern and female choice in guppies. *Behavioral Ecology*, 3, 346–351.

Hudson, P. & Greenman, J. (1998). Competition mediated by parasites: biological and theoretical progress. *Trends in Ecology and Evolution*, 13, 387–390.

Hudson, P. J., Rizzoli, A., Grenfell, B. T., Heesterbeek, H. & Dobson, A. P. (2001). *The Ecology of Wildlife Diseases*. Oxford: Oxford University Press.

Iida, Y. & Mizokami, A. (1996). Outbreaks of coldwater disease in wild ayu and pale chub. *Fish Pathology*, 31, 157–164.

Johansen, L. H., Jensen, I., Mikkelsen, H., *et al.* (2011). Disease interaction and pathogens exchange between wild and farmed fish populations with special reference to Norway. *Aquaculture*, 315, 167–186.

Johnsen, B. O. & Jensen, A. J. (1991). The gyrodactylus story in Norway. *Aquaculture*, 98, 289–302.

Johnsen, B. O. & Jensen, A. J. (1994). The spread of furunculosis in salmonids in Norwegian rivers. *Journal of Fish Biology*, 45, 47–55.

Johnson, P. T. J. & Paull, S. H. (2011). The ecology and emergence of diseases in fresh waters. *Freshwater Biology*, 56, 638–657.

Johnson, S. C., Treasurer, J. W., Bravo, S., Nagasawa, K. & Kabata, Z. (2004). A review of the impact of parasitic copepods on marine aquaculture. *Zoological Studies*, 43, 229–243.

Jones, P. G., Hammell, K. L., Dohoo, I. R. & Revie, C. W. (2012). Effectiveness of emamectin benzoate for treatment of *Lepeophtheirus salmonis* on farmed Atlantic salmon *Salmo salar* in the Bay of Fundy, Canada. *Diseases of Aquatic Organisms*, 102, 53–64.

Jorgensen, A., Torp, K., Bjorland, M. A. & Poppe, T. T. (2011). Wild Arctic char *Salvelinus alpinus* and trout *Salmo trutta*: hosts and reservoir of the salmonid pathogen *Spironucleus salmonicida* (Diplomonadida; Hexamitidae). *Diseases of Aquatic Organisms*, 97, 57–63.

Karvonen, A., Rintamaki, P., Jokela, J. & Valtonen, E. T. (2010). Increasing water temperature and disease risks in aquatic systems: climate change increases the risk of some, but not all, diseases. *International Journal for Parasitology*, 40, 1483–1488.

Kelly, D. W., Paterson, R. A., Townsend, C. R., Poulin, R. & Tompkins, D. M. (2009a). Has the introduction of brown trout altered disease patterns in native New Zealand fish? *Freshwater Biology*, 54, 1805–1818.

Kelly, D. W., Paterson, R. A., Townsend, C. R., Poulin, R. & Tompkins, D. M. (2009b). Parasite spillback: a neglected concept in invasion ecology? *Ecology*, 90, 2047–2056.

Kelly, D. W., Poulin, R., Tompkins, D. M. & Townsend, C. R. (2010a). Synergistic effects of glyphosate formulation and parasite infection on fish malformations and survival. *Journal of Applied Ecology*, 47, 498–504.

Kelly, D. W., Thomas, H., Thieltges, D. W., Poulin, R. & Tompkins, D. M. (2010b). Trematode infection causes malformations and population effects in a declining New Zealand fish. *Journal of Animal Ecology*, 79, 445–452.

Kortet, R., Vainikka, A. & Taskinen, J. (2002). Epizootic cutaneous papillomatosis in roach *Rutilus rutilus*: sex and size dependence, seasonal occurrence and between-population differences. *Diseases of Aquatic Organisms*, 52, 185–190.

Kozubikova, E., Filipova, L., Kozak, P., *et al.* (2009). Prevalence of the Crayfish plague pathogen *Aphanomyces astaci* in invasive American crayfishes in the Czech Republic. *Conservation Biology*, 23, 1204–1213.

Krkosek, M. & Hilborn, R. (2011). Sea lice (*Lepeophtheirus salmonis*) infestations and the productivity of pink salmon (*Oncorhynchus gorbuscha*) in the Broughton Archipelago, British Columbia, Canada. *Canadian Journal of Fisheries and Aquatic Sciences*, 68, 17–29.

Krkošek, M., Lewis, M. A., Morton, A., Frazer, L. N. & Volpe, J. P. (2006). Epizootics of wild fish induced by farm fish. *Proceedings of the National Academy of Sciences of the USA*, 103, 15506–15510.

Krkošek, M., Connors, B., Mages, P., *et al.* (2011a). Fish farms, parasites, and predators: implications for salmon population dynamics. *Ecological Applications*, 21, 897–914.

Krkošek, M., Connors, B. M., Morton, A., *et al.* (2011b). Effects of parasites from salmon farms on productivity of wild salmon. *Proceedings of the National Academy of Sciences of the United States of America*, 108, 14700–14704.

Krkosek, M., Ashander, J., Frazer, & Lewis, M. A. (2013a). Allee effect from parasite spillback. *American Naturalist*, 182, 640–652.

Krkosek, M., Revie, C., Gargan, P., *et al.* (2013b). Impact of parasites on salmon recruitment in the Northeast Atlantic Ocean. *Proceedings of the Royal Society B*, 280, 20122359.

Kurath, G. & Winton, J. (2011). Complex dynamics at the interface between wild and domestic viruses of finfish. *Current Opinion in Virology*, 1, 73–80.

Lafferty, K. D. (1997). Environmental parasitology: what can parasites tell us about human impacts on the environment? *Parasitology Today*, 13, 251–255.

Lafferty, K. D. (2009). The ecology of climate change and infectious diseases. *Ecology*, 90, 888–900.

Lafferty, K. D. & Gerber, L. R. (2002). Good medicine for conservation biology: the intersection of epidemiology and conservation theory. *Conservation Biology*, 16, 593–604.

Lafferty, K. D. & Morris, A. K. (1996). Altered behavior of parasitised killifish increases susceptibility to predation by bird final hosts. *Ecology*, 77, 1390–1397.

Lenormand, T. & Raymond, M. (1998). Resistance management: the stable zone strategy. *Proceedings of the Royal Society B*, 265, 1985–1990.

Leung, T. L. F. & Bates, A. E. (2013). More rapid and severe disease outbreaks for aquaculture at the tropics: implications for food security. *Journal of Applied Ecology*, 50, 215–222.

Lipsitch, M., Bergstrom, C. T. & Levin, B. R. (2000). The epidemiology of antibiotic resistance in hospitals: paradoxes and prescriptions. *Proceedings of the National Academy of Sciences of the United States of America*, 97, 1938–1943.

Luebke, R. W., Hodson, P. V., Faisal, M., *et al.* (1997). Aquatic pollution-induced immunotoxicity in wildlife species. *Fundamental and Applied Toxicology*, 37, 1–15.

Macnab, V. & Barber, I. (2012). Some (worms) like it hot: fish parasites grow faster in warmer water, and alter host thermal preferences. *Global Change Biology*, 18, 1540–1548.

Mages, P. A. & Dill, L. M. (2010). The effect of sea lice (*Lepeophtheirus salmonis*) on juvenile pink salmon (*Oncorhynchus gorbuscha*) swimming endurance. *Canadian Journal of Fisheries and Aquatic Sciences*, 67, 2045–2051.

McCallum, H. I., Kuris, A., Harvell, C. D., *et al.* (2004). Does terrestrial epidemiology apply to marine systems? *Trends in Ecology and Evolution*, 19, 585–591.

McDonald, D. L., Bonner, T. H., Brandt, T. M. & Trevino, G. H. (2006). Size susceptibility to trematode-induced mortality in the endangered fountain darter (*Etheostoma fonticola*). *Journal of Freshwater Ecology*, 21, 293–299.

Mesa, M. G., Poe, T. P., Maule, A. G. & Schreck, C. B. (1998). Vulnerability to predation and physiological stress responses in juvenile chinook salmon (*Oncorhynchus tshawytscha*) experimentally infected with *Renibacterium salmoninarum*. *Canadian Journal of Fisheries and Aquatic Sciences*, 55, 1599–1606.

Meyer, F. P. 1991. Aquaculture disease and health management. *Journal of Animal Science*, 69, 4201–4208.

Meyer, J. L., Sale, M. J., Mulholland, P. J. & Poff, N. L. (1999). Impacts of climate change on aquatic ecosystem functioning and health. *Journal of the American Water Resources Association*, 35, 1373–1386.

Miller, K. M., Li, S., Kaukinen, K. H., *et al.* (2011). Genomic signatures predict migration and spawning failure in wild Canadian salmon. *Science*, 331, 214–217.

Miller, M. P. & Vincent, E. R. (2008). Rapid natural selection for resistance to an introduced parasite of rainbow trout. *Evolutionary Applications*, 1, 336–341.

Mitchell, A. J., Overstreet, R. M., Goodwin, A. E. & Brandt, T. M. (2005). Spread of an exotic fish-gill trematode: a far-reaching and complex problem. *Fisheries*, 30, 11–16.

Murray, A. G. & Peeler, E. J. (2005). A framework for understanding the potential for emerging diseases in aquaculture. *Preventive Veterinary Medicine*, 67, 223–235.

Nese, L. & Enger, O. (1993). Isolation of *Aeromonas salmonicida* from salmon lice *Lepeophtheirus salmonis* and marine plankton. *Diseases of Aquatic Organisms*, 16, 79–81.

Nikolic, N., Butler, J. R. A., Baglinière, J. L., *et al.* (2009). An examination of genetic diversity and effective population size in Atlantic salmon populations. *Genetics Research*, 91, 395–412.

Obrien, S. J., Roelke, M. E., Marker, L., *et al.* (1985). Genetic basis for species vulnerability in the cheetah. *Science*, 227, 1428–1434.

Okamura, B., Hartikainen, H., Schmidt-Posthaus, H. & Wahli, T. (2011). Life cycle complexity, environmental change and the emerging status of salmonid proliferative kidney disease. *Freshwater Biology*, 56, 735–753.

Ondrackova, M., Simkova, A., Civanova, K., Vyskocilova, M. & Jurajda, P. (2012). Parasite diversity and microsatellite variability in native and introduced populations of four *Neogobius* species (Gobiidae). *Parasitology*, 139, 1493–1505.

Packer, C., Holt, R. D., Hudson, P. J., Lafferty, K. D. & Dobson, A. P. (2003). Keeping the herds healthy and alert: implications of predator control for infectious disease. *Ecology Letters*, 6, 797–802.

Paterson, R. A., Townsend, C. R., Poulin, R. & Tompkins, D. M. (2011). Introduced brown trout alter native acanthocephalan infections in native fish. *Journal of Animal Ecology*, 80, 990–998.

Peacock, S., Connors, B., Krkosek, M., Irvine, J. & Lewis, M. A. (2014). Can reduced predation offset negative effects of sea louse parasites on chum salmon? *Proceedings of the Royal Society B*, 281, 20132913.

Pedersen, A. B., Jones, K. E., Nunn, C. L. & Altizer, S. (2007). Infectious diseases and extinction risk in wild mammals. *Conservation Biology*, 21, 1269–1279.

Persson, L. (1997). Competition, predation and environmental factors as structuring forces in freshwater fish communities: Sumari (1971) revisited. *Canadian Journal of Fisheries and Aquatic Sciences*, 54, 85–88.

Peterson, R. O., Thomas, N. J., Thurber, J. M., Vucetich, J. A. & Waite, T. A. (1998). Population limitation and the wolves of Isle Royale. *Journal of Mammalogy*, 79, 828–841.

Petterson, E., Sandberg, M. & Santi, N. (2009). Salmonid alphavirus associated with *Lepeophtheirus salmonis* (Copepoda: Caligidae) from Atlantic salmon, *Salmo salar* L. *Journal of Fish Diseases*, 32, 477–479.

Poulin, R. (1992). Toxic pollution and parasitism in freshwater fish. *Parasitology Today*, 8, 58–61.

Poulin, R. (2006). Global warming and temperature-mediated increases in cercarial emergence in trematode parasites. *Parasitology*, 132, 143–151.

Poulin, R. (2010). Parasite manipulation of host behavior: an update and frequently asked questions. *Advances in the Study of Behavior*, 41, 151–186.

Poulin, R. & Fitzgerald, G.J. (1989). Shoaling as an anti-ectoparasite mechanism in juvenile sticklebacks (*Gasterosteus* spp). *Behavioral Ecology and Sociobiology*, 24, 251–255.

Poulin, R., Paterson, R. A., Townsend, C. R., Tompkins, D. M. & Kelly, D. W. (2011). Biological invasions and the dynamics of endemic diseases in freshwater ecosystems. *Freshwater Biology*, 56, 676–688.

Prophete, C., Carlson, E. A., Li, Y., *et al.* (2006). Effects of elevated temperature and nickel pollution on the immune status of Japanese medaka. *Fish & Shellfish Immunology*, 21, 325–334.

Pulkkinen, K., Suomalainen, L. R., Read, A. F., *et al.* (2010). Intensive fish farming and the evolution of pathogen virulence: the case of columnaris disease in Finland. *Proceedings of the Royal Society B*, 277, 593–600.

Quinn, T. P. (2005). *The Behavior and Ecology of Pacific Salmon and Trout*. Seattle, WA: University of Washington Press.

Rader, R. B., Belk, M. C. & Keleher, M. J. (2003). The introduction of an invasive snail (*Melanoides tuberculata*) to spring ecosystems of the Bonneville Basin, Utah. *Journal of Freshwater Ecology*, 18, 647–657.

Rauque, C. A., Viozzi, G. P. & Semenas, L. G. (2003). Component population study of *Acanthocephalus tumescens* (Acanthocephala) in fishes from Lake Moreno, Argentina. *Folia Parasitologica*, 50, 72–78.

Ray, R. A. & Bartholomew, J. L. (2013). Estimation of transmission dynamics of the *Ceratomyxa shasta* actinospore to the salmonid host. *Parasitology*, 140, 907–916.

Richards, E. L., van Oosterhout, C. & Cable, J. (2010). Sex-specific differences in shoaling affect parasite transmission in guppies. *PLoS ONE*, 5, e13285.

Roche, D. G., Leung, B., Franco, E. F. M. & Torchin, M. E. (2010). Higher parasite richness, abundance and impact in native versus introduced cichlid fishes. *International Journal for Parasitology*, 40, 1525–1530.

Sanjayan, M. A., Crooks, K., Zegers, G. & Foran, D. (1996). Genetic variation and the immune response in natural populations of pocket gophers. *Conservation Biology*, 10, 1519–1527.

Schmitz, O. & Nudds, T. (1994). Parasite-mediated competition in deer and moose: how strong is the effect of meningeal worm on moose? *Ecological Applications*, 4, 91–103.

Seppala, O., Karvonen, A. & Valtonen, E. T. (2004). Parasite-induced change in host behaviour and susceptibility to predation in an eye fluke–fish interaction. *Animal Behaviour*, 68, 257–263.

Smiley, J. E., Okihiro, M. S., Drawbridge, M. A. & Kaufmann, R. S. (2012). Pathology of ocular lesions associated with gas supersaturation in white seabass. *Journal of Aquatic Animal Health*, 24, 1–10.

Smith, M. J., Telfer, S., Kallio, E. R., *et al.* (2009). Host–pathogen time series data in wildlife support a transmission function between density and frequency dependence. *Proceedings of the National Academy of Sciences of the United States of America*, 106, 7905–7909.

Stephens, P. A. & Sutherland, W. J. (1999). Consequences of the Allee effect for behaviour, ecology and conservation. *Trends in Ecology and Evolution*, 14, 401–405.

Strange, R. N. & Scott, P. R. (2005). Plant disease: a threat to global food security. *Annual Review of Phytopathology*, 43, 83–116.

Strecker, U. (2006). The impact of invasive fish on an endemic *Cyprinodon* species flock (Teleostei) from Laguna Chichancanab, Yucatan, Mexico. *Ecology of Freshwater Fish*, 15, 408–418.

Sures, B. & Knopf, K. (2004). Individual and combined effects of cadmium and 3,3′, 4,4′, 5-pentachlorobiphenyl (PCB 126) on the humoral immune response in European eel (*Anguilla anguilla*) experimentally infected with larvae of *Anguillicola crassus* (Nematoda). *Parasitology*, 128, 445–454.

Taraschewski, H. (2006). Hosts and parasites as aliens. *Journal of Helminthology*, 80, 99–128.

Taugbol, T., Skurdal, J. & Hastein, T. (1993). Crayfish plague and management strategies in Norway. *Biological Conservation*, 63, 75–82.

Thieltges, D. W., Jensen, K. T. & Poulin, R. (2008). The role of biotic factors in the transmission of free-living endohelminth stages. *Parasitology*, 135, 407–426.

Thompson, R. C. A., Lymbery, A. J. & Smith, A. (2010). Parasites, emerging disease and wildlife conservation. *International Journal for Parasitology*, 40, 1163–1170.

Thrall, P. H., Antonovics, J. & Hall, D. W. (1993). Host and pathogen coexistence and sexually-transmitted vector-borne diseases characterised by frequency-dependent disease transmission. *American Naturalist*, 142, 543–552.

Thrall, P. H., Antonovics, J. & Dobson, A. P. (2000). Sexually transmitted diseases in polygynous mating systems: prevalence and impact on reproductive success. *Proceedings of the Royal Society B*, 267, 1555–1563.

Tompkins, D. M., Dunn, A. M., Smith, M. J. & Telfer, S. (2011). Wildlife diseases: from individuals to ecosystems. *Journal of Animal Ecology*, 80, 19–38.

Torchin, M. E., Lafferty, K. D., Dobson, A. P., McKenzie, V. J. & Kuris, A. M. (2003). Introduced species and their missing parasites. *Nature*, 421, 628–630.

Valtonen, E. T., Holmes, J. C. & Koskivaara, M. (1997). Eutrophication, pollution, and fragmentation: effects on parasite communities in roach (*Rutilus rutilus*) and perch (*Perca fluviatilis*) in four lakes in central Finland. *Canadian Journal of Fisheries and Aquatic Sciences*, 54, 572–585.

Van Oosterhout, C., Smith, A. M., Hanfling, B., *et al.* (2007). The guppy as a conservation model: implications of parasitism and inbreeding for reintroduction success. *Conservation Biology*, 21, 1573–1583.

Wagner, E. J., Bartley, M., Arndt, R., Oplinger, R. W. & Routledge, M. D. (2012). Comparison of hatchery and field performance between a whirling-disease-resistant strain and the ten sleep strain of rainbow trout. *Journal of Aquatic Animal Health*, 24, 110–120.

Walker, P. J. & Winton, J. R. (2010). Emerging viral diseases of fish and shrimp. *Veterinary Research*, 41, 51.

Westcott, J. D., Revie, C. W., Griffin, B. L. & Hammell, K. L. (2010). Evidence of sea lice *Lepeophtheirus salmonis* tolerance to emamectin benzoate in New Brunswick Canada. Sea Lice 2010–8th International Sea Lice Conference, Victoria, BC.

Non-indigenous fishes and their role in freshwater fish imperilment

M. JAKE VANDER ZANDEN, NICOLAS W. R. LAPOINTE
AND MICHAEL P. MARCHETTI

8.1 INTRODUCTION

The current state of freshwater biodiversity, and that of freshwater fishes in particular, can only be described as dismal (Dudgeon *et al.*, 2006; Burkhead, 2012; Chapters 1 and 2). Freshwater extinction rates have been estimated to be around 1000× above background extinction rates (Ricciardi & Rasmussen, 1999; Burkhead, 2012). Using North America as an example, a recent American Fisheries Society assessment indicates that 39% of North American freshwater fish taxa are imperilled: 230 are vulnerable, 190 are threatened, 280 are endangered and 61 are extinct or extirpated (Jelks *et al.*, 2008). Global assessments paint the same general picture (www.iucn.org). What are the factors ultimately responsible for this situation? The objective of this chapter is to examine the role of non-indigenous fishes (hereafter NIF) in the decline and imperilment of native freshwater fishes. We provide a short primer to clarify a few key terms relating to invasion biology (Box 8.1).

A biological invasion should not be viewed as an event, but rather a process comprised of several successive stages (Figure 8.1). The non-indigenous species introduction process begins with uptake of individuals, transport to a new area (*transport outside of native range*), and subsequently release into the wild (*introduction to the wild*). These released individuals may then establish a self-sustaining population in the new area (*establishment*), and an established population may increase in abundance and expand its geographic range (*spread*). Typically it is only when a species becomes abundant and/or widespread that it is

Conservation of Freshwater Fishes, eds G. P. Closs, M. Krkosek and J. D. Olden. Published by Cambridge University Press. © Cambridge University Press 2016.

Box 8.1. Terminology relating to non-indigenous species

Studies tend to use a wide assortment of poorly defined and often confusing terms to represent the introduction and establishment of species outside of their presumed native or historical geographic range (Davis, 2009). While the terms invasive, non-native, non-indigenous and exotic are all commonly used, we adopt the term non-indigenous fishes (NIFs) in this chapter to refer to any fish species occurring in the wild outside of its natural range or potential natural range of dispersal, irrespective of the original country of origin or political jurisdiction. By using this general definition, we do not differentiate between species from another country/continent, transplants to new regions within a continent, or even translocations among watersheds. We also note that a wide variety of non-indigenous species that are not fish may have both direct and indirect impacts on native freshwater fishes and aquatic ecosystems, but that in this chapter we will limit our treatment to the effects of NIFs.

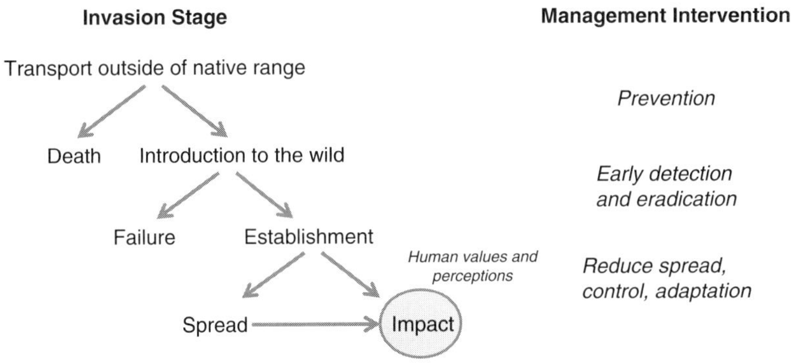

Figure 8.1 Successive stages of the biological invasion process, and the appropriate or relevant management responses along the successive stages. Our diagram highlights the fact that both occurrence and spread can contribute to invasive species impact, and that human values and perceptions also play a large role when judging impact.

perceived to cause ecological or economic harm (*impact*), such that the species is considered invasive (Lockwood *et al.*, 2013). Human values and perceptions greatly influence how and whether a non-indigenous species is viewed as having undesirable consequences. Alternatively, an introduced species population may remain small and localised in its distribution, with few or no tangible ecological or economic effects. To pass through the invasion stages shown in Figure 8.1, a species must overcome a suite of barriers to advancement. At each stage, there are ecological, environmental and stochastic forces that can either facilitate or hinder transition to subsequent stages. These may include the characteristics of the species (e.g. reproductive rate, physiological tolerance), environmental conditions (e.g. temperature or salinity match) and the nature of the introduction event (e.g. timing of introduction, number of individuals moved). Identifying the stages and associated barriers of the invasion process allows for an explicit recognition that human beings can act as either inhibitors or facilitators at each stage.

Introduced species are widely recognised as an important component of global environmental change, and often have local consequences for populations, communities, food webs and ecosystems (Davis, 2009; Lockwood *et al.*, 2013). The key question from the conservation perspectives is, 'how often do they result in the imperilment and extinction of native fishes?' While there is relatively little debate surrounding the contributing role of NIF introductions in the decline of freshwater fishes (Lassuy, 1995; Richter *et al.*, 1997; Dudgeon *et al.*, 2006; Jelks *et al.*, 2008), it is often difficult or impossible to isolate the causal role of a single factor such as introduced fishes. This is because native species and ecosystems are often subject to multiple human impacts simultaneously, and the role of individual factors can be nearly impossible to tease apart. One approach has been to ask whether NIFs function as a 'passenger' or 'driver' of biodiversity loss (MacDougall & Turkington, 2005; Light & Marchetti, 2007). To illustrate the idea of 'passenger', habitat degradation may produce conditions that are no longer suitable for native species, but suitable for more tolerant non-natives. In such a case, a non-indigenous species could simply replace a native species due to changing habitat, even if there is no direct biotic interaction such as predation or competition. At the other extreme, non-natives can have direct predatory or competitive impacts on native species, such that the non-native is a causal 'driver' of native species declines. Differentiating between these two options is often challenging from correlational and observational data (typically what is available). In reality, the decline and extinction of native species typically involves multiple interacting drivers. For example, habitat degradation may simultaneously favour non-native species and produce declines in native species. Expansion of non-native species could further alter ecosystem properties, thereby causing additional declines in the native species, while also producing direct ecological impacts, such as predation on, or competition with, native species (Bauer, 2012).

While we acknowledge that NIFs typically interact with other human impacts on ecosystems and native fishes, the goal of this chapter is to systematically examine the role of NIFs as a driver of freshwater fish imperilment. To this end, we employ regional examples and case studies to illustrate key points.

8.2 TRANSPORT AND INTRODUCTION OF NON-INDIGENOUS FRESHWATER FISHES

In this section, we address the first two stages of the invasion process (Figure 8.1): transport beyond the native range and introduction into the wild. Humans transport vast numbers of live organisms outside of their native range. Reasons for transport are varied and include the desire to create new fisheries, surround themselves with familiar organisms, accidental transport and simple negligence. Current estimates by the US Geological Survey document 697 unique fish introduced to both the fresh and brackish waters of the US (USGS, 2013). Of these, approximately 42% originate from other countries, and 54% are native to the USA but were introduced outside their native range (USGS, 2013). Freshwater fishes have been introduced to the USA from all continents except Antarctica. The rate of introduction of NIFs accelerated rapidly in the middle of the twentieth century (Allen & Flecker, 1993; Crivelli, 1995; Fuller *et al.*, 1999), but appears to have stabilised since 2000 (USGS, 2013).

While there are several ways to categorise the transport and introduction of NIFs, we classify NIF introductions according to the introduction pathway, which is the mode or reason by which the species was introduced. The broadest level classification is a simple dichotomy of intentional versus unintentional introductions. Intentional introduction refers to the import of organisms to a new area for an express purpose; these are often important to humans, and sometimes involve considerable effort and expenditure. Thus, we typically have relatively good records of intentional events. By contrast, unintentional introductions are an unintended consequence of the intentional transfer of some good or service, such that we typically have much poorer knowledge of these.

8.2.1 Intentional introductions

The intentional transport and introduction of organisms by humans has been occurring since ancient times (Wilson *et al.*, 2009). Here

we consider five general categories of intentional fish introductions: food-related releases, game and bait fishes, aquarium releases,
biocontrol and scientific/conservation releases.

8.2.1.1 Food-related releases

Fishes are commonly introduced into new areas as a food source, both
for aquaculture and to create or improve wild-capture fisheries. For
example, tilapia (*Tilapia* spp. and *Oreochromis* spp.) have been introduced into over 90 countries for a wide variety of reasons, most recently
to create new fisheries (Canonico *et al.*, 2005). Tilapia aquaculture has
grown exponentially in the past few decades, and is valued at many billions of dollars (www.fao.org). While providing much-needed protein
and employment to many developing areas, they are also highly invasive and exert major ecological impacts (Canonico *et al.*, 2005). One
of the earliest known examples of fish translocation is that of common carp (*Cyprinus carpio*) (Balon, 1995). Native to the waters of the
Ponto-Caspian in central Asia, common carp were transported to China
and later Europe for pond aquaculture dating back 2000 years (Copp
et al., 2005). Common carp have since been widely introduced to inland
waters throughout the world (Moyle, 2002; Koehn, 2004).

At present, the majority of global aquaculture production consists of NIFs, predominantly carps, tilapia, salmonids and catfishes.
Unfortunately, aquaculture fishes often escape facility confines, or are
intentionally released into nearby streams and waterways, and sometimes become successfully established (Canonico *et al.*, 2005; Kolar *et al.*,
2005; Weyl, 2008). An example of this is the farming of Atlantic salmon (*Salmo salar*), where the introduction and frequent escape of farmed
fishes along the Atlantic and Pacific coastlines is an enormous problem
(Naylor *et al.*, 2001). It has been estimated that up to 40% of 'wild-caught'
Atlantic salmon fished in the North Atlantic and > 90% fished from the
Baltic Sea are of farmed origin (Naylor *et al.*, 2001). In addition, more
than half a million individual Atlantic salmon raised in aquaculture facilities on the west coast of North America escaped captivity between 1987
and 1997 (Naylor *et al.*, 2001). At the present time, Atlantic salmon have
been found in 77 rivers and streams across British Columbia, and have
been observed to spawn in some locations (Naylor *et al.*, 2001).

8.2.1.2 Game and bait fishes

Perhaps the largest category of NIFs are those introduced for game, not-
ing that game and food fishes are sometimes one and the same. Until

recent decades, fisheries management in North America and elsewhere has been virtually synonymous with artificial propagation and translocation of fishes, which are often non-indigenous. Many NIF introductions in North America during the nineteenth and twentieth centuries involved intentional stocking by management agencies, and were aimed at 'improving' the native biota, expanding sportfishing, or providing forage for native or other introduced game fishes (Rahel, 1997). Settlers of the western US in the nineteenth and twentieth centuries preferred the familiar game species of the eastern US such as brook trout (*Salvelinus fontinalis*), black basses (*Micropterus* spp.) and catfishes (*Ictalurus* spp.) over the native fish fauna. NIFs were widely introduced throughout the western US, and many have successfully established (Moyle, 2002). Perhaps the best example of an introduced game fish is the virtually cosmopolitan rainbow trout (*Oncorhynchus mykiss*). This salmonid is native to streams and rivers of the western US, but as of 1996, has been introduced to every continent except Antarctica and more than 100 countries and oceanic islands worldwide (Lever, 1996; Cambray, 2003).

Coincident with the movement of game fishes is the importation of forage and bait fishes. Fishes used as bait are often released at the end of the day or fishing trip (Litvak & Mandrak, 1993). Species such as red shiner (*Cyprinella lutrensis*), golden shiner (*Notemigonus crysoleucas*) and fathead minnow (*Pimephales promelas*) now have broad distributions in the USA because of widespread transport and release due to their popularity as bait fishes (USGS, 2013). Ludwig and Leitch (1996) combined mail surveys, baitfish samples, interviews and literature data to quantify the role of bait bucket transfer in the dispersal of NIFs. They found bait bucket transfer to be a major vector for the dispersal of NIFs.

8.2.1.3 Aquarium pet releases

Many fishes have been introduced intentionally through the release of aquarium pets. The number of individuals and species involved in this pathway can be staggering. Radonski et al. (1984) estimate that up to 1500 species of freshwater fish were regularly imported into the USA for the aquarium trade in the early 1980s, although Fuller et al. (1999) suspect this to be a severe underestimate. Although aquarium fish release rates are not well known, Gertzen et al. (2008) estimated that approximately 10,000 fishes were released to the wild annually in areas surrounding Montréal, Canada, with most releases involving a handful of common aquarium species such as goldfish (*Carassius auratus*) and guppies (*Poecilia reticulata*). A similar result was reported for the Pacific

Northwest of North America by Strecker *et al.* (2011). Pet releases are an underestimated source of NIF introductions, and its importance as a pathway for NIFs is expected to grow in the future, especially as the global climate warms and the probability of establishment of aquarium fishes increases.

8.2.1.4 Biocontrol

Non-native fish introductions for biocontrol are often overlooked. Known natural enemies of pest species are introduced with an expectation that the enemy will reduce or control the population of the pest. In this case, pests have historically included native species considered harmful to humans, but also non-native species that cause economic or environmental harm. Some early biocontrol organisms were fishes. For example, the western mosquitofish (*Gambusia affinis*) has been widely introduced for mosquito control in the western US and around the world (Lever, 1996).

8.2.1.5 Research and conservation

Fishes are sometimes transported to new areas to conserve or protect the species. This includes organisms brought to zoos, aquaria, or wildlife preserves, as well as endangered species moved to novel natural environments for their immediate protection and propagation. For example, the US Endangered Species Act (ESA) allows for the establishment of 'experimental' populations outside of the species range. The Devil's Hole pupfish (*Cyprinodon diabolis*) was introduced to the Amargosa Pupfish Station in 1983 (Sigler & Sigler, 1987) because the sole wild population was threatened by groundwater pumping (Soltz & Naiman, 1978). It was subsequently returned to its native habitat, although there now remain several additional non-native populations of the species.

The transport and introduction of species for conservation purposes has recently become an issue under the guise of assisted colonisation/ migration. Assisted colonisation is a conservation strategy whereby a species is intentionally introduced to new locations, presumably because native populations are threatened by environmental change (Hoegh-Guldberg *et al.*, 2008; Ricciardi & Simberloff, 2009). While this may ultimately be a viable environmental management strategy in some cases, it should be recognised that assisted colonisation is essentially the purposeful transport and release of a non-native species, for which the ecological consequences will be unpredictable (Ricciardi & Simberloff, 2009; Olden *et al.*, 2011).

8.2.2 Unintentional introductions

Many fishes are transported unintentionally as a by-product of the movement of other goods or services; for example, fish are frequently transported as hitchhikers within cargo holds or ballast water. By their nature,
unintentional introductions are difficult to evaluate in terms of their overall importance, although recent evidence suggests they are quite prevalent
(Molnar *et al.*, 2008). The best-known example of unintentional transport involves movement of species in ship ballast tanks. Large transport
ships often pump harbour water into ballast tanks to balance their cargo
load. At the receiving port, much of this water and associated organisms
(invertebrates, fishes, larvae) are unceremoniously dumped into the
water. The significance of this pathway as a source of non-native fishes
has grown with global trade and the size and speed of vessels (Carlton,
2000). It has been estimated that > 10,000 species may be transported
within ship ballast water on any given day (Carlton, 1999). For example,
round goby (*Neogobius melanostomus*) was transported to the Laurentian
Great Lakes via ship ballast water, and subsequently spread throughout
the Great Lakes and beyond via ballast water and by natural dispersal
through connected waterways (Kornis *et al.*, 2012). In other situations,
fishes are unintentionally transported within water diverted for human
use. An example that combines both ballast water and water diversion
pathways is the Shimofuri goby (*Tridentiger bifasciatus*), a small Japanese
estuarine fish. The Shimofuri goby was initially introduced into the San
Francisco Bay Estuary via ballast water during the mid-1980s (Matern,
2001; Moyle, 2002). It was then transported inland over 500 km, and
then across a mountain range into the Los Angeles basin by water export
through the California State Water Project (Matern, 2001).

We summarise the relative importance of different vectors of NIF
transport for several regions around the world in Table 8.1. The two most
important categories for NIFs are food-related and non-food-related,
highlighting the utilitarian purpose of most NIF introductions.
Unintentional transport was generally of moderate importance for
freshwater NIFs across these diverse regions. In contrast, unintentional
introductions tend to be the dominant vector for global marine organisms, indicating that the forces driving the proliferation of non-native
species differs among biomes and ecosystems. While a relatively small
fraction of NIFs are unintentional introductions (USGS, 2013), the relative importance of unintentional introductions has generally increased
in recent decades (Ricciardi, 2006; Keller *et al.*, 2009). Recent recognition of the undesirable consequences of many species introductions in

Table 8.1 A summary of documented NIF introductions, according to the reason/pathway of introduction for different regions of the world. Values presented are numbers of NIFs for each region, while the percentage of the total pool for each region is shown in parentheses. The last column shows pathways for global marine organisms for comparison.

Reason/pathway for introduction	US freshwater fish[a]	California freshwater fish[b]	Great Britain[c]	Australia[d]	Spain[e]	Global marine organisms[f]
Food-related and game species	190 (31%)	49 (69%)	7 (39%)	30 (34%)	18 (72%)	141 (43%)
Non-food related	202 (33%)	5 (7%)	10 (56%)	29 (33%)	6 (24%)	20 (6%)
Biocontrol	35 (6%)	2 (3%)	0 (0%)	2 (2%)	1 (4%)	0 (0%)
Research and conservation	35 (6%)	2 (3%)	0 (0%)	15 (17%)	0 (0%)	0 (0%)
Transportation related (unintentional)	154 (25%)	13 (18%)	1 (6%)	10 (12%)	0 (0%)	282 (86%)
Unknown	0 (0%)	0 (0%)	0 (0%)	9 (10%)	0 (0%)	0 (0%)
Totals	616	71	18	87	25	329

Notes

[a]Based on data from the US Geological survey, 2013. Nonindigenous Aquatic Species Database summary graphs. Available from: http://nas.er.usgs.gov/xxxx.

[b]Based on data in Moyle (2002).

[c]Based on data in Keller *et al.* (2009).

[d]Based on data in Lintermans (2004) (note that some organisms are identified with multiple modes of transport so the percentages add up to greater than 100%).

[e]Based on data in Elvira and Almodovar (2001).

[f]Based on data from Molnar *et al.* (2008) (note that some organisms are identified with multiple modes of transport so the percentages add up to greater than 100%).

the developed world has led to policies that have curtailed intentional stocking of NIFs. Yet the rise of global trade and transport in recent decades has greatly increased opportunities for accidental introductions.

8.2.3 Many are called, few are chosen

Of the large number of NIF introductions, only a subset of the species ultimately establishes self-sustaining populations. For example, Courtenay and Williams (1992) noted that, of 99 exotic species occurrence records in the continental USA, 46 had self-sustaining populations. While the oft-cited 'tens rule' (Williamson, 1996) predicts that roughly 10% of introduced species establish and 10% of those spread, Jeschke and Strayer (2005) report much higher rates of establishment and spread (~50%) for vertebrates intentionally introduced between North America and Europe. Nevertheless, the point is that a species may be transported and introduced, but environmental conditions are often not suitable or favourable for maintaining a self-sustaining population. This is particularly the case at high latitudes, where cold winter temperatures limit survival for warm-water species. Even where conditions are potentially suitable, NIF colonists can face an uphill battle. The small numbers of individuals transported into a novel environment may have a difficult time finding food, habitat, or mates, while avoiding predators. Inbreeding can also be a concern. Not surprisingly, repeated introductions or introductions of a large number of individuals (i.e. increased propagule pressure) can increase the likelihood of establishment (Kolar & Lodge, 2001).

Among established NIFs, the majority have relatively limited geographic distributions, while a handful of species are geographically widespread (Fuller et al., 1999; Gido & Brown, 1999; Rahel, 2000; Schade & Bonar, 2005). This highly right-skewed pattern is evident in Figure 8.2A, which shows that the vast majority of NIFs inhabiting the USA inhabit a small number of watersheds, while a very small number of species are widespread. The list of the most widespread North American species is dominated by intentionally stocked game fishes (Rahel, 2000). On other continents, a similar set of NIF species commonly occurs, with some variation as a result of climatic suitability. For example, an extensive basin-by-basin review of freshwater fishes in South America reports many of the usual suspects: tilapia (*Tilapia* spp., *Oreochromis* spp.), brown trout (*Salmo trutta*), brook trout (*Salvelinus fontinalis*), common carp and largemouth bass (*Micropterus salmoides*) (Barletta et al., 2010). Even though large numbers of NIFs are transported and introduced, only a subset of these are capable of establishing self-sustaining populations.

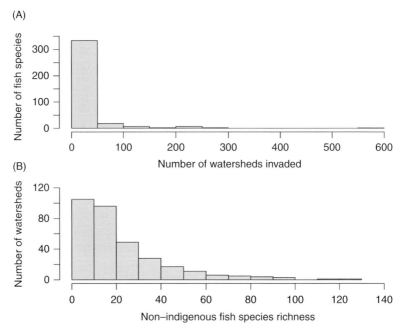

Figure 8.2 Frequency histograms showing NIF spatial distributions and species richness at the watershed level. (A) The number of watersheds in the US invaded by each individual NIF species known in the US. Most NIF species occur in a small number of watersheds, while a small number of NIFs are geographically widespread. (B) NIF species richness across these same watersheds. Most watersheds contain relatively few NIFs, while a small number of watersheds have high NIF species richness. Watersheds are USGS 6-digit HUCs (J.D. Olden, unpublished data).

A relatively small number of NIFs, predominantly game fishes, have become geographically widespread. Across US watersheds, NIF species richness is also right-skewed, with most watersheds containing relatively few NIF species and only a handful of watersheds containing large numbers of NIFs (Figure 8.2B). Despite this skewed distribution of species richness, NIFs dominate regional fish faunas in many human-dominated regions of the world, particularly in the Mediterranean and dryland habitats (Leprieur *et al.*, 2008; Marr *et al.*, 2010).

8.3 ECOLOGICAL IMPACTS AND RESPONSES

Established NIFs have the potential to affect native fishes, either directly or indirectly. In this section we consider only the ecological impacts of NIFs and ignore socioeconomic considerations, given that our focus is the conservation of freshwater fishes. Although impacts of NIFs are at times considered dichotomous (i.e. benign versus high-impact), they are more accurately seen as a continuum (Lapointe *et al.*, 2012). In a practical sense, impacts of a NIF are often difficult to characterise. Although direct interactions with native species are sometimes quantified, indirect effects can be subtle and difficult to detect. It has been argued that a loss of biological or genetic diversity through species extirpations is required for large ecological impacts (Gozlan *et al.*, 2010), yet there are a suite of other effects that NIFs can have on aquatic species and ecosystems. Many freshwater systems are essentially habitat 'islands' because there is minimal connectivity among systems, and many species are adapted to localised conditions. This leaves freshwater communities vulnerable to the impacts of NIFs, particularly where the native taxa lack an evolutionary history of competition or predation with other species (Cox & Lima, 2006; Kuehne & Olden, 2012).

8.3.1 Mechanisms of NIF impacts on native fishes

There are a variety of potential mechanisms through which NIFs can impact native fishes. Although detailed elsewhere (Gozlan *et al.*, 2010; Leunda, 2010; Cucherousset & Olden, 2011; Ribeiro & Leunda, 2012), we briefly discuss mechanisms of impact here.

8.3.1.1 Predation and direct mortality

NIFs often prey directly on adults, juveniles and eggs of native species (Witte *et al.*, 1992b; Chaichana & Jongphadungkiet, 2012). For example, the introduction of Nile perch (*Lates niloticus*) into Lake Victoria is widely regarded as the direct cause of the decline or extinction of native cichlid species. There are many examples of direct predation as an important mechanism (Ross, 1991), particularly in the case of predator-naïve native prey species. Other forms of direct mortality can also have important population-level consequences. For example, parasitic NIFs such as sea lamprey (*Petromyzon marinus*) are known to kill some native fishes (Smith & Tibbles, 1980). In addition, the behaviour of some NIFs can lead to the destruction of native fish nests and eggs. For example, non-native rainbow trout are known to construct redds on top of native salmonid redds (i.e. superimposition), damaging and destroying native eggs in the process (Taniguchi *et al.*, 2000).

8.3.1.2 Competition

Species introductions may cause a shift or decline in a prey community that negatively affects native species as a result of resource competition (Post & Cucin, 1984; Ross, 1991). Additionally, NIFs can compete with native species for space, shelter, or mates (Marchetti, 1999; Fleming *et al.*, 2000). In response, native species may alter their behaviour, including aggression, food consumption, dominance hierarchies, habitat use, diel activity patterns and emigration (Ross, 1991; Fraser & Gilliam, 1992; Marchetti, 1999; Cucherousset *et al.*, 2007; Nasmith *et al.*, 2010). For example, native brown trout in France changed their diet in the presence of introduced brook trout (Cucherousset *et al.*, 2007), and dace (*Phoxinus* spp.) became more active at crepuscular periods and adopted a more benthic-oriented behaviour in lakes stocked with non-native trout in Alberta, Canada (Nasmith *et al.*, 2010). The fitness consequences of competition may include decreased growth and fecundity (Fraser & Gilliam, 1992; Marchetti, 1999; Irons *et al.*, 2007; Britton *et al.*, 2010). Overall, resource competition with NIFs is an important mechanism underlying native fish declines (Ross, 1991).

8.3.1.3 Parasites and disease

The introduction of novel parasites and diseases is often inexorably linked to the introduction of NIF hosts. Novel parasites and diseases can be directly detrimental to native fishes (Vitule *et al.*, 2009). For example, common carp in the Czech Republic were affected by *Bothriocephalus gowkongensis*, a tapeworm carried by introduced grass carp (*Ctenopharyngodon idella*) (Lusk *et al.*, 2010). It is also possible for NIFs to alter the population dynamics of existing native parasites by acting as an alternate host, although such effects may either exacerbate or alleviate their impact on native species (Poulin *et al.*, 2011).

8.3.1.4 Ecosystem alterations

Aquatic habitats can undergo ecosystem-level changes in response to NIFs. Species such as goldfish (*Carassium auratus*) that feed in soft substrates can uproot macrophytes and re-suspend sediments, leading to increases in turbidity (Richardson *et al.*, 1995). Grass carp feed directly on macrophytes (Bain, 1993) and thereby remove habitat for other aquatic species. Feeding and excretion by NIFs can alter nitrogen and phosphorus cycling and availability in aquatic systems (Schindler *et al.*, 1993; Vanni, 2002). The ecosystem-level changes that come with NIF introductions can have a range of secondary consequences. Due to the complexity of interactions between many NIF and native species and ecosystems, many of the broader ecosystem-level implications of fish introductions and game-fish stocking remain poorly known (Eby *et al.*, 2006).

8.3.2 Responses of native fishes to NIF introductions

8.3.2.1 *Genetic effects*

The potential genetic effects of NIFs are a critical conservation concern, particularly the loss of genetic uniqueness via hybridisation (Hitt *et al.*, 2003). Similarly, genetic introgression can occur when a non-native strain is stocked where a native population of the same species exists (Ferguson, 1990; Garcia-Marin *et al.*, 1999), or if fish escape from a hatchery and breed with natives. Such losses are a concern for fish conservation, but are restricted to cases where introduced species or strains are closely related to native populations. Perhaps more pervasively, NIFs can affect genetic diversity by contributing to the local extirpation of populations, which reduces native genetic diversity. Additionally, NIFs can exert strong selective pressures on native fishes, leading to evolutionary responses (Roberge *et al.*, 2008). Such responses could lead to the development of novel traits and thus increase the genetic diversity of native species (Schlaepfer *et al.*, 2011), but the development of novel traits in response to NIFs is more realistically viewed as a case of 'one step forward, two steps back' from a conservation perspective.

8.3.2.2 *Population and species effects*

NIFs commonly contribute to population declines in native species, which may ultimately lead to population extirpations or extinction (Ross, 1991). For instance, declines in commercially important native species in the Gargalheiras Reservoir, Brazil were attributed to the establishment of Nile tilapia (Attayde *et al.*, 2011). Studies of biodiversity decline have generally focused on species-level declines, imperilment and extinction. With the exception of highly localised endemics, species declines are the by-product of the extirpation of many individual populations (Hughes *et al.*, 1997). Rates of population loss are not well known, but are estimated to be 1000× higher than species extinction rates (Hughes *et al.*, 1997). Studies that have examined long-term freshwater fish community change indicate high rates of local population loss. Using historical data for rivers in Mexico, Mercado-Silva *et al.* (2006a) reported native species population-level loss rates of 10–30% per decade and a concomitant 10–20% increase per decade in non-native species. Other studies report similar patterns of local extirpation of native populations, and increasing incidence of non-native species (Gido *et al.*, 2010). Native species undergo a gradual range-wide decline as local populations blink out. This phenomenon is analogous to the concept of extinction debt (Tilman *et al.*, 1994), in that species declines may not be easily observed, and regional or global extinction occurs only after a prolonged period of individual population losses.

As populations are locally extirpated, the risk of species extinction increases. Introduced species were the second leading factor (after habitat alteration), contributing to 68% of North American fish extinctions (Miller *et al.*, 1989). Similarly, Lassuy (1995) found that of the 92 freshwater fish species listed under the US Endangered Species Act, species introductions were a contributing factor for 70% of the taxa. Of these, 73% of cases involved species related to sport fishing (game, forage or bait fish), with centrarchids, ictalurids and salmonids leading the list. This same general finding, that NIFs are a major contributor to fish imperilment and extinction, has been reported elsewhere (Richter *et al.*, 1997; Wilcove *et al.*, 1998; Harrison & Stiassny, 1999). In contrast, a study of endangered fishes in Canada found that invasive species were notably less important as they were implicated for only 27% of the imperilled taxa (Venter *et al.*, 2006). Similarly non-native species were suggested as a main factor in the decline of only 15% of rare species across the Laurentian Great Lakes (Mandrak & Cudmore, 2010). Equivalent figures are lacking for many regions, owing in part to limited research in developing nations (Fazey *et al.*, 2005); however, non-native species are implicated in the decline of 29% of threatened neotropical fishes in South America (Rodriguez, 2001).

8.3.2.3 *Community, food web and ecosystem effects*
NIFs can cause profound changes in freshwater fish communities. Introduced predators can eliminate native fish species and assemblages in lakes (Chapleau *et al.*, 1997; Findlay *et al.*, 2000; Trumpickas *et al.*, 2011; Aiken *et al.*, 2012). Examples of assemblage-level impacts include the Nile perch (*Lates niloticus*) impacts on the endemic cichlids in Lake Victoria (Witte *et al.*, 1992a, 1992b) or the peacock bass (*Cichla*) introductions into neotropical reservoirs (Pelicice & Agostinho, 2009). Additionally, introduced salmonids have caused declines in native galaxiids across their range (McDowall, 2006; Habit *et al.*, 2010). In many aquatic systems around the globe, entire food webs are increasingly

dominated by NIFs (Ricciardi, 2006; Leprieur *et al.*, 2008; Marr *et al.*, 2010). While this can often result in increased local species richness, their presence strongly contributes to biotic homogenisation, a process whereby disjunct communities or biota become more similar over time due to the combined effects of native species loss and NIF introductions (Rahel, 2000; Olden *et al.*, 2010).

NIFs can alter food webs and ecosystem processes across trophic levels and ecosystem boundaries. For example, the introduction of non-native smallmouth bass (*Micropterus dolomieu*) and rock bass (*Ambloplites rupestris*) into Ontario lakes caused dramatic declines in native littoral prey fishes via direct predation (Vander Zanden *et al.*, 1999). This had unexpected negative effects on the native top predator, lake trout (*Salvelinus namaycush*), which are highly dependent on littoral prey fish in the absence of their preferred pelagic forage fish prey (Vander Zanden *et al.*, 1999). The ecological effect of NIFs can propagate across several trophic levels, as demonstrated by studies in New Zealand streams, where brown trout introduction has altered insect communities and increased periphyton standing-stock (Flecker & Townsend, 1994; Simon & Townsend, 2003). In some cases, NIFs have been stocked to induce a system-wide trophic cascade for the purpose of improving water quality and clarity (Scharf, 2007). In addition, food web effects can sometimes cross ecosystem boundaries. In Japanese streams, introduced rainbow trout produced a suite of effects including reduced aquatic insect emergence, increased periphyton growth, and severed food web linkages between streams and the surrounding riparian ecosystem (Baxter *et al.*, 2004).

8.3.3 Generalities about impact

There is a great deal of uncertainty regarding the relative impacts of NIFs. It seems that a large percentage of impacts are caused by only a few species. However, it is unclear whether other NIFs are relatively benign or simply have poorly documented impacts. Such uncertainty can be ascribed to a number of factors including: differences in impact among regions, a lack of baseline data, the lack of impact assessments, and potential threshold, cumulative and synergistic effects. The impact of a non-native species depends on the distribution and abundance of both native and introduced species, and the response of the recipient biota and ecosystem. Impacts are likely to be heterogeneous across the introduced range (Ricciardi, 2003). For instance, species introduced to ecosystems near the limits of their physiological tolerances are unlikely to have major impacts. Such variability leads to difficulties in broadly classifying NIFs as high- or low-impact invaders. Furthermore, pre-invasion data are generally unavailable (Cambray, 2003; Gherardi, 2007), and even when baseline data exist, impacts can be difficult to identify in the face of other environmental stressors and species introductions. These can involve cumulative or synergistic effects (Ricciardi, 2001), making it difficult to tease out the effects of a given species introduction.

Impacts also vary through time. There are often lag times between establishment and impact (Crooks, 2005; Richardson & Pysek, 2006). For instance, introduced predators with long generation times may not have significant impacts until some population threshold is breached (Schoen *et al.*, 2012). Impacts may only be realised after NIFs adapt to their new environment (Ricciardi *et al.*, 2013). Alternatively, NIFs can remain relatively benign until ecosystem conditions change (e.g. through climate change or the introduction of a new, facilitative species), leading Courtenay (1993) to describe non-indigenous species as potential ecological time bombs. Many introduced populations have not fully colonised the total extent of suitable habitat in their non-native range (Mercado-Silva *et al.*, 2006b; Papes *et al.*, 2011). Strayer (2010) described this situation as an 'invasion debt', likely to be paid over time as species spread. The period before the debt is 'paid' may be protracted for freshwater systems, as their spread at times requires secondary introductions. It is also possible that impacts may attenuate after reaching an initial peak if the NIF population eventually declines or because no further impacts on a native species are possible after the species has been extirpated (Strayer *et al.*, 2006).

8.3.3.1 *Traits of high-impact species*

Improved understanding of the traits of high-impact NIFs could lead to better predictability and inform risk assessments (Chapter 2); however, few generalities exist for NIFs because impacts depend in part on the characteristics of recipient ecosystems. Habitat suitability (e.g. on a latitudinal gradient) may not be equivalent across the introduced range of a NIF, and introduced piscivores may have a greater impact in systems lacking a top predator relative to communities with native piscivores. Interactions among NIFs and native ecosystems can be complex. Nevertheless, a few commonalities do exist in the traits of high-impact NIFs. Some of the most notable impacts of NIFs are the result of introduced piscivores. Introductions of predatory sport fishes are the main threat to endemic fishes in the western USA (Minckley & Deacon, 1991). Primary consumers may also have strong impacts, whereas omnivores are often relatively benign (Moyle & Light, 1996; Ribeiro *et al.*, 2008). Broad physiological tolerance is also often associated with impact (Kolar & Lodge, 2002; Marchetti *et al.*, 2004; Ribeiro *et al.*, 2008), and studies have found large maximum body size and small egg size to be important (Kolar & Lodge, 2002; Marchetti *et al.*, 2004). Aside from species traits, taxonomic uniqueness relative to recipient communities, prior invasion success, and association with humans also correspond with impacts for NIFs (Marchetti *et al.*, 2004; Ricciardi & Atkinson, 2004; Ribeiro *et al.*, 2008). Species risk assessments provide a useful tool for prioritising and identifying potentially harmful NIFs (Ricciardi & Rasmussen, 1998; Lodge *et al.*, 2006).

8.4 NIFS AND INTERACTIONS WITH OTHER DRIVERS

Although this chapter is meant to examine the role of NIFs in the decline of native fishes, it is important to consider species interactions within the broader context of habitat and environmental change, which brings us back to the passenger versus driver dichotomy discussed earlier (Light & Marchetti, 2007). While often presented as a dichotomy, native species decline may commonly be the result of interactions among biotic and environmental drivers, such that NIF may simultaneously be both a driver of native species declines and a passenger of other environmental changes. This framework is analogous in many ways to biotic homogenisation, which emphasises the role of species introductions, native species loss, and habitat alterations in increasing the similarity of biotas over time (Rahel, 2002). At the landscape scale, the consequence is that initially distinctive fish faunas are replaced by a handful of tolerant and ubiquitous fish species (common carp, tilapia, rainbow trout, etc.). As a result, beta diversity, reflective of the uniqueness of local faunas, is reduced. Below, we highlight some of the well-known drivers that interact with NIFs.

8.4.1 Impoundments and altered flow regimes

The creation of impoundments tends to favour NIFs at the expense of native species (Marchetti *et al.*, 2004; Havel *et al.*, 2005; Johnson *et al.*, 2008; Mims & Olden, 2013). The twentieth century has witnessed the construction of 40,000 large dams and untold millions of small dams worldwide, transforming many river systems into series of impounded reservoirs (Chapter 4). In addition to creating slackwater conditions where water formerly flowed, dams block fish movements, eliminate floodplains and habitat complexity, disconnect fishes from spawning grounds, and alter downstream physical features, water temperatures, water quality and flow regimes (Baxter, 1977; Ligon *et al.*, 1995; Chapter 4). NIFs adapted to these novel, lacustrine conditions are often introduced or otherwise colonise impoundments. Downstream, flow regimes and habitats are also altered, including changes in the timing of peak- and low-flow events, water-level fluctuation patterns and thermal regimes (Poff *et al.*, 2007; Mims & Olden, 2013). Such changes tend to favour NIFs, often leading to severe declines in native fishes (Marchetti & Moyle, 2001), a well-known example being the declining endemic fishes of the Colorado River (Olden *et al.*, 2006).

8.4.2 Climate change

Climate change will have a suite of consequences for freshwater ecosystems, as well as human use and management of these systems (Rahel & Olden, 2008; Chapter 3). For example, increased surface water temperature is allowing a northward expansion of temperature-limited warm-water fishes such as smallmouth bass and other centrarchids (Sharma *et al.*, 2007). The northward expansion of warm-water fishes is already under way, with pronounced impacts on native fishes and lake food webs (Vander Zanden *et al.*, 1999). Using IPCC climate change scenarios, Sharma *et al.* (2009) identified thousands of lakes vulnerable to smallmouth bass invasion in response to projected climate change, serving as an example of how climate change might facilitate the spread of NIFs (Rahel & Olden, 2008).

8.4.3 Other drivers

A range of other drivers (e.g. pollution, sedimentation, stream channelisation, excessive surface and groundwater extraction, and over-harvest) can interact with NIFs to cause native species declines. As an example, the African Rift Lake, Lake Victoria, experienced the probable extinction of several hundred endemic cichlid fishes, and offers an illustrative case study. The lake experienced massive watershed erosion, shoreline development, and urbanisation during the twentieth century, eventually leading to cyanobacterial blooms, hypoxia, an approximately fivefold decline in water clarity, and infestation by introduced water hyacinth

(*Eichhomia crassipes*), a highly invasive macrophyte. Nile perch (*Lates niloticus*) was introduced into Lake Victoria in 1956, but it was more than 20 years before the population increased dramatically. Although predation by Nile perch undoubtedly played a central role in the loss of native cichlids from Lake Victoria, ultimately untangling the effects of interacting drivers of biodiversity loss may not be possible (Gurevitch & Padilla, 2004). This holds true for other complex case studies such the Laurentian Great Lakes (Christie, 1974; Shuter & Mason, 2001; Madenjian *et al.*, 2002). In short, many aquatic ecosystems have been subject to a barrage of interacting stressors. In many cases, a single stressor is unlikely to cause the extirpation of a native species. It may be that when multiple stressors act in concert we tend to see NIF establishment, wholesale food web changes and native species extirpation. Examples involving multiple interacting ecological stressors may or may not involve non-native species, although in light of the expanding scope and magnitude of their spread and impact, they often do. There is a need to develop a better understanding of the interacting roles of introduced species, habitat alteration and other potential factors as drivers of native species decline (Light & Marchetti, 2007). It appears that at many spatial scales, anthropogenic impacts, NIFs and native species imperilment are inextricably linked (Leprieur *et al.*, 2008).

8.5 MANAGEMENT STRATEGIES

The main point of this chapter is that NIFs are often an important contributing factor in the decline and imperilment of native freshwater fish species and communities. NIFs are introduced for a variety of reasons, and can impact native species in multiple ways. In a general sense, the introduction of NIFs is inexorably tied to human activities (Leprieur *et al.*, 2008), and is part of a wider suite of environmental transformations that are associated with human-driven environmental change. While teasing apart the ultimate drivers of declining native species and communities is a challenge, managing and preventing impacts on native species and ecosystems is undoubtedly a far greater challenge. In this section, we examine management strategies in response to NIFs and their impacts. In reality, it is not only NIFs, but non-native species generally (i.e. both fishes and non-fishes), that may invoke environmental management responses aimed at reducing impacts on native fishes, as well as other environmental concerns (economic, aesthetic and broader ecosystem integrity impacts).

8.5.1 Prevention and containment

Perhaps it is self-evident that preventing undesired NIF introductions is likely to be the most cost-effective way of minimising their adverse impacts (Leung *et al.*, 2002). One key prevention strategy in many developed nations has been to limit or ban the import and introduction of presumed high-risk species (Ruesink *et al.*, 1995). This generally involves targeting potentially harmful species, as identified through risk assessments based on species traits as discussed above (Lodge *et al.*, 2006). Efforts to halt pest species at the border have the potential to be effective, although this approach is often not straightforward in light of the diverse pathways and reasons for NIF introductions, and the potential for conflicting values and demands. In considering an intentional introduction, one could simply weigh the expected costs and benefits, and act accordingly. For example, introducing tilapia into a newly built reservoir may be viewed as undesirable or high-risk from an ecological point of view (Canonico *et al.*, 2005). On the other hand, economic pressures to create new fisheries are likely to trump ecological concerns. Notably, if NIF introductions threaten the provisioning of ecosystem services such as water quality and recreation, economic and ecological concerns may align. Unfortunately, the value of ecosystem services remains poorly recognised.

There are many examples of where an undesirable species is in a region and spreading. In such cases, the management aim is to contain the species and prevent their further spread. As an example, rainbow smelt (*Osmerus mordax*) appeared in the Laurentian Great Lakes in the 1920s, and have subsequently appeared in hundreds of smaller inland lakes in the region as a result of illegal transport via bait-bucket spread and angler carelessness (Evans & Loftus, 1987). Rainbow smelt have documented strong negative impacts on native game species (Evans & Loftus, 1987; Hrabik *et al.*, 1998; Mercado-Silva *et al.*, 2007). Whether species such as rainbow smelt continue to spread will likely depend on whether anglers recognise and understand their adverse impacts, and alter their behaviour accordingly.

The issue of containment is at the core of the management debate surrounding Asian carp (primarily *Hypophthalmichthys nobilis* and *H. molitrix*) in North America. Asian carp escaped southern US aquaculture facilities in the 1970s, and subsequently dispersed throughout much of the Mississippi River basin (Kolar *et al.*, 2005). Asian carp eventually reached the Chicago Sanitary and Ship canal. This canal is an important shipping and navigational link between the Laurentian

Great Lakes and Mississippi River basins, and also represents an artificial breach of the biogeographic barrier that historically separated these basins. Although an electric barrier has been built in an attempt to halt fish movement through the Chicago Sanitary and Ship canal, Asian carp are now at the cusp of moving into and establishing in the Great Lakes drainage basin (Jerde *et al.*, 2011). While shipping canals have increased the connectivity of transportation networks, they have unfortunately also increased connectivity for aquatic biota (Rahel, 2007).

8.5.2 Control and eradication

In many situations, NIFs have established and exert negative impacts on native species. Eradication is an attractive management option, particularly where recolonisation from surrounding populations is unlikely. Eradication has sometimes been achieved in small, self-contained ecosystems. For example, introduced brown trout and brook trout have been eradicated through over-harvest in numerous high elevation Sierra Nevada (USA) lakes, allowing rapid recovery of native community structure and the endangered mountain yellow-legged frogs (*Rana muscosa*) (Vredenburg, 2004; Knapp *et al.*, 2007). Rotenone has also been used to successfully eradicate unwanted NIFs (Britton *et al.*, 2011).

Often eradication is not practical, and management efforts aim to simply control NIF populations with the goal of minimising future spread, or reducing abundance and impacts. A well-known example is the multi-million dollar sea lamprey (*Petromyzon marinus*) control programme in the Laurentian Great Lakes aimed at protecting the current mix of native and non-native fisheries (Smith & Tibbles, 1980). Introduced lake trout (*Salvelinus namaycush*) in Yellowstone Lake, Wyoming are predators on native cutthroat trout (*Oncorhynchus clarki bouvieri*). Intensive lake trout removal efforts have reduced their abundance and impact, allowing native salmonids to persist (Ruzycki *et al.*, 2003). Similarly, native Lahontan cutthroat trout (*Oncorhynchus clarki henshawi*) populations appear to have increased in response to removal of introduced brook trout at Independence Lake, California (Scoppettone *et al.*, 2012).

8.5.3 Restoration

NIFs and habitat alteration are inextricably linked. A key question is whether habitat restoration or invasive species control is the best option to achieve native species and ecosystem recovery. In Ash Meadows, NV, efforts to restore spring habitats back to something resembling their pre-disturbance condition shifted fish composition from non-native back to dominance by native species (Scoppettone *et al.*, 2005). In contrast, a replicated large-scale stream restoration experiment found that native fish species showed a greater recovery in response to exotic removal than to flow restoration (Marks *et al.*, 2010; Chapter 4, Box 4.2). Stream restoration efforts, including dam removals, allow restoration of natural flow regimes and habitats and tend to strongly favour native species (Kanehl *et al.*, 1997; Marchetti & Moyle, 2001). The flip side is that river restoration also tends to increase connectivity and thus may facilitate invasive species dispersal, highlighting the two counteracting sides of increasing connectivity (Rahel, 2007; Chapter 10).

8.5.4 Management targeting

In light of the magnitude and complexity of NIFs and their impacts, there is an urgent need to more effectively prioritise and target management activities, so as to minimise the spread and impacts of NIFs. Doing so requires that we ask questions such as the following. Which NIFs are of greatest concern? Which pathways or vectors are of greatest concern? Which sites or locations are most at risk or vulnerable to future spread and impact? Which management activities are most cost-effective in reducing impacts? Ecological and economic approaches such as species risk assessments (Kolar & Lodge, 2002), ecological niche modelling (Vander Zanden & Olden, 2008) and benefit–cost studies (Naylor, 2000) fall beyond the scope of this chapter, but are useful tools for improving the decision-making process and targeting and directing management actions more effectively.

8.5.5 Potential conservation value of NIFs

While the undesirable impacts of NIFs have been widely described, NIFs can also provide conservation benefits by facilitating or serving as prey for native species, or playing a substitute ecological role formerly played by an extinct species (Schlaepfer *et al.*, 2011). While few of the examples of conservation benefits noted by Schlaepfer *et al.* involved NIFs, the reality is that NIFs are increasingly widespread, and will only continue to expand their range and their ecological role in freshwater ecosystems. Freshwater food webs will increasingly include multiple interacting introduced species. This is the real-world context under which future ecosystem management decisions will need to be made. Decisions to control or eradicate NIFs will need to carefully consider their broader role in ecosystems (Zavaleta *et al.*, 2001).

8.6 SUMMARY

Although fishes have been translocated outside of their native range for millennia, the rate of fish introductions has rapidly accelerated in parallel with the globalisation of travel and trade. There are many ultimate causes for the transport and introduction of NIFs to new waters, and as such, efforts to address the problem need to be specific to the pathways and underlying reasons for their introduction. As a result, it is not likely that there will ever be a silver-bullet solution. Interestingly, the underlying reasons for fish introductions have changed over time in many areas. For example, in much of the developed world, sanctioned stocking of game fish to new waters has declined, while aquarium releases and illegal transport of game fish by anglers appears to be on the rise.

Perhaps the most troubling feature of NIFs is their permanence once they have established. In contrast with chemical pollutants, which often diminish over time, invasive species often exhibit secondary spread following their initial establishment in a region (Vander Zanden & Olden, 2008). For many NIFs, the outlook for continued spread and impact is high, in that there is much suitable but currently unoccupied habitat (Mercado-Silva *et al.*, 2006b). Preventing the introduction of new potentially harmful species and containing already established species is the key to minimising their adverse impacts. At the same time, some NIFs may be disfavoured in changing climates (Rahel & Olden, 2008). Furthermore, NIFs are the new reality, as ecosystems are increasingly compromised by multiple interacting NIFs. In light of this, we need to continue to find creative ways to manage ecosystems, and hopefully protect native species and ecosystems in spite of the presence, and occasional dominance, of NIFs (Rosenzweig, 2003). We must also recognise that NIFs interact with other anthropogenic drivers, and that factors such as habitat loss, water quality, flow regimes, climate and harvest mediate the interactions between native and non-native fishes. We can expect rapidly changing ecosystems in the coming decades, which will continue to pose novel challenges as we attempt to conserve native fishes, and the ecosystems that sustain them.

8.7 DISCUSSION QUESTIONS

1. How might researchers tease apart the impacts of NIFs on native freshwater fishes from other factors such as habitat loss and degradation?

2. How might researchers identify the indirect effects of NIFs on native freshwater fishes?

3. Should conservationists focus on conserving native freshwater fish species, or protecting ecosystem function and processes?

4. If biotic homogenisation continues for freshwater fishes around the world, what are some possible global consequences? Are there ways to forestall this end?

5. Because we are unlikely to rid freshwaters of NIFs, how might we best approach the management and conservation of native freshwater fishes?

8.8 ACKNOWLEDGEMENTS

The authors thank Pam Fuller at the USGS for her kind assistance with introduced fish numbers for the USA. We also thank Jereme Gaeta, Julian Olden, Gretchen Hansen and an anonymous reviewer for providing helpful comments on the manuscript.

8.9 REFERENCES

Aiken, J. K., Findlay, C. S. & Chapleau, F. (2012). Long-term assessment of the effect of introduced predatory fish on minnow diversity in a regional protected area. *Canadian Journal of Fisheries and Aquatic Sciences*, 69, 1798–1805.

Allen, J. D. & Flecker, A. S. (1993). Biodiversity conservation in running waters. *Bioscience*, 43, 32–43.

Attayde, J. L., Brasil, J. & Menescal, R. A. (2011). Impacts of introducing Nile tilapia on the fisheries of a tropical reservoir in North-eastern Brazil. *Fisheries Management and Ecology*, 18, 437–443.

Bain, M. B. (1993). Assessing impacts of introduced aquatic species – grass carp in large systems. *Environmental Management*, 17, 211–224.

Balon, E. K. (1995). Origin and domestication of the wild carp, *Cyprinus carpio* – from Roman gourmets to the swimming flowers. *Aquaculture*, 129, 3–48.

Barletta, M., Jaureguizar, A. J., Baigun, C., *et al.* (2010). Fish and aquatic habitat conservation in South America: a continental overview with emphasis on neotropical systems. *Journal of Fish Biology*, 76, 2118–2176.

Bauer, J. T. (2012). Invasive species: 'back-seat drivers' of ecosystem change? *Biological Invasions*, 14, 1295–1304.

Baxter, C. V., Fausch, K. D., Murakami, M. & Chapman, P. L. (2004). Fish invasion restructures stream and forest food webs by interrupting reciprocal prey subsidies. *Ecology*, 85, 2656–2663.

Baxter, R. M. (1977). Environmental effects of dams and impoundments. *Annual Review of Ecology and Systematics*, 8, 255–283.

Britton, J. R., Davies, G. D. & Harrod, C. (2010). Trophic interactions and consequent impacts of the invasive fish *Pseudorasbora parva* in a native aquatic foodweb: a field investigation in the UK. *Biological Invasions*, 12, 1533–1542.

Britton, J. R., Gozlan, R. E. & Copp, G. H. (2011). Managing non-native fish in the environment. *Fish and Fisheries*, 12, 256–274.

Burkhead, N. M. (2012). Extinction rates in North American freshwater fishes, 1900–2010. *Bioscience*, 62, 798–808.

Cambray, J. A. (2003). Impact on indigenous species biodiversity caused by the globalization of alien recreational freshwater fisheries. *Hydrobiologia*, 500, 217–230.

Canonico, G. C., Arthington, A., McCrary, J. K. & Thieme, M. L. (2005). The effects of introduced tilapias on native biodiversity. *Aquatic Conservation – Marine and Freshwater Ecosystems*, 15, 463–483.

Carlton, J. T. (1999). The scale and ecological consequences of biological invasions in the world's oceans. In *Invasive Species and Biodiversity Management*. Sundlund, O. T., Shei, P. J. & Viken, A. (Eds). Dordecht, Netherlands: Kluwer Academic.

Carlton, J. T. (2000). Global change and biological invasions in the oceans. In *Invasive Species in a Changing World*. Mooney, H. A. & Hobbs, R. J. (Ed). Washington, DC: Island Press.

Chaichana, R. & Jongphadungkiet, S. (2012). Assessment of the invasive catfish *Pterygoplichthys pardalis* (Castelnau, 1855) in Thailand: ecological impacts and biological control alternatives. *Tropical Zoology*, 25, 173–182.

Chapleau, F., Findlay, C. S. & Szenasy, E. (1997). Impact of piscivorous fish introductions on fish species richness of small lakes in Gatineau Park, Quebec. *Ecoscience*, 4, 259–268.

Christie, W. J. (1974). Changes in the fish species composition of the Great Lakes. *Journal of the Fisheries Research Board of Canada*, 31, 827–854.

Copp, G. H., Bianco, P. G., Bogutskaya, N. G., *et al.* (2005). To be, or not to be, a non-native freshwater fish? *Journal of Applied Ichthyology*, 21, 242–262.

Courtenay, W. R. & Williams, J. D. (1992). Dispersal of exotic species from aquaculture sources, with emphasis on freshwater fishes. In *Dispersal of Living Organisms into Aquatic Ecosystems*. College Park, MD: University of Maryland Sea Grant Program.

Courtenay, W. R. J. (1993). Biological pollution through fish introductions. In *Biological Pollution: The Control and Impact of Invasive Exotic Species*. Mcknight, B. N. (Ed.). Indianapolis, IN: Indiana Academy of Science.

Cox, J. G. & Lima, S. L. (2006). Naivete and an aquatic–terrestrial dichotomy in the effects of introduced predators. *Trends in Ecology and Evolution*, 21, 674–680.

Crivelli, A. J. (1995). Are fish introductions a threat to endemic fresh-water fishes in the northern Mediterranean region. *Biological Conservation*, 72, 311–319.

Crooks, J. A. (2005). Lag times and exotic species: the ecology and management of biological invasions in slow-motion. *Ecoscience*, 12, 316–329.

Cucherousset, J. & Olden, J. D. (2011). Ecological impacts of nonnative freshwater fishes. *Fisheries*, 36, 215–230.

Cucherousset, J., Aymes, J. C., Santoul, F. & Cereghino, R. (2007). Stable isotope evidence of trophic interactions between introduced brook trout *Salvelinus fontinalis* and native brown trout *Salmo trutta* in a mountain stream of south-west France. *Journal of Fish Biology*, 71, 210–223.

Davis, M. A. (2009). *Invasion Biology*. New York, NY: Oxford Univeristy Press.

Dudgeon, D., Arthington, A. H., Gessner, M. O., *et al.* (2006). Freshwater biodiversity: importance, threats, status and conservation challenges. *Biological Reviews*, 81, 163–182.

Eby, L. A., Roach, W. J., Crowder, L. B. & Stanford, J. A. (2006). Effects of stocking-up freshwater food webs. *Trends in Ecology and Evolution*, 21, 576–584.

Elvira, B. & Almodóvar, A. (2001). Freshwater fish introductions in Spain: facts and figures at the beginning of the 21st century. *Journal of Fish Biology*, 59, 323–331.

Evans, D. O. & Loftus, D. H. (1987). Colonization of inland lakes in the Great Lakes region by rainbow smelt, *Osmerus mordax*: their freshwater niche and effects on indigenous fishes. *Canadian Journal of Fisheries and Aquatic Sciences*, 44(Suppl. 2), 249–266.

Fazey, I., Fischer, J. & Lindenmayer, D. B. (2005). Who does all the research in conservation biology? *Biodiversity and Conservation*, 14, 917–934.

Ferguson, M. M. (1990). The genetic impact of introduced fishes on native species. *Canadian Journal of Zoology*, 68, 1053–1057.

Findlay, C. S., Bert, D. G. & Zheng, L. (2000). Effect of introduced piscivores on native minnow communities in Adirondack lakes. *Canadian Journal of Fisheries and Aquatic Sciences*, 57, 570–580.

Flecker, A. S. & Townsend, C. R. (1994). Community-wide consequences of trout introductions in New Zealand streams. *Ecological Applications*, 4, 798–807.

Fleming, I. A., Hindar, K., Mjolnerod, I .B., *et al.* (2000). Lifetime success and interactions of farm salmon invading a native population. *Proceedings of the Royal Society B, Biological Sciences*, 267, 1517–1523.

Fraser, D. F. & Gilliam, J. F. (1992). Nonlethal impacts of predator invasion – facultative suppression of growth and reproduction *Ecology*, 73, 959–970.

Fuller, P. F., Nico, L. G. & Williams, J. D. (1999). *Nonindigeous Fishes Introduced into Inland Waters of the United States*. Bethesda, MD: American Fisheries Society Special Publication 27.

Garcia-Marin, J. L., Sanz, N. & Pla, C. (1999). Erosion of the native genetic resources of brown trout in Spain. *Ecology of Freshwater Fish*, 8, 151–158.

Gertzen, E., Familiar, O. & Leung, B. (2008). Quantifying invasion pathways: fish introductions from the aquarium trade. *Canadian Journal of Fisheries and Aquatic Sciences*, 65, 1265–1273.

Gherardi, F. (2007). Measuring the impact of freshwater NIS: what are we missing? In *Biological Invaders in Inland Waters: Profiles, Distribution, and Threats*. Dordrecht, The Netherlands: Springer.

Gido, K. B. & Brown, J. H. (1999). Invasion of North American drainages by alien fish species. *Freshwater Biology*, 42, 387–399.

Gido, K. B., Dodds, W. K. & Eberle, M. E. (2010). Retrospective analysis of fish community change during a half-century of landuse and streamflow changes. *Journal of the North American Benthological Society*, 29, 970–987.

Gozlan, R. E., Britton, J. R., Cowx, I. & Copp, G. H. (2010). Current knowledge on non-native freshwater fish introductions. *Journal of Fish Biology*, 76, 751–786.

Gurevitch, J. & Padilla, D. K. (2004). Are invasive species a major cause of extinctions? *Trends in Ecology and Evolution*, 19, 470–474.

Habit, E., Piedra, P., Ruzzante, D. E., *et al.* (2010). Changes in the distribution of native fishes in response to introduced species and other anthropogenic effects. *Global Ecology and Biogeography*, 19, 697–710.

Harrison, I. J. & Stiassny, M. L. J. (1999). The quiet crisis: a preliminary listing of the freshwater fishes of the world that are extinct or 'missing in action'. In *Extinctions in Near Time*. New York, NY: Kluwer Academic/Plenum.

Havel, J. E., Lee, C. E. & Vander Zanden, M. J. (2005). Do reservoirs facilitate invasions into landscapes? *BioScience*, 55, 518–522.

Hitt, N. P., Frissell, C. A., Muhlfeld, C. C. & Allendorf, F. W. (2003). Spread of hybridization between native westslope cutthroat trout, *Oncorhynchus clarki lewisi*, and nonnative rainbow trout, *Oncorhynchus mykiss*. *Canadian Journal of Fisheries and Aquatic Sciences*, 60, 1440–1451.

Hoegh-Guldberg, O., Hughes, L., Mcintyre, S., *et al.* (2008). Assisted colonization and rapid climate change. *Science*, 321, 345–346.

Hrabik, T. R., Magnuson, J. J. & Mclain, A. S. (1998). Predicting the effects of rainbow smelt on native fishes in small lakes: evidence from long-term research on two lakes. *Canadian Journal of Fisheries and Aquatic Sciences*, 55, 1364–1371.

Hughes, J. B., Daily, G. C. & Ehrlich, P. R. (1997). Population diversity: Its extent and extinction. *Science*, 278, 689–692.

Irons, K. S., Sass, G. G., McCelland, M. A. & Stafford, J. D. (2007). Reduced condition factor of two native fish species coincident with invasion of non-native Asian carps in the Illinois River, USA – is this evidence for competition and reduced fitness? *Journal of Fish Biology*, 71, 258–273.

Jelks, H. L., Walsh, S. J., Burkhead, N. M., *et al.* (2008). Conservation status of imperiled North American freshwater and diadromous fishes. *Fisheries*, 33, 372–407.

Jerde, C. L., Mahon, A. R., Chadderton, W. L. & Lodge, D. M. (2011). 'Sight-unseen' detection of rare aquatic species using environmental DNA. *Conservation Letters*, 4, 150–157.

Jeschke, J. M. & Strayer, D. L. (2005). Invasion success of vertebrates in Europe and North America. *Proceedings of the National Academy of Science (USA)*, 102, 7198–7202.

Johnson, P. T., Olden, J. D. & Vander Zanden, M. J. (2008). Dam invaders: impoundments facilitate biological invasions in freshwaters. *Frontiers in Ecology and the Environment*, 6, 357–63.

Kanehl, P. D., Lyons, J. & Nelson, J. E. (1997). Changes in the habitat and fish community of the Milwaukee River, Wisconsin, following removal of the Woolen Mills dam. *North American Journal of Fisheries Management*, 17, 387–400.

Keller, R. P., Zu Ermgassen, P. S. & Aldridge, D. C. (2009). Vectors and timing of freshwater invasions in Great Britain. *Conservation Biology*, 23, 1526–1534.

Knapp, R. A., Boiano, D. M. & Vredenburg, V. T. (2007). Removal of nonnative fish results in population expansion of a declining amphibian (mountain yellow-legged frog, *Rana muscosa*). *Biological Conservation*, 135, 11–20.

Koehn, J. D. (2004). Carp (*Cyprinus carpio*) as a powerful invader in Australian waterways. *Freshwater Biology*, 49, 882–894.

Kolar, C. S. & Lodge, D. M. (2001). Progress in invasion biology: predicting invaders. *Trends in Ecology and Evolution*, 16, 199–204.

Kolar, C. S. & Lodge, D. M. (2002). Ecological predictions and risk assessment for alien fishes in North America. *Science*, 298, 1233–1236.

Kolar, C. S., Chapman, D. C., Courtenay, W. R. J., *et al.* (2005). *Asian Carps of the Genus Hypophthalmichthys (Pisces, Cyprinidae) – A Biological Synopsis and Environmental Risk Assessment.* US Fish and Wildlife Service.

Kornis, M. S., Mercado-Silva, N. & Vander Zanden, M. J. (2012). Twenty years of invasion: a review of round goby *Neogobius melanostomus* biology, spread and ecological implications. *Journal of Fish Biology*, 80, 235–285.

Kuehne, L. M. & Olden, J. D. (2012). Prey naivety in the behavioural responses of juvenile Chinook salmon (*Oncorhynchus tshawytscha*) to an invasive predator. *Freshwater Biology*, 57, 1126–1137.

Lapointe, N. W. R., Pendleton, R. M. & Angermeier, P. L. (2012). A comparison of approaches for estimating relative impacts of nonnative fishes. *Environmental Management*, 49, 82–95.

Lassuy, D. R. (1995). Introduced species as a factor in extinction and endangerment of native fish species. *American Fisheries Society Symposium*, 15, 391–396.

Leprieur, F., Beauchard, O., Blanchet, S., Oberdorff, T. & Brosse, S. (2008). Fish invasions in the world's river systems: when natural processes are blurred by human activities. *PLoS Biology*, 6, 404–410.

Leunda, P. M. (2010). Impacts of non-native fishes on Iberian freshwater ichthyofauna: current knowledge and gaps. *Aquatic Invasions*, 5, 239–262.

Leung, B., Lodge, D. M., Finnoff, D., *et al.* (2002). An ounce of prevention or a pound of cure: bioeconomic risk analysis of invasive species. *Proceedings of the Royal Society of London Series B*, 269, 2407–2413.

Lever, C. (1996). *Naturalized Fishes of the World*. London: Academic Press.

Light, T. & Marchetti, M. P. (2007). Distinguishing between invasions and habitat changes as drivers of diversity loss among California's freshwater fishes. *Conservation Biology*, 21, 434–446.

Ligon, F. K., Dietrich, W. E. & Trush, W. J. (1995). Downstream ecological effects of dams. *Bioscience*, 45, 183–192.

Lintermans, M. (2004). Human-assisted dispersal of alien freshwater fish in Australia. *New Zealand Journal of Marine and Freshwater Research*, 38, 481–501.

Litvak, M. K. & Mandrak, N. E. (1993). Ecology of freshwater baitfish use in Canada and the United States. *Fisheries*, 18, 6–13.

Lockwood, J. L., Hoopes, M. F. & Marchetti, M. P. (2013). *Invasion Ecology*, second edition. Oxford: Wiley-Blackwell Publishing.

Lodge, D. M., Williams, S., MacIsaac, H. J., *et al.* (2006). Biological invasions: Recommendations for US policy and management. *Ecological Applications*, 16, 2035–2054.

Ludwig, H. R. & Leitch, J. A. (1996). Interbasin transfer of aquatic biota via angler's bait buckets. *Fisheries*, 21, 14–18.

Lusk, S., Lusková, V. & Hanel, L. (2010). Alien fish species in the Czech Republic and their impact on the native fish fauna. *Folia Zoologica*, 59, 57–72.

Macdougall, A. S. & Turkington, R. (2005). Are invasive species the drivers or passengers of change in degraded ecosystems? *Ecology*, 86, 42–55.

Madenjian, C. P., Fahnenstiel, G. L., Johengen, T. H., *et al.* (2002). Dynamics of the Lake Michigan food web, 1970–2000. *Canadian Journal of Fisheries and Aquatic Sciences*, 59, 536–753.

Mandrak, N. E. & Cudmore, B. (2010). The fall of native fishes and the rise of non-native fishes in the Great Lakes basin. *Aquatic Ecosystem Health and Management*, 13, 255–268.

Marchetti, M. P. (1999). An experimental study of competition between the native Sacramento perch (*Archoplites interruptus*) and introduced bluegill (*Lepomis macrochirus*). *Biological Invasions*, 1, 55–65.

Marchetti, M. P. & Moyle, P. B. (2001). Effects of flow regime on fish assemblages in a regulated California stream. *Ecological Applications*, 11, 530–539.

Marchetti, M. P., Moyle, P. B. & Levine, R. (2004). Alien fishes in California watersheds: characteristics of successful and failed invaders. *Ecological Applications*, 14, 587–596.

Marks, J. C., Haden, G. A., O'Neill, M. & Pace, C. (2010). Effects of flow restoration and exotic species removal on recovery of native fish: lessons from a dam decommissioning. *Restoration Ecology*, 18, 934–943.

Marr, S. M., Marchetti, M. P., Olden, J. D., *et al.* (2010). Freshwater fish introductions in mediterranean-climate regions: are there commonalities in the conservation problem? *Diversity and Distributions*, 16, 606–619.

Matern, S. A. (2001). Using temperature and salinity tolerances to predict the success of the shimofuri goby, a recent invader into California. *Transactions of the American Fisheries Society*, 130, 592–599.

McDowall, R. M. (2006). Crying wolf, crying foul, or crying shame: alien salmonids and a biodiversity crisis in the southern cool-temperate galaxioid fishes? *Reviews in Fish Biology and Fisheries*, 16, 233–422.

Mercado-Silva, N., Lyons, J., Díaz-Pardo, E., *et al.* (2006a). Long-term changes in the fish assemblage of the Laja River, Guanajuato, central Mexico. *Aquatic Conservation – Marine and Freshwater Ecosystems*, 16, 533–546.

Mercado-Silva, N., Olden, J. D., Maxted, J. T., *et al.* (2006b). Forecasting the spread of invasive rainbow smelt in the Laurentian Great Lakes region of North America. *Conservation Biology*, 20, 1740–1749.

Mercado-Silva, N., Gilbert, S., Sass, G. G., Roth, B. M. & Vander Zanden, M. J. (2007). Impact of rainbow smelt (*Osmerus mordax*) invasion on walleye (*Sander vitreus*) recruitment in Wisconsin lakes. *Canadian Journal of Fisheries and Aquatic Sciences*, 64, 1543–1550.

Miller, R. R., Williams, J. D. & Williams, J. E. (1989). Extinctions of North American fishes during the past century. *Fisheries*, 14, 22–38.

Mims, M. C. & Olden, J. D. (2013). Fish assemblages respond to altered flow regimes via ecological filtering of life history strategies. *Freshwater Biology*, 58, 50–62.

Minckley, W. L. & Deacon, J. E. (Eds). (1991). *Battle Against Extinction: Native Fish Management in the American West*. Tuscon, AZ: University of Arizona Press.

Molnar, J. L., Gamboa, R. L., Revenga, C. & Spalding, M. D. (2008). Assessing the global threat of invasive species to marine biodiversity. *Frontiers in Ecology and the Environment*, 6, 485–492.

Moyle, P. B. (2002). *Inland Fishes of California*. Berkeley, CA: University of California Press.

Moyle, P. B. & Light, T. (1996). Biological invasions of fresh water – empirical rules and assembly theory. *Biological Conservation*, 78, 149–161.

Nasmith, L. E., Tonn, W. M., Paszkowski, C. A. & Scrimgeour, G. J. (2010). Effects of stocked trout on native fish communities in boreal foothills lakes. *Ecology of Freshwater Fish*, 19, 279–289.

Naylor, R. L. (2000). The economics of alien species invasions. In *Invasive Species in a Changing World*. Mooney, H. & Hobbs, R. (Eds). Washington, DC: Island Press.

Naylor, R. L., Williams, S. L. & Strong, D. R. (2001). Aquaculture – a gateway for exotic species. *Science*, 294, 1655–1656.

Olden, J. D., Poff, N. L. & Bestgen, K. R. (2006). Life-history strategies predict fish invasions and extirpations in the Colorado River Basin. *Ecological Monographs*, 76, 25–40.

Olden, J. D., Kennard, M. J., Leprieur, F., *et al.* (2010). Conservation biogeography of freshwater fishes: recent progress and future challenges. *Diversity and Distributions*, 16, 496–513.

Olden, J. D., Kennard, M. J., Lawler, J. J. & Poff, N. L. (2011). Challenges and opportunities in implementing managed relocation for conservation of freshwater species. *Conservation Biology*, 25, 40–47.

Papes, M., Sallstrom, M., Asplund, T. R. & Vander Zanden, M. J. (2011). Invasive species research to meet the needs of resource management and planning. *Conservation Biology*, 25, 867–872.

Pelicice, F. M. & Agostinho, A. A. (2009). Fish fauna destruction after the introduction of a non-native predator (*Cichla kelberi*) in a neotropical reservoir. *Biological Invasions*, 11, 1789–1801.

Poff, N. L., Olden, J. D., Merritt, D. M. & Pepin, D. M. (2007). Homogenization of regional river dynamics by dams and global biodiversity implications. *Proceedings of the National Academy of Sciences of the United States of America*, 104, 5732–5737.

Post, J. R. & Cucin, D. (1984). Changes in the benthic community of a small precambrian lake following the introduction of yellow perch, *Perca flavescens*. *Canadian Journal of Fisheries and Aquatic Sciences*, 41, 1496–1501.

Poulin, R., Paterson, R. A., Townsend, C. R., Tompkins, D. M. & Kelly, D. W. (2011). Biological invasions and the dynamics of endemic diseases in freshwater ecosystems. *Freshwater Biology*, 56, 676–688.

Radonski, G. C., Prosser, N. S., Martin, R. G. & Stroud, R. H. (1984). Exotic fishes and sport fishing. In *Distribution, Biology, and Management of Exotic Fishes*. Courtenay, W.R. & Stauffer, J.R. (Eds). Baltimore, MD: The Johns Hopkins University Press.

Rahel, F. J. (1997). From Johnny Appleseed to Dr Frankenstein: changing values and the legacy of fisheries management. *Fisheries*, 22, 8–9.

Rahel, F. J. (2000). Homogenization of fish faunas across the United States. *Science*, 288, 854–856.

Rahel, F. J. (2002). Homogenization of freshwater faunas. *Annual Review of Ecology and Systematics*, 33, 291–315.

Rahel, F. J. (2007). Biogeographic barriers, connectivity and homogenization of freshwater faunas: it's a small world after all. *Freshwater Biology*, 52, 696–710.

Rahel, F. J. & Olden, J. D. (2008). Assessing the effects of climate change on aquatic invasive species. *Conservation Biology*, 22, 521–533.

Ribeiro, F., Elvira, B., Collares-Pereira, M. J. & Moyle, P. B. (2008). Life-history traits of non-native fishes in Iberian watersheds across several invasion stages: a first approach. *Biological Invasions*, 10, 89–102.

Ribeiro, F. & Leunda, P. M. (2012). Non-native fish impacts on Mediterranean freshwater ecosystems: current knowledge and research needs. *Fisheries Management and Ecology*, 19, 142–156.

Ricciardi, A. (2001). Facilitative interactions among aquatic invaders: Is an 'invasion meltdown' occurring in the Great Lakes? *Canadian Journal of Fisheries and Aquatic Sciences*, 58, 2513–2525.

Ricciardi, A. (2006). Patterns of invasion in the Laurentian Great Lakes in relation to changes in vector activity. *Diversity and Distributions*, 12, 425–433.

Ricciardi, A. (2003). Predicting the impacts of an introduced species from its invasion history: an empirical approach applied to zebra mussel invasions. *Freshwater Biology*, 48, 972–981.

Ricciardi, A. & Atkinson, S. K. (2004). Distinctiveness magnifies the impact of biological invaders in aquatic ecosystems. *Ecology Letters*, 7, 781–784.

Ricciardi, A. & Rasmussen, J. B. (1998). Predicting the identity and impact of future biological invaders: a priority for aquatic resource management. *Canadian Journal of Fisheries and Aquatic Sciences*, 55, 1759–1765.

Ricciardi, A. & Rasmussen, J. B. (1999). Extinction rates of North American freshwater fauna. *Conservation Biology*, 13, 1220–1222.

Ricciardi, A. & Simberloff, D. (2009). Assisted colonization is not a viable conservation strategy. *Trends in Ecology and Evolution*, 24, 248–253.

Ricciardi, A., Hoopes, M. F., Marchetti, M. P. & Lockwood, J. L. (2013). Progress towards understanding the ecological impacts of nonnative species. *Ecological Monographs*, 83, 263–282.

Richardson, D. M. & Pysek, P. (2006). Plant invasions: merging the concepts of species invasiveness and community invasibility. *Progress in Physical Geography*, 30, 409–431.

Richardson, M. J., Whoriskey, F. G. & Roy, L. H. (1995). Turbidity generation and biological impacts of an exotic fish *Carassius auratus*, introduced into shallow seasonally anoxic ponds *Journal of Fish Biology*, 47, 576–585.

Richter, B. D., Braun, D. P., Mendelson, M. A. & Master, L. L. (1997). Threats to imperiled freshwater fauna. *Conservation Biology*, 11, 1081–1093.

Roberge, C., Blanchet, S., Dodson, J. J., Guderley, H. & Bernatchez, L. (2008). Disturbance of social hierarchy by an invasive species: a gene transcription study. *Plos ONE*, 3.

Rodriguez, J. P. (2001). Exotic species introductions into South America: an underestimated threat? *Biodiversity and Conservation*, 10, 1983–1996.

Rosenzweig, M. L. (2003). *Win–Win Ecology*. Oxford:, Oxford University Press.

Ross, S. T. (1991). Mechanisms structuring stream fish assemblages: are there lessons from introduced species? *Environmental Biology of Fishes*, 30, 359–368.

Ruesink, J. L., Parker, I. M., Groom, M. J. & Kareiva, P. M. (1995). Reducing the risks of nonindigenous species introductions. *BioScience*, 45, 465–477.

Ruzycki, J. R., Beauchamp, D. A. & Yule, D. L. (2003). Effects of introduced lake trout on native cutthroat trout in yellowstone lake. *Ecological Applications*, 13, 23–37.

Schade, C. B. & Bonar, S. A. (2005). Distribution and abundance of nonnative fishes in streams of the Western United States. *North American Journal of Fisheries Management*, 25, 1386–1394.

Scharf, W. (2007). Biomanipulation as a useful water quality management tool in deep stratifying reservoirs. *Hydrobiologia*, 583, 21–42.

Schindler, D. E., Kitchell, J. F., He, X., *et al.* (1993). Food-web structure and phosphorus cycling in lakes. *Transactions of the American Fisheries Society*, 122, 756–772.

Schlaepfer, M. A., Sax, D. F. & Olden, J. D. (2011). The potential conservation value of non-native species. *Conservation Biology*, 25, 428–437.

Schoen, E. R., Beauchamp, D. A. & Overman, N. C. (2012). Quantifying latent impacts of an introduced piscivore: pulsed predatory inertia of lake trout and decline of kokanee. *Transactions of the American Fisheries Society*, 141, 1191–1206.

Scoppettone, G. G., Rissler, P. H., Gourley, C. & Martinez, C. (2005). Habitat restoration as a means of controlling non-native fish in a Mojave Desert oasis. *Restoration Ecology*, 13, 247–256.

Scoppettone, G. G., Rissler, P. H., Shea, S. P. & Somer, W. (2012). Effect of brook trout removal from a spawning stream on an adfluvial population of lahontan cutthroat trout. *North American Journal of Fisheries Management*, 32, 586–596.

Sharma, S., Jackson, D. A., Minns, C. K. & Shuter, B. J. (2007). Will northern fish populations be in hot water because of climate change? *Global Change Biology*, 13, 2052–2064.

Sharma, S., Jackson, D. A. & Minns, C. K. (2009). Quantifying the effects of climate change and invasive species on native species. *Ecography*, 32, 517–525.

Shuter, B. J. & Mason, D. M. (2001). Exotic invertebrates, food-web disruption, and lost fish production: understanding impacts of dreissenid and cladoceran invaders on lower-lakes fish communities and forecasting invasion impacts on upper-lakes fish communities. Board of Technical Experts, Great Lakes Fishery Commission.

Sigler, W. F. & Sigler, J. W. (1987). *Fishes of the Great Basin: A Natural History*, Reno, NV: University of Nevada Press.

Simon, K. S. & Townsend, C. R. (2003). Impacts of freshwater invaders at different levels of ecological organisation, with emphasis on salmonids and ecosystem consequences. *Freshwater Biology*, 48, 982–994.

Smith, B. R. & Tibbles, J. J. (1980). Sea lamprey (*Petromyzon marinus*) in Lakes Huron, Michigan, and Superior: history of invasion and control, 1936–78. *Canadian Journal of Fisheries and Aquatic Sciences*, 37, 1780–1801.

Soltz, D. L. & Naiman, R. J. (1978). *The Natural History of Native Fishes in the Death Valley System*. Los Angeles, CA: Natural History Museum of Los Angeles County.

Strayer, D. L. (2010). Alien species in fresh waters: ecological effects, interactions with other stressors, and prospects for the future. *Freshwater Biology*, 55, 152–174.

Strayer, D. L., Eviner, V. T., Jeschke, J. M. & Pace, M. L. (2006). Understanding the long-term effects of species invasions. *Trends in Ecology and Evolution*, 21, 645–651.

Strecker, A. L., Campbell, P. M. & Olden, J. D. (2011). The aquarium trade as an invasion pathway in the Pacific Northwest. *Fisheries*, 36, 74–85.

Taniguchi, Y., Miyake, Y., Saito, T., Urabe, H. & Nakano, S. (2000). Redd superimposition by introduced rainbow trout, *Oncorhynchus mykiss*, on native charrs in a Japanese stream. *Ichthyological Research*, 47, 149–156.

Tilman, D., May, R. M., Lehman, C. L. & Nowak, M. A. (1994). Habitat destruction and the extinction debt. *Nature*, 371, 65–66.

Trumpickas, J., Mandrak, N. E. & Ricciardi, A. (2011). Nearshore fish assemblages associated with introduced predatory fishes in lakes. *Aquatic Conservation – Marine and Freshwater Ecosystems*, 21, 338–347.

USGS. (2013). Nonindigenous Aquatic Species Database summary graphs. http://nas.er.usgs.gov (downloaded 27 February2013).

Vander Zanden, M. J. & Olden, J. D. (2008). A management framework for preventing the secondary spread of aquatic invasive species. *Canadian Journal of Fisheries and Aquatic Science*, 65, 1512–1522.

Vander Zanden, M. J., Casselman, J. M. & Rasmussen, J. B. (1999). Stable isotope evidence for the food web consequences of species invasions in lakes. *Nature*, 401, 464–467.

Vanni, M. J. (2002). Nutrient cycling by animals in freshwater ecosystems. *Annual Review of Ecology and Systematics*, 33, 341–370.

Venter, O., Brodeur, N. N., Nemiroff, L., *et al.* (2006). Threats to endangered species in Canada. *Bioscience*, 56, 903–910.

Vitule, J. R. S., Freire, C. A. & Simberloff, D. (2009). Introduction of non-native freshwater fish can certainly be bad. *Fish and Fisheries*, 10, 98–108.

Vredenburg, V. T. (2004). Reversing invasive species effects: experimental removal of introduced fish leads to rapid recovery of a declining frog. *Proceedings of the National Academy of Science (USA)*, 101, 7646–7650.

Weyl, O. L. F. (2008). Rapid invasion of a subtropical lake fishery in central Mozambique by Nile tilapia, *Oreochromis niloticus* (Pisces: Cichlidae). *Aquatic Conservation – Marine and Freshwater Ecosystems*, 18, 839–851.

Wilcove, D. S., Rothstein, D., Dubow, J., Phillips, A. & Losos, E. (1998). Quantifying threats to imperiled species in the United States. *Bioscience*, 48, 607–615.

Williamson, M. (1996). *Biological Invasions.* New York, NY: Chapman and Hall.

Wilson, J. R. U., Dormontt, E. E., Prentis, P. J., Lowe, A. J. & Richardson, D. M. (2009). Something in the way you move: dispersal pathways affect invasion success. *Trends in Ecology and Evolution,* 24, 136–144.

Witte, F., Goldschmidt, T., Goudswaard, P. C., *et al.* (1992a). Species extinction and concomitant ecological changes in Lake Victoria. *Netherlands Journal of Zoology,* 42, 214–232.

Witte, F., Goldschmidt, T., Wanink, J. *et al.* (1992b). The destruction of an endemic species flock – quantitative data on the decline of the Haplochromine cichlids of Lake Victoria. *Environmental Biology of Fishes,* 34, 1–28.

Zavaleta, E. S., Hobbs, R. J. & Mooney, H. A. (2001). Viewing invasive species removal in a whole-ecosystem context. *Trends in Ecology and Evolution,* 16, 454–459.

Riparian management and the conservation of stream ecosystems and fishes

MARK S. WIPFLI AND JOHN S. RICHARDSON

9.1 ECOLOGICAL IMPORTANCE OF RIPARIAN HABITATS TO FRESHWATER ECOSYSTEMS

Riparian areas are the terrestrial environment adjacent to water that both influences and is influenced by the aquatic feature (Gregory *et al.*, 1991; Naiman *et al.*, 2010). Riparian areas along streams provide shade, sources of wood and organic matter, contribute to bank stability, filter sediments, take up excess nutrients from groundwater inputs, and other key processes that protect freshwaters (e.g. Naiman *et al.*, 2010; Richardson & Danehy, 2007; Figure 9.1). Riparian areas also increase biodiversity through habitat complexity and close juxtaposition of aquatic and terrestrial environments (Quinn *et al.*, 2004; Naiman *et al.*, 2010). Alterations to riparian areas, despite their small area relative to the landscape, have disproportionate effects on habitats and fish communities (Naiman *et al.*, 2010; Wipfli & Baxter, 2010). Key habitat losses and alterations are derived from modification of riparian areas by reducing instream habitat complexity (Bilby & Ward, 1989; Fausch & Northcote, 1992; Naiman *et al.*, 2010), diminishing the productive basis of freshwater food webs (Belsky *et al.*, 1999; Quinn *et al.*, 2004), increasing nutrient, contaminant and sediment intrusion (Muscutt *et al.*, 1993; Daniels & Gilliam, 1996; Nguyen *et al.*, 1998; Waters, 1999).

Riparian and freshwater ecosystems are typically tightly coupled, especially in their natural states, and the linkages that couple them frequently exert strong influence on their associated invertebrate and fish fauna (e.g. Gregory *et al.*, 1991; Naiman *et al.*, 2010). Riparian habitats,

Conservation of Freshwater Fishes, eds G. P. Closs, M. Krkosek and J. D. Olden. Published by Cambridge University Press. © Cambridge University Press 2016.

and the condition of these habitats, further plays a key role in the ecology of these fresh waters, influencing critical processes such as water, nutrient and sediment delivery and dynamics; prey resources for fish and other consumers, and other organic materials exchanged between aquatic and terrestrial habitats (Nakano *et al.*, 1999; Naiman *et al.*, 2010); light and water temperature dynamics that in turn affect food web processes and fish metabolism and growth; aquatic physical habitat (wood); and terrestrial consumers that prey upon fishes (Bisson & Bilby, 1998; Naiman *et al.*, 2010; Wipfli & Baxter, 2010). These processes in turn directly or indirectly influence fishes in freshwater systems (Wang *et al.*, 2001; Pusey & Arthington, 2003; Allan, 2004; Richardson *et al.*, 2010a).

9.1.1 Key physical, chemical and biological interactions

Riparian plant community composition and structure, and the extent of plant cover associated with water bodies, can influence the hydrograph, including the duration and magnitude of discharge following storm events (Tabacchi *et al.*, 2000; McGlynn & McDonnell, 2003; Meador & Goldstein, 2003). Discharge and associated processes (e.g. flow, turbidity and sediment dynamics) in turn affect the quantity and quality of aquatic habitats, and the food webs that support fishes (Dunbar *et al.*, 2010; Wellington *et al.*, 2010; Jones *et al.*, 2012; Chapter 6). Water delivery through riparian areas to streams, lakes and wetlands also affects nutrient and sediment delivery and storage (Meador & Goldstein, 2003). Nutrient dynamics indirectly affect fishes through influencing aquatic food web processes, including stream periphyton and heterotrophic bacteria dynamics (Rier & Stevenson, 2002). These basal trophic levels (algae and organic matter) are consumed by invertebrates, which form the trophic basis for many fish populations (Winterbourn, 1990). Many freshwater fishes eat biofilm and detritus directly as well (Pringle & Hamazaki, 1997; Katano *et al.*, 2003), so sediment loads and nutrient shifts that reduce algal production will impact some fishes directly. Loss of riparian plants, including a reduction in community structural complexity, can accelerate run-off to water bodies, impacting resident wood and therefore aquatic habitats, and nutrient dynamics (Meador & Goldstein, 2003).

The structure of the riparian plant community influences radiative inputs through shading (Figure 9.1). Light inputs have several influences on streams and fish habitat. Solar radiation affects primary production in streams, thermal dynamics, and fish perception of their

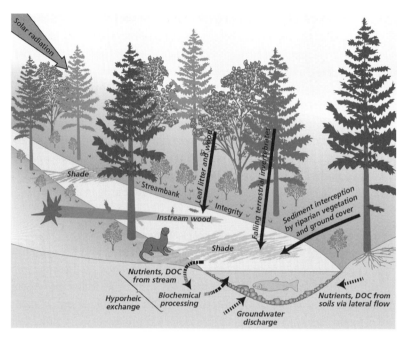

Figure 9.1 Linkages between the riparian zone and streams at the reach scale. DOC, dissolved organic carbon. Modified from original drawing by Eric Leinberger, University of British Columbia.

food resources (Diehl, 1988; Fraser & Metcalfe, 1997). Algal production is positively related to light (e.g. Kiffney *et al.*, 2003, 2004). In deciduous forests, the springtime flush of leaves can depress light inputs to < 3% of the pre-leaf period and reduce algal productivity to less than 25% of the pre-leaf period (Hill *et al.*, 2001). Similar changes can be seen along streams as stream width increases and canopy closure decreases (Finlay *et al.*, 2011; Sakamaki & Richardson, 2013). Land uses that reduce riparian canopy cover changes radiative and allochthonous inputs to streams. This can change aquatic-produced food resources from allochthonous-dependent invertebrates to autochthonous-dependent invertebrates (Bisson & Bilby, 1998), which would be detrimental to food-limited fishes if this trophic shift renders the prey community less accessible or less preferred.

As stated above, light is important to fish foraging success; visual foragers depend on light for capturing prey (Jönsson *et al.*, 2011; Erős *et al.*, 2012). Many fish are unable to forage under low light conditions, but their abilities vary among species, ages, trophic positions and even seasonally (Rader *et al.*, 2007). The scattering of light through riparian canopies can also affect foraging efficiency as demonstrated by lower capture success of cutthroat trout in areas with vegetation above the stream surface, in contrast to reaches with reduced canopy cover (Wilzbach *et al.*, 1986).

Water temperature is a major controlling factor in invertebrate and ectothermic vertebrate populations, with all species having specific temperature tolerance limits for growth, reproduction and survival (Eaton & Scheller, 1996; Zeigler *et al.*, 2013). Riparian canopy cover plays a crucial role in regulating temperature, primarily from shading, especially in small streams (Poole & Berman, 2001; Moore *et al.*, 2005). Loss or reduction of riparian vegetation, or less channel incision, can mean direct habitat loss for fishes when water temperatures exceed tolerance limits for fish (Leach *et al.*, 2012).

Riparian vegetation plays a large role in bank stability (Millar, 2000) and limits the mobility of sediments, particularly fine particles that often impair aquatic habitats (Waters, 1999). Riparian forest canopies help shield soil surfaces from direct strike from falling raindrops as rain hits the foliage and limbs as they pass through the canopy, reducing impact and rainsplash, and again directly at the soil surface from leaf litter previously fallen from the canopies (Young & Wiersma, 1973; Morgan, 2009). Leaves form a protective layer on soil surfaces, reducing erosion (Sayer, 2006), thereby reducing the movement of fine sediments into streams that pose threats to stream biota (Wellington *et al.*, 2010; Jones *et al.*, 2012; Chapter 6). Riverbed fine sediment can change dominance in aquatic invertebrate communities from surface dwellers to burrowers, which are largely unavailable to drift-feeding fishes such as salmonids (Suttle *et al.*, 2004). Plant litter in riparian habitats plays other important roles too. Litter is an inherent part of nutrient and carbon cycling and regulates soil microclimates (Sayer, 2006). Soil fertility and nutrient cycling influences aquatic biota as nutrients make their way to streams affecting primary production and the consumers that primary producers support (Winterbourne, 1990; Rier & Stevenson, 2002).

Wood provided to aquatic habitats by riparian forests plays a key role in aquatic ecosystem structure and function. Large wood (historically termed large woody debris, defined as > 10 cm diameter and > 1 m long) has been shown to exert many controlling influences on streams, affecting channel form, influencing routing and storage of organic matter and sediment (Gomi *et al.*, 2002; Gurnell *et al.*, 2002; Montgomery *et al.*, 2003), and modifying the movement and transformation of nutrients

(Bisson *et al.*, 1987). Wood influences the physical characteristics of streams and riparian habitats, affecting in-stream fish habitat formation and maintenance (Maser & Sedell, 1994; Culp *et al.*, 1996; Bilby *et al.*, 1998; Dolloff & Warren, 2003; Whiteway *et al.*, 2010), pool area and volume (Whiteway *et al.*, 2010) and cover for fishes (Kail *et al.*, 2007; Antón et al., 2011; Baillie *et al.*, 2013). The distribution of large wood in streams largely depends on channel size (Bilby & Ward, 1989; Swanson *et al.*, 1982). In headwater channels, large wood exhibits a random distribution reflecting the pattern and rate of large wood recruitment because average peak flows cannot normally move the debris (although large flows may). This is in sharp contrast to large channels where large wood is distributed in clumps (log jams, which often also include small wood), due to transport and subsequent aggregation. Wood in streams often supports higher densities of fishes (Whiteway *et al.*, 2010; Kano *et al.*, 2013), although not for all species (Langford *et al.*, 2012), and generally greater fish biodiversity as documented in tropical streams (Wright & Flecker, 2004; Kano *et al.*, 2013). Some tropical fishes live in holes in wood (e.g. Power, 2003). Riparian vegetation provides cover for fish and the absence of this cover influences fish behaviour, pushing fish deeper in the water column when cover is less abundant (Bugert *et al.*, 1991; Kano *et al.*, 2013).

9.1.2 Resource subsidies from riparian habitats

Riparian vegetation can provide a large portion of the energy that fuels stream food webs, an example of a cross-ecosystem resource subsidy (Wallace *et al.*, 1997; Richardson *et al.*, 2010b; Wipfli & Baxter, 2010; Figure 9.1). Plant litter (leaves, twigs, seeds and fruiting structures) drops into streams, providing a source of fixed carbon that is converted by microbes into food for invertebrates, amphibians and some fishes (e.g. Webster *et al.*, 1999; Richardson *et al.*, 2005). In very small forested streams, organic matter inputs can contribute the largest fraction of the energy budget to a stream (Fisher & Likens, 1973). The quality of plant litter varies widely depending on plant species (e.g. coniferous vs. deciduous) and plant part (e.g. leaves vs. stems) (Richardson *et al.*, 2004). For example, some nitrogen fixers (e.g. alder) are naturally high in nitrogen (N) and provide a highly digestible, high N content organic matter source for detritivores (Fyles & Fyles, 1993; Motomori *et al.*, 2001; Leroy & Marks, 2006). As leaf litter is higher-quality detritus, it is processed and utilised by consumers more readily than woody material, other factors being equal (Hofer & Richardson, 2007). These terrestrial–freshwater linkages are crucial for stream communities. England and Rosemond (2004) demonstrated that reductions in riparian forest canopy cover weakened terrestrial–aquatic linkages via reductions in stream particulate organic matter, subsequently affecting crucial energy flow to consumers (Wallace *et al.*, 1997).

In addition to wood in streams having a large role in creating physical habitat for fishes, it also has a direct biological role. Wood offers a substratum for benthic invertebrate production (Anderson *et al.*, 1978; Wallace & Benke, 1984; Johnson *et al.*, 2003), providing a source of carbon for biofilms, which are grazed by various guilds of invertebrates and fishes that feed on biofilms (Pringle & Hamazaki, 1997; Katano *et al.*, 2003). Plant species type affects the longevity of wood in streams, and tree density and size class influence the amount of wood that enters streams and their residency once there. Conifers on average decay more slowly (Spänhoff & Meyer, 2004) and are often much larger than deciduous counterparts, at least in western North America.

Riparian plant cover also influences the amount, timing and type of terrestrial prey (invertebrates) that enter water bodies as prey for freshwater fishes (Mason & MacDonald, 1982; Wipfli, 1997; Nakano *et al.*, 1999; Richardson & Danehy, 2007). The quantities of terrestrial invertebrate subsidies to water bodies vary globally. Terrestrial prey organisms often contribute a large proportion (greater than half in many cases) of the diet of freshwater fishes during much of the year in streams in parts of North America and Japan (Wipfli, 1997; Nakano *et al.*, 1999; Kawaguchi & Nakano, 2001; Baxter *et al.*, 2005), but may be less important in other places with different riparian conditions – New Zealand tussock grassland streams, for example (Edwards & Huryn, 1996). There is some evidence supporting the hypothesis that these terrestrial subsidies affect local fish abundance in streams (Kawaguchi *et al.*, 2003). Further, the amount and type of riparian vegetation cover affects terrestrial prey subsidies to streams, and therefore the overall prey budget for fishes, and could be especially crucial in habitats where fishes are food limited (Allan *et al.*, 2003; Kawaguchi *et al.*, 2003). Red alder houses and provides more terrestrial invertebrates than co-occurring hemlock and spruce along Alaskan streams (Allan *et al.*, 2003), and sycamore provides more terrestrial prey to fishes than oak and ash in selected English streams (Mason & MacDonald, 1982). Riparian disturbances that affect riparian canopy structure, plant abundance and species composition will have an impact on stream invertebrates, and thus on the food resources for fishes, with subsequent impacts on resident fishes where fish are food-limited.

There are other examples connecting resource subsidies from the forest beyond the immediate riparian area to fish productivity. Plant litter leachate makes its way into subsurface and ground water (Lu *et al.*, 2014) and is used by bacteria in the stream biofilm on substrate surfaces, and subsequently consumed by a range of species, including invertebrates (e.g. Ledger & Hildrew, 2001) and fishes (Flecker & Taylor, 2004).

Much more distant and obscure habitat and ecosystem connections can also affect stream fishes in ways that have become better understood in the last couple of decades. Vannote *et al.* (1980) highlighted the reliance of downstream communities on upstream riparian habitats and processes, illustrating how community structure and productivity may be linked to and dependent upon very distant riparian habitats and the plant communities they support. Subsidies from upstream habitats, including small, high-gradient headwater streams where fish often do not occur (due to lack of suitable habitat), are important for freshwater fishes (Wipfli & Gregovich, 2002). Both terrestrial invertebrates originating from surrounding forests and aquatic prey produced in fishless headwater channels drift via surface water flow and are carried to fish habitats further downstream. Both prey types, sourced from upland habitats, are important prey items for fishes. And again, riparian vegetation type (e.g. conifer vs. deciduous) strongly influences the amount of prey that is carried by stream flow, with deciduous-canopied streams providing 2–4× more prey delivery per water volume to fishes than coniferous species (Piccolo & Wipfli, 2002; Wipfli & Musslewhite, 2004) throughout the temperate rainforests of western North America. Wallace *et al.* (1997) demonstrated the dependence of headwater invertebrate communities on riparian organic matter subsidies, which can indirectly affect downstream fishes via invertebrate transport from upland habitats.

9.1.3 Other linkages to the riparian area

Freshwater fishes are also indirectly affected by riparian plant cover through the piscivorous predators that these terrestrial habitats support. Piscivorous birds that prey on freshwater fishes use riparian plants for perching, nesting and roosting (Baker & Baker, 2002). For instance, kingfishers in southern Tanzania depend on riparian canopies for perching while foraging, with some species preferring high perches and others lower perches, suggesting that riparian forest structural complexity can influence the foraging success of avian piscivores (Bonnington *et al.*, 2008). Some cormorants in Great Britain utilise trees for nesting, feed on a wide range of fish species, and are viewed as having serious potential to impact fisheries in some cases (Kirby *et al.*, 1996). The type of riparian cover influences the quality of habitat for avian predators (Monadjem, 1996) and therefore their abundance and performance, in turn affecting predation risk of fishes. Protection of riparian areas and their vegetation can also be a significant contributor to the conservation of a variety of piscivorous mammals, including river otters (e.g. Sepulveda *et al.*, 2007).

Some aquatic organisms modify stream and riparian habitats in ways that facilitate habitat creation for other species. For instance, beavers create slow-water habitats in streams that are important to some fishes and amphibians (e.g. Wright *et al.*, 2002; Stevens *et al.*, 2007). Beaver ponds and associated off-channel habitats are key rearing habitats for some species of salmonids (Pollock *et al.*, 2004). These changes also lead to changes in riparian areas, as flooded trees provide wildlife habitat and lead to early successional vegetation along edges of ponded areas. Returning adult salmon, and other species, can constitute a nutritional source to riparian areas. Carcasses are carried by scavengers to the riparian areas, and signals of marine-derived nitrogen have been detected in riparian plants (Mathewson *et al.*, 2003; Drake & Naiman, 2007) and can contribute to biodiversity and productivity of riparian plant communities (Reimchen *et al.*, 2003; Hocking & Reynolds, 2011).

Crickets and grasshoppers from forests and fields around streams may be infected by a parasite that has an obligate freshwater stage, the horsehair worm (Sato *et al.*, 2011). When the insects are parasitised, the worms alter the host's behaviour and drive it to find water where the worms breed. The hapless insects are doomed to die as the parasite escapes into the water and fishes take advantage of the former parasite host as prey. In one very detailed study, 60% of the annual energy budget of Japanese trout was supplied by this parasite-mediated resource subsidy (Sato *et al.*, 2011).

9.2 RIPARIAN HABITAT ALTERATIONS AND IMPACTS

Riparian habitats, especially floodplains, arc often very vulnerable to human impacts because they are desirable areas for humans for urbanisation and agriculture (Tockner & Stanford, 2002; Chapter 2). They are often the most productive areas within watersheds for agriculture and forestry, and therefore historically targeted for crop and livestock production and timber harvesting (Décamps *et al.*, 1988; Tockner & Stanford, 2002; Figure 9.2). Floodplains also provide access for navigation and recreation, fishing, waste disposal and other human needs (Tockner & Stanford, 2002; Thoms, 2003; Opperman *et al.*, 2009; Naiman *et al.*, 2010). However, this urbanisation and resource development almost always decouples natural stream–riparian processes, and negatively impacts stream–riparian linkages (Naiman *et al.*, 2010; Reid *et al.*, 2011). Such impacts most commonly include loss of riparian vegetation, or at least a major shift of plant community type, loss of riparian vegetation species and structural diversity. These shifts away from more natural states invariably affect run-off and flood magnitudes (Moore & Wondzell, 2005), nutrient (Feller, 2005) and sediment delivery to, and routing through, water bodies (May & Gresswell, 2003), affecting fishes through loss of habitat quality and quantity as previously discussed (Dawson, 2002; Whiteway *et al.*, 2010; Baillie *et al.*, 2013; Kano *et al.*, 2013). This typically changes the food web structure and productivity that fish rely upon (Winterbourn, 1990; Bisson & Bilby, 1998; Dunbar *et al.*, 2010).

Forestry and agriculture lead to dramatic changes in riparian plant cover, changing the flow of allochthonous plant material quality and quantity to water bodies that drive benthic invertebrate communities, and the flow of terrestrial prey subsidies to water bodies, both impacting the amount and type of prey available for freshwater fishes (Wallace *et al.*, 1997; Piccolo & Wipfli, 2002; Naiman *et al.*, 2010). Furthermore, livestock with access to streams and riparian habitats dramatically impact fishes and their habitats. Water quality and aquatic habitat are degraded from trampling and increased sedimentation, defecation, and the loss of vegetation cover and diversity (Agouridis *et al.*, 2005; Armour *et al.*, 1991). This degradation impacts not only fishes and their habitats directly (Peterson *et al.*, 2010), but also the lower trophic levels on which they depend (Ranganath *et al.*, 2009; Hagen *et al.*, 2010; Bergfur *et al.*, 2012).

Invasive riparian plant species are a growing threat to freshwater ecosystems (e.g. tamarisk, tallow, Russian olive, Japanese knotweed – e.g. Lecerf *et al.*, 2007; Kominoski *et al.*, 2013). For instance, European bird cherry in riparian zones in Alaska outcompetes and displaces native plants, which in turn support fewer herbivorous insects than native plants (Roon, 2011). Fewer invertebrates in riparian plant communities likely translate into fewer terrestrial prey

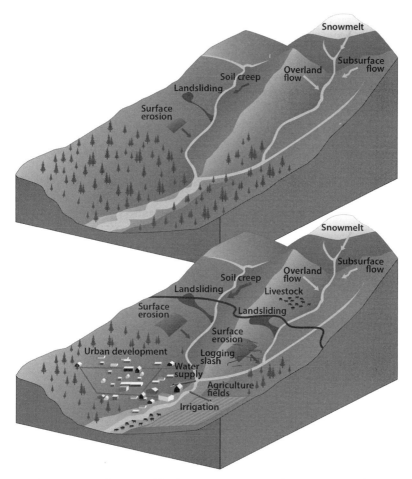

Figure 9.2 Catchment and landscape connections to the stream network influencing the population and community processes affecting stream fishes.

subsidies to water bodies, thus impacting fishes in places where fish may be food-limited (Allan *et al.*, 2003). Other invasive plants including pasture grass (*Brachiaria mutica*) and water hyacinth (*Eichhornia* spp.) impact hydrologic processes and aquatic habitat in streams in tropical Australia (Perna *et al.*, 2012). In many instances, non-native species, such as eucalyptus, radiata pine and Sitka spruce, are purposefully planted along riparian areas for timber production, to the detriment of stream food webs (e.g. Neal *et al.*, 2004; Canhoto *et al.*, 2013). Invasive plants and animals continue to spread and degrade aquatic habitats, impacting fish and fish habitats throughout the world (Lodge *et al.*, 2006; Pejchar & Mooney, 2009; Strayer, 2010), but also hold conservation value of their own (Schlaepfer *et al.*, 2011).

9.2.1 Catchment-scale processes

Processes occurring at and near the terrestrial–water interface are often occurring at a much broader spatial scale – often at the catchment and riverscape scales – and the consequences for water bodies of these processes, including disturbance-related processes, are frequently magnified according to their spatial extent and nature, and catchment properties (Leopold & Marchand, 1968; Bilby *et al.*, 1998; Benda *et al.*, 2003; Meador & Goldstein, 2003; Allan, 2004; Benda & Bigelow, 2014). Livestock grazing, timber harvesting and urbanisation are good examples (Figure 9.2). The broader the spatial scale, typically the greater the influence on streams (Benda & Sias, 2003; Benda & Bigelow, 2014).

Hynes (1975) noted: 'in every respect, the valley rules the stream', highlighting the close linkage between catchments and the riverscapes that drain them. Land cover at the catchment scale, including forest type, influences stream water chemistry (Muscutt *et al.*, 1993; Daniels & Gilliam, 1996; Johnson *et al.*, 1997; Correll, 1998; Nguyen *et al.*, 1998; Compton *et al.*, 2003; Czarnomski *et al.*, 2008) and water chemistry affects freshwater foodwebs, productivity and biodiversity of aquatic biota, including fishes (Rosenberg & Resh, 1993; Wetzel & Likens, 2000; Di Giulio & Hinton, 2008). Timber harvesting within catchments results in loss of large wood in streams, effects usually lasting for many decades, and is broadly known to severely impact fishes in forested catchments (Bilby *et al.*, 1998; Bojsen & Barriga, 2002; Benda & Sias, 2003; Wright & Flecker, 2004). Run-off and sediment movement and loading within catchments, whether naturally or from anthropogenic causes such as agriculture or forestry, generally holds ill consequences for stream fishes and other aquatic biota, even though the source can be a great distance upstream from the site of impact (Meador & Goldstein, 2003; Allan, 2004; Benda & Bigelow, 2014).

Connectivity throughout a catchment helps maintain populations of fishes (Chapter 2). Many species of fish require seasonal movements among habitats and connectivity of habitats throughout catchments for persistence of metapopulations (Fagan, 2002). Network connectivity enables the transport of materials – large wood, detritus and invertebrates (Benda & Sias, 2003; Wipfli *et al.*, 2007). This connectivity can be disrupted for all or parts of the year, and human activities taking place within catchments often contribute to fragmentation of suitable freshwater habitat for fishes (Fagan, 2002; Wipfli *et al.*, 2007; McKay *et al.*, 2013). Loss of connectivity also impacts other processes (nutrient transport and cycling) and movement of other biota (Pringle, 2003; Freeman *et al.*, 2007).

Catchment processes are likely to overwhelm reach-scale processes depending in part on how much of the catchment is upstream and the spatial extent of the source (Roni & Beechie, 2013). For instance, much of the management (e.g. riparian area protection, fencing, restoration) of freshwaters is done at the reach scale, but may not be successful without regard to the rest of the catchment (Beechie *et al.*, 2010). Although catchment-scale processes extend beyond what we discuss here, and can have profound influences on instream processes and biota, we have kept our treatment of them limited to maintain the key focus of this chapter, i.e. riparian zones.

9.3 FUTURE PROSPECTS

Increased recognition and appreciation over the years of the critical functions provided to fish and their supporting ecosystems by riparian habitats (e.g. Gregory *et al.*, 1991; Pusey & Arthington, 2003; Lovell & Sullivan, 2006) has led to improved riparian management yet with room for further improvement. Limiting livestock access to aquatic and riparian habitats has led to increased habitat quality in most instances (Sarr, 2002; Agouridis *et al.*, 2005; Miller *et al.*, 2010), but can have negative consequences, for example in situations including where invasive *Salix* spp. in Australia dominates river banks and supports fewer terrestrial arthropods for insectivorous consumers (Greenwood *et al.*, 2004). Enforcing protective riparian buffers in crop agriculture and in forestry settings has also generally led to improved aquatic habitat quality, with benefits extending to fishes (Richardson *et al.*, 2012).

Long-term changes in landscapes, including conversion to other uses (agriculture, urban, etc.) may lead to reduced capacity for resilience in catchments, reach-scale riparian areas and their associated freshwater ecosystems, with potentially adverse impacts on freshwater fishes (Lapointe *et al.*, 2014). For instance, some species are being slowly reduced in abundance as a result of changes to landscapes and the connectivity discussed above. The reduced global numbers of some species will diminish their potential to recolonise habitats where they have been lost (e.g. Fagan *et al.*, 2002). These effects can lead to reduced resilience of fish populations directly or through food web effects mediated through their prey or productivity of local ecosystems. Considerable evidence indicates that local losses of biodiversity lead to reduced rates of ecological functions and loss of resilience to natural and anthropogenic disturbances (Downing *et al.*, 2012). Recovery following disturbance may not proceed at historic rates when pressed by ongoing stressors, scarcity of colonists and changing forcing functions, such as climate.

Global change will bring new and unpredictable challenges to riparian management. Warming has been shown to change species distributions, with warmer-climate species expanding to higher altitudes and latitudes globally (Parmesan & Yohe, 2003; Comte *et al.*, 2013), and to change riparian plant community composition (Ström *et al.*, 2011). Other changes to riparian vegetation due to human management and planting can alter the nature and value of riparian areas (e.g. Kominoski *et al.*, 2013). For instance, the rapid decline in eastern hemlock in the eastern United States (Webster *et al.*, 2012), or the extensive planting of eucalyptus and radiata pine may alter the productive capacity of streams in such areas (e.g. Davies & Boulton, 2009). Plantations of exotic species in riparian areas can alter habitat for freshwater species other than fish, such as frogs, invertebrates and others (Parris & Lindenmayer, 2004). These shifts in plant and animal distributions will likely affect stream–riparian linkages as plant species, for example, occupy riparian habitats that were historically inaccessible, potentially outcompeting and displacing resident species.

9.4 KEY KNOWLEDGE GAPS AND FUTURE RESEARCH DIRECTIONS

Many jurisdictions have adopted some version of maintenance of linear buffers of vegetation as a tool to protect freshwaters (Richardson *et al.*, 2012). However, innovation in the application of conservation measures is needed. The allocation of protective measures along riparian areas of streams is an improvement over the past, but it is by no means clear that the amount of protection is sufficient to meet conservation targets, especially under extreme conditions. There have been a surprising dearth of well-designed studies to evaluate the adequacy of current Best Management Practices for stream fishes (e.g. Lee *et al.*, 2004) and there is an acquiescence to use of the numbers derived from the expert opinions developed in the FEMAT (Forest Ecosystem Management Assessment Team) process in the USA (FEMAT, 1993), despite a shortage of actual tests. This need for innovation is also true of restoration activities (Roni & Beechie, 2013), many of which have dubious value, but few projects are carefully assessed for their success or their efficiency at meeting targets, the latter of which are often not explicit.

Much of the biodiversity and ecosystem function literature argues that loss of biodiversity impairs the functioning and the resilience of ecosystems (e.g. Downing *et al.*, 2012). Resilience may be lost as freshwater species are lost (e.g. Schindler *et al.*, 2010). There are few empirical studies to demonstrate these effects, and research that addresses concerns about the effects of species loss on the functioning and resilience of ecosystems is needed.

Intact riparian ecosystems help protect stream fishes and other biota from stressors (e.g. nutrient and sediment run-off discussed earlier). Multiple stressors may show additive or non-additive effects in streams in a cumulative effects context (e.g. Wagenhoff *et al.*, 2011). Many freshwater fish populations are subject to a range of stressors (Chapter 6), but how these interact, from antagonistic to additive to synergistic, will require empirical testing and a mechanistic framework for their evaluation, especially within a riparian management context. Prediction of the effects, given the range of possible interactions as riparian conditions vary, can only be dealt with when the mechanisms of the physico-chemical system can be anticipated.

Riparian habitats provide key resources to stream fishes, in terms of quality and quantity of both habitat and food (Naiman *et al.*, 2010; Wipfli & Baxter, 2010). Natural resource development, agriculture, urban sprawl, pollution, invasive species and climate change all pose threats to riparian ecosystems, riparian–freshwater linkages, and the fishes that rely upon the habitats and processes within these ecosystems. Riparian management over the past several decades has converged on protection of fixed-width, linear buffers along streams for forestry (Richardson *et al.*, 2012), and has also been applied to agricultural and urban landscapes, although this has also limited innovation. A one-size-fits-all approach to riparian management does not take into account the natural variability among riverine ecosystems and habitat types and their reciprocal linkages that ultimately shape their communities and support fishes and other ecosystem services. It can be argued that riparian areas are so essential to many stream fishes that they should be considered critical habitat in the sense of endangered species conservation (Richardson *et al.*, 2010a). We still do not fully understand nor are we able to predict the quantitative value of riparian areas to particular species of fish, and especially not to whole communities.

However, there is little doubt about the qualitative contributions of the processes linking riparian areas to stream ecosystems. It is clear that riparian habitats are crucial for the productivity and biodiversity of freshwater fishes, and adequate protections and careful resource management are essential for long-term sustainability of fish ecosystems, fishes and fisheries resources.

9.5 DISCUSSION QUESTIONS

1. How will changes in the amounts of refuge habitats, past changes and current changes interact to affect resilience of stream fish populations?
2. Is it really possible to separate direct abiotic stresses due to management of riparian areas and catchment forest harvesting from the indirect effects mediated through food web interactions?
3. How do we convince funders and regulators that more extensive sampling over longer time frames is needed to understand fish populations?
4. From a manager's point of view, how can one deal with undesired variation in numbers in a fish population in response to riparian management?
5. How would increasing rates of productivity or other ecosystem functions impacted by riparian management affect species diversity of stream fishes?

9.6 REFERENCES

Agouridis, C. T., Workman, S. R., Warner, R. C. & Jennings, G. D. (2005). Livestock grazing management impacts on stream water quality: a review. *Journal of the American Water Resources Association*, 41, 591–606.

Allan, J. D. (2004). Landscapes and riverscapes: the influence of land use on stream ecosystems. *Annual Review of Ecology, Evolution, and Systematics*, 35, 257–284.

Allan, J. D., Wipfli, M. S., Caouette, J. P., Prussian, A. & Rodgers, J. (2003). Influence of streamside vegetation on inputs of terrestrial invertebrates to salmonid food webs. *Canadian Journal of Fisheries and Aquatic Sciences*, 60, 309–320.

Anderson, N. H., Sedell, J. R., Roberts, L. M. & Triska, F. J. (1978). The role of aquatic invertebrates in processing of wood debris in coniferous forest streams. *American Midland Naturalist*, 100, 64–82.

Antón, A., Elosegi, A., García-Arberas, L., Díez, J. & Rallo, A. (2011). Restoration of dead wood in Basque stream channels: effects on brown trout population. *Ecology of Freshwater Fish*, 20, 461–471.

Armour, C. L., Duff, D. A. & Elmore, W. (1991). The effects of livestock grazing on riparian and stream ecosystems. *Fisheries*, 16, 7–11.

Baillie, B. R., Hicks, B. J., Den Heuvel, M. R., Kimberley, M. O. & Hogg, I. D. (2013). The effects of wood on stream habitat and native fish assemblages in New Zealand. *Ecology of Freshwater Fish*, 22, 553–566.

Baker, N. E. & Baker, E. (2002). *Important Bird Areas in Tanzania: A First Inventory*. Dar es Salaam, Tanzania: Wildlife Conservation Society of Tanzania.

Baxter, C. V., Fausch, K. D. & Saunders, W. C. (2005). Tangled webs: reciprocal flows of invertebrate prey link streams and riparian zones. *Freshwater Biology*, 50, 201–220.

Beechie, T. J., Sear, D. A., Olden, J. D., *et al.* (2010). Process-based principles for restoring river ecosystems. *Bioscience*, 60, 209–222.

Belsky, A., Matzke, A. & Uselman, S. (1999). Survey of livestock influences on stream and riparian ecosystems in the western United States. *Journal of Soil and Water Conservation*, 54, 419–431.

Benda, L. E. & Bigelow, P. (2014). On the patterns and processes of wood in northern California streams. *Geomorphology*, 209, 79–97.

Benda, L. E. & Sias, J. C. (2003). A quantitative framework for evaluating the mass balance of in-stream organic debris. *Forest Ecology and Management*, 172, 1–16.

Benda, L. E., Miller, D. J., Sias, J. C., *et al.* (2003). Wood recruitment processes and wood budgeting. In *The Ecology and Management of Wood in World Rivers*. Bethesda, MD: American Fisheries Society, pp. 49–73.

Bergfur, J., Demars, B. O. L., Stutter, M. I., Langan, S. J. & Friberg, N. (2012). The tarland catchment initiative and its effect on stream water quality and macroinvertebrate indices. *Journal of Environmental Quality*, 41, 314–321.

Bilby, R. E. & Ward, J. W. (1989). Changes in characteristics and function of woody debris with increasing size of streams in western Washington. *Transactions of the American Fisheries Society*, 118, 368–378.

Bilby, R., Bisson, P. & Naiman, R. (1998). Function and distribution of large woody debris. In *River Ecology and Management: Lessons from the Pacific Coastal Ecoregion*. New York, NY: Springer-Verlag, pp. 373–398.

Bisson, P. A. & Bilby, R. E. (1998). Organic matter and trophic dynamics. In *River Ecology and Management: Lessons from the Pacific Coastal Ecoregion*. New York, NY: Springer, pp. 373–398.

Bisson, P. A., Bilby, R. E., Bryant, M. D., *et al.* (1987). Large woody debris in forested streams in the Pacific Northwest: past, present, and future. In *Streamside Management: Forestry and Fishery Interactions*. University of Washington, Seattle, WA: Institute of Forest Resources, pp. 143–190.

Bojsen, B. H. & Barriga, R. (2002). Effects of deforestation on fish community structure in Ecuadorian Amazon streams. *Freshwater Biology*, 47, 2246–2260.

Bonnington, C., Weaver, D. & Fanning, E. (2008). The habitat preference of four kingfisher species along a branch of the Kilombero River, Southern Tanzania. *African Journal of Ecology*, 46, 424–427.

Bugert, R. M., Bjornn, T. C. & Meehan, W. R. (1991). Summer habitat use by young salmonids and their responses to cover and predators in a small southeast Alaska stream. *Transactions of the American Fisheries Society*, 120, 474–485.

Canhoto, C., Calapez, R., Goncalves, A. L. & Moreira-Santos, M. (2013). Effects of eucalyptus leachates and oxygen on leaf-litter processing by fungi and stream invertebrates. *Freshwater Science*, 32, 411–424.

Compton, J. E., Church, M. R., Larned, S. T. & Hogsett, W. E. (2003). Nitrogen export from forested watersheds in the Oregon Coast Range: the role of N_2 fixing red alder. *Ecosystems*, 6, 773–785.

Comte, L., Buisson, L., Daufresne, M. & Grenouillet, G. (2013). Climate-induced changes in the distribution of freshwater fish: observed and predicted trends. *Freshwater Biology*, 58, 625–639.

Correll, D. L. (1998). The role of phosphorus in the eutrophication of receiving waters: a review. *Journal of Environmental Quality*, 27, 261–266.

Culp, J. M., Scrimgeour, G. J. & Townsend, G. D. (1996). Simulated fine woody debris accumulations in a stream increase rainbow trout fry abundance. *Transactions of the American Fisheries Society*, 125, 472–479.

Czarnomski, N. M., Dreher, D. M., Snyder, K. U., Jones, J. A. & Swanson, F. J. (2008). Dynamics of wood in stream networks of the Western Cascades Range, Oregon. *Canadian Journal of Forest Research*, 38, 2236–2248.

Daniels, R. & Gilliam, J. (1996). Sediment and chemical load reduction by grass and riparian filters. *Soil Science Society of America Journal*, 60, 246–251.

Davies, J. N. & Boulton, A. J. (2009). Great house, poor food: effects of exotic leaf litter on shredder densities and caddisfly growth in 6 subtropical Australian streams. *Journal of the North American Benthological Society*, 28, 491–503.

Dawson, K. (2002). Fish kill events and habitat losses of the Richmond River, NSW Australia: an overview. *Journal of Coastal Research*, 36, 216–221.

Décamps, H., Fortune, M., Gazelle, F. & Pautou, G. (1988). Historical influence of man on the riparian dynamics of a fluvial landscape. *Landscape Ecology*, 1, 163–173.

Di Giulio, R. T. & Hinton, D. E. (2008). *The Toxicology of Fishes*. Boca Raton, FL: Taylor and Francis Group.

Diehl, S. (1988). Foraging efficiency of three freshwater fishes: effects of structural complexity and light. *Oikos*, 207–214.

Dolloff, C. A. & Warren Jr, M. L. (2003). Fish relationships with large wood in small streams. *American Fisheries Society Symposium*, 37, 179–193.

Downing, A. S., van Nes, E. H., Mooij, W. M., & Scheffer, M. (2012). The resilience and resistance of an ecosystem to a collapse of diversity. *PloS ONE*, 7, e46135.

Drake, D. C. & Naiman, R. J. (2007). Reconstruction of pacific salmon abundance from riparian tree-ring growth. *Ecological Applications*, 17, 1523–1542.

Dunbar, M. J., Pedersen, M. L., Cadman, D., *et al.* (2010). River discharge and local-scale physical habitat influence macroinvertebrate LIFE scores. *Freshwater Biology*, 55, 226–242.

Eaton, J. G. & Scheller, R. M. (1996). Effects of climate warming on fish thermal habitat in streams of the United States. *Limnology and Oceanography*, 41, 1109–1115.

Edwards, E. D. & Huryn, A. D. (1996). Effect of riparian land use on contributions of terrestrial invertebrates to streams. *Hydrobiologia*, 337, 151–159.

England, L. E. & Rosemond, A. D. (2004). Small reductions in forest cover weaken terrestrial-aquatic linkages in headwater streams. *Freshwater Biology*, 49, 721–734.

Erős, T., Gustafsson, P., Greenberg, L. A. & Bergman, E. (2012). Forest–stream linkages: effects of terrestrial invertebrate input and light on diet and growth of brown trout (*Salmo trutta*) in a boreal forest stream. *PloS ONE*, 7, e36462.

Fagan, W. F. (2002). Connectivity, fragmentation, and extinction risk in dendritic metapopulations. *Ecology*, 83, 3243–3249.

Fagan, W. F., Unmack, P. J., Burgess, C., & Minckley, W. L. (2002). Rarity, fragmentation, and extinction risk in desert fishes. *Ecology*, 83, 3250–3256.

Fausch, K. D. & Northcote, T. G. (1992). Large woody debris and salmonid habitat in a small coastal British Columbia stream. *Canadian Journal of Fisheries and Aquatic Sciences*, 49, 682–693.

Feller, M. C. (2005). Forest harvesting and streamwater inorganic chemistry in western North America: a review. *Journal of the American Water Resources Association*, 41, 785–811.

FEMAT (Forest Ecosystem Management Assessment Team). (1993). *Forest Ecosystem Management: An Ecological, Economic, and Social Assessment.* Portland, OR: US Forest Service, US Department of Commerce, National Oceanic and Atmospheric Administration, National Marine Fisheries Service, United States Bureau of Land Management, & US Fish and Wildlife Service.

Finlay, J. C., Hood, J. M., Limm, M. P., *et al.* (2011). Light-mediated thresholds in stream-water nutrient composition in a river network. *Ecology*, 92, 140–150.

Fisher, S. G. & Likens, G. E. (1973). Energy flow in bear brook, New Hampshire: an integrative approach to stream ecosystem metabolism. *Ecological Monographs*, 43, 421–439.

Flecker, A. S. & Taylor, B. W. (2004). Tropical fishes as biological bulldozers: density effects on resource heterogeneity and species diversity. *Ecology*, 85, 2267–2278.

Fraser, N. & Metcalfe, N. (1997). The costs of becoming nocturnal: feeding efficiency in relation to light intensity in juvenile Atlantic salmon. *Functional Ecology*, 11, 385–391.

Freeman, M. C., Pringle, C. M. & Jackson, C. R. (2007). Hydrologic connectivity and the contribution of stream headwaters to ecological integrity at regional scales. *Journal of the American Water Resources Association*, 43, 5–14.

Fyles, J. W. & Fyles, I. H. (1993). Interaction of Douglas-fir with red alder and salal foliage litter during decomposition. *Canadian Journal of Forest Research*, 23, 358–361.

Gomi, T., Sidle, R. C. & Richardson, J. S. (2002). Understanding processes and downstream linkages of headwater systems headwaters differ from downstream reaches by their close coupling to hillslope processes, more temporal and spatial variation, and their need for different means of protection from land use. *Bioscience*, 52, 905–916.

Greenwood, H., O'Dowd, D. J. & Lake, P. S. (2004). Willow (*Salix rubens*) invasion of the riparian zone in south-eastern Australia: reduced abundance and altered composition of terrestrial arthropods. *Diversity and Distributions*, 10, 485–492.

Gregory, S. V., Swanson, F. J., McKee, W. A. & Cummins, K. W. (1991). An ecosystem perspective of riparian zones. *Bioscience*, 41, 540–551.

Gurnell, A. M., Piegay, H., Swanson, F. J. & Gregory, S. V. (2002). Large wood and fluvial processes. *Freshwater Biology*, 47, 601–619.

Hagen, E. M., Mctammany, M. E., Webster, J. R. & Benfield, E. F. (2010). Shifts in allochthonous input and autochthonous production in streams along an agricultural land-use gradient. *Hydrobiologia*, 655, 61–77.

Hill, W. R., Mulholland, P. J. & Marzolf, E. R. (2001). Stream ecosystem responses to forest leaf emergence in spring. *Ecology*, 82, 2306–2319.

Hocking, M. D. & Reynolds, J. D. (2011). Impacts of salmon on riparian plant diversity. *Science*, 331, 1609–1612.

Hofer, N. & Richardson, J. S. (2007). Comparisons of the colonisation by invertebrates of three species of wood, alder leaves, and plastic 'leaves' in a temperate stream. *International Review of Hydrobiology*, 92, 647–655.

Hynes, H. B. N. (1975). The stream and its valley. *Verhandlungen des Internationalen Verein Limnologie*, 19, 1–15.

Johnson, L. B., Richards, C., Host, G. E. & Arthur, J. W. (1997). Landscape influences on water chemistry in midwestern stream ecosystems. *Freshwater Biology*, 37, 193–208.

Johnson, L. B., Breneman, D. H. & Richards, C. (2003). Macroinvertebrate community structure and function associated with large wood in low gradient streams. *River Research and Applications*, 19, 199–218.

Jones, J. I., Murphy, J. F., Collins, A. L., *et al.* (2012). The impact of fine sediment on macroinvertebrates. *River Research and Applications*, 28, 1055–1071.

Jönsson, M., Hylander, S., Ranåker, L., Nilsson, P. A. & Brönmark, C. (2011). Foraging success of juvenile pike *Esox lucius* depends on visual conditions and prey pigmentation. *Journal of Fish Biology*, 79, 290–297.

Kail, J., Hering, D., Muhar, S., Gerhard, M. & Preis, S. (2007). The use of large wood in stream restoration: experiences from 50 projects in Germany and Austria. *Journal of Applied Ecology*, 44, 1145–1155.

Kano, Y., Miyazaki, Y., Tomiyama, Y., *et al.* (2013). Linking mesohabitat selection and ecological traits of a fish assemblage in a small tropical stream (Tinggi River, Pahang Basin) of the Malay Peninsula. *Zoological Science*, 30, 178–184.

Katano, O., Aonuma, Y., Nakamura, T. & Yamamoto, S. (2003). Indirect contramensalism through trophic cascades between two omnivorous fishes. *Ecology*, 84, 1311–1323.

Kawaguchi, Y. & Nakano, S. (2001). Contribution of terrestrial invertebrates to the annual resource budget for salmonids in forest and grassland reaches of a headwater stream. *Freshwater Biology*, 46, 303–316.

Kawaguchi, Y., Taniguchi, Y. & Nakano, S. (2003). Terrestrial invertebrate inputs determine the local abundance of stream fishes in a forested stream. *Ecology*, 84, 701–708.

Kiffney, P. M., Richardson, J. S. & Bull, J. P. (2003). Responses of periphyton and insects to experimental manipulation of riparian buffer width along forest streams. *Journal of Applied Ecology*, 40, 1060–1076.

Kiffney, P. M., Richardson, J. S. & Bull, J. P. (2004). Establishing light as a causal mechanism structuring stream communities in response to experimental manipulation of riparian buffer width. *Journal of the North American Benthological Society*, 23, 542–555.

Kirby, J. S., Holmes, J. S. & Sellers, R. M. (1996). Cormorants *Phalacrocorax carbo* as fish predators: an appraisal of their conservation and management in Great Britain. *Biological Conservation*, 75, 191–199.

Kominoski, J. S., Shah, J. J. F., Canhoto, C., *et al.* (2013). Forecasting functional implications of global changes in riparian plant communities. *Frontiers in Ecology and the Environment*, 11, 423–432.

Langford, T. E. L., Langford, J. & Hawkins, S. J. (2012). Conflicting effects of woody debris on stream fish populations: implications for management. *Freshwater Biology*, 57, 1096–1111.

Lapointe, N. W. R., Cooke, S. J., Imhof, J. G., *et al.* (2014). Principles for ensuring healthy and productive freshwater ecosystems that support sustainable fisheries. *Environmental Reviews*, 22, 110–134.

Leach, J. A., Moore, R. D., Hinch, S. G. & Gomi, T. (2012).Estimation of forest harvesting-induced stream temperature changes and bioenergetic consequences for cutthroat trout in a coastal stream in British Columbia, Canada. *Aquatic Sciences*, 74, 427–441.

Lecerf, A., Patfield, D., Boiche, A., *et al.* (2007). Stream ecosystems respond to riparian invasion by Japanese knotweed (*Fallopia japonica*). *Canadian Journal of Fisheries and Aquatic Sciences*, 64, 1273–1283.

Ledger, M. E. & Hildrew, A. G. (2001). Growth of an acid-tolerant stonefly on epilithic biofilms from streams of contrasting pH. *Freshwater Biology*, 46, 1457–1470.

Lee, P., Smyth, C. & Boutin, S. (2004). Quantitative review of riparian buffer width guidelines from Canada and the United States. *Journal of Environmental Management*, 70, 165–180.

Leopold, L. B. & O'Brien Marchand, M. (1968). On the quantitative inventory of the riverscape. *Water Resources Research*, 4, 709–717.

Leroy, C. J. & Marks, J. C. (2006). Litter quality, stream characteristics and litter diversity influence decomposition rates and macroinvertebrates. *Freshwater Biology*, 51, 605–617.

Lodge, D. M., Williams, S., MacIsaac, H. J., *et al.* (2006). Biological invasions: recommendations for US policy and management. *Ecological Applications*, 16, 2035–2054.

Lovell, S. T. & Sullivan, W. C. (2006). Environmental benefits of conservation buffers in the united states: evidence, promise, and open questions. *Agriculture, Ecosystems & Environment*, 112, 249–260.

Lu, Y. H., Bauer, J. E., Canuel, E. A., *et al.* (2014). Effects of land use on sources and ages of inorganic and organic carbon in temperate headwater streams. *Biogeochemistry*, 119, 275–292.

Maser, C. & Sedell, J. R. (1994). *From the Forest to the Sea: The Ecology of Wood in Streams, Rivers, Estuaries, and Oceans*. Delray Beach, FL: St. Lucie Press.

Mason, C. & MacDonald, S. (1982). The input of terrestrial invertebrates from tree canopies to a stream. *Freshwater Biology*, 12, 305–311.

Mathewson, D., Hocking, M. & Reimchen, T. (2003). Nitrogen uptake in riparian plant communities across a sharp ecological boundary of salmon density. *BMC Ecology*, 3, 4.

May, C. L. & Gresswell, R. E. (2003). Processes and rates of sediment and wood accumulation in headwater streams of the Oregon Coast Range, USA. *Earth Surface Processes and Landforms*, 28, 409–424.

McGlynn, B. L. & McDonnell, J. J. (2003). Quantifying the relative contributions of riparian and hillslope zones to catchment runoff. *Water Resources Research*, 39, 1310.

McKay, S. K., Schramski, J. R., Conyngham, J. N. & Fischenich, J. C. (2013). Assessing upstream fish passage connectivity with network analysis. *Ecological Applications*, 23, 1396–1409.

Meador, M. R. & Goldstein, R. M. (2003). Assessing water quality at large geographic scales: relations among land use, water physicochemistry, riparian condition, and fish community structure. *Environmental Management*, 31, 504–517.

Millar, R. G. (2000). Influence of bank vegetation on alluvial channel patterns. *Water Resources Research*, 36, 1109–1118.

Miller, J., Chanasyk, D., Curtis, T., Entz, T. & Willms, W. (2010). Influence of streambank fencing with a cattle crossing on riparian health and water quality of the lower Little Bow River in southern Alberta, Canada. *Agricultural Water Management*, 97, 247–258.

Monadjem, A. (1996). Habitat associations of birds along the Sabie River, South Africa. *African Journal of Ecology*, 34, 75–78.

Montgomery, D. R., Collins, B. D., Buffington, J. M. & Abbe, T. B. (2003). Geomorphic effects of wood in rivers. *American Fisheries Society Symposium*, 37, 21–47.

Moore, R. D. & Wondzell, S. M. (2005). Physical hydrology and the effects of forest harvesting in the Pacific Northwest: a review. *Journal of the American Water Resources Association*, 41, 763–784.

Moore, R. D., Spittlehouse, D. L. & Story, A. (2005). Riparian microclimate and stream temperature response to forest harvesting: a review. *Journal of the American Water Resources Association*, 41, 813–834.

Morgan, R. P. C. (2009). *Soil Erosion and Conservation*. Malden, MA: Wiley-Blackwell.

Motomori, K., Mitsuhashi, H. & Nakano, S. (2001). Influence of leaf litter quality on the colonization and consumption of stream invertebrate shredders. *Ecological Research*, 16, 173–182.

Muscutt, A., Harris, G., Bailey, S. & Davies, D. (1993). Buffer zones to improve water quality: a review of their potential use in UK agriculture. *Agriculture, Ecosystems & Environment*, 45, 59–77.

Naiman, R. J., Decamps, H. & McClain, M. E. (2010). *Riparia: Ecology, Conservation, and Management of Streamside Communities*. Burlington, MA: Academic Press.

Nakano, S., Miyasaka, H. & Kuhara, N. (1999). Terrestrial–aquatic linkages: riparian arthropod inputs alter trophic cascades in a stream food web. *Ecology*, 80, 2435–2441.

Neal, C., Ormerod, S. J., Langan, S. J., Nisbet, T. R. & Roberts, J. (2004). Sustainability of UK forestry: contemporary issues for the protection of freshwaters, a conclusion. *Hydrology and Earth System Sciences*, 8, 589–595.

Nguyen, M., Sheath, G., Smith, C. & Cooper, A. (1998). Impact of cattle treading on hill land: 2. Soil physical properties and contaminant runoff. *New Zealand Journal of Agricultural Research*, 41, 279–290.

Opperman, J. J., Galloway, G. E., Fargione, J., *et al.* (2009). Sustainable floodplains through large-scale reconnection to rivers. *Science*, 326, 1487–1488.

Parmesan, C. & Yohe, G. (2003). A globally coherent fingerprint of climate change impacts across natural systems. *Nature*, 421, 37–42.

Parris, K. M. & Lindenmayer, D. B. (2004). Evidence that creation of a *Pinus radiata* plantation in south-eastern Australia has reduced habitat for frogs. *Acta Oecologica*, 25, 93–101.

Pejchar, L. & Mooney, H. A. (2009). Invasive species, ecosystem services and human well-being. *Trends in Ecology & Evolution*, 24, 497–504.

Perna, C. N., Cappo, M., Pusey, B. J., Burrows, D. W. & Pearson, R. G. (2012). Removal of aquatic weeds greatly enhances fish community richness and diversity: an example from the Burdekin River Floodplain, tropical Australia. *River Research and Applications*, 28, 1093–1104.

Peterson, D. P., Rieman, B. E., Young, M. K. & Brammer, J. A. (2010). Modeling predicts that redd trampling by cattle may contribute to population declines of native trout. *Ecological Applications*, 20, 954–966.

Piccolo, J. J. & Wipfli, M. S. (2002). Does red alder (*Alnus rubra*) in upland riparian forests elevate macroinvertebrate and detritus export from headwater streams to downstream habitats in south-eastern Alaska? *Canadian Journal of Fisheries and Aquatic Sciences*, 59, 503–513.

Pollock, M. M., Pess, G. R. & Beechie, T. J. (2004). The importance of beaver ponds to coho salmon production in the Stillaguamish River Basin, Washington, USA. *North American Journal of Fisheries Management*, 24, 749–760.

Poole, G. C. & Berman, C. H. (2001). An ecological perspective on in-stream temperature: natural heat dynamics and mechanisms of human-caused thermal degradation. *Environmental Management*, 27, 787–802.

Power, M. E. (2003). Life cycles, limiting factors, and behavioral ecology of four loricariid catfishes in a Panamanian stream. *Catfishes*, 2, 581–600.

Pringle, C. M. (2003). What is hydrologic connectivity and why is it ecologically important? *Hydrological Processes*, 17, 2685–2689.

Pringle, C. M. & Hamazaki, T. (1997). Effects of fishes on algal response to storms in a tropical stream. *Ecology*, 78, 2432–2442.

Pusey, B. J. & Arthington, A. H. (2003). Importance of the riparian zone to the conservation and management of freshwater fish: a review. *Marine and Freshwater Research*, 54, 1–16.

Quinn, J. M., Boothroyd, I. K. & Smith, B. J. (2004). Riparian buffers mitigate effects of pine plantation logging on New Zealand streams: 2. Invertebrate communities. *Forest Ecology and Management*, 191, 129–146.

Rader, R. B., Belish, T., Young, M. K. & Rothlisberger, J. (2007). The scotopic visual sensitivity of four species of trout: a comparative study. *Western North American Naturalist*, 67, 524–537.

Ranganath, S. C., Hession, W. C. & Wynn, T. M. (2009). Livestock exclusion influences on riparian vegetation, channel morphology, and benthic macroinvertebrate assemblages. *Journal of Soil and Water Conservation*, 64, 33–42.

Reid, M. A., Ogden, R. & Thoms, M. C. (2011). The influence of flood frequency, geomorphic setting and grazing on plant communities and plant biomass on a large dryland floodplain. *Journal of Arid Environments*, 75, 815–826.

Reimchen, T., Mathewson, D., Hocking, M., Moran, J. & Harris, D. (2003). Isotopic evidence for enrichment of salmon-derived nutrients in vegetation, soil, and insects in riparian zones in coastal British Columbia. *American Fisheries Society Symposium*, 34, 59–70.

Richardson, J. S. & Danehy, R. J. (2007). A synthesis of the ecology of headwater streams and their riparian zones in temperate forests. *Forest Science*, 53, 131–147.

Richardson, J. S., Shaughnessy, C. R. & Harrison, P. G. (2004). Litter breakdown and invertebrate association with three types of leaves in a temperate rainforest stream. *Archiv für Hydrobiologie*, 159, 309–325.

Richardson, J. S., Bilby, R. E. & Bondar, C. A. (2005). Organic matter dynamics in small streams of the Pacific Northwest. *Journal of the American Water Resources Association*, 41, 921–934.

Richardson, J. S., Taylor, E., Schluter, D., Pearson, M. & Hatfield, T. (2010a). Do riparian zones qualify as critical habitat for endangered freshwater fishes? *Canadian Journal of Fisheries and Aquatic Sciences*, 67, 1197–1204.

Richardson, J. S., Zhang, Y. X. & Marczak, L. B. (2010b). Resource subsidies across the land–freshwater interface and responses in recipient communities. *River Research and Applications*, 26, 55–66.

Richardson, J. S., Naiman, R. J. & Bisson, P. A. (2012). How did fixed-width buffers become standard practice for protecting freshwaters and their riparian areas from forest harvest practices? *Freshwater Science*, 31, 232–238.

Rier, S. T. & Stevenson, R. J. (2002). Effects of light, dissolved organic carbon, and inorganic nutrients on the relationship between algae and heterotrophic bacteria in stream periphyton. *Hydrobiologia*, 489, 179–184.

Roni, P. & Beechie, T. (2013). Introduction to restoration: key steps for designing effective programs and projects. In *Stream and Watershed Restoration: A Guide to Restoring Riverine Processes and Habitats*. Chichester: Wiley-Blackwell, pp. 1–10.

Roon, D. A. (2011). Ecological effects of invasive European birdcherry on salmonid food webs in Alaska streams. Masters of Science Thesis. Fairbanks, AK: University of Alaska Fairbanks.

Rosenberg, D. M. & Resh, V. H. (1993). *Freshwater Biomonitoring and Benthic Macroinvertebrates*. New York, NY: Chapman & Hall.

Sakamaki, T. & Richardson, J. S. (2013). Nonlinear variation of stream–forest linkage along a stream-size gradient: an assessment using biogeochemical proxies of in-stream fine particulate organic matter. *Journal of Applied Ecology*, 50, 1019–1027.

Sarr, D. A. (2002). Riparian livestock exclosure research in the western United States: a critique and some recommendations. *Environmental Management*, 30, 516–526.

Sato, T., Watanabe, K., Kanaiwa, M., *et al.* (2011). Nematomorph parasites drive energy flow through a riparian ecosystem. *Ecology*, 92, 201–207.

Sayer, E. J. (2006). Using experimental manipulation to assess the roles of leaf litter in the functioning of forest ecosystems. *Biological Reviews*, 81, 1–31.

Schindler, D. E., Hilborn, R., Chasco, B., *et al.* (2010). Population diversity and the portfolio effect in an exploited species. *Nature*, 465, 609–612.

Schlaepfer, M. A., Sax, D. F. & Olden, J. D. (2011). The potential conservation value of non-native species. *Conservation Biology*, 25, 428–437.

Sepúlveda, M. A., Bartheld, J. L., Monsalve, R., Gómez, V. & Medina-Vogel, G. (2007). Habitat use and spatial behaviour of the endangered southern river otter (*Lontra provocax*) in riparian habitats of Chile: conservation implications. *Biological Conservation*, 140, 329–338.

Spänhoff, B. & Meyer, E. I. (2004). Breakdown rates of wood in streams. *Journal of the North American Benthological Society*, 23, 189–197.

Stevens, C. E., Paszkowski, C. A. & Foote, A. L. (2007). Beaver (*Castor canadensis*) as a surrogate species for conserving anuran amphibians on boreal streams in Alberta, Canada. *Biological Conservation*, 134, 1–13.

Strayer, D. L. (2010). Alien species in fresh waters: ecological effects, interactions with other stressors, and prospects for the future. *Freshwater Biology*, 55, 152–174.

Ström, L., Jansson, R., Nilsson, C., Johansson, M. E. & Xiong, S. (2011). Hydrologic effects on riparian vegetation in a boreal river: an experiment testing climate change predictions. *Global Change Biology*, 17, 254–267.

Suttle, K. B., Power, M. E., Levine, J. M. & McNeely, C. (2004). How fine sediment in riverbeds impairs growth and survival of juvenile salmonids. *Ecological Applications*, 14, 969–974.

Swanson, F. J., Gregory, S., Sedell, J. & Campbell, A. (1982). Land–water interactions: the riparian zone. In *Analysis of Coniferous Forest Ecosystems in the Western United States*. Stroudsburg, PA: Hutchinson Ross Publishing Co., pp. 267–291.

Tabacchi, E., Lambs, L., Guilloy, H., *et al.* (2000). Impacts of riparian vegetation on hydrological processes. *Hydrological Processes*, 14, 2959–2976.

Thoms, M. C. (2003). Floodplain–river ecosystems: lateral connections and the implications of human interference. *Geomorphology*, 56, 335–349.

Tockner, K. & Stanford, J. A. (2002). Riverine flood plains: present state and future trends. *Environmental Conservation*, 29, 308–330.

Vannote, R. L., Minshall, G. W., Cummins, K. W., Sedell, J. R. & Cushing, C. E. (1980). The river continuum concept. *Canadian Journal of Fisheries and Aquatic Sciences*, 37, 130–137.

Wagenhoff, A., Townsend, C. R., Phillips, N. & Matthaei, C. D. (2011). Subsidy-stress and multiple-stressor effects along gradients of deposited fine sediment and dissolved nutrients in a regional set of streams and rivers. *Freshwater Biology*, 56, 1916–1936.

Wallace, J. B. & Benke, A. C. (1984). Quantification of wood habitat in sub-tropical coastal-plain streams. *Canadian Journal of Fisheries and Aquatic Sciences*, 41, 1643–1652.

Wallace, J. B., Eggert, S. L., Meyer, J. L. & Webster, J. R. (1997). Multiple trophic levels of a forest stream linked to terrestrial litter inputs. *Science*, 277, 102–104.

Wang, L., Lyons, J., Kanehl, P. & Bannerman, R. (2001). Impacts of urbanization on stream habitat and fish across multiple spatial scales. *Environmental Management*, 28, 255–266.

Waters, T. F. (1999). Long-term trout production dynamics in Valley Creek, Minnesota. *Transactions of the American Fisheries Society*, 128, 1151–1162.

Webster, J. R., Benfield, E. F., Ehrman, T. P., *et al.* (1999). What happens to allochthonous material that falls into streams? A synthesis of new and published information from Coweeta. *Freshwater Biology*, 41, 687–705.

Webster, J. R., Morkeski, K., Wojculewski, C. A., *et al.* (2012). Effects of hemlock mortality on streams in the southern Appalachian Mountains. *American Midland Naturalist*, 168, 112–131.

Wellington, C. G., Mayer, C. M., Bossenbroek, J. M. & Stroh, N. A. (2010). Effects of turbidity and prey density on the foraging success of age 0 year yellow perch *Perca flavescens. Journal of Fish Biology*, 76, 1729–1741.

Wetzel, R. G. & Likens, G. E. (2000). Inorganic nutrients: nitrogen, phosphorus, and other nutrients. In *Limnological Analyses*. New York, NY: Springer Science, pp. 85–112.

Whiteway, S. L., Biron, P. M., Zimmermann, A., Venter, O. & Grant, J. W. A. (2010). Do in-stream restoration structures enhance salmonid abundance? A meta-analysis. *Canadian Journal of Fisheries and Aquatic Sciences*, 67, 831–841.

Wilzbach, M. A., Cummins, K. W. & Hall, J. D. (1986). Influence of habitat manipulations on interactions between cutthroat trout and invertebrate drift. *Ecology*, 67, 898–911.

Winterbourn, M. (1990). Interactions among nutrients, algae and invertebrates in a New Zealand mountain stream. *Freshwater Biology*, 23, 463–474.

Wipfli, M. S. (1997). Terrestrial invertebrates as salmonid prey and nitrogen sources in streams: contrasting old-growth and young-growth riparian forests in southeastern Alaska, USA. *Canadian Journal of Fisheries and Aquatic Sciences*, 54, 1259–1269.

Wipfli, M. S. & Baxter, C. V. (2010). Linking ecosystems, food webs, and fish production: subsidies in salmonid watersheds. *Fisheries*, 35, 373–387.

Wipfli, M. S. & Gregovich, D. P. (2002). Export of invertebrates and detritus from fishless headwater streams in southeastern Alaska: implications for downstream salmonid production. *Freshwater Biology*, 47, 957–969.

Wipfli, M. S. & Musslewhite, J. (2004). Density of red alder (*Alnus rubra*) in headwaters influences invertebrate and detritus subsidies to downstream fish habitats in Alaska. *Hydrobiologia*, 520, 153–163.

Wipfli, M. S., Richardson, J. S. & Naiman, R. J. (2007). Ecological linkages between headwaters and downstream ecosystems: transport of organic matter, invertebrates, and wood down headwater channels. *Journal of the American Water Resources Association*, 43, 72–85.

Wright, J. P. & Flecker, A. S. (2004). Deforesting the riverscape: the effects of wood on fish diversity in a Venezuelan piedmont stream. *Biological Conservation*, 120, 439–447.

Wright, J. P., Jones, C. G. & Flecker, A. S. (2002). An ecosystem engineer, the beaver, increases species richness at the landscape scale. *Oecologia*, 132, 96–101.

Young, R. A. & Wiersma, J. (1973). The role of rainfall impact in soil detachment and transport. *Water Resources Research*, 9, 1629–1636.

Zeigler, M. P., Brinkman, S. F., Caldwell, C. A., *et al.* (2013). Upper thermal tolerances of Rio Grande cutthroat trout under constant and fluctuating temperatures. *Transactions of the American Fisheries Society*, 142, 1395–1405.

Fragmentation, connectivity and fish species persistence in freshwater ecosystems

KEITH B. GIDO, JAMES E. WHITNEY,
JOSHUAH S. PERKIN AND THOMAS F. TURNER

10.1 OVERVIEW

Fragmentation poses one of the greatest threats to freshwater fish bio-diversity (Nilsson *et al.*, 2005; Reidy-Liermann *et al.*, 2012). Whereas damming of large rivers is perhaps the most obvious form of fragmentation (e.g., Nilsson *et al.*, 2005), smaller, semipermeable barriers such as road crossings (Perkin & Gido, 2012) or water withdrawals that dry sections of a river network (Falke *et al.*, 2011) also pose a conservation challenge. In glacial regions, lakes that are naturally connected through waterways are increasingly being isolated by summer evaporation and groundwater loss (Baki *et al.*, 2012). Climate and land-use changes also isolate populations in headwater reaches by increasing temperatures (Rahel *et al.*, 1996) or drying of streams (Falke *et al.*, 2011) in downstream reaches. Finally, barriers can form when the occurrence of a species, such as a large predator, inhibits the movement of prey through a dispersal corridor (Fraser *et al.*, 1995). This severing of connectivity in aquatic habitats affects species persistence through multiple stressors (Chapters 4 and 6) including limiting dispersal necessary to fulfil important life stages, exacerbating negative species interactions, and inhibiting recolonisation following disturbance. Barriers to movement isolate small populations leading to reduced genetic diversity (Chapter 16) and potentially compromise long-term population persistence (e.g. Wofford *et al.*, 2005).

In this chapter, we discuss how fragmentation disrupts dispersal and migration of freshwater fishes and the long-term consequences for population diversity and stability. We begin with a global overview

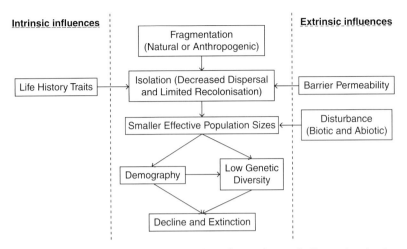

Figure 10.1 Conceptual model describing factors driving decline and extinction of species following fragmentation. Both intrinsic and extrinsic factors can regulate the degree of isolation following fragmentation. Small populations also are more prone to stochastic disturbance events and low genetic diversity can reduce reproductive performance and the ability of a population to adapt to altered conditions.

of the problem followed by a review of theoretical and empirical methods for quantifying the effects of fragmentation on population viability. We conclude with a discussion of conservation challenges along with future research and management recommendations. The primary tenet of our review is that persistence of species in fragmented systems is dependent on the nature of barriers to dispersal and ecological traits of species, particularly their ability to complete critical life-history stages within fragmented habitats (Figure 10.1). We often refer to the terms fragmentation, isolation and connectivity. Whereas there are instances where these might be used interchangeably, we consider fragmentation to represent habitats that have been partitioned into smaller habitats and by extension result in smaller populations. The terms connectivity and isolation refer to the ability or lack of ability, respectively, of fishes to disperse into or out of particular habitats.

10.2 FRAGMENTATION OF LARGE RIVERS, STREAMS AND LAKES

Fragmentation of large rivers is linked to declining biodiversity, but it is often difficult to separate the effects of fragmentation with concurrent alterations to river flow and other abiotic factors (e.g. temperature regimes, geomorphology). Barriers on large rivers range from reservoir dams tens of metres high to water diversion structures less than 1 m high. Damming of rivers is a relatively recent phenomenon, in that most large impoundments are less than 100 years old, with most dams constructed since the 1950s (World Commission on Dams, 2000; Chapter 4). Regardless, there have been massive declines in large river fishes that require long-distance dispersal (Chapter 11). Most notable are diadromous fishes (e.g. salmonids, eels, lampreys, striped bass), because barriers on large rivers inhibit movement between critical spawning, rearing and adult foraging habitats. Potamodromous fishes (e.g. sturgeon, suckers) that move considerable distances within freshwater are similarly affected by fragmentation. Mature adults of both diadromous and potamodromous fishes typically move upstream to spawn while early life stages either passively (e.g. pelagic eggs and larvae) or actively (e.g. smolt) move downstream. Dams on large rivers are typically impassable to fishes at all life stages in the upstream direction unless specifically designed to allow passage. Downstream movement at some or all life stages (i.e. drifting eggs and larvae, juveniles, adults) is more likely, albeit with increased mortality that is dependent on the outflow structure (Coutant & Whitney, 2000; Agostinho *et al.*, 2007). Reservoirs also create barriers to dispersal that inhibit movement among tributaries entering a single reservoir (Falke & Gido, 2006; Franssen, 2012; Hudman & Gido, 2013), and there are numerous examples of species extirpations upstream of impoundments (e.g. Winston *et al.*, 1991; Matthews & Marsh-Matthews, 2007). Another consequence of dams and other water management practices is that many large rivers have been disconnected from their floodplains, which provide vital habitats for many species of fish (e.g. Welcomme, 1979; Goulding, 1980; Lowe-McConnell, 1987; Kingsford, 2000). The construction of levees, channelisation of rivers for barge passage and altered flow regimes are all culprits in reducing or eliminating lateral connectivity to off-channel habitats. Just as in-stream obstructions limit access to vital spawning habitats, reduced lateral connectivity with off-channel habitats has restricted the recruitment of many large river species (e.g. Quist & Spiegel, 2012; Chapter 12).

Human-engineered barriers in small streams are typically represented by road crossings and diversion structures. Natural barriers that impede dispersal such as waterfalls, beaver dams, dry stream reaches and unsuitable meso-habitats are also common (Roberts & Angermeier, 2007). Like larger river barriers, obstructions in smaller systems can completely eliminate dispersal, but others are semipermeable and allow passage (up- or downstream) under periods of high flow. The effects of semipermeable barriers are much harder to predict (Norman *et al.*, 2009; Anderson *et al.*, 2012). Extinctions caused by abiotic (drying) or biotic (predation) factors should be higher in small streams than large rivers because the habitat volume is lower (Horwitz, 1978; Schlosser, 1982). Barriers in small streams will limit recolonisation dynamics if populations are extirpated in these hydrologically dynamic reaches above barriers (Winston *et al.*, 1991; Luttrell *et al.*, 1999). Although not as well quantified, food web interactions might also intensify in small isolated habitats if densities are high and prey refugia are lacking, particularly during periods of drought (Magoulick & Kobza, 2003; Creed, 2006).

Natural lakes, primarily those connected by seepage streams, are also threatened by fragmentation. Research in such lakes provides strong evidence for the role of fragmentation in structuring fish communities (e.g. Tonn & Magnuson, 1982; Tonn *et al.*, 1990; Hershey *et al.*, 1999; Olden *et al.*, 2001). Tonn and Magnuson (1982) demonstrated that piscivorous fish were only present in lakes with low winter dissolved oxygen concentrations if those lakes maintained connectivity to refugia lakes and, if present, piscivores limited the occurrence of prey species. Englund *et al.* (2009) found similar trade-offs between abiotic conditions and presence of predators for isolated lakes in Sweden. Olden *et al.* (2001) further showed that fishes in isolated seepage lakes (i.e. large distances from potential sources) were more susceptible to local extinction and recolonisation rates were slower than in more connected lakes. These studies illustrate how lake isolation alters fish community structure through either abiotic (low dissolved oxygen) or biotic (predation) disturbances. Moreover, there appears to be a trade-off between the presence of refugia habitat for predators and the presence and diversity of prey fishes and other aquatic organisms. For example, Scheffer *et al.* (2006) reviewed the association between isolation and diversity of a variety of plants and animals and found that lake isolation limits fish diversity but increases diversity of other plants and animal taxa (e.g. amphibians and insects).

In summary, fish communities have typically evolved in a network of highly connected habitats, with the exception of headwater species in habitats isolated by natural waterfalls or ephemeral stream reaches. Stochastic environmental conditions associated with climate are more likely to structure fish communities rather than biotic interactions in lotic systems (Grossman *et al.*, 1982). Fragmentation not only reduces habitat availability, but also isolates populations that historically dispersed over much greater distances and may increase the chances and intensity of biotic interactions. Alternatively, species that evolved in lentic systems have adapted in more isolated and potentially under more intense biotic interactions than fishes in lotic systems (Jackson *et al.*, 2001). Increasing isolation of lakes should have the greatest impact on small, shallow lakes that are connected to other lakes, but other anthropogenic activities, such as non-native species or eutrophication, are potentially of greater concern for resident lake species.

10.3 CONCEPTUAL FRAMEWORK FOR UNDERSTANDING FRAGMENTATION EFFECTS ON FISHES

The basic principles of the Theory of Island Biogeography (MacArthur & Wilson, 1967) elegantly describe how dividing populations into smaller, more-isolated habitats results in higher rates of extinction and less opportunity for colonisation. These principles can be applied to naturally isolated lakes or river basins, as evidenced by the strong relationship between freshwater fish diversity and habitat area (Eadie & Keast, 1984; Hugueny *et al.*, 2011). However, diversity–area relationships result from both extinction and speciation events occurring over evolutionary time (Smith, 1981), while human-engineered fragmentation occurs over much shorter time scales. Thus, mechanisms for reductions in diversity might differ between natural and human-influenced fragmentation of habitats. Moreover, if barriers restrict but do not eliminate movement, reductions in diversity and abundance may not be evident due to a threshold response, or there will be lags in the response to fragmentation (Branco *et al.*, 2012). A mechanistic understanding of declining fish abundance and diversity in fragmented habitats begins by recognising that rates of decline are influenced by extrinsic (e.g. barrier type, disturbance regime) as well as intrinsic (e.g. life-history traits of species) properties of a system (Figure 10.1). When these factors limit dispersal and recolonisation, smaller populations become increasingly susceptible to extinction. This 'extinction vortex' (*sensu* Gilpin & Soulé, 1986) is exacerbated by an increase in genetic drift, a corresponding decrease in genetic variance and increase in genetic load (i.e. fewer individuals in the population are near optimal fitness), and concomitant reductions in population viability (Gaggiotti, 2003).

10.3.1 Intrinsic factors influencing fragmentation

Metapopulation theory (Levins, 1969) also provides a mechanistic framework for assessing the consequences of fragmentation. Most freshwater fish populations do not establish the theoretical balance between genetic drift and gene flow because they fluctuate in local abundance and periodically go extinct in response to environmental fluctuations or other factors. Nevertheless, numerous studies invoke a metapopulation framework to conceptualise the dynamics of freshwater fishes (e.g. Fagan, 2002; Falke & Fausch, 2010). In metapopulation theory, local extinction is balanced by recolonisation and the genetic outcome depends on the genetic characteristics of the colonists (Whitlock & McCauley, 1990). For example, if colonists come as many propagules from a number of upstream localities to a downstream confluence (as in a dendritic model with mass effects), we would expect the newly colonised population to have higher genetic diversity, on average, than any one donor population (Morrissey & deKerchove, 2009). Conversely, if colonists come from a single, geographically adjacent locality, then we might expect lowered genetic diversity, on average, compared to remaining populations within the stream network. In general, the number of colonists and the number of distinct populations from which they originate determines overall genetic diversity under a metapopulation model with extinction and recolonisation (Slatkin, 1985; Whitlock & McCauley, 1990). When the pool of colonists is restricted, theory predicts that fragmentation will lower genetic diversity for the entire metapopulation roughly proportional to the extent that barrier placement limits the geographic scope (and genetic diversity) of the colonist pool.

Because the most significant human-mediated fragmentation events occurred within the last century, genetic drift, gene flow and natural selection should be more important forces than mutation as agents of change of genetic diversity on a recently fragmented riverscape. This is because mutation can take thousands of years to appreciably alter gene frequencies. If mutation can be safely ignored, genetic drift will reduce heterozygosity and genetic diversity at a rate that is inversely proportional to the population (census) size that has become completely isolated (i.e. closed to migration and gene flow) due to fragmentation. Thus, for a small population, detectable changes in allele frequencies will arise in a few generations, but in a very large population it may take hundreds of generations for reductions in diversity to be detectable.

Life-history and behavioural traits of species are useful in predicting their sensitivity to fragmentation. Species with complex life cycles requiring dispersal from rearing, feeding and spawning habitats are highly sensitive to fragmentation (e.g. Rieman & Dunham, 2000; Morita & Yamamoto, 2002; Chapters 2 and 11). A common phenomenon of stream fish is that individuals within populations tend to either be resident or migrant (e.g. Gowan *et al.*, 1994; Skalski & Gilliam, 2000; Rodriguez, 2002). Salmonids provide examples of populations with dual movement strategies, and not surprisingly, fragmentation selects for resident forms at the expense of migratory forms (e.g. Morita & Yamamoto, 2002). By contrast, Humphries *et al.* (1999) proposed the 'low-flow recruitment hypothesis' and noted that some species of large-river fishes are capable of sustaining populations in systems that have been disconnected from floodplain habitats. Body size might also be a predictor of species sensitivity to fragmentation. Blanchet *et al.* (2010) found that fish with large and intermediate body size lost genetic diversity to drift more rapidly than small-bodied species following human-induced fragmentation. They acknowledged that body size, per se, may not be the only trait responsible for fragmentation sensitivity, but it may correlate with numerous other vital ecological traits such as movement behaviour and feeding ecology. Other species with reproductive strategies requiring that eggs and larvae drift for an extended period of time also are sensitive to fragmentation (e.g. Dudley & Platania, 2007). Whereas many rheophilic (flowing water) specialists decline in fragmented rivers, fragmentation has less influence on eurytopic (generalist) species and limnophilic (standing water) specialists (Musil *et al.*, 2012). It is not surprising that some species persist in fragmented habitats because most fragments are large relative to the home range of those species and can support large populations (e.g. Reid *et al.*, 2008; Hudman & Gido, 2013).

10.3.2 Extrinsic factors influencing fragmentation

Extrinsic factors such as the disturbance regime of the system, fragment size, or barrier permeability influence population responses to fragmentation. Adverse effects on fish communities are more likely to occur in small fragments isolated with impermeable barriers in systems with frequent disturbance events. In river systems, species are likely to have different thresholds for the size of river fragments in which they can persist. For example, pelagic-broadcast spawning fishes in the Great

Plains, USA, appear to need > 100 river kilometres (rkm), but some species may require up to 300 rkm to persist (Perkin & Gido, 2011). This is in contrast to some benthic-dwelling fish where movement is restricted to < 5 km (e.g. Lamphere & Blum, 2012). Increasing the number of river obstacles, which also reduces fragment sizes and the probability of passage, was found to have a negative effect on young of year fishes in European rivers (Musil *et al.*, 2012). However, when barriers allow some movement both up- and downstream of the structure (i.e. through fishways), there can be limited influence on population genetic structure (e.g. Reid *et al.*, 2008). Understanding the link between intrinsic and extrinsic properties of freshwater systems is clearly necessary to predict species responses to fragmentation.

In a fragmented system consisting of two populations of equal population size, the overall restorative effect of gene flow depends on barrier permeability, the prevailing direction of migration (e.g. whether symmetric or asymmetric), and species-specific life-history features that influence vagility and migration rates between populations (e.g. body size, egg size, fecundity). In Figure 10.2, we present a conceptual model that is an extension of that presented in Figure 10.1 to develop predictions for the relative importance of genetic drift, gene flow and natural selection, as a function of fragment size and barrier permeability. This model assumes that a hypothetical focal population is continuously and uniformly distributed in space, which is subsequently fragmented by a barrier into two subpopulations. Additionally, we assume that all individuals complete their life histories in the fragmented subpopulations and are otherwise 'idealised' as described in standard population genetic models (i.e. sex ratio equal, Poisson variance in reproductive success, random union of gametes, etc. – this means that population size (N) and N_e are equal). Thus, the model only explicitly considers extrinsic effects on genetic diversity. Under conditions of uniform distribution, fragment size relates directly to the effective population size (N_e) and predicts the response to genetic drift. Barrier permeability relates to the probability that migrants are exchanged between subpopulations, and that they breed in the recipient population. To facilitate discussion, we identified four quadrants with different general responses to fragment size and permeability. Quadrant 1 represents small and isolated fragments, where genetic drift is predicted to be the overriding force that changes allele frequencies each generation. At extremely small values, inbreeding effects on viability are highly likely. In quadrant 2, migration is the most important factor that impinges on allele frequencies

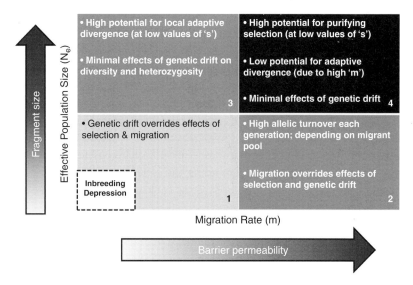

Figure 10.2 A conceptual model of evolutionary response to fragmentation. Three parameters are tracked in the model. The *x*-axis tracks increasing barrier permeability and the genetic effective migration rate (m). The *y*-axis tracks fragment size and the genetic effective population size (N_e) of a fragmented subpopulation. Numbers in the lower right-hand corner of each cell identify quadrants that facilitate discussion, but we envision a continuous gradient of *m* and N_e as barrier permeability and fragment size increase. The selection coefficient (s) is tracked in each quadrant. At maximum values of permeability and fragment size in quadrant 4, the fragmented subpopulations approach panmixia. At minimum values in quadrant 1, fragmented populations are subject to genetic drift and inbreeding. Diversifying natural selection across subpopulations is predicted to be more likely in quadrant 3 because of low gene flow but large N_e. Gene flow resists divergent selection in quadrant 4 unless the selection coefficient is very large.

in a small recipient subpopulation, and the overall effects depend on genetic characteristics of the subpopulation from which migrants are derived. This is analogous to a source–sink dynamic (e.g. Waits *et al.*, 2008). Quadrant 3 represents isolated fragments that are large enough to preclude strong negative effects on diversity from genetic drift. In this case, relatively small selection coefficients may drive diversification across isolated fragments, depending on the nature of selective gradients. Finally, quadrant 4 represents a case where large fragmented subpopulations exchange many migrants through a barrier with high permeability. Subpopulations would approach panmixia at extreme values. Diversifying selection would be countered by high gene flow, but relatively small selection coefficients could drive purifying selection under these conditions. Metapopulation persistence also decreases when dispersal opportunity in simulated riverscapes is reduced from bidirectional movement to only unidirectional movement (Jager et al., 2001; Fagan, 2002), which frequently occurs when dams limit dispersal to only occur in a downstream direction (Agostinho et al., 2007).

Our model is formulated to examine symmetric migration among subpopulations, but persistent unidirectional downstream movement through a barrier promotes asymmetric gene flow. If gene flow is strictly asymmetric, then the upstream population contributes and the downstream population receives migrants each generation. With respect to our model, this puts the upstream population in quadrants 1 or 3, with the effects of drift depending on fragment size alone, and the recipient population in quadrants 2 or 4. Under these conditions, we would expect diminished genetic diversity upstream versus downstream, all else being equal (Wares & Pringle, 2008). Furthermore, there is the potential for strong interactions of extrinsic and intrinsic factors in shaping diversity across this simple landscape. For example, for fish species with drifting eggs or larvae, barriers may magnify advective downstream losses of propagules thereby lowering genetic diversity overall (Alò & Turner, 2005). Population models and empirical data suggest that the depletion of reproductive productivity by unidirectional advection can have species-wide impacts on diversity (Wares & Pringle, 2008; Pringle et al., 2011). Lakes and ponds generally do not have persistent unidirectional flow, but physical forces such as water convection and wind can facilitate fish movement, and thereby affect gene flow rates and genetic diversity. Lakes could be modelled using the two-subpopulation case with symmetrical migration, unless physical transport forces facilitate asymmetrical movement.

10.4 QUANTIFYING CONNECTIVITY

A first step in understanding the effects of fragmentation on freshwater fish communities is to quantify the relationship between connectivity and genetic, population and community structure. Connectivity can generally be classified as structural or functional (Taylor et al., 1993). Structural connectivity refers to physical properties of aquatic habitats that facilitate or inhibit exchange among habitat units or patches. For example, structural connectivity includes characteristics of a network such as distance between habitats and permeability of barriers to movement. Functional connectivity represents the movement rates of materials or biota among habitats that influence ecosystem processes and community structure. The classic example of functional connectivity in rivers comes from the River Continuum Concept (Vannote *et al.*, 1980), where ecosystem processes in upstream habitats cascade to influence downstream habitats. Similarly, the migration of fishes through a network of connected habitats might influence ecological interactions such as predator–prey relations (Woodford & McIntosh, 2010), transport of nutrients (Flecker *et al.*, 2010) and population genetic structure and evolutionary potential (Morrissey & de Kerchove, 2009). In this section, we review methods of quantifying structural and functional connectivity, as well as reviewing current literature illustrating the strong relationship between these aspects of connectivity.

10.4.1 Measuring structural connectivity

Efforts to quantify structural connectivity of freshwater habitats have expanded greatly in the past two decades (Fullerton *et al.*, 2010). Wiens (2002) conceptualised how principles of landscape ecology could be integrated into stream networks and emphasised the importance of connectivity in mediating landscape processes. He and many others (e.g. Grant *et al.*, 2007; Erös *et al.*, 2012) argued that streams and lakes might be dissected into a network of habitat patches with differing degrees of connectivity. This framework allowed important advances in understanding the constraints of fragmentation within aquatic systems, but there is still debate in how to quantify patches and connections among patches. A succinct definition of a patch is difficult to make because patches vary with life-history traits and life stages of fishes (Schlosser & Angermeier, 1995; Fullerton *et al.*, 2010) and with scale of analysis (Frissell *et al.*, 1986; Fausch *et al.*, 2002). In general, patches are defined as areas that are relatively homogeneous (in physical structure or community composition) and differ from their immediate surroundings. However, the degree to which a species (or life stage within a species) perceives heterogeneity is determined by their temporal movement patterns and resource use. For freshwater systems, patches in a riverscape have been defined from small microhabitats (e.g. grains of sand) up to entire watersheds (Winemiller *et al.*, 2010; Erös *et al.*, 2012). For our consideration of fragmented fish populations, relevant patch sizes range from discrete reaches (e.g. segments between tributaries) of streams to entire watersheds or lakes. Key characteristics of patches that are relevant to fragmentation include: size, distribution within the landscape, juxtaposition, diversity, duration, and physical processes that affect patch formation (Pringle *et al.*, 1988). Some patches are easily delimited, such as a series of lakes connected by streams. Boundaries between patches in streams can be less distinct (e.g. longitudinal gradients in physico-chemical features), with some authors arguing that connections between patches, such as stream confluence zones can, in themselves, provide important habitats or patches (e.g. Grant *et al.*, 2007). Others, for ease of modelling, consider confluences as transition zones through which fishes simply move to get to other patches (e.g. Erös *et al.*, 2011). Finally, structural connectivity can be naturally dynamic in some regions, including arid biomes where dewatering is a natural feature of landscapes and acts to structure stream fish communities by altering functional connectivity (Beatty *et al.*, 2009; Jaeger & Olden, 2012; Rolls *et al.*, 2012). Despite these complexities, most would agree that aquatic systems have discrete boundaries (shoreline, stream confluences, etc.) and can be partitioned into patches of discrete habitat. This approach has paved a way forward in quantifying structural connectivity and developing indices of fragmentation.

Quantifying structural connectivity in aquatic habitats can range from simple linear arrangement of habitat patches to complex networks of connected patches where patches and linkages vary in quality. Assessing barriers to movement along a linear arrangement of patches is useful for identifying small-scale movement patterns, such as pool-to-pool movement of stream fishes (Roberts & Angermeier, 2007). Structural connectivity might also consider lateral connectivity to floodplain or other off-channel habitats (Ward *et al.*, 1999; Fullerton *et al.*, 2010). More complex patch-based spatial graphs were recently recognised as a useful approach to quantifying structural connectivity among a series of connected patches including lakes and stream networks (Erös *et al.*, 2012). This approach is derived from graph theory where patches are nodes connected by links, and graphs can be used to generate matrices that represent dispersal probability between any patch in the network (Figure 10.3). In streams, graphs and associated matrices can be refined to consider the hierarchical nature of stream networks including larger segments (heterogeneous nodes) downstream and unidirectional flow (heterogeneous links) among patches that might cause directional bias in dispersal (Grant *et al.*, 2007; Grant, 2011; Erös *et al.*, 2012). Incorporating fragmentation is done by increasing the resistance (i.e. decreasing transition probabilities) to dispersal among patches

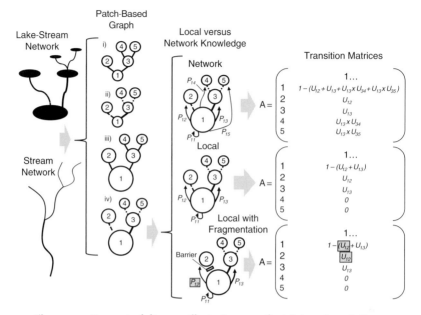

Figure 10.3 Conceptual diagram illustrating use of patch-based graph theory to model fish dispersal in fragmented lake–stream and stream networks. Lakes or the longitudinal length of stream between confluences are considered as patches (Erös *et al.*, 2011), which can be converted into patch-based graphics assuming (i) homogeneous node and link resolution (i.e. nodes = circles of equal size, links = lines of equal character), (ii) homogeneous node resolution and heterogeneous link resolution, (iii) homogeneous link resolution and heterogeneous node resolution, and (iv) heterogeneous node and link resolution (Erös *et al.*, 2012). Within patch-based graphics, fish dispersal can be limited to local or network knowledge during simulations using spatially explicit transition probabilities (P_{xx}) between patches (Padgham & Webb, 2010; Neeson *et al.*, 2011). Fragmentation is modelled by adjusting transition probabilities within deterministic matrices to incorporate barrier location, barrier permeability and node quality (Padgham & Webb, 2010). For transition matrices, only transition probabilities and portions of matrices associated with dispersal from node 1 are shown for brevity.

(Padgham & Webb, 2010; Perkin *et al.*, 2013). Other, related approaches include circuit theory that attempts to quantify and visualise structural (and functional) connectivity among patches (McRae *et al.*, 2008). This approach considers the analogous use of resistance (barriers), current (dispersal rate) and voltage (genetic difference among populations) to describe movement of individuals (and genes) among habitat patches. Other measures and approaches are reviewed in Storfer *et al.* (2010).

Patch-based spatial graphs and resulting matrices allow quantification of structural connectivity of patches. Complemented with indices of habitat availability, graph or network based analyses can be used to indicate the level of fragmentation within hierarchically structured aquatic systems and help prioritise systems based on structural connectivity. An example of a simple index of longitudinal connectivity in a stream network is provided by Cote *et al.* (2009). Their dendritic connectivity index (DCI) includes dividing the total stream network length into sections isolated by barriers and computing a weighted average of connectivity within and among all sections. The DCI can be adapted to represent life histories for diadromous fishes (requiring connectivity to the ocean) and potamodromous fishes (requiring connectivity throughout a watershed), and in both cases produces a value of 100 in the absence of fragmentation (i.e. complete natural connectivity) that declines toward zero based on the number, placement and permeability of barriers. A greater DCI value corresponds with isolation of a smaller fraction of the riverscape. Although a potential criticism of the DCI is that assignment of a static value of permeability is unrealistic because barrier permeability is generally temporally dynamic, approaches for measuring dynamic permeability have recently been developed (Bourne *et al.*, 2011; Anderson *et al.*, 2012) and empirical evidence suggests this is an informative measure of structural connectivity (e.g. Perkin & Gido, 2012). The DCI and related measures of longitudinal connectivity can be employed to evaluate cost–benefit approaches to assessing fragmentation and reconnecting habitats (O'Hanley & Tomberlin, 2005). In such approaches, the cost of removing or modifying a barrier to allow passage is weighed against a set budget as well as the potential gain in habitat, measured as the longitudinal length of stream reconnected following barrier removal (Kemp & O'Hanley, 2010; O'Hanley, 2011; O'Hanley *et al.*, 2013). With such advances has come an increasing recognition of the importance of assigning ecologically relevant permeability to individual barriers because highly permeable barriers might not block fish movement (Pepino *et al.*, 2012) and existing methods for measuring permeability might not represent interpretations by fishes (Mahlum *et al.*, 2014).

10.4.2 Measuring functional connectivity

In most cases, quantifying structural connectivity is used to infer functional connectivity, but does not guarantee functional connectivity.

Thus, associating a break in structural connectivity with either direct observations of dispersal of individuals or indirect measures of connectivity that infer dispersal through analysis of spatial variation in genetic, population or community structure analysis. Arguably the best approach to directly measure functional connectivity among fish populations or communities is through the marking or tracking of individuals throughout a network of habitat patches (e.g. Warren & Pardew, 1998). Unfortunately, this can be labour-intensive and tracking the movement of small individuals is difficult because of tag-size limitations, increased handling stress, as well as sheer abundance and low survivorship, which requires large numbers to be tagged to assure adequate number of recaptured fishes. Coarse approximation of functional connectivity can be assessed with some indirect methods. For example, analyses of spatial autocorrelation (e.g. distance–decay relationships) of species abundance or community structure after statistically removing environmental effects can provide indirect evidence of connectivity, in that species demonstrating greater spatial autocorrelation are inferred to be more connected (Olden *et al.*, 2001; Hitt & Angermeier, 2008; Shurin *et al.*, 2009). A promising approach to evaluate functional connectivity, which is not constrained by the size of the study organism, is through the use of population genetics.

Population geneticists have developed a suite of indices, focused predominantly on genetic dissimilarity or Bayesian assignment tests, to infer functional connectivity across distinct localities in a network of stream or lake sampling localities. The traditional and most widely used metric is F_{ST} which is usually computed pairwise among populations to yield a symmetrical matrix of genetic distance. Wright's (1978) approximation is used to convert F_{ST} to $N_e m$, which is the product of the genetic effective population size (N_e) and the per-generation migration rate (m). More recently developed metrics, based on Approximate Bayesian Computation (ABC) procedures (e.g. BAYESASS (Wilson & Rannala, 2003); BAPS (Corander *et al.*, 2004); GENECLASS2 (Piry *et al.*, 2004)) allow indirect estimation of the fraction of genotypes assigned as migrants, m (where $m_{ij} \neq m_{ji}$), at each locality (where values of m_{ij} comprise the elements of the migration matrix, M). In this case, ABC approaches allow estimation of asymmetric migration (e.g. Fraser *et al.*, 2007). Once a pairwise matrix is determined, landscape genetics approaches (Manel *et al.*, 2003; Mullen *et al.*, 2010) are used to evaluate the relationship of the F_{ST} or migration matrix (M) and abiotic and biotic features of the stream network.

One of the main challenges to using genetic methods for estimation of functional connectivity is that classical population genetics theory (e.g. Wright, 1939) is based on an assumption that spatial and temporal patterns of genetic variation are shaped by the interplay of genetic drift, migration and selection that are in equilibrium with respect to each other. In other words, classical theory assumes that all populations are stable in size, persistent in time and are connected by migration that occurs at a constant rate. This assumption is clearly violated in metapopulations that are subject to local extinction and recolonisation. However, metapopulation theory that accounts for non-equilibrium dynamics has been well developed over the last 40 years and ABC approaches permit interpretation in an array of populations where extinction and recolonisation are important. For populations that fluctuate in size, sampling the same localities across time steps also permits evaluation of temporal genetic stability in terms of variance in effective population size and migration rates (Wang & Whitlock, 2003). Parameters estimated via temporal approaches are based on allele frequency changes between time steps and give contemporary rather than historical insight into processes that shape allele frequencies (Waples, 1989). Historical drainage connectivity can override modern connectivity and affect estimates of F_{ST}, $N_e m$ and m based on genetic data (Poissant *et al.*, 2005; Turner & Robison, 2006; Sterling *et al.*, 2012). Given some of the problems with using genetic data alone to estimate connectivity and that genetic data can only measure connectivity following the successful reproduction of individuals, it is prudent to conduct simultaneous ecological and genetic studies (Lowe & Allendorf, 2010). Ecological and genetic approaches are highly complementary because they (can) use the same currency (e.g. abundance and effective population sizes, dispersal and effective migration rate, respectively) and together give insights into processes that drive distribution and abundance on the landscape that neither approach can offer alone.

10.4.3 Linking structural and functional connectivity

Not surprisingly, there is strong evidence to support the link between structural connectivity and changes in functional connectivity. Dams and the associated impoundments have caused the decline of fishes including salmonids (Morita & Yamamoto, 2002), cyprinids (Winston *et al.*, 1991), catostomids (Catalano *et al.*, 2007) and many others (Chapter 4). Fishes inhabiting stream segments isolated upstream of reservoirs have unique genetic structure (Hudman & Gido, 2013), including alterations

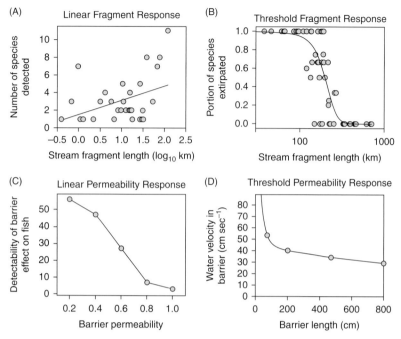

Figure 10.4 Empirical examples of fish responses to fragmentation. These examples illustrate a linear response to fragment length by fish communities in streams of New York (A; Bain & Wine, 2010), a threshold response to fragment length by pelagic-broadcast spawning fish communities in the Great Plains (B; Perkin & Gido, 2011), a linear response in the power to detect effect of a barrier on fish dispersal across a range of permeability values (C; Pepino *et al.*, 2012) and an exhaustion-threshold (i.e. length a fish can swim at a given water velocity before exhaustion) response by fish to the water velocity flowing through a barrier of particular length (D; Neary, 2012).

to phenotypic expression, suggesting evolutionary (i.e. adaption to novel environments) consequences of fragmentation (Haas *et al.*, 2010; Franssen, 2012; Sterling *et al.*, 2012). Consequently, increased attention is being devoted toward understanding the ecological effects of attenuated structural connectivity (Fullerton *et al.*, 2010). Based on recent works, it is clear that the size and isolation of stream fragments created by in-stream barriers have important consequences for stream fish community structure, but these responses can be linear or non-linear (Figure 10.4). For example, Bain and Wine (2010) found a linear relationship between stream fragment length and species richness in highly fragmented streams of New York (Figure 10.4A) and a threshold-like response in brook trout extirpation among small fragments. Similarly, Perkin and

Gido (2011) documented threshold responses in extirpation of multiple cyprinids among smaller stream fragments distributed throughout the Great Plains (Figure 10.4B). In terms of fragment isolation, Pepino *et al.* (2012) found a linear decline in fish dispersal as road–stream culvert permeability declined (Figure 10.4C), whereas Neary (2012) reviewed the use of fish passage models such as exhaustion-threshold curves to evaluate road–stream permeability (Figure 10.4D).

Results such as those presented in Figure 10.4 suggest stream fish communities respond to the combined effects of the number, placement and permeability of barriers distributed throughout a watershed. In general, we expect genetic structure to be the most responsive to barrier permeability, whereas population and community dynamics likely respond to both permeability and fragmentation (Figure 10.5). Thus community or population responses to fragmentation are generally strongest among watersheds characterised by many barriers with low permeability, in turn creating a large number of small stream fragments (Figure 10.5A). Under this scenario, any level of fragmentation by barriers with decreased permeability (relative to a natural stream channel) will elicit a response of the fish community, but the response will depend on the magnitude of alteration. However, given the hierarchical organisation of patches in stream networks, emergent properties such as threshold responses to fragment size or barrier permeability are common in fragmented riverscapes (discussed earlier; Grant *et al.*, 2007; Perkin *et al.*, 2013). Such threshold patterns associated with fragment size (Figure 10.5B) and barrier permeability (Figure 10.5C) can result in responses of greater magnitude than expected based on the number of barriers alone. Examples of thresholds for barrier permeability might include the migration of a small number of individuals necessary to rescue genetic diversity; whereas thresholds for fragment size might occur for species with pelagic eggs (e.g. Leslie *et al.*, 1982) that require minimum length of stream for successful reproduction. In reality, the synergistic effects of threshold responses to both fragment size and barrier permeability contribute to community and population responses that are not predicted by the presence or number of barriers alone (Figure 10.5D). The latter scenario suggests many populations or communities will persist under low to moderate levels of fragmentation, but the nature of the response will depend on how traits of species of interest interact with the number, placement and permeability of barriers (Branco *et al.*, 2012; Pepino *et al.*, 2012; Perkin *et al.*, 2013).

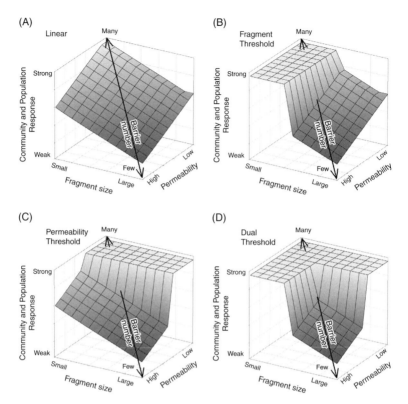

Figure 10.5 Conceptual models describing potential fish community or population responses to changes in stream habitat connectivity. (A) Responses are strongest among watersheds characterised by many barriers with low permeability and consequently small fragment sizes. Alteration to community structure might be greater than expected based on the number of barriers alone because fish interpretations of riverscapes result in threshold responses to (B) fragment size or (C) barrier permeability. (D) In real-world riverscapes, low permeability among many barriers contributes to smaller fragment sizes so that communities effectively respond to barrier number, position (contributing to fragment size) and permeability.

10.5 MODELLING RESPONSES TO FRAGMENTATION

More detailed and realistic modelling exercises are helpful in understanding metapopulation persistence, genetic structure and gene flow, and the interactions of intrinsic and extrinsic forces that shape organism and community responses to fragmentation. Using metapopulation models, Fagan (2002) revealed that fragmentation has fundamentally different consequences for bifurcating dendritic networks compared to systems with linear geometries (e.g. only considering the mainstem of a large river), in that multiple fragmentation events in a dendritic network produced a greater number of small fragments. This result has drastic consequences for the persistence of fish metapopulations in dendritic river systems, because extinction risk increases as fragment (and presumably population) size decreases (Lande, 1993). The influence of fragment size on population persistence was further illustrated by the theoretical studies of Jager et al. (2001) and Morita and Yokota (2002) that investigated the influence of fragmentation and fragment size on white sturgeon (*Acipenser transmontanus*) and white-spotted charr (*Salvelinus leucomaenis*), respectively. These studies showed that decreasing fragment size decreased the probability of metapopulation persistence. In addition, metapopulation persistence also decreases when dispersal opportunity in simulated riverscapes is reduced from bidirectional movement to only unidirectional movement (Jager et al., 2001; Fagan, 2002), which frequently occurs when dams limit dispersal to a downstream-only direction. Allowing bidirectional movement via the construction of fish passages, however, does not automatically increase metapopulation persistence. Jager (2006a) found that upstream movement around barriers for a white sturgeon metapopulation in a simulated riverscape was only beneficial if subsequent downstream movement was restricted for larger fish (through screening) or allowed safe passage. Similarly, the longitudinal positioning of long and short fragments interspersed among multiple barriers may also impact the influence passage construction has on the persistence of simulated metapopulations, with passageways generally being most beneficial when there is a long segment of available habitat upstream of many barriers (Jager, 2006a,b). Finally, Jager et al. (2001) found that as the number of impoundments in a simulated river network increased, genetic diversity within theoretical white sturgeon populations eroded, simulating the genetic impact impoundments may have on real-world populations. Modelling exercises such as these provide hypotheses on how fish populations and communities will respond to different levels of fragmentation.

10.6 CONSERVATION CHALLENGES AND OPPORTUNITIES

Different barrier types, fragment lengths and traits of species create unique challenges for conservation. Impoundments are perhaps the greatest hurdle in that they are expensive to remove, have a large value to society (e.g., Sethi et al., 2004; Chapter 4), and might prevent the spread of non-native species (Jackson & Pringle, 2010). Nevertheless, efforts such as the Elwha Dam removal (Wunderlich *et al.*, 1994) demonstrate that this is a feasible management option, particularly for dams with limited function or in need of repair (e.g. Gregory *et al.*, 2002; Stanley & Doyle, 2003; Chapter 4). Moreover, restoration efforts that reconnect isolated habitats have generally been successful at improving fish populations (Roni *et al.*, 2012), and a number of studies show rapid colonisation of former impoundment sites and upstream areas by migratory and resident fishes following dam removal (e.g. Hart *et al.*, 2002; Catalano *et al.*, 2007). Although dam removal is necessary for river restoration, logistical and societal challenges make it essential to prioritise the removal of dams (Hitt *et al.*, 2012). Fortunately, there are an increasing number of studies and options for prioritising dam removal that conservation organisations can draw from (Fullerton *et al.*, 2010). If dam removal is not practical, increasing connectivity through the construction of fish passage devices or mechanical movement of fishes through barriers might be alternative management solutions (e.g. Walters *et al.*, 2014).

The removal or modification of smaller barriers such as road crossing culverts and low-head dams is much more feasible than for large dams (O'Hanley, 2011). Removing low-head dams has been successful at increasing connectivity of habitats used by anadromous (Burdick & Hightower, 2006) and potamodromous (Pepino *et al.*, 2012) species. The effects of poorly designed road crossings on freshwater fish assemblages are well documented (Gibson *et al.*, 2005) and most prevalent in smaller, headwater streams characterised by greater slopes (Poplar-Jeffers *et al.*, 2009). Bridges across larger streams generally span the entire channel and are less likely to cause fragmentation (see Perkin & Gido, 2012 supplemental material). The potentially large number of roads crossing small streams creates a challenge in the sheer number of barriers in need of renovation (Forman & Alexander, 1998). Similar to dam removal, a prioritisation scheme will be necessary to identify those barriers that cause the greatest extent of fragmentation and therefore require renovation. To this end, metrics such as the DCI, as described above, are potential tools for prioritisations (Cote *et al.*, 2009; Fullerton *et al.*, 2010).

An emerging problem with efforts to mitigate fragmentation is the trade-off that increasing connectivity to reduce extinction risk of native species facilitates invasion by non-native species (Fausch *et al.*, 2009; Jackson & Pringle, 2010; Chapter 4). Conservation decisions will have to balance the probabilities that non-natives will disperse into a system if barriers are removed (Clarkson *et al.*, 2012) and the predicted impact of those species (Rahel, 2013). Understanding the latter is particularly complex, as many species invasions are benign, but some are devastating (Chapter 8). The most comprehensive examples of purposeful fragmentation involve management and conservation of native and endemic inland trout species. For these species, an important threat to persistence is competition and/or hybridisation with non-natives, usually introduced rainbow, brown or brook trout (Fausch *et al.*, 2009). To mitigate this threat, an impassable barrier to upstream movement is typically constructed, followed by removal of non-native species (Novinger & Rahel, 2003; Fausch *et al.*, 2009). Once non-natives are removed, wild or hatchery-reared natives are repatriated to these fragmented, renovated reaches. The management trade-off is that barrier construction excludes non-natives but necessarily isolates the native population from gene flow, and thereby increases vulnerability to extinction through stochastic demographic events, localised catastrophic disturbance and inbreeding depression (Roberts *et al.*, 2013). This trade-off extends to co-occurring species, and may disrupt production and linkages of energy and nutrient flow, which further disrupts local fish abundance.

10.7 FUTURE DIRECTION AND CHALLENGES

Fragmentation alone has negative consequences for freshwater fishes, but isolating populations and communities is intensified by other human-mediated disturbances such as increased intensity of drought via climate change, poor water quality, and invasive species (Hoagstrom *et al.*, 2011; Cooke *et al.*, 2012). Both conceptual and mathematical frameworks are available to assess the consequences of fragmentation (e.g. Fausch *et al.*, 2009; Padgham & Webb, 2010). Future efforts to build on these conceptual frameworks should incorporate technological advances as well as inclusion of other potentially important intrinsic and extrinsic factors influencing fish communities. For example, technological advances in mapping aquatic habitats and quantifying barrier permeability will help quantify structural connectivity in fragmented lakes, rivers and streams (Chin *et al.*, 2008; Park *et al.*, 2008). However, it will also be important to test measures of structural connectivity with empirical measurements of dispersal and ultimately population persistence. Such measurements will be facilitated through advances in genetic analyses and advanced tracking methods (e.g. Fraser *et al.*, 2004; Gido & Jackson, 2010; Winemiller *et al.*, 2010).

A potentially important but understudied source of fragmentation is due to biotic factors that inhibit or promote movement of fishes through natural or artificial corridors. Gilliam and Fraser (2001) noted the pattern of reduced abundance of small fishes in mainstem habitats compared to small tributaries might be caused by predator avoidance in the mainstem. Although predators could fragment habitat for these smaller fishes, dispersal was dependent on intrinsic (body size) and extrinsic (water level) factors and suggested predation as an agent of fragmentation was complex. More research is needed to identify the role of individual behaviour and biotic interactions in fragmented systems. For example, the construction of instream barriers that alter habitat upstream and downstream might attract predators that inhibit passage, even if the structure itself is passable. The potential of fish passageways to become 'ecological traps' by luring populations to unsuitable habitats needs to be evaluated so that facilities can be properly constructed (Pelicice & Agostinho, 2008, Chapter 4). Although this is of particular concern for impounded waters that create lentic habitats that provide habitat for large predatory fishes (e.g. Gillette *et al.*, 2005), concentrations of fishes below dams can also attract predators, such as sea lions in coastal drainages (Keefer *et al.*, 2012). These indirect or unexpected consequences of fragmentation certainly need to be considered when prescribing management recommendations.

Adverse effects of climate change on fish communities (see Chapter 3) also are exacerbated by fragmentation because increases in disturbance frequency and altered thermal regimes will shrink suitable habitat of fishes (Rahel *et al.*, 1996), and barriers will impede movement or escape from potentially inhospitable conditions. Moreover, in arid and semi-arid regions, increased frequency of drying will force fishes downstream into impoundments where they have to contend with lentic habitats and species. Identifying the constraints that climate change will have on fragmented habitats is a necessary area of future research.

Finally, the evolutionary consequences of living in fragmented habitats are not well understood because a number of factors (intrinsic and extrinsic) and forces (e.g. genetic drift, selection and gene flow) are acting simultaneously to shape organism response to altered habitats and reduced connectivity. Historically, the rate at which natural fragmentation reduced connectivity of populations supported increased diversification in evolutionary time (Dias *et al.*, 2012). Recent fragmentation, however, has resulted in rapid loss of connectivity, causing decreased population viability in ecological time and ultimately decreased diversity in evolutionary time. Reductions in genetic diversity and dispersal constraints will create selective environments that are quite different than in connected habitats. Smaller, less-mobile individuals might have a selective advantage over larger more mobile individuals (Bain & Wine, 2010). Trophic interactions might change as individuals dependent on movement are unable to access foraging habitats (e.g. oceans, large rivers) in a fragmented system. Similarly, reproductive ecology and migration patterns of fishes are likely to change. Movement of migrant individuals will be highly constrained and if there is a trade-off in reproductive success, those individuals might be selected against in fragmented habitats. Comparative landscape genetic studies with multiple species analysed simultaneously are rare for freshwater fishes, but the promise of this approach is well appreciated (Wares, 2002; Urban & Skelly, 2006). One outcome is that intrinsic differences in life history and ecology have important consequences for the distribution of genetic diversity in space and time (e.g. Waples, 1987; Avise, 1992; Tibbets et al., 1996; Turner & Trexler, 1998; Whiteley et al., 2004; Burridge et al., 2008) and that life history can interact with landscape features to determine overall population structure within species (Neville et al., 2006) and differences among species (e.g. Manier & Arnold, 2006; Turner et al., 2006). Current research across species is beginning to identify traits that predict responses to fragmentation, but clearly more research is needed to refine these predictions.

In conclusion, as the human population increases, the need to traverse aquatic habitats with structures that potentially constrain movement of fishes will increase. Major strides have been made to research and effectively design structures to allow dispersal of fishes (e.g. Stuart & Mallen-Cooper, 1999; Pelicice & Agostinho, 2008). Given that the frequency of major dam construction has declined in many countries, and that smaller barriers can be modified to allow fish passage, current conservation goals should begin a trend of restoring connectivity of aquatic habitats on global, national and watershed scales. Additional advances have been put forward in the literature to prioritise barriers for removal or renovation. These tools consider both the location of barriers in a network of habitats, the permeability of existing barriers and the intrinsic characteristics of the fish populations impacted by those barriers. These cumulative efforts can inform natural resource agencies, but there are still large gaps in understanding the context in which barriers influence dynamics of fish populations and communities. As with many conservation problems, empirical data describing how barriers will influence natural populations lags well behind theoretical expectations. Thus, researchers should embrace emerging technological advances in conservation genetics and our ability to track fish movement to gather critical data on habitat needs and responses of fishes to fragmentation. Using demographic and genetic data in concert (e.g. genetic mark–recapture approaches; relatedness and behavioural ecology) offers great potential for synthesis and prediction of ecological and evolutionary outcomes of fragmentation on fish populations and communities, and may help overcome shortcomings of each method when used alone (e.g. Kanno *et al.*, 2014).

10.8 DISCUSSION QUESTIONS

1. What intrinsic characteristics (traits) of species make them vulnerable to fragmentation?
2. What extrinsic characteristics of water bodies (streams or lakes) influence how fragmentation affects fish communities?
3. Are demographic or genetic responses more important predictors of long-term population persistence following fragmentation?
4. Why does local species diversity of freshwater fish communities generally decline following fragmentation? Are there scenarios where fragmentation can increase species diversity?
5. What potentially confounding effects, such as habitat degradation, are typically associated with fragmentation?

10.9 ACKNOWLEDGEMENTS

Financial support for KBG and TFT was provided by the Bureau of Reclamation Water Smart Program. JEW was supported by the Department of Education GAANN scholarship and JSP was supported by the Kansas State University Division of Biology. Thoughtful comments on an earlier draft of this chapter were provided by the Aquatic Journal Club at Kansas State University. We thank Nick Bond and an anonymous reviewer for thoughtful comments that greatly improved the presentation of these materials.

10.10 REFERENCES

Alò, D. & Turner, T. F. (2005). Effects of habitat fragmentation on effective population size in the endangered Rio Grande silvery minnow. *Conservation Biology*, 19, 1138–1148.
Anderson, G. B., Freeman, M. C., Freeman, B. J., *et al.* (2012). Dealing with uncertainty when assessing fish passage through culvert road crossings. *Environmental Management*, 50, 462–477.

Agostinho, A. A., Marques, E. E., Agostinho, C. S., *et al.* (2007). Fish ladder of Lajeado Dam: migrations on one-way routes? *Neotropical Ichthyology*, 5, 121–130.

Avise, J. C. (1992). Molecular population structure and the biogeographic history of a regional fauna: a case history with lessons for conservation biology. *Oikos*, 63, 62–76.

Bain, M. B. & Wine, M. L. (2010). Testing predictions of stream landscape theory for fish assemblages in highly fragmented watersheds. *Folia Zoologica*, 59, 231–239.

Baki, A. B., Zhu, D. Z., Courtice, G. & Rajaratnam, N. (2012). Exploring the potential fish pathways in an experimental rocky ramp type nature-like fish pass. In *9th International Symposium on Ecohydraulics 2012 Proceedings*. Mader, H. & Kraml, J. (editors).

Beatty, R. J., Rahel, F. J. & Hubert, W. A. (2009). Complex influences of low-head dams and artificial wetlands on fishes in a Colorado River tributary system. *Fisheries Management and Ecology*, 16, 457–467.

Blanchet, S., Rey, O., Etienne, R., Lek, S. & Loot, G. (2010). Species-specific responses to landscape fragmentation: implications for management strategies. *Evolutionary Applications*, 3, 291–304.

Bourne, C. M., Kehler, D. G., Wiersma, Y. F. & Cote, D. (2011). Barriers to fish passage and barriers to fish passage assessments: the impact of assessment methods and assumptions on barrier identification and quantification of watershed connectivity. *Aquatic Ecology*, 45, 389–403.

Branco, P., Segurado, P., Santos, J. M., Pinheiro, P. & Ferreira, M. T. (2012). Does longitudinal connectivity loss affect the distribution of freshwater fish? *Ecological Engineering*, 48, 70–78.

Burdick, S. M. & Hightower, J. E. (2006). Distribution of spawning activity by anadromous fishes in an Atlantic slope drainage after removal of a low-head dam. *Transaction of the American Fisheries Society*, 135, 1290–1300.

Burridge, C. P., Craw, D., Fletcher, D. & Waters, J. M. (2008). Geological dates and molecular rates: fish DNA sheds light on time dependency. *Molecular Biology and Evolution*, 25, 624–633.

Catalano, M. J., Bozek, M. A. & Pellet, T. D. (2007). Effects of dam removal on fish assemblage structure and spatial distributions in the Baraboo River, Wisconsin. *North American Journal of Fisheries Management*, 27, 519–530.

Chin, A., Laurencio, L. R. & Martinez, A. E. (2008). The hydrologic importance of small- and medium-sized dams: examples from Texas. *The Professional Geographer*, 60, 238–251.

Clarkson, R. W., Marsh, P. C. & Dowling, T. E. (2012). Population prioritization for conservation of imperiled warmwater fishes in an arid-region drainage. *Aquatic Conservation: Marine and Freshwater Ecosystems*, 22, 498–510.

Cooke, S. J., Paukert, C. P. & Hogan, Z. (2012). Endangered river fish: factors hindering conservation and restoration. *Endangered Species Research*, 17, 179–191.

Corander, J., Waldmann, P., Marttinen, P. & Sillanpa, M. J. (2004). BAPS 2: enhanced possibilities for the analysis of genetic population structure. *Bioinformatics*, 20, 2363–2369.

Cote, D., Kehler, D. G., Bourne, C. & Wiersma, Y. F. (2009). A new measure of longitudinal connectivity for stream networks. *Landscape Ecology*, 24, 101–113.

Coutant, C. C. & Whitney, R. R. (2000). Fish behavior in relation to passage through hydropower turbines: a review. *Transactions of the American Fisheries Society*, 129, 351–380.

Creed, R. P. (2006). Predator transitions in stream communities: a model and evidence from field studies. *Journal of the North American Benthological Society*, 25, 533–544.

Dias, M. S., Cornu, J. F., Oberdorff, T., Lasso, C. A. & Tedesco, P. A. (2012). Natural fragmentation in river networks as a driver of speciation for freshwater fishes. *Ecography*, 35, 1–7.

Dudley, R. K. & Platania, S. P. (2007). Flow regulation and fragmentation imperil pelagic-spawning riverine fishes. *Ecological Applications*, 17, 2074–2086.

Eadie, J. M. & Keast, A. (1984). Resource heterogeneity and fish species diversity in lakes. *Canadian Journal of Zoology*, 62, 1689–1695.

Englund, G., Johansson, F., Olofsson, P., Salonsaari, J. & Öhman, J. (2009). Predation leads to assembly rules in fragmented fish communities. *Ecology Letters*, 12, 663–671.

Erös, T., Schmera, D. & Schick, R. S. (2011). Network thinking in riverscape conservation: a graph-based approach. *Biological Conservation*, 144, 184–192.

Erös, T., Olden, J. D., Schick, R. S., Schmera, D. & Fortin, M. J. (2012). Characterizing connectivity relationships in freshwaters using patch-based graphs. *Landscape Ecology*, 27, 303–317.

Fagan, W. F. (2002). Connectivity, fragmentation, and extinction risk in dendritic metapopulations. *Ecology*, 83, 3243–3249.

Falke, J. A. & Fausch, K. D. (2010). From metapopulations to metacommunities: linking theory with empirical observations of the spatial population dynamics of stream fishes. In *Community Ecology*

Of Stream Fishes: Concepts, Approaches, And Techniques. Bethesda, MD: American Fisheries Society Symposium 73, pp. 207–234.

Falke, J. A. & Gido, K. B. (2006). Effects of reservoir connectivity on stream fish assemblages in the Great Plains. *Canadian Journal of Fisheries and Aquatic Sciences*, 63, 480–493.

Falke, J. A., Fausch, K. D., Magelky, R., *et al.* (2011). The role of groundwater pumping and drought in shaping ecological futures for stream fishes in a dryland river basin of the western Great Plains, USA. *Ecohydrology*, 4, 682–697.

Fausch, K. D., Torgersen, C. E., Baxter, C. V. & Li, H. W. (2002). Landscapes to riverscapes: bridging the gap between research and conservation of stream fishes. *BioScience*, 52, 483–498.

Fausch, K. D., Rieman, B. E., Dunham, J. B., Young, M. K. & Peterson, D. P. (2009). Invasion versus isolation: tradeoffs in managing native salmonids with barriers to upstream movement. *Conservation Biology*, 23, 859–870.

Flecker, A. S., McIntyre, P. B., Moore, J. W., *et al.* (2010). Migratory fishes as material and process subsidies in riverine ecosystems. In *Community Ecology of Stream Fishes: Concepts, Approaches, and Techniques*. Bethesda, MD: American Fisheries Society Symposium 73, pp. 559–592.

Forman, R. T. & Alexander, L. E. (1998). Roads and their major ecological effects. *Annual Reviews in Ecology and Systematics*, 29, 207–257.

Franssen, N. R. (2012). Genetic structure of a native cyprinid in a reservoir-altered stream network. *Freshwater Biology*, 57, 155–165.

Fraser, D. F., Gilliam, J. F. & Yip-Hoi, T. (1995). Predation as an agent of population fragmentation in a tropical watershed. *Ecology*, 76, 1461–1472.

Fraser, D. J., Lippe, C. & Bernatchez, L. (2004). Consequences of unequal population size, asymmetric gene flow and sex-biased dispersal on population structure in brook charr (*Salvelinus fontinalis*). *Molecular Ecology*, 13, 67–80.

Fraser, D. J., Hansen, M. M., Ostergaard, S., *et al.* (2007). Comparative estimation of effective population sizes and temporal gene flow in two contrasting population systems. *Molecular Ecology*, 16, 3866–3889.

Frissell, C. A., Liss, W. J., Warren, C. E. & Hurley, M. D. (1986). A hierarchical framework for stream habitat classification: viewing streams in a watershed context. *Environmental Management*, 10, 199–214.

Fullerton, A. H., Burnett, K. M., Steel, E. A., *et al.* (2010). Hydrological connectivity for riverine fish: measurement challenges and research opportunities. *Freshwater Biology*, 55, 2215–2237.

Gaggiotti, O. E. (2003). Genetic threats to population persistence. *Annales Zoologici Fennici*, 40, 155–168.

Gibson, R. J., Haedrich, R. L. & Wernerheim, C. M. (2005). Loss of fish habitat as a consequence of inappropriately constructed stream crossings. *Fisheries*, 30, 10–17.

Gido, K. B. & Brown, J. H. (1999). Invasion of North American drainages by alien fish species. *Freshwater Biology*, 42, 387–399.

Gido, K. B. & Jackson, D. A. (Eds). (2010). *Community Ecology of Stream Fishes: Concepts, Approaches, and Techniques*. Bethesda, MD: American Fisheries Society, Symposium 73.

Gillette, D. P., Tiemann, J. S., Edds, D. R. & Wildhaber, M. L. (2005). Spatiotemporal patterns of fish assemblage structure in a river impounded by low-head dams. *Copeia*, 2005, 539–549.

Gilliam, J. F. & Fraser, D. F. (2001). Movement in corridors: enhancement by predation threat, disturbance, and habitat structure. *Ecology*, 82, 258–273.

Gilpin, M. E. & Soulé, M. E. (1986). Minimum viable populations: processes of species extinction. In *Conservation Biology: The Science of Scarcity and Diversity*. Soulé, M. E. (Ed.). Sunderland, MA: Sinauer, pp. 19–34.

Goulding, M. (1980). *The fishes and the Forest. Explorations in Amazonian Natural History*. Berkeley, CA: University of California Press.

Gowan, C., Young, M. K., Fausch, K. D. & Riley, S. C. (1994). Restricted movement in stream salmonids: a paradigm lost? *Canadian Journal of Fisheries and Aquatic Sciences*, 51, 2626–2637.

Grant, E. H. C. (2011). Structural complexity, movement bias, and metapopulation extinction risk in dendritic ecological networks. *Journal of the North American Benthological Society*, 30, 252–258.

Grant, E. H. C., Lowe, W. H. & Fagan, W. F. (2007). Living in the branches: population dynamics and ecological processes in dendritic networks. *Ecology Letters*, 10, 165–175.

Gregory, S., Li, H. & Li, J. (2002). The conceptual basis for ecological responses to dam removal. *BioScience*, 52, 713–723.

Grossman, G. D., Moyle, P. B. & Whitaker, J. O. (1982). Stochasticity in structural and functional characteristics of an Indiana stream fish assemblage – a test of community theory. *American Naturalist*, 120, 423–454.

Haas, T. C., Blum, M. J. & Heins, D. C. (2010). Morphological responses of stream fish to water impoundment. *Ecology Letters*, 6, 803–806.

Hart, D. D., Johnson, T. E., Bushaw-Newton, K. L., *et al.* (2002). Dam removal: challenges and opportunities for ecological research and river restoration. *BioScience*, 52, 669–681.

Hershey, A. E., Gettel, G. M., McDonald, M. E., *et al.* (1999). A geomorphic-trophic model for landscape control of Arctic lake food webs. *BioScience*, 49, 887–897.

Hitt, N. P. & Angermeier, P. L. (2008). Evidence for fish dispersal from spatial analysis of stream network topology. *Journal of the North American Benthological Society*, 27, 304–320.

Hitt, N. P., Eyler, S. & Wofford, J. E. B. (2012). Dam removal increases American eel abundance in distant headwater streams. *Transactions of the American Fisheries Society*, 141, 1171–1179.

Hoagstrom, C. W., Brooks, J. E. & Davenport, S. R. (2011). A large-scale conservation perspective considering endemic fishes of the North American plains. *Biological Conservation*, 144, 21–34.

Horwitz, R. J. (1978). Temporal variability patterns and the distributional patterns of stream fishes. *Ecological Monographs*, 48, 307–321.

Hudman, S. P. & Gido, K. B. (2013). Multi-scale effects of impoundments on genetic structure of creek chub (*Semotilus atromaculatus*) in the Kansas River basin. *Freshwater Biology*, 58, 441–453.

Hugueny, B., Movellan, A. & Belliard, J. (2011). Habitat fragmentation and extinction rates within freshwater fish communities: a faunal relaxation approach. *Global Ecology and Biogeography*, 20, 449–463.

Humphries, P., King, A. J. & Koehn, J. D. (1999). Fish, flows and flood plains: links between freshwater fishes and their environment in the Murray–Darling River system, Australia. *Environmental Biology of Fishes*, 56, 129–151.

Jackson, C. R. & Pringle, C. M. (2010). Ecological benefits of reduced hydrologic connectivity in intensively developed landscapes. *BioScience*, 60, 37–46.

Jackson, D. A., Peres-Neto, P. R. & Olden, J. D. (2001). What controls who is where in freshwater fish communities – the roles of biotic, abiotic, and spatial factors. *Canadian Journal of Fisheries and Aquatic Sciences*, 58, 157–170.

Jaeger, K. L. & Olden, J. D. (2012). Electrical resistance sensor arrays as a means to quantify longitudinal connectivity of rivers. *River Research and Applications*, 28, 1843–1852.

Jager, H. I. (2006a). Chutes and ladders and other games we play with rivers. I. Simulated effects of upstream passage on white sturgeon. *Canadian Journal of Fisheries and Aquatic Sciences*, 63, 165–175.

Jager, H. I. (2006b). Chutes and ladders and other games we play with rivers. II. Simulated effects of translocation on sturgeon. *Canadian Journal of Fisheries and Aquatic Sciences*, 63, 176–185.

Jager, H. I., Chandler, J. A., Lepla, K. B. & Winkle, W. V. (2001). A theoretical study of river fragmentation by dams and its effect on white sturgeon populations. *Environmental Biology of Fishes*, 60, 347–361.

Kanno, Y., Letcher, B. H., Coombs, J. A., Nislow, K. H. & Whiteley, A. R. (2014). Linking movement and reproductive history of brook trout to assess habitat connectivity in a heterogeneous stream network. *Freshwater Biology*, 59, 142–154.

Keefer, M. L., Stansell, R. J., Tackley, S. C., *et al.* (2012). Use of radiotelemetry and direct observations to evaluate sea lion predation on adult pacific salmonids at Bonneville Dam. *Transactions of the American Fisheries Society*, 141, 1236–1251.

Kemp, P. S. & O'Hanley, J. R. (2010). Procedures for evaluating and prioritizing the removal of fish passage barriers: a synthesis. *Fisheries Management and Ecology*, 17, 297–322.

Kingsford, R. T. (2000). Ecological impacts of dams, water diversions and river management on floodplain wetlands in Australia. *Austral Ecology*, 25, 109–127.

Lamphere, B. A. & Blum, M. J. (2012). Genetic estimates of population structure and dispersal in a benthic stream fish. *Ecology of Freshwater Fish*, 21, 75–86.

Lande, R. (1993). Risk of population extinction from demographic and environmental stochasticity and random catastrophes. *The American Naturalist*, 142, 911–927.

Letcher, B. H., Nislow, K. H., Coombs, J. A., O'Donnell, M. J. & Dubreuil, T. L. (2007). Population response to habitat fragmentation in a stream-dwelling brook trout population. *PLoS ONE*, 2, e1139.

Leslie, A. J. Jr, Van Dyke, J. M., Nall, L. E. & Miley, W. W. II. (1982). Current velocity for transport of grass carp eggs. *Transactions of the American Fisheries Society*, 111, 99–101.

Levins, R. (1969). Some demographic and genetic consequences of environmental heterogeneity for biological control. *Bulletin of the Entomological Society of America*, 15, 237–240.

Liermann, C. R., Nilsson, C., Robertson, J. & Ng, Y. (2012). Implications of dam obstruction for global freshwater fish diversity. *BioScience*, 62, 539–548.

Lowe, W. H. & Allendorf, F. W. (2010). What can genetics tell us about population connectivity? *Molecular Ecology*, 19, 3038–3051.

Lowe-McConnell, R. H. (1987). *Ecological Studies in Tropical Fish Communities*. New York, NY: Cambridge University Press.

Luttrell, G. R., Echelle, A. A., Fisher, W. L. & Eisenhour, D. J. (1999). Declining status of two species of the *Macrhybopsis aestivalis* complex (Teleostei: Cyprinidae) in the Arkansas River Basin and related effects of reservoirs and barriers to dispersal. *Copeia*, 1999, 981–989.

MacArthur, R. H. & Wilson, E. O. (1967). *The Theory of Island Biogeography*. Princeton, NJ: Princeton University Press.

Magoulick, D. D. & Kobza, R. M. (2003). The role of refugia for fishes during drought: a review and synthesis. *Freshwater Biology*, 48, 1186–1198.

Mahlum, S., Cote, D., Wiersma, Y. F., Kehler, D. & Clarke, K. D. (2014). Evaluating the barrier assessment technique derived from FishXing software and the upstream movement of brook trout through road culverts. *Transactions of the American Fisheries Society*, 143, 39–48.

Manel, S., Schwartz, M. K., Luikart, G. & Taberlet, P. (2003). Landscape genetics: combining landscape ecology and population genetics. *Trends in Ecology and Evolution*, 18, 189–197.

Manier, M. K. & Arnold, S. J. (2006). Ecological correlates of population genetic structure: a comparative approach using a vertebrate metacommunity. *Proceedings of the Royal Society – Biological Sciences*, 273, 3001–3009.

Matthews, W. J. & Marsh-Matthews, E. (2007). Extirpation of red shiner in direct tributaries of Lake Texhoma (Oklahoma–Texas): a cautionary case history from a fragmented river–reservoir system. *Transactions of the American Fisheries Society*, 136, 1041–1062.

McRae, B. H., Dickson, B. G., Keitt, T. H. & Shah, V. B. (2008). Using circuit theory to model connectivity in ecology, evolution, and conservation. *Ecology*, 89, 2712–2724.

Morita, K. S. & Yokota, A. (2002). Population viability of stream-resident salmonids after habitat fragmentation: a case study with white-spotted charr (*Salvelinus leucomaenis*) by an individual based model. *Ecological Modelling*, 155, 85–94.

Morita, K., Yamamoto, S. & Hoshino, N. (2000). Extreme life-history change of white-spotted charr (*Salvelinus leucomaenis*) after damming. *Canadian Journal of Fisheries and Aquatic Sciences*, 57, 1300–1306.

Morrissey, M. B. & de Kerckhove, D. T. (2009). The maintenance of genetic variation due to asymmetric gene flow in dendritic metapopulations. *American Naturalist*, 174, 875–889.

Moyle, P. B., Li, H. W. & Barton, B. A. (1986). The Frankenstein effect: impact of introduced fishes on native fishes in North America. In *Fish Culture in Fisheries Management*. Stroud, R. H. (Ed.). Bethesda, MD: American Fisheries Society, pp. 415–426.

Mullen, L. B., Woods, H. A., Schwartz, M. K., Sepulveda, A. J. & Lowe, W. H. (2010). Scale-dependent genetic structure of the Idaho giant salamander (*Dicamptodon aterrimus*) in stream networks. *Molecular Ecology*, 19, 898–909.

Musil, J., Horky, P., Slavik, O., Zboril, A. & Horka, P. (2012). The response of the young of the year fish to river obstacles: functional and numerical linkages between dams, weirs, fish habitat guilds and biotic integrity across large spatial scale. *Ecological Indicators*, 23, 634–640.

Neary, V. S. (2012). Binary fish passage models for uniform and nonuniform flows. *River Research and Applications*, 28, 418–428.

Neeson, T. M., Wiley, M. J., Adlerstein, S. A. & Riolo, R. L. (2011). River network structure shapes interannual feedbacks between adult sea lamprey migration and larval habitation. *Ecological Modelling*, 222, 3181–3192.

Neville, H. M., Dunham, J. B. & Peacock, M. M. (2006). Landscape attributes and life history variability shape genetic structure of trout populations in a stream network. *Landscape Ecology*, 21, 901–916.

Nilsson, C., Reidy, C. A., Dynesius, M. & Revenga, C. (2005). Fragmentation and flow regulation of the world's large river systems. *Science*, 308, 405–408.

Nislow, K. H., Hudy, M., Letcher, B. H. & Smith, E. P. (2011). Variation in local abundance and species richness of stream fishes in relation to dispersal barriers: implications for management and conservation. *Freshwater Biology*, 56, 2135–2144.

Norman, J. R., Hagler, M. H., Freeman, M. C. & Freeman, B. J. (2009). Application of a multistate model to estimate culvert effects on movement of small fishes. *Transactions of the American Fisheries Society*, 138, 826–838.

Novinger, D. C. & Rahel, F. J. (2003). Isolation management with artificial barriers as a conservation strategy for cutthroat trout in headwater streams. *Conservation Biology*, 17, 772–781.

O'Hanley, J. R. (2011). Open rivers: barrier removal planning and the restoration of free-flowing rivers. *Journal of Environmental Management*, 92, 3112–3120.

O'Hanley, J. R. & Tomberlin, C. (2005). Optimizing the removal of small fish passage barriers. *Environmental Modelling and Assessment*, 10, 85–98.

O'Hanley, J. R., Wright, J., Deibel, M., Fedora, M. A. & Soucy, C. L. (2013). Restoring stream habitat connectivity: a proposed method for prioritizing the removal of resident fish passage barriers. *Journal of Environmental Management*, 125, 19–27.

Olden, J. D., Jackson, D. A. & Peres-Neto, P. R. (2001). Spatial isolation and fish communities in drainage lakes. *Oecologia*, 127, 572–585.

Padgham, M. & Webb, J. A. (2010). Multiple structural modifications to dendritic ecological networks produce simple responses. *Ecological Modelling*, 221, 2537–2545.

Park, D., Sullivan, M., Bayne, E. & Scrimgeour, G. (2008). Landscape-level stream fragmentation caused by hanging culverts along roads in Alberta's boreal forest. *Canadian Journal of Forest Research*, 38, 566–575.

Pelicice, F. M. & Agostinho, A. A. (2008). Fish-passage facilities as ecological traps in large neotropical rivers. *Conservation Biology*, 22, 180–188.

Pepino, M., Rodriguez, M. A. & Magnan, P. (2012). Fish dispersal in fragmented landscapes: a modelling framework for quantifying the permeability of structural barriers. *Ecological Applications*, 22, 1435–1445.

Perkin, J. S. & Gido, K. B. (2011). Stream fragmentation thresholds for a reproductive guild of Great Plains fishes. *Fisheries*, 36, 371–383.

Perkin, J. S. & Gido, K. B. (2012). Fragmentation alters stream fish community structure in dendritic ecological networks. *Ecological Applications*, 22, 2176–2187.

Perkin, J. S., Gido, K. B., Al-Ta'Ani, O. & Scoglio, C. (2013). Simulating fish dispersal in stream networks fragmented by multiple road crossings. *Ecological Modelling*, 257, 44–56.

Piry, S., Alapetite, A., Cornuet, J. M., et al. (2004). GENECLASS 2: a software for genetic assignment and first-generation migrant detection. *Journal of Heredity*, 95, 536–539.

Poissant, J., Knight, T. W. & Ferguson, M. M. (2005). Nonequilibrium conditions following landscape rearrangement: the relative contribution of past and current hydrological landscapes on the genetic structure of a stream-dwelling fish. *Molecular Ecology*, 14, 1321–1331.

Poplar-Jeffers, I. O., Petty, J. T., Anderson, J. T., et al. (2009). Culvert replacement and stream habitat restoration: implications from brook trout management in an Application Watershed, U.S.A. *Restoration Ecology*, 17, 404–413.

Pringle, C. M., Naiman, R. J., Bretschko, G., et al. (1988). Patch dynamics in lotic systems: the stream as a mosaic. *Journal of the North American Benthological Society*, 7, 503–524.

Pringle, J. M., Blakeslee, A. M. H., Byers, J. E. & Roman, J. (2011). Asymmetric dispersal allows an upstream region to control population structure throughout a species' range. *Proceedings of the National Academy of Sciences USA*, 108, 15288–15293.

Quist, M. C. & Spiegel, J. R. (2012). Population demographics of catostomids in large river ecosystems: effects of discharge and temperature on recruitment dynamics and growth. *River Research and Applications*, 28, 1567–1586.

Rahel, F. J. (2013). Intentional fragmentation as a management strategy in aquatic systems. *BioScience*, 63, 362–372.

Rahel, F. J., Keleher, C. J. & Anderson, J. L. (1996). Potential habitat loss and population fragmentation for cold water fish in the North Platte River drainage of the Rocky Mountains: response to climate warming. *Limnology and Oceanography*, 41, 1116–1123.

Reid, S. M., Wilson, C. C., Mandrak, N. E. & Carl, L. M. (2008). Population structure and genetic diversity of black redhorse (*Moxostoma duquesnei*) in a highly fragmented watershed. *Conservation Genetics*, 9, 531–546.

Rieman, B. E. & Dunham, J. B. (2000). Metapopulations and salmonids: a synthesis of life history patterns and empirical observations. *Ecology of Freshwater Fish*, 9, 51–64.

Roberts, J. H. & Angermeier, P. L. (2007). Movement responses of stream fishes to introduced corridors of complex cover. *Transactions of the American Fisheries Society*, 136, 971–978.

Roberts, J. J., Fausch, K. D., Peterson, D. P. & Hooten, M. B. (2013). Fragmentation and thermal risks from climate change interact to affect persistence of native trout in the Colorado River basin. *Global Change Biology*, 19, 1383–1398.

Rodriguez, M. A. (2002). Restricted movement in stream fish: the paradigm is incomplete, not lost. *Ecology*, 83, 1–13.

Rolls, R. J., Leigh, C. & Sheldon, F. (2012). Mechanistic effects of low-flow hydrology on riverine ecosystems: ecological principles and consequences of alteration. *Freshwater Science*, 31, 1163–1186.

Roni, P., Beechie, T., Schmutz, S. & Muhar, S. (2012). Prioritization of watersheds and restoration projects. In *Stream and Watershed Restoration: A Guide to Restoring Riverine Processes and Habitats*. Roni, P. and Beechie, T. (Eds). Chichester: John Wiley and Sons, pp. 189–214.

Scheffer, M., Van Geest, G. J., Zimmer, K., *et al.* (2006). Small habitat size and isolation can promote species richness: second-order effects on biodiversity in shallow lakes and ponds. *Oikos*, 112, 227–231.

Schlosser, I. J. (1982). Fish community structure and function along two habitat gradients in a headwater stream. *Ecological Monographs*, 52, 395–414.

Schlosser, I. J. & Angermeier, P. L. (1995). Spatial variation in demographic processes of lotic fishes: conceptual models, empirical evidence, and implications for conservation. *American Fisheries Society Symposium*, 17, 392–401.

Sethi, S. A., Selle, A. R., Doyle, M. W., Stanley, E. H. & Kitchel, H. E. (2004). Response of unionid mussels to dam removal in Koshkonong Creek, Wisconsin (USA). *Hydrobiologia*, 525, 1–3.

Shurin, J. B., Cottenie, K. & Hillebrand, H. (2009). Spatial autocorrelation and dispersal limitation in freshwater organisms. *Oecologia*, 159, 151–159.

Skalski, G. T. & Gilliam, J. F. (2000). Modeling diffusive spread in a heterogeneous population: a movement study with stream fish. *Ecology*, 81, 1685–1700.

Slatkin, M. (1985). Gene flow in natural populations. *Annual Review of Ecology and Systematics*, 16, 393–430.

Smith, J. M. (1981). Macroevolution. *Nature*, 289, 13–14.

Stanley, E. H. & Doyle, M. W. (2003). Trading off: the ecological effects of dam removal. *Frontiers in Ecology and the Environment*, 1, 15–22.

Sterling, K. A., Reed, D. H., Noonan, B. P. & Warren, M. L. (2012). Genetic effects of habitat fragmentation and population isolation on *Etheostoma raneyi* (Percidae). *Conservation Genetics*, 13, 859–872.

Storfer, A., Murphy, M. A., Spear, S. F., Holderegger, R. & Waits, L. P. (2010). Landscape genetics: where are we now? *Molecular Ecology*, 19, 3496–3514.

Stuart, I. G. & Mallen-Cooper, M. (1999). An assessment of the effectiveness of a vertical-slot fishway for non-salmonid fish at a tidal barrier on a large tropical/subtropical river. *Regulated Rivers: Research and Management*, 15, 575–590.

Taylor, P. D., Fahrig, L. & Merriam, G. (1993). Connectivity is a vital element of landscape structure. *Oikos*, 68, 571–573.

Tibbets, C. A. & Dowling, T. E. (1996). Effects of intrinsic and extrinsic factors on population fragmentation in three species of North American minnows (Teleostei: Cyprinidae). *Evolution*, 50, 1280–1292.

Tonn, W. M. & Magnuson, J. J. (1982). Patterns in the species composition and richness of fish assemblages in northern Wisconsin lakes. *Ecology*, 63, 1149–1166.

Tonn, W. M., Magnuson, J. J., Rask, M. & Toivonen, J. (1990). Intercontinental comparison of small-lake fish assemblages – the balance between local and regional processes. *American Naturalist*, 136, 345–375.

Turner, T. F., Osborne, M. J., Moyer, G. R., Benavides, M. A. & Alo, D. (2006). Life history and environmental variation interact to determine effective population to census size ratio. *Proceedings of the Royal Society – Biological Sciences*, 273, 3065–3073.

Turner, T. F. & Robison, H. W. (2006). Genetic diversity of the Caddo Madtom, *Noturus taylori*, with comments on factors that promote genetic divergence in fishes endemic to the Ouachita highlands. *Southwestern Naturalist*, 51, 338–345.

Turner, T. F. & Trexler, J. C. (1998). Ecological and historical associations of gene flow in darters (Teleostei: Percidae). *Evolution*, 52, 1781–1801.

Urban, M. C. & Skelly, D. K. (2006). Evolving metacommunities: toward and evolutionary perspective on metacommunities. *Ecology*, 87, 1616–1626.

Vannote, R. L., Minshall, G. W., Cummins, K. W., Sedell, J. R. & Cushing, C. E. (1980). The river continuum concept. *Canadian Journal of Fisheries and Aquatic Sciences*, 37, 130–137.

Waits, E. R., Bagley, M. J., Blum, M. J., McCormick, F. H. & Lazorchak, J. M. (2008). Source–sink dynamics sustain central stonerollers (*Campostoma anomalum*) in a heavily urbanized catchment. *Freshwater Biology*, 53, 2061–2075.

Walters, D. M., Zuellig, R. E., Crockett, H. J., *et al.* (2014). Barriers impeded upstream spawning migration of flathead chub. *Transactions of the American Fisheries Society*, 143, 17–25.

Wang, J. L., & Whitlock, M. C. (2003). Estimating effective population size and migration rates from genetic samples over space and time. *Genetics*, 163, 429–446.

Waples, R. S. (1987). A multispecies approach to the analysis of gene flow in marine shore fishes. *Evolution*, 41, 385–400.

Waples, R. S. (1989). Temporal variation in allele frequencies – testing the right hypothesis. *Evolution*, 43, 1236–1251.

Ward, J. V., Tockner, K. & Schiemer, F. (1999). Biodiversity of floodplain river ecosystems: ecotones and connectivity. *Regulated Rivers: Research and Management*, 15, 125–139.

Wares, J. P. (2002). Community genetics in the Northwestern Atlantic intertidal. *Molecular Ecology*, 11, 1131–1144.

Wares, J. P. & Pringle, J. M. (2008). Drift by drift: effective population size is limited by advection. *BMC Evolutionary Biology*, 8, 235.

Warren, M. J. Jr & Pardew, M. G. (1998). Road crossings as barriers to small-stream fish movement. *Transactions of the American Fisheries Society*, 127, 637–644.

Welcomme, R. L. (1979). *The fisheries Ecology of floodplain Rivers*. London: Longman.

Whiteley, A. R., Spruell, P. & Allendorf, F. W. (2004). Ecological and life history characteristics predict population genetic divergence of two salmonids in the same landscape. *Molecular Ecology*, 13, 3675–3688.

Whitlock, M. C. & McCauley, D. E. (1990). Some population genetic consequences of colony formation and extinction – genetic correlations within founding groups. *Evolution*, 44, 1717–1724.

Wiens, J. A. (2002). Riverine landscapes: taking landscape ecology into the water. *Freshwater Biology*, 47, 501–515.

Wilson, G. A. & Rannala, B. (2003). Bayesian inference of recent migration rates using multilocus genotypes. *Genetics* 163, 1177–1191.

Winemiller, K. O., Flecker, A. S. & Hoeinghaus, D. J. (2010). Patch dynamics and environmental heterogeneity in lotic ecosystems. *Journal of the North American Benthological Society*, 29, 84–99.

Winston, M. R., Taylor, C. M. & Pigg, J. (1991). Upstream extirpation of four minnow species due to damming of a prairie stream. *Transactions of the American Fisheries Society*, 120, 98–105.

Wofford, J. E. B., Gresswell, R. E. & Banks, M. A. (2005). Influence of barriers to movement on within-watershed genetic variation of coastal cutthroat trout. *Ecological Applications*, 15, 628–637.

Woodford, D. J. & McIntosh, A. R. (2010). Evidence of source-sink metapopulations in a vulnerable native galaxiid fish driven by introduced trout. *Ecological Applications*, 20, 967–977.

World Commission on Dams. (2000). *Dams and Development: A New Framework for Decision-Making*. London: EarthScan Publications.

Wright, S. (1939). The distribution of self-sterility alleles in populations. *Genetics*, 24, 538–552.

Wright, S. (1978). *Evolution and the Genetics of Populations, Volume 4: Variability Within and Among Natural Populations (Evolution and the Genetics of Populations)*. Chicago, IL: University of Chicago Press.

Wunderlich, R. C., Winter, B. D. & Meyer, J. H. (1994). Restoration of the Elwah River ecosystem. *Fisheries*, 19, 11–19.

11

Conservation of migratory fishes in freshwater ecosystems

PETER B. MCINTYRE, CATHERINE REIDY LIERMANN,
EVAN CHILDRESS, ELLEN J. HAMANN, J. DEREK HOGAN,
STEPHANIE R. JANUCHOWSKI-HARTLEY,
AARON A. KONING, THOMAS M. NEESON, DANIEL L. OELE
AND BRENDA M. PRACHEIL

Migratory fishes are natural wonders. For many people, the term migratory fish evokes images of salmon audaciously jumping at waterfalls as they return to their own riverine birthplace to spawn after years of growth in the ocean, but freshwater fishes actually show a broad spectrum of migration strategies. Migratory fishes include small species – three-spined sticklebacks that spawn in coastal streams around the northern Pacific and gobies that move from the ocean into tropical island streams by climbing waterfalls (McDowall, 1988) – as well as some of the largest freshwater fishes in the world, such as the Mekong dog-eating catfish and the Chinese paddlefish (Stone, 2007). Aside from migratory habits, these species have few shared characteristics; they encompass numerous evolutionary lineages, enormous differences in life history, and every possible direction and distance of migration. Biologists treat migratory freshwater fishes as a functional group because their life-history strategy revolves around long-distance movement between ecosystems in a perilous quest to take advantage of both high-quality breeding sites and bountiful feeding areas. As humans have physically blocked fish migrations, degraded breeding and feeding grounds and relentlessly harvested migrants for their flesh and roe, many populations have declined or been extirpated. This chapter will provide an overview of fundamental and applied research that is helping to guide efforts to conserve migratory freshwater fishes.

For practical purposes, we define migratory behaviour as the synchronized movement of a substantial proportion of a population between

Conservation of Freshwater Fishes, eds G. P. Closs, M. Krkosek and J. D. Olden. Published by Cambridge University Press. © Cambridge University Press 2016.

distinct habitats, which is repeated through time within or across generations. Modern definitions of fish migrations typically recognise both the adaptive benefits of migrating and individual variation in executing the general strategy (see McDowall, 1988; Lucas & Baras, 2001). Not every individual must move, the timing may vary somewhat from year to year, and the motive for migrating may include seeking refuge from harsh conditions in addition to breeding and feeding. Nonetheless, in most cases, migration is critical to individual fitness and population persistence because it enables specialised use of different habitats for growth and reproduction. Where their migration routes are blocked or key habitats are lost, migratory fishes often suffer rapid and catastrophic losses.

Human appropriation and degradation of the Earth's freshwater ecosystems (Vörösmarty et al., 2010; Carpenter et al., 2011) have transformed this reliance on multiple habitats into a detriment for many migratory fishes. Historically, travelling spectacular distances along the same routes swum by their ancestors was sufficiently advantageous to give rise to migrations of over 1400 km among salmon in the Snake River of the USA, whitefish in the Omul River of Russia, sturgeon in Europe's Danube River, and various catfishes in Asia's Mekong River and South America's Amazon River (McKeown, 1984; Lucas & Baras, 2001). Today, these extreme migrants are especially vulnerable to changes in environmental conditions specifically because their long-distance movements expose them to many risks. Barriers and habitat loss have reduced many migratory fish populations, and even entire species. Emblematic cases include the sturgeons and paddlefishes – an ancient lineage of which all extant species appear on the IUCN's Red List of endangered species (Pikitch et al., 2005) – as well as numerous salmon populations along the coasts of North America and Europe that now depend on stringent protections to avoid extirpation (Parrish et al., 1998; Ruckelshaus et al., 2002). The growing literature on declines of migratory fishes across the world suggests that a conservation crisis point is at hand, meriting urgent attention (Chapter 1).

The thesis of this chapter is that conserving migratory fishes is a global challenge that requires balancing numerous biological, social and economic factors. This scenario is not unique to fishes (Wilcove & Wikelski, 2008), but river networks impose special constraints on conservation approaches. Unfortunately, scientific research has long focused on just a handful of migratory fishes. Pioneering books on fish migrations dealt primarily with salmon, eels, shad and sturgeon (e.g. Meck, 1916; Hasler, 1966; Harden Jones, 1968), and taxonomic and geographic coverage have expanded only in the last few decades (e.g. McKeown, 1984; McDowall, 1988; Lucas & Baras, 2001). Similarly, conservation and management efforts have focused disproportionately on a modest number of high-profile migratory taxa such as salmon and sturgeons (e.g. Ruckelshaus *et al.*, 2002; Landsman *et al.*, 2011), while far more species remain obscure despite their imperilment. Economically valuable migrants can sometimes serve as umbrella species whose conservation will also improve the prospects for other migratory fishes (e.g. Limburg & Waldman, 2009), but many regions are losing migrations quietly without the benefit of such rallying-point species.

We will begin with an overview of the range of migratory strategies and the diversity of freshwater fishes that migrate. With this broad perspective in mind, we will discuss the conservation predicament created by losses of both key habitats and the connections among them. We will then explore some of the ramifications of declining migrations for metapopulation dynamics, ecosystem processes and human fisheries. Finally, we will discuss some key practical considerations that complicate conservation efforts, including cost-efficiency of restoring connectivity, societal demand for new dams and roads that create barriers, risks of species invasions and contaminant transport, and the tendency to focus on economically valuable species. An exhaustive review of any one of these topics is beyond the scope of this chapter; our goal is to introduce a variety of viewpoints on the challenge of conserving migratory freshwater fishes.

11.1 ECOLOGICAL CLASSES OF MIGRATORY FRESHWATER FISHES

There are four major classes of migration patterns among fishes that depend on freshwater ecosystems for at least part of their life cycle (Myers, 1949; McDowall, 1988). *Potamodromy* describes migrations that occur entirely within freshwaters. Most potamodromous species spend part of the year in either large rivers or lakes, and their migrations take them upstream into tributaries (or occasionally between zones of the same lake). The other three forms of freshwater migratory behaviour are classified as *diadromy* (Figure 11.1) because they involve switching between freshwater and saltwater environments, which demands a range of physiological adaptations. Diadromous species that breed in saltwater are *catadromous*; the young travel to freshwater ecosystems to grow and mature before returning to saltwater to complete their life

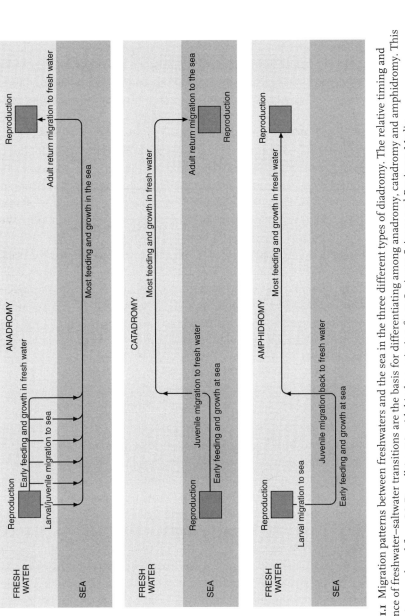

Figure 11.1 Migration patterns between freshwaters and the sea in the three different types of diadromy. The relative timing and importance of freshwater–saltwater transitions are the basis for differentiating among anadromy, catadromy and amphidromy. This diagram is reproduced from McDowall (1997), with kind permission from Springer Science and Business Media.

cycle. Among species that breed in freshwaters, some diadromous fishes use marine waters as their primary growth environment (*anadromy*), while others remain in saltwater only briefly prior to returning to fresh-water (*amphidromy*).

A second major dimension of life-history variation among fresh-water migratory fishes regards the potential number of breeding bouts in a fish's lifetime, known as *parity*. A large majority of migratory species are capable of breeding repeatedly (*iteroparity*), including almost all potamodromous fishes. In contrast, some migratory species invest all of their resources in one large reproductive bout that is inevitably followed by death (*semelparity*). Semelparous migrants include all lampreys and anguillid eels as well as some salmonids, clupeids (shad and herring) and sticklebacks. Despite the fame of Pacific salmon for mass mortality after spawning, the majority of salmonid species are at least partially iteroparous (Quinn, 2005).

Convergent evolution of each of the four migration strategies and two parity states in many lineages attests to the advantages conferred by stereotyped migrations (McDowall, 1988, 1997). For most potamodromous and diadromous fishes, the underlying motivation for large-scale migrations is to take advantage of prime breeding and feeding areas that are geographically distinct, much the same as for migratory birds and mammals (Baker, 1978). Some fish migrations are also used to seek refuge from inhospitable environmental conditions (Lucas & Baras, 2001), although these movements often align with feeding or breeding cycles as well. In all anadromous fishes and most potamodromous species, feeding and growth occur in a relatively large ecosystem with high productivity (ocean, lake or large river), and breeding is concentrated in smaller, less-productive streams and rivers. In contrast, catadromous species achieve most of their growth in a small ecosystem (stream, lake) before returning to the ocean to spawn. Amphidromous species both breed and grow in freshwater environments, but retain a marine larval phase that enables dispersal and colonisation (McDowall, 1988).

Strong latitudinal patterns in the frequency of each diadromous strategy underscore the match of the growth phase to the more productive environment; catadromous and amphidromous fishes predominate in the tropics where rivers and lakes are more productive than most marine habitats, while anadromous species take advantage of the high productivity of oceans compared to rivers and lakes at higher latitudes (Gross *et al.*, 1988). At the same time, the less-productive environment is thought to feature less risk of predation for young fishes, making it appealing as a nursery area despite lower potential growth rates (McDowall, 1988; Lucas & Baras, 2001). The importance of ecosystem productivity and predation presumably also govern the direction of potamodromous fish migrations, although the most advantageous direction to migrate likely depends on both dietary and reproductive requirements. For example, most of the large minnows and alestid fishes in African lakes use the lacustrine environment for growth but move into tributaries to spawn (Leveque, 1997). Intriguingly, large tropical rivers often contain some species that move upstream to breed in headwaters and others that move downstream to breed in floodplains and deltas – these fish may literally migrate past each other (Lucas & Baras, 2001).

The ecology of potamodromous fishes has received far less attention than that of diadromous species, despite the fact that the majority of the world's migratory freshwater fishes are potamodromous. The total number of diadromous species worldwide is estimated to be around 250 (McDowall, 1997), which is comparable to the sum of potamodromous species documented in case studies of just the Mekong River (165 species; Baran & Myschowoda, 2009), Canada (34 species; Lucas & Baras, 2001) and South American rivers (at least 67 species; Carolsfeld *et al.*, 2004). Roughly one-third of Canadian freshwater fish species are potamodromous (Lucas & Baras, 2001) and most major tropical rivers include at least a few dozen species (e.g. Welcomme, 1985; Carolsfeld *et al.*, 2004). Moreover, potamodromous species dominate fish biomass in many tropical (e.g. Carolsfeld *et al.*, 2004; McIntyre *et al.*, 2008; Ziv *et al.*, 2012) and temperate (Klingler *et al.*, 2003; Cooke *et al.*, 2005) rivers during their migrations, suggesting that they play essential roles in ecosystems.

The conservation of migratory freshwater fishes is a global challenge. Diadromous fishes use (or once used) virtually every coastal river basin on Earth (McDowall, 1988), and potamodromous species can be found in most inland river networks and lakes as well as many coastal drainages. Unfortunately, there exists neither a comprehensive map of diversity patterns among the world's migratory freshwater fishes nor a systematic assessment of their endangerment. These undertakings would be of great value for research and conservation, but the diversity patterns summarised above provide context for understanding why migratory fishes are so often imperilled.

11.2 TRIPLE JEOPARDY: DEPENDENCE ON MULTIPLE HABITATS AND THEIR INTERCONNECTIONS

Animals have evolved complex life histories to allow utilisation of multiple environments that collectively maximise lifetime reproductive success by taking advantage of the opportunities (and avoiding the costs) afforded by each place (Wilbur, 1980). In some cases, the spatial dimension of this transition is trivial; for instance, most insects and amphibians undergo a radical metamorphosis from larval to adult morphotypes, but the spatial distance between their respective habitats is often small. Nonetheless, such transitions expose these species to a form of double jeopardy; threats in either of their two critical habitats can lead to extirpation, even if the other habitat remains pristine (e.g. Todd *et al.*, 2011). Migratory animals face this same double jeopardy, but also a third major challenge (i.e. triple jeopardy) due to the necessity of suitable conditions in the migration corridors connecting their often-distant feeding and breeding habitats (Wilcove & Wikelski, 2008). Compared to aerial migrations of birds or insects, migratory mammals and fishes experience an especially stringent form of triple jeopardy in that their pathway itself is often physically blocked. Thus, conserving fish migrations requires ensuring appropriate conditions in the habitats used as breeding grounds and feeding areas, but also protecting migratory corridors that are not obstructed by dams, road crossings or other barriers (Figure 11.2).

The importance of recognising this triple jeopardy scenario is underscored by cases where migratory freshwater fishes have declined due to changes in just one of the three essential habitats. For instance, restored spawning grounds in California's Shasta River attract anadromous coho salmon, but low flow and high temperatures lead to mortality of young fish before they are ready to migrate to the ocean (Jeffres & Moyle, 2012). That losses of spawning or nursery grounds will compromise population viability is obvious, but degradation of feeding habitats can be just as problematic. Long-term variation in salmon populations in the northern Pacific Ocean indicate that climatic control of marine productivity drives fluctuations in anadromous fish stocks (Finney *et al.*, 2002; Irvine & Fukuwaka, 2011), thereby illustrating both the importance of feeding areas and the necessity of accounting for human impacts on the marine realm when seeking to conserve diadromous species. Striped bass declines in Chesapeake Bay through the 1970s illustrate a different type of anthropogenic influence on feeding areas; this downturn was driven

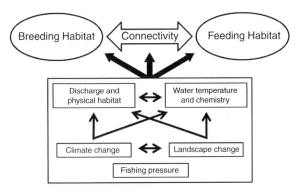

Figure 11.2 Conceptual model of the triple jeopardy faced by migratory fishes. Population declines can be driven by loss or degradation of breeding habitat, feeding habitat, or the migration corridor that connects breeding and feeding habitats. Major classes of anthropogenic change, and the resulting stressor pathways, leading to habitat loss or degradation are indicated in the lower box.

by overfishing in the marine waters where they spend years growing to adult size (Richards & Rago, 1999).

Both breeding and feeding habitats in rivers have been compromised by widespread alterations of substrate, flow, temperature and chemistry (Vörösmarty *et al.*, 2010; Carpenter *et al.*, 2011) that lead to mismatches with migratory life-history strategies that have evolved to suit the natural properties of rivers. In particular, critical life-history events such as spawning and larval rearing are generally timed to match patterns of discharge and temperature (Poff *et al.*, 1997; Lytle & Poff, 2004). For example, the American paddlefish relies on temperature, flow and turbidity cues indicative of spring high-flows to ensure that river conditions are conducive to larval growth (Jennings & Zigler, 2009). Alteration of these spawning cues has had demonstrable negative effects on paddlefish recruitment (Pracheil *et al.*, 2009). Moreover, when migrating upstream, many species depend on appropriate channel complexity and hydrology to create low-energy corridors for efficient swimming, but human activities can eliminate these favourable conditions (McElroy *et al.*, 2012).

Intensive alteration of the landscape associated with agriculture, urbanisation and forestry leads to degradation of breeding and feeding habitats through physical changes (e.g. substrate and hydrology) as well as shifts in temperature and chemistry that are not directly visible yet can have profound effects (Allan & Castillo, 2007). Migratory fishes often travel long distances to reach productive feeding areas, but the detrital, algal, invertebrate and prey fish resources that they seek may have been diminished by changing substrate, flow and chemistry. In breeding habitats, substrate composition is critical for the spawning and early life history of many migratory fishes. For example, lamprey larvae bury themselves in soft, fine sediments on stream bottoms, and changes in sediment stability or composition can lead to larval mortality and eventually population declines (Renaud, 1997; Smith *et al.*, 2012). Pacific salmon require a different kind of substrate; their large eggs must be buried in gravel to protect them from predators and ensure adequate water flow for oxygenation (Quinn, 2005). Land-use changes can dramatically elevate watershed inputs of sediment that lead to burial of coarse substrates in silt (Allan & Castillo, 2007), while dams starve downstream areas of sediment as they trap particles within impoundments (Vörösmarty *et al.*, 2003). Either of these changes can disrupt the availability of food resources for young fishes, which must feed intensively to survive and grow. Human activities may also boost the exposure of fry to predators through species introductions and habitat alteration. For instance, predation by non-native trout on young humpback chubs – a native migratory species at risk of extinction – within the Grand Canyon of the Colorado River is facilitated by clear-water conditions below dams (Marsh & Douglas, 1997).

Although maintaining the integrity of both breeding and feeding grounds is essential, so too is ensuring unobstructed migration corridors connecting these habitats. The general consequences of dams and fragmentation are discussed in detail elsewhere in this book (Chapters 4 and 10); hence, we will only outline a few key issues relevant to migratory fishes here. Dams, weirs and road culverts that act as barriers to fish migrations are pervasive in the world's rivers (Reidy Liermann *et al.*, 2012). Millions of dams block fish movement through river networks (Lehner *et al.*, 2011), yet culverts and other structures associated with road crossings are many times more numerous. For instance, there are 38 times as many road crossings as dams in tributaries of the North American Great Lakes (Januchowski-Hartley *et al.*, 2013), and a substantial proportion of these road crossings partially or entirely block fish movement (Januchowski-Hartley *et al.*, 2014). Yet, even as dams and culverts are gaining attention for their detrimental effects on stream-resident and migratory fish species in developed countries, new dams, weirs and road crossings continue to proliferate in poorer countries to meet development and economic needs

(Kareiva, 2012; Laurance & Balmford, 2013). Unfortunately, even progressive discussions of the consequences of infrastructure expansion sometimes fail to consider an aquatic perspective (e.g. Laurance & Balmford, 2013).

The impact of dams and culverts on the downstream migration of fry and adults can be just as damaging as blocking upstream movement. Small dams and culverts probably pose few challenges for moving downstream during high flows, although they may become problematic when discharge is low (Benstead *et al.*, 1999). However, passing over the spillway or through the turbines of a large dam can impart severe injury or death due to high-velocity contact or acute turbulence (Coutant & Whitney, 2000). Safe passage downstream is further complicated by variation among species and life stages in their position and orientation in the water column (Coutant & Whitney, 2000; Williams *et al.*, 2012). Screening systems (Williams, 2008) and surface bypass routes (Johnson & Dauble, 2006) have been developed to divert juvenile salmonids away from the turbines of dams in the Columbia River system (USA), but these methods may not be broadly applicable to other diadromous or potamodromous fishes (Cada, 1990; Coutant & Whitney, 2000). Sensor-packed 'robo-fish' are now providing insights into the experience of larger fishes passing over dams or through hydroelectric turbines; each route often leads to immense instantaneous accelerations (> 95 times gravity) due to direct physical contact with the spillway (100% incidence) or turbine blade (62% incidence; Duncan, 2013). These examples and many others (Chapter 4) illustrate the importance of designing safe passage methods for fishes of all sizes that face dams and road culverts while migrating up- and downstream.

Dams, weirs and road culverts can also exert indirect effects on migratory fishes by changing ecosystem structure and functioning. Large dams alter thermal regimes, hydraulics and fluxes of sediment and nutrients for tens of kilometres downstream (Ward & Stanford, 1983; Poff *et al.*, 2007). In turn, these abiotic alterations drive changes in substrate and physico-chemical characteristics (Ligon *et al.*, 1995), as well as lateral connectivity between the channel and floodplains (Bayley, 1995). Attendant ecological changes may include decreased recruitment of riparian vegetation (Scott *et al.*, 1996), altered fish community structure (Bain *et al.*, 1988; Perkin & Gido, 2012) and decreased community resilience to disturbances (Sedell *et al.*, 1990). Moreover, large dams disrupt the very environmental signals (e.g. discharge, temperature) that many fishes use as migratory cues (Lehner *et al.*, 2011). Overall, migratory fishes that must rely on fragmented river systems experience wildly different conditions than those typical of the free-flowing rivers in which their migratory life histories evolved.

In addition to dependence on multiple habitats and their connections, migratory species often cross political boundaries. As a result of such trans-jurisdictional movements, it is challenging to incorporate the entire geography and life history of migratory species into management decisions (e.g. Pracheil *et al.*, 2012). Where fish move between areas overseen by multiple governing bodies that differ in their priorities and constituencies, they will often be subject to contrasting regulations (e.g. catch limits) and practices (e.g. habitat restoration). The difficulty of establishing, maintaining and funding coherent management strategies is likely to increase dramatically with the number of jurisdictions involved. Thus, institutions that operate at the actual scale of migrations could improve conservation outcomes by aligning the actions of disparate players. For example, the Great Lakes Fishery Commission, which coordinates federal and state governments across the North American Great Lakes, has helped unify the management of migratory lake sturgeon (Welsh, 2004).

The summary above has merely scratched the surface with regard to the ways and places in which loss of breeding habitat, productive feeding grounds and riverine connections have negatively affected freshwater migratory fishes. The salient point is that degradation of any one of these three key habitats can be sufficient in itself to eliminate a migratory population. Accounting for this triple jeopardy scenario can help to ensure that interventions target the habitats that most limit long-term prospects for migratory fish populations.

11.3 ECOLOGICAL AND SOCIETAL IMPLICATIONS OF DECLINING FRESHWATER FISH MIGRATIONS

The ramifications of population declines of migratory fishes extend well beyond their local ecosystem. The reproductive success of individual migrants contributes not only to their own population, but also to persistence of their species as a whole. As the diversity and population sizes of migratory fishes are diminished, ripple effects are often felt across entire ecosystems based on shifts in the rates of fundamental processes like primary production and organic decomposition. Beyond the river channel, human populations often depend on migratory fishes for sustenance, commerce and recreation. Reducing or eliminating fish migration runs can rob communities of a key food supply as well as an integral dimension of their cultural identity. We now address each of these implications of declining migrations in turn.

11.3.1 Long-term conservation in a metapopulation context

Practical constraints often require that conservation efforts focus on just a subset of a species' range, but metapopulation theory suggests that these localised efforts can have long-term benefits for the species as a whole (Hanski, 1999). A metapopulation is a network of patchily distributed populations that are connected via dispersal or migration of individuals – a scenario that could apply to most migratory fish species. The exchange of genes and individuals stabilises the dynamics and enhances the persistence of both individual populations (Brown & Kodric-Brown, 1977) and the metapopulation as a whole (Hanski, 1999), partly by reducing the deleterious effects of inbreeding, mutation load and genetic drift. Conversely, if any one population in the network dwindles or is extirpated, loss of its genetic diversity can reduce the effective size of the entire metapopulation, thereby increasing genetic drift and risk of extinction (Frankham, 1995). These and other aspects of fish metapopulation dynamics are more fully discussed in Chapter 10.

Metapopulation dynamics are likely to characterise many migratory fishes because migrations divide the life cycle into phases of spatial segregation and potential mixing among populations. Among diadromous species, catadromous fishes often have the opportunity to mix while spawning in marine waters, whereas riverine breeding areas used by anadromous and amphidromous species are isolated (although straying occurs; e.g. Hamann & Kennedy, 2012). A similar dichotomy exists among potamodromous fishes; some species migrate to many distinct feeding areas in headwaters but mix freely while spawning downstream in floodplains (e.g. prochilodontids in South American rivers; Collins et al., 2013), while others share feeding grounds in lakes or large river channels but scatter into separate tributaries to breed (e.g. migrants in the North American Great Lakes; Landsman et al., 2011). We might speculate that genetic exchange among populations that converge upon shared breeding areas would enhance metapopulation persistence, but also diminish the likelihood of intraspecific differentiation like that observed in some salmonids (Quinn, 2005) and clupeids (Leggett & Carscadden, 1978). In any case, comparisons of metapopulation structure, genetic exchange and breeding site fidelity across diadromous and potamodromous species would clarify the degree to which conservation strategies need to be tailored to particular migration patterns.

In keeping with metapopulation theory, sockeye salmon fisheries in Alaska show regional stability despite interannual variability in the size of each constituent population (Schindler *et al.*, 2010). The key insight is that while the number of breeders in each river fluctuates considerably, the independence of these dynamics in space and time effectively stabilises the aggregate catch. This model system exemplifies the benefits of conserving a portfolio of populations where mixing and rescue processes can occur via natural connections. Conversely, metapopulation models and research on resident stream fishes underscore the ease of losing ecological connectivity in river systems due to inherent limitations on dispersal pathways in dendritic systems (Gotelli & Taylor, 1999; Fagan, 2002). By extension, the benefits of restoring feeding and breeding areas – or alleviating barriers to movement between them – are expected to be substantial for metapopulations of migratory fishes.

11.3.2 Ecosystem consequences of declining migrations

Looking beyond fishes themselves, ecosystem dynamics often depend upon influxes of energy and nutrients delivered by fish migrations. Indeed, anadromous fishes have been proposed as keystone species for terrestrial vertebrates (Willson & Halupka, 1995). A growing body of research is elucidating the many pathways by which fish migrations can drive ecosystem processes via material subsidies, disturbance, food web interactions and effects on ecosystem processes (Flecker *et al.*, 2010). This work demonstrates that loss of migratory fishes can greatly alter the character of aquatic ecosystems, thereby adding another motive for conservation efforts.

Studies on the ecosystem roles of migratory fish originated with and continue to focus particularly on anadromous salmon, which achieve the majority of their growth in the marine environment but generally die after spawning in freshwater streams and lakes. During spawning they have two major effects on stream ecosystems: disturbance of the substrate during nest-building (Moore *et al.*, 2004) and addition of nutrients from excretion and decomposing carcasses and eggs (Naiman *et al.*, 2002). Fish-derived nutrients are incorporated by algae (Wipfli *et al.*, 1998), invertebrates (Hicks *et al.*, 2005) and resident fishes (Rinella *et al.*, 2012), and are even transferred to terrestrial vegetation (Bilby *et al.*,

2003). This subsidy can have a positive feedback on salmon fisheries as juveniles benefit from the nutrients provided by spawning adults (Bilby et al., 2003; Rinella et al., 2012). These effects are context-dependent; nutrient inputs are most important when substrate size is large and therefore less prone to disturbance, when run size is large relative to ecosystem size, and when background nutrient levels are low (Janetski et al., 2009; Flecker et al., 2010).

There is growing evidence that excretion by live adults is also an important pathway for subsidies that complements inputs from decaying salmon carcasses (Tiegs et al., 2011). This finding raises the possibility that iteroparous species also provide substantial nutrient subsidies despite high spawner survival rates. For instance, suckers are iteroparous, potamodromous species that are abundant across much of North America (Cooke et al., 2005), and their large migrations can constitute a major component of stream nutrient dynamics (Hall, 1972; Flecker et al., 2010; Childress et al., 2014). Similarly, studies on anadromous clupeids also show evidence of nutrient subsidies (Browder & Garman, 1994; Walters et al., 2009; Hanson et al., 2010). Ecosystem responses to nutrient transfers by iteroparous fishes are poorly documented, but the example provided by semelparous salmon provides ample motivation for further research.

A broad range of consumers benefit from fish migrations either through direct consumption of migrants or indirect dependence on the productivity arising from ecosystem subsidies. Bald eagles have greater reproductive success when nesting near rivers with spawning migrations (Gerrard et al., 1975). Bear populations, which transfer salmon nutrients far into terrestrial ecosystems (Hilderbrand et al., 1999), are sensitive to decreases in salmon availability (Levi et al., 2012). Even the abundance and diversity of songbirds increase in the presence of salmon migrations as they capitalise on the salmon subsidy indirectly via increases in the emergence of aquatic insects (Christie & Reimchen, 2008). In large neotropical rivers, migrations of juvenile fish from productive nursery habitats to oligotrophic streams help to support piscivorous fishes (Winemiller & Jepsen, 2004).

In addition to providing nutrients and energy, migratory fish can alter material processing. *Prochilodus* and their relatives are major iteroparous migrants in most river systems of South America, and they play a key role in ecosystem dynamics by fastidiously consuming sediment from the bottom. Their feeding alters algal and invertebrate assemblages on rocky substrates (Flecker, 1996), dramatically affects several dimensions of carbon cycling (Taylor *et al.*, 2006) and contributes substantially to nitrogen and phosphorus recycling (McIntyre *et al.*, 2007). On the tropical island of Puerto Rico, dams frequently block access to streams for migratory fishes and shrimps, which leads to substantially higher biomass of algae, detritus and invertebrates upstream (Greathouse *et al.*, 2006). These examples underscore the strength and variety of ecosystem-level effects of losing migratory fishes (Flecker *et al.*, 2010).

Ultimately, the conservation of major freshwater fish migrations should be seen as a necessary component of protecting the integrity of aquatic ecosystems, rather than being a purely fish-centric effort. An extensive web of species depends on the material subsidies and activities of migratory fishes, and this has been well demonstrated for various anadromous species as well as a few potamodromous counterparts. Coverage of catadromous species from an ecosystem perspective is distinctly lacking, which raises interesting questions about whether certain marine ecosystems are meaningfully fertilised by the spawning activities of mullets or anguillid eels.

11.3.3 Human consequences of lost migratory fisheries

Freshwater fisheries are gaining increasing recognition as an anthropocentric reason to conserve migratory fishes (Beard *et al.*, 2011). Migratory species constitute a large proportion of inland fish catch in many regions of the world, the loss of which can have significant nutritional, economic and social consequences. Reliable global data on the magnitude of fishery catches is lacking for potamodromous and amphidromous species, but 360,000 T of diadromous fishes were caught in 2010 (~5% of total inland catches; FAO, 2012). Riverine catches of diadromous species represent the bulk of reported harvest for many temperate nations: > 80% each for the USA, Canada and Russia. Diadromous catches are lower in tropical nations where rivers are the more productive habitat, peaking at ~10% for Bangladesh and ~7% for India. Although limited data quality precludes any strong conclusions, tropical diadromous species – most of which are catadromous or amphidromous – may simply not produce enough biomass from any single river to support a major fishery. Low catadromous fish catches could also reflect historical overfishing. In the case of anguillid eels, for example, culinary traditions in many cultures motivate continued fishing pressure even as eel populations become increasingly endangered (Crook, 2010).

Available data from select inland fisheries indicate that potamodromous species comprise a large proportion of catches. For instance, 40–70% of the fish biomass harvested from the Mekong River basin is believed to be migratory species (Barlow *et al.*, 2008; Baran & Myschowoda, 2009), most of which are potamodromous. With the Mekong's annual fishery valued between $4.3 and $7.8 billion on retail markets (Hortle, 2009), the loss of these migrations would be economically devastating. Moreover, river fisheries provide up to 80% of animal protein consumed by people in the lower Mekong; hence, declining migrations could create a crisis in food security. These Mekong data highlight the high stakes involved in migratory fish conservation even from a strictly human perspective, and additional examples exist throughout the world (Limburg & Waldman, 2009; Welcomme *et al.*, 2010; Beard *et al.*, 2011).

Despite their dire consequences for migratory fishes (e.g. Agostinho *et al.*, 2008), new dams are under development for the mainstem and large tributaries of rivers that support many of the world's greatest freshwater fisheries (Kareiva, 2012). Even as dam removals begin to outnumber dam construction projects in North America and Europe, many additional large dams are proposed for rivers in Asia, Africa and South America where migratory species represent a major portion of the fish fauna (e.g. Finer & Jenkins, 2012; Ziv *et al.*, 2012). For example, channel obstruction by the Three Gorges and Gezhouba dams on the Yangtze River in China is blamed for endangerment of the anadromous Reeve's shad, Chinese sturgeon, Yangtze sturgeon and the now-extinct Chinese paddlefish (Dudgeon, 2010). Estimates from a variety of tropical rivers suggest that the contribution of potamodromous species to fisheries is disproportionately high compared to their richness (Welcomme, 1985; Carolsfeld *et al.*, 2004; Ziv *et al.*, 2012). Unfortunately, reliance of local people on fish migrations as a low-input, low-cost source of protein sometimes seems to be overlooked in evaluating proposals for new dams.

Reservoir fisheries are often cited as compensation for the adverse impacts of dams on riverine fishes (e.g. Costa-Pierce, 1997), but losses of migratory fisheries caused by dams may rival or exceed the reservoir yields that have been gained. For example, average diadromous catches reported by FAO for the last 10 years (FAO, 2012) exceed aggregate yields estimated for reservoirs in India, Russia and many other Asian and European countries, signifying that all of these reservoir fisheries together could not compensate for loss of diadromous catches (Reidy

Liermann *et al.*, in preparation). Unfortunately, our ability to quantify the human consequences of lost migratory fisheries is limited by lack of high-resolution catch data, especially in developing nations (Welcomme *et al.*, 2010). Despite recent calls for economic and nutritional valuation of inland fisheries (Beard *et al.*, 2011), lack of worldwide mapping and valuation of migratory fisheries precludes understanding of the full ramifications of their loss.

11.4 SOCIETAL DIMENSIONS OF CONSERVATION

Keeping in mind the diverse rationales for conserving migratory fishes, we now turn to the societal considerations that influence conservation actions. Some of the most complex decisions involve weighing the costs and benefits of dams and road crossings, which directly serve human interests. Spatial prioritisation methods (see Moilanen *et al.*, 2009) can be used to balance conservation and societal needs when selecting dam and road crossing remediation projects as well as the placement of new structures. In contemplating barrier removals, decision-makers must also account for legitimate ecological costs arising from species invasions and chemical pollutants. Finally, conservation approaches developed for one taxon or environmental setting are often applied much more broadly, sometimes with disappointing or even negative results. We conclude this section by discussing the importance of tailoring conservation efforts to the local migratory fauna.

11.4.1 Managing dams – old, new and impending

After millennia of dam construction to serve societal needs (Chapter 4), two contrasting outlooks have emerged among nations of the world, each having distinct implications for conserving migratory fishes. In countries that had built many large dams by the mid-twentieth century, there is an accelerating trend towards decommissioning and removing dams that is driven by financial, safety and environmental concerns (Stanley & Doyle, 2003). Many of these dams have reached an age at which maintenance costs rival or outweigh costs of removal (Doyle & Havlick, 2009). Amplifying cost issues are regulatory shifts; for instance, the European Union's Water Framework Directive mandates sustainable management of river networks, including free passage of migratory fishes and environmentally sensitive flow regimes (Acreman & Ferguson, 2010). In the USA, the Federal Energy Regulatory Commission (FERC) is increasingly addressing obstruction of migrations for endangered species, even requiring complete dam removal in a few cases. For instance, in 1997 FERC mandated removal of a 160-year-old dam at the mouth of the Kennebec River in Maine with the goal of restoring habitat access for seven species of anadromous fishes, including the endangered shortnose sturgeon (Bowman, 2002). Shortnose sturgeon numbers are now rising in the Kennebec River (Bain *et al.*, 2007), and FERC has subsequently required the removal of two large dams on the Elwha River in Washington to restore breeding habitat access for endangered anadromous salmon (Brenkman *et al.*, 2012). The wave of dam removals in North America and Europe is likely to expand in the coming years (Chapter 4).

Most developing countries exhibit a very different outlook on dam projects; new hydropower dams are seen as essential for meeting projected electricity and irrigation demands as their populations and standard-of-living increase. In many of these regions, new hydropower generation is likely to impose high costs on diverse assemblages of migratory fishes (Chapter 4). For example, dams in China's Yangtze River have eliminated access to over 99% of the historical spawning grounds of the anadromous Chinese sturgeon, which is critically endangered (Wang *et al.*, 2012). Proposed dams in Brazil and Southeast Asia would carve up the two most species-rich rivers on Earth, the Amazon (Finer & Jenkins, 2012) and the Mekong (Ziv *et al.*, 2012), where migratory species are both diverse and essential for fisheries.

There is emerging consensus within the conservation community that successfully protecting and restoring freshwater fish migrations will require adaptive management, using lessons learned from previous dam projects to inform those currently under way (Beck *et al.*, 2012; Kareiva, 2012). In most developed nations, decades of research provide guidance on best practices and needed regulations; acting on this knowledge could allow protection and recovery of many migratory fishes. Unfortunately, in the tropical rivers that are home to many more species, we generally lack a full accounting of the number and biomass of migratory fishes, let alone an understanding of their lifelong habitat needs, migration phenology and contribution to human food security. Without such data, the full consequences of dams for migratory fishes are difficult to predict beyond the basic generalisation that losses are likely. Nonetheless, incomplete information does not justify overlooking the societal and environmental implications of new dams; while some projects will deservedly pass muster, environmental impact assessments should be required to account for lost migrations and fisheries, useful lifetime of reservoirs and long-term maintenance costs.

11.4.2 Prioritising restoration and conservation efforts

In this era of simultaneous barrier removals and construction, efforts to safeguard migratory freshwater fishes while giving due consideration to sociopolitical and economic issues can benefit from quantitative tools that identify cost-effective strategies. Spatially explicit prioritisation methods (Moilanen *et al.*, 2009) can be tailored to identify locations where removing dams or road culverts would maximise gains in fish habitat access throughout a river network, or where the placement of new barriers would have minimal effects on habitat connectivity. For both removals and additions of barriers, prioritisation analyses should consider economic costs, site-specific social and political factors, and the distribution of resident and migratory species (Kuby *et al.*, 2005; Kemp & O'Hanley, 2010; Null & Lund, 2012). Such multifactor decision-making scenarios are complex, and realistic recommendations often require more input data than are available (e.g. Kuby *et al.*, 2005). Nonetheless, decision support software is becoming ever more accessible and power-ful, reducing the logistical barriers to exploring quantitative prioritisa-tion approaches (Chapter 14).

As with other aspects of migratory fish conservation, data availabil-ity is a major constraint on the prioritisation of barrier management strategies. For large river basins, lakes or coastal regions, mapping the locations of dams and road culverts can be daunting because georef-erenced data are often unavailable or require labour-intensive collation and refinement (Januchowski-Hartley *et al.*, 2013). Even after locating all potential barriers to migrations, assessing their passability to par-ticular species can be a challenge (e.g. Pepino *et al.*, 2012). Bridges (fully passable) and tall dams (completely impassable without specialised passage structures) are end-members of the passability spectrum, and most small dams and road culverts are somewhere in between. In many cases, detailed field surveys are required to determine whether particu-lar fishes can move through these smaller structures. The passability of a structure depends on its size and construction details (e.g. baffles to create low-flow zones), the local environmental context (e.g. stream gradient), hydrological conditions during the migration season, and the fish's swimming abilities (Kemp & O'Hanley, 2010). With data from a representative subset of barriers, it is possible to calibrate models that predict the passability of additional structures based on readily derived variables such as landscape slope and stream size (Januchowski-Hartley et al., 2014).

To supplement data on the location and passability of each structure, quantifying the amount of upstream habitat and the costs of alternative barrier removal actions can facilitate restoration prioritisation (Kuby et al., 2005; Kemp & O'Hanley, 2010). Spatial data on the distribution of historical and present-day habitats both up- and downstream and the ecological condition of these habitats can be difficult to find. However, coarse spatial data on landscape stressors or other proxies can be substituted for field data on actual stream habitat condition when necessary (e.g. Hermoso et al., 2011; Linke et al., 2012). Explicit consideration of economic costs and political constraints will, in most cases, yield more cost-effective recommendations than prioritisation analyses that do not account for these real-world issues (e.g. Carwardine et al., 2008; O'Hanley, 2011). Unfortunately, data on economic and political constraints are usually even scarcer than ecological data. Thus, methods for estimating these factors and incorporating them into spatial prioritisations and conservation planning remain a growing area of research. Even if truly comprehensive analyses are not possible, accounting for at least a subset of the factors outlined above can boost conservation effectiveness compared to ad hoc selection of remediation projects (Kuby et al., 2005; Kemp & O'Hanley, 2010).

For migratory fishes, the value of spatial prioritisation analyses derives from the facts that there are typically more potential barrier remediation projects than society can pay for, the costs and benefits of individual projects vary widely, and gains from each project are contingent on other projects undertaken within the same river network (Chapter 14). Initial efforts could focus on barriers that no longer serve any identifiable purpose for society (low social cost) or are cheap to remove (low economic cost), as these targets often engender minimal trade-offs (Stanley & Doyle, 2003; Doyle & Havlick, 2009). However, low-cost projects may not be worth including in an overall restoration portfolio if they only make a small amount of habitat accessible. For example, there may be little habitat gain from removing dams or culverts lying above or just below a large dam. Moreover, a series of dams, weirs and road culverts along a river channel can collectively block most migratory fishes even if each structure is partially passable when considered alone (O'Hanley, 2011; Neeson et al., 2015). Spatial prioritisation models can be designed to account for these fish-oriented factors while also balancing economic costs, political complexity, diverse management objectives and stake-holder attitudes toward both fishes and barriers (Johnson & Graber, 2002). As a result, prioritisation models have become an invaluable tool for selecting portfolios of restoration projects that could conserve migratory fishes in a cost-effective way (e.g. Kuby *et al.*, 2005; Null & Lund, 2012; Ziv *et al.*, 2012; Neeson *et al.*, 2015).

11.4.3 When not to remove barriers: containing invasive species and pollution

River networks are managed for many values other than fish migrations, and even from a conservation perspective there are sometimes ecological reasons to retain existing barriers despite their detrimental effects on migrations (Stanley & Doyle, 2003; Jackson & Pringle, 2010; McLaughlin *et al.*, 2012). Foremost among these are cases where barriers prevent the movement up- or downstream of invasive species that have arrived at the site after the barrier was put in place. Potential invaders are sometimes migratory species themselves, as in the case of potamodromous sea lampreys in the North American Great Lakes or anadromous Atlantic salmon that have escaped from aquaculture facilities on the Pacific coast of Chile. In such situations, creating facilities where all migrating fishes are trapped and then invaders can be culled may be a feasible – albeit costly – option. This approach relies on swimways or lift systems to route fish into a holding area where screening allows passage of only the desirable species. The US Fish and Wildlife Service has used such 'trap-and-sort' facilities to control sea lampreys for decades in tributaries of the Great Lakes (Klingler *et al.*, 2003), and lift systems have become common at hydropower dams on large rivers worldwide even though their effectiveness at native fish passage is questionable (Brown *et al.*, 2013). Automated sorting has not yet been feasible, but might one day facilitate selective upstream passage, although the challenge of downstream passage for adults and fry would remain.

In specific cases, semipermeable barriers can enable upstream migrations by select fishes while blocking unwanted species. The sea lamprey control programme in the Great Lakes again exemplifies this approach; seasonal barriers have been used at some sites, while others use permanent low-head dams to block lampreys but allow salmonids to pass (Lavis *et al.*, 2003). The latter approach embodies a very specific value system; exotic Pacific salmon are able to leap small dams readily, while both invasive sea lampreys and a wide variety of native migratory fishes (catostomids, percids, esocids) are denied access (Velez-Espino *et al.*, 2011). Interestingly, eel-ladders designed to enable catadromous anguillid eels to overcome barriers are now being tested as another selective means of trapping invasive sea lampreys (Lavis *et al.*, 2003).

Perhaps more common are situations where anthropogenic barriers limit range expansions by non-migratory invaders, giving management agencies pause when assessing restoration projects (McLaughlin *et al.*, 2012). Invasive species are often talented dispersers, and removing a dam or other barrier may open up a large amount of new habitat to them. The feared invader usually occurs downstream, where it is poised for upstream expansion into a river network upon removal of a barrier (e.g. Fausch *et al.*, 2009). Problematic invaders include more than just fishes; parasites and diseases may also be of concern when barrier removals would open new routes for infection to spread (McLaughlin *et al.*, 2012).

Downstream and upstream movement of chemical contaminants is another major ecological concern when considering barrier remov-als (Jackson & Pringle, 2010; McLaughlin *et al.*, 2012). In industrial-ised countries, the most common worry is downstream release of sediments contaminated with toxic chemicals (e.g. metals, polychlorin-ated biphenyls) as a legacy of historical pollution. In addition, trapping sediment behind dams can foster nitrogen removal through denitrifi-cation, thereby alleviating nitrogen-loading in agricultural landscapes (Powers *et al.*, 2013). Moreover, migratory fishes have a unique capacity to serve as an upstream vector for contaminants. For instance, salmon bioaccumulate toxic chemicals while feeding in the ocean and the North American Great Lakes, and subsequently boost concentrations of these toxins in stream biota when they die after spawning (e.g. Baker *et al.*, 2009; Janetski *et al.*, 2012). Thus, fish migrations themselves can com-plicate contaminant containment strategies by carrying pollution against the current, particularly when semelparous species are involved.

How should we balance these concerns against the benefits of restor-ing fish migrations when considering barrier removals? The negative effects of dams and road culverts on migratory species have often been deemed less pressing than the benefits of preventing the spread of inva-sive species or contaminants, particularly when no gamefish species are at stake. In reality, these are complicated scenarios that involve multiple hypothetical elements and implicit value judgements that are rarely dis-cussed fully and openly. Thus, it is understandable that the spectre of spreading invasive species and contaminants could trump the potential benefits of restored native fish migrations, favouring the maintenance of existing barriers. For a case study of a structured decision-making approach to this type of scenario, we recommend the US Fish and Wildlife Service's pilot analysis of barrier removal options in the Bad River, a tributary of Lake Superior (Patronski *et al.*, 2009). In that example, the status quo of barrier maintenance was favoured after considering a wide variety of alternatives. Unfortunately, we are not aware of any quantitative analyses at regional (or larger) scales for which data were available on both conservation benefits (e.g. Northeast Aquatic Connectivity Project; Martin & Apse, 2011) and conservation costs of barrier removals, let alone an analysis that also incorporates economic and social management objectives at large scales.

11.4.4 Clarifying our values: what species are we managing for?

Designing sensible conservation strategies for migratory fishes must begin with specifying what species we are managing for, because the benefits of particular restoration or protection efforts will vary among species. In rivers, lakes and oceans with numerous freshwater migratory species, a broad range of life-history strategies, migration phenologies and habitat needs are invariably present. Despite this diversity, management usually focuses on a subset of species that have high commercial and recreational value. This tacit value judgement is rarely discussed openly, yet it can direct barrier management, technological research investment and field assessments towards these select species, potentially to the detriment of other migrants.

In North America and Europe, the bias has often been toward salmonid-centric management of barriers unless endangered status gives special priority to other taxa. For instance, the largest dam removal projects in the USA have focused on coastal rivers where Pacific salmon (Elwha River, Washington) or Atlantic salmon (Kennebec River and Penobscot River, Maine) are the focal beneficiaries, although other diadromous species may also profit (Limburg & Waldman 2009). Operators of large dams have made modest shifts in many countries toward 'environmental flows' that better replicate the timing of high and low flows during an annual cycle (Bunn & Arthington, 2002; Poff *et al.*, 2007), but major experiments in restoring natural flow regimes have rarely targeted migratory fishes explicitly.

Mitigating the impacts of dams on migratory fishes often focuses on development of passage structures, but this substantial literature deals primarily with salmonids (Roscoe & Hinch, 2010; Katopodis &

Williams, 2012). Unfortunately, the design parameters of fishways for salmon, which are unusual in both their athleticism and semelparity, are a poor match to the needs of other migratory species found in the same rivers (Stuart & Mallen-Cooper, 1999; Agostinho et al., 2007; Mallen-Cooper & Brand, 2007). In the Columbia River basin, declines of the once-abundant anadromous Pacific lamprey reflect the incompatibility of their anguilliform swimming mechanics with the fast, turbulent flows that attract salmon (Moser et al., 2002). Failures of salmon-oriented fishways have been so profound for American shad, which were once a dominant anadromous migrant in eastern North America, that dam removals have been suggested as the only viable restoration approach (Brown et al., 2013). Despite these shortcomings, salmon passage technology has been used in many rivers where no salmon – or even ecological analogues of salmon – are present (e.g. Figure 11.3), leading to a variety of failures worldwide (Mallen-Cooper & Brand, 2007). Comparatively high migratory biomass, species diversity, variable hydrology and floodplain extent in tropical rivers distinguish their fish passage challenges from those of temperate coastal rivers (Oldani & Baigún, 2002; Pelicice & Agostinho, 2008; Baumgartner et al., 2011). Addressing all of these issues using a single structure for upstream and downstream passage represents a formidable challenge (Godinho & Kynard, 2008), but is essential as major new dams are being planned for many tropical rivers.

A parallel overgeneralisation of salmonid results to other migratory fishes may apply to concern about contaminant transport. Most of the published papers on this topic address semelparous salmonids as vectors. While salmon are indeed ideal bioaccumulators by virtue of their large body size, piscivorous diet and lipid-rich tissues, they are also unusual among migratory species in being semelparous. Iteroparous migrants are less likely to introduce substantial quantities of bioaccumulated chemicals to their breeding grounds, both because they do not die after spawning and because they are generally lower in tissue lipids and create smaller eggs (which can concentrate lipophilic compounds) than salmonids. Thus, the evidence supporting fears about the delivery of toxic chemicals by migratory fishes applies primarily to coastal drainages where semelparous salmonids are native, as well as the introduced ranges of Pacific salmon in the North American Great Lakes, southern South America and New Zealand (Correa & Gross, 2008) and Atlantic salmon in western Canada, Chile and Tasmania (Thorstad et al., 2008). Further research is needed to determine whether contamination

Figure 11.3 Pak Moon Dam fish ladder in Thailand. Spanning the Mun River just 5 km upstream of its confluence with the Mekong River, the Pak Mun Dam has been named among the world's most ecologically and socially damaging hydropower projects. To bypass the dam, fish must swim through a minimum of 44 baffles, which requires a climb of 15 vertical metres across a 92-m horizontal distance (i.e. a 16% grade). Few potamodromous species have been reported to successfully traverse the ladder, which was modelled after ladder designs for anadromous salmon. Photo: Aaron A. Koning.

concerns apply equally to the iteroparous species that constitute the majority of the world's migratory fishes.

We believe that a broader value system is requisite to conserving migratory freshwater fishes. If management and conservation efforts focus primarily on species with direct economic value derived from recreational and commercial fishing, we are likely to see continued erosion of the world's ichthyofaunal diversity even if priority fisheries remain stable. Low economic value should not be equated with low conservation value as we attempt to manage new and old barriers, minimise degradation of water quality and quantity (Vörösmarty *et al.*, 2010; Carpenter *et al.*, 2011) and limit overfishing (Allan *et al.*, 2005). Rather, conservation strategies should be designed to meet the requirements of all migratory species in an area. The resulting efforts are unlikely to conflict with the needs of non-migratory species, and applying a more inclusive value system to migratory fish conservation efforts will ultimately benefit most freshwater species through greater habitat protections and improved analytical tools (Kemp & O'Hanley, 2010).

11.5 FUTURE RESEARCH DIRECTIONS

Much work remains for both researchers and practitioners interested in conserving migratory freshwater fishes around the world. While successes may not always be cheered by the public at large, our review indicates that fish migrations provide diverse benefits to species, ecosystems and humanity. The metapopulation structure of migratory species means that conserving each local population also contributes to species persistence at larger spatial and temporal scales, as well as buffering regional fishery potential (Schindler et al., 2010). The future of migratory fishes and their contributions to ecosystem functioning and human well-being will depend on forward-looking policies as well as management informed by past lessons and ongoing research.

We conclude by touching on four issues that we believe are especially important for advancing conservation research on migratory freshwater fishes. First, it is imperative that the development of fish-passage technologies – ladders, lifts, bypass channels, and the like – address a wider range of fish species and background hydrologies (Roscoe & Hinch, 2010; Baumgartner et al., 2011), as well as tackling downstream passage in addition to upstream access. We must avoid unthinking extrapolation of lessons from one region or taxon to others. Instead, a concerted effort is needed to engage the hydropower industry, water users, government regulators and conservation organisations in testing fish-passage methods in each region where new dams are contemplated or existing ones fail to demonstrate adequate passage. Developing efficient passage structures for a diversity of migratory fishes will surely be challenging and expensive (Lucas & Baras, 2001; Bunt et al., 2012; Thiem et al., 2013), and will require engineers to 'think like a fish' to match their approaches to the traits of the migratory fauna at a given site (Williams et al., 2012). After commissioning, the ability of these structures to effectively pass migrants should be evaluated against a standard of maintaining viable populations of all migratory species (da Silva et al., 2012; Pompeu et al., 2012). This approach could encourage local adaptive management at a particular dam as well as fostering the transfer of effective designs to projects elsewhere in a judicious manner (Bunt et al., 2012; Katopodis & Williams, 2012).

Second, major conceptual and practical challenges arise from the fact that many fish migrations encompass multiple ecological and political zones. Conserving migratory populations requires large-scale habitat protections and fishery regulations supported by coordinated actions, incentives and funding across agencies and interest groups (Alvarez-Romero *et al.*, 2011; Pracheil *et al.*, 2012). We must also recognise that the condition of key habitats is itself a reflection of numerous stressors, each of which may be driven by multiple human activities occurring across disparate spatial scales (e.g. Vörösmarty *et al.*, 2010). Thus, improving habitat quality for migratory fishes will often necessitate actions beyond the migration corridor itself. Attribution of responsibility for population declines is complicated by the triple-jeopardy nature of migratory life histories. For instance, low returns of chinook salmon from the Snake River compared to populations downstream have been attributed to mortality at hydroelectric dams during outmigration, but poor survival in the marine realm may in fact be responsible (Rechisky *et al.*, 2013). Identifying the bottlenecks that limit rehabilitation of migratory fish populations is an essential step, but causality is often contentious. Ultimately, successful conservation at appropriate spatial scales will often require interjurisdictional political agreements that account for the key details of species ecology, threats arising from diverse sources and social and economic considerations. As researchers, we must be prepared for long-term engagement in this complex process.

Third, we see an important role for basic research on the significance of the iteroparity–semelparity and diadromy–potamodromy life-history spectra for conservation strategies. For instance, no theoretical models or meta-analyses have addressed whether anthropogenic pressures more strongly affect species showing one life history compared to others. Natural selection favours semelparity when adult mortality is high but juvenile mortality is low (Crespi & Teo, 2002), whereas iteroparity is advantageous when adult survival is high but juveniles have low odds of survival. To the extent that migration barriers, fisheries and other factors boost mortality of adult migrants without affecting young fish, humans may be selecting for semelparous species or for maximal reproductive investment by first-time breeders within iteroparous species (Crespi & Teo, 2002). Conversely, if degradation of spawning or nursery habitats affects juveniles but not adults, then long-lived iteroparous species may perform best. With regard to migrations between freshwater and saltwater, landlocked populations of at least a few diadromous species have arisen due to strong selection pressure by dams (e.g. Tsunagawa & Arai, 2009) and pollution (e.g. Bleackley *et al.*, 2009), and growing evidence of natural plasticity in migratory behaviour (e.g. Feutry *et al.*, 2012; Hogan *et al.*, 2014; Huey *et al.*, 2014) suggests that many diadromous species have the capacity to adjust their migration strategy. Anadromous salmonids also show fishery-induced evolution in a variety of traits (Hard *et al.*, 2008). Potamodromous species can experience reduced gene flow and population differentiation due to barriers between habitats (e.g. Wollebaek *et al.*, 2011) and intensive fisheries can drive large shifts in body size (e.g. Taylor *et al.*, 2006), and perhaps even favour loss of migratory behaviour (Theriault *et al.*, 2008). In each of these cases, elucidating differences among life-history strategies in their effect on susceptibility to anthropogenic stressors could help direct conservation efforts toward species and regions that are most at risk.

Finally, climate change research represents an essential frontier in the conservation of migratory fishes. Migrants may be strongly affected by climate change through shifts in phenology, habitat suitability and connections between habitats. Migrations are closely tied to seasonal environmental cues such as water temperature and discharge, which reflect regional climate. Shifts in migration phenology have already been attributed to climate change for a variety of species including shad, Pacific salmon (Quinn & Adams, 1996), Atlantic salmon (Juanes *et al.*, 2004), northern pike and smelt (Ahas & Aasa, 2006). Whether these phenological shifts reflect successful tracking of changes in the optimal timing of migrations, or instead turn out to be maladaptive, remains unclear. Migrations are timed to maximise resource availability or reduce mortality based on long-term averages of environmental conditions (Quinn & Adams, 1996); hence, the growth or reproductive success of migrants may be depressed when cues in one habitat that trigger the onset of migration become misaligned with the timing of optimal conditions in the destination habitat (Visser & Both, 2005). In some cases, the very existence of key habitats (e.g. floodplains) and the passability of corridors connecting them may be affected by altered precipitation patterns under climate change. For example, amphidromous Hawaiian gobies migrate from the ocean into streams whose hydrology fluctuates wildly between low baseflow and flood events (Radtke & Kinzie, 1996). Many of these streams already suffer reduced flow due to stream diversions for human use (Brasher, 2003), and recent and projected reductions in rainfall across the Hawaiian archipelago are likely to both reduce stream habitat and diminish migration corridors providing access to headwaters

Stressors:	Exploitation	Dams, water use	Urbanisation
Habitat:	Pristine	Stressed	Poor
Fisheries:	Plentiful	Declining	Trivial

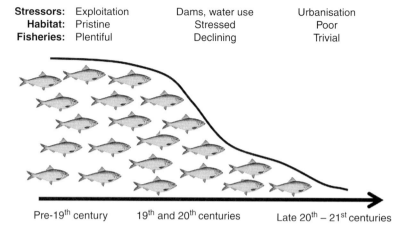

Pre-19th century   19th and 20th centuries   Late 20th – 21st centuries

Figure 11.4 Historical overview of declines in diadromous fishes of northeastern North America. Fisheries drove initial population declines, followed by habitat degradation and loss, and finally further habitat loss and climate change. Throughout this process, loss of institutional and societal memory about the past abundance and importance of migratory fish complicates setting conservation goals. This diagram is adapted from Limburg and Waldman (2009), with permission from Oxford University Press.

(Walter *et al.*, 2012). Taken together, shifts in migration phenology, habitat availability and connectivity under ongoing climate change are sure to complicate efforts to conserve migratory freshwater fishes, necessitating an adaptive approach to management (see Stein *et al.*, 2013).

As researchers and practitioners work together to resolve these and other challenges in conserving freshwater fish migrations, the perceptual problem of shifting baselines cannot be ignored (Humphries & Winemiller, 2009; Limburg & Waldman, 2009): remembering (or even imagining) the benefits of substantially larger fish migrations during a bygone era is difficult (Figure 11.4). This limitation of human perspectives undermines conservation efforts of all sorts, but neither decision makers nor the public can be faulted for it when we are frequently reminded that invasive species, habitat destruction, diseases and other changes make ecosystems of the past unrecoverable. To combat the shifting baseline challenge, we should continue to make use of iconic migratory fishes such as Pacific salmon as standard-bearers, but also doggedly highlight the fact that substantial fish migrations once occurred annually in rivers close to home for almost everyone. Broadening awareness of long-term declines, ecosystem importance and fisheries benefits of migratory fishes is essential for sustaining remaining migrations and restoring those that have been lost. Overall, we are optimistic that the ongoing expansion of both public appreciation and scientific expertise will enable more effective conservation of migratory freshwater fishes worldwide.

11.6 DISCUSSION QUESTIONS

1. How many migratory fish species are in your area, and which major classes of migration do they represent?
2. Which is the most common barrier to fish migrations in your area: dams or road culverts?
3. Is it more problematic to inhibit adult fish from reaching upstream spawning grounds, or young fish from migrating back downstream?
4. When both desirable native fishes and harmful invasive species are migratory, which should have priority in management efforts?
5. How can we compare the ecological benefits of restoring migrations to the economic costs of removing barriers and improving habitats?
6. What would happen to sport fisheries if migratory fishes continue to be extirpated? How about subsistence fisheries?

11.7 ACKNOWLEDGEMENTS

We are grateful for financial support of our research by the Upper Midwest and Great Lakes Landscape Conservation Cooperative, University of Wisconsin Water Resources Institute (WR11R002), Wisconsin Department of Natural Resources, National Science Foundation (DEB-1030242, DEB- 1115025), US Environmental Protection Agency (GL00E00553-0), US Department of Defense (SERDP RC-1646), University of Michigan Water Center, The Nature Conservancy, Packard Fellowship in Science and Engineering, and Mustard Seed Foundation.

11.8 REFERENCES

Acreman, M. C. & Ferguson, A. J. D. (2010). Environmental flows and the European water framework directive. *Freshwater Biology*, 55, 32–48.
Agostinho, C. S., Agostinho, A. A., Pelicice, F., Almeida, D. A. A. & Marques, E. E. (2007). Selectivity of fish ladders: a bottleneck in Neotropical fish movement. *Neotropical Ichthyology*, 5, 205–213.
Agostinho, A. A., Pelicice, F. M. & Gomes, L. C. (2008). Dams and the fish fauna of the Neotropical region: impacts and management related to diversity and fisheries. *Brazilian Journal of Biology*, 68, 1119–1132.
Ahas, R. & Aasa, A. (2006). The effects of climate change on the phenology of selected Estonian plant, bird and fish populations. *International Journal of Biometeorology*, 51, 17–26.
Allan, J. D. & Castillo, M. M. (2007). *Stream Ecology: Structure and Function of Running Waters*, second edition. New York, NY: Springer.

Allan, J. D., Abell, R., Hogan, Z. E. B., *et al.* (2005). Overfishing of inland waters. *BioScience*, 55, 1041–1051.

Alvarez-Romero, J. G., Pressey, R. L., Ban, N. C., *et al.* (2011). Integrated land–sea conservation planning: the missing links. *Annual Reviews of Ecology, Evolution, and Systematics*, 42, 381–409.

Baker, M. R., Schindler, D. E., Holtgrieve, G. W. & St. Louis, V. L. (2009). Bioaccumulation and transport of contaminants: migrating sockeye salmon as vectors of mercury. *Environmental Science and Technology*, 43, 8840–8846.

Baker, R. R. (1978). *The Evolutionary Ecology of Animal Migration*. New York, NY: Holmes and Meier.

Bain, M. B., Finn, J. T. & Brooke, H. E. (1988). Streamflow regulation and fish community structure. *Ecology*, 69, 382–392.

Bain, M. B., Haley, N., Peterson, D. L., *et al.* (2007). Recovery of a US endangered fish. *PLoS ONE*, 2(1), e168.

Baran, E. & Myschowoda, C. (2009). Dams and fisheries in the Mekong Basin. *Aquatic Ecosystem Health and Management*, 12, 227–234.

Barlow, C., Baran, E., Halls, A. & Kshatriya, M. (2008). How much of the Mekong fish catch is at risk from mainstem dam development? *Catch and Culture*, 14, 16–21.

Baumgartner, L. J., Marsden, T., Singhanouvong, D., *et al.* (2011). Using an experimental *in situ* fishway to provide key design criteria for lateral fish passage in tropical rivers: a case study from the Mekong River, Central Lao PDR. *River Research and Applications*, 28, 1217–1229.

Bayley, P. B. (1995). Understanding large river–floodplain ecosystems. *BioScience*, 45, 153–158.

Beard, T. D., Arlinghaus, R., Cooke, S. J., *et al.* (2011). Ecosystem approach to inland fisheries: research needs and implementation strategies. *Biology Letters*, 7, 481–483.

Beck, M. W., Claassen, A. H. & Hunt, P. J. (2012). Environmental and livelihood impacts of dams: common lessons across development gradients that challenge sustainability. *International Journal of River Basin Management*, 10, 73–92.

Benstead, J. P., March, J. G., Pringle, C. M. & Scatena, F. N. (1999). Effects of a low-head dam and water abstraction on migratory tropical stream biota. *Ecological Applications*, 9, 656–668.

Bilby, R. E., Beach, E. W., Fransen, B. R., Walter, J. K. & Bisson, P. A. (2003). Transfer of nutrients from spawning salmon to riparian vegetation in western Washington. *Transactions of the American Fisheries Society*, 132, 733–745.

Bleackley, N. A., Landman, M. J. & Ling, N. (2009). Ecology of common bully (*Gobiomorphus cotidianus*) in the Tarawera and Rangitaiki rivers: isolation by inland distance or anthropogenic discharge? *New Zealand Journal of Marine and Freshwater Research*, 43, 889–899.

Bowman, M. B. (2002). Legal perspectives on dam removal. *BioScience*, 52, 739–747.

Brasher, A. M. D. (2003). Impacts of human disturbances on biotic communities in Hawaiian streams. *BioScience*, 53, 1052–1060.

Brenkman, S. J., Duda, J. J., Torgersen, C. E., *et al.* (2012). A riverscape perspective of Pacific salmonids and aquatic habitats prior to large-scale dam removal in the Elwha River, Washington, USA. *Fisheries Management and Ecology*, 19, 36–53.

Browder, R. G. & Garman, G. C. (1994). Increased ammonium concentrations in a tidal freshwater stream during residence of migratory clupeid fishes. *Transactions of the American Fisheries Society*, 123, 993–996.

Brown, J. H. & Kodric-Brown, A. (1977). Turnover rates in insular biogeography – effect of immigration on extinction. *Ecology*, 58, 445–449.

Brown, J. J., Limburg, K. E., Waldman, J. R., *et al.* (2013). Fish and hydropower on the U.S. Atlantic coast: failed fisheries policies from half-way technologies. *Conservation Letters*, 6, 280–286.

Bunn, S. E. & Arthington, A. H. (2002). Basic principles and ecological consequences of altered flow regimes for aquatic biodiversity. *Environmental Management*, 30, 492–507.

Bunt, C. M., Castro-Santos, T. & Haro, A. (2012). Performance of fish passage structures at upstream barriers to migration. *River Research and Applications*, 28, 457–478.

Cada, G. F. (1990). A review of studies relating to the effects of propeller-type turbine passage on fish early life stages. *North American Journal of Fisheries Management*, 10, 418–426.

Carolsfeld, J., Harvey, B., Ross, C. & Baer, A. (2004). *Migratory Fishes of South America: Biology, Fisheries, and Conservation Status*. Ottawa: International Development Research Centre.

Carpenter, S. R., Stanley, E. & Vander Zanden, M. J. (2011). State of the world's freshwater ecosystems: physical, chemical, and biological changes. *Annual Review of Environment and Resources*, 36, 75–99.

Carwardine, J., Wilson, K. A., Watts, M., *et al.* (2008). Avoiding costly conservation mistakes: the importance of defining actions and costs in spatial priority setting. *PLoS ONE*, 3, e2586.

Childress, E., Allan, J. D. & McIntyre, P. B. (2014). Nutrient subsidies from native migratory fish enhance productivity in Great Lakes tributaries. *Ecosystems*, 17, 522–534.

Christie, K. S. & Reimchen, T. E. (2008). Presence of salmon increases passerine density on Pacific Northwest streams. *The Auk*, 125, 51–59.

Collins, S., Bickford, N., McIntyre, P. B., *et al.* (2013). Population structure of a Neotropical migratory fish: contrasting perspectives from genes and otolith microchemistry. *Transactions of the American Fisheries Society*, 142, 1192–1201.

Cooke, S. J., Bunt, C. M., Hamilton, S. J., *et al.* (2005). Threats, conservation strategies, and prognosis for suckers (Catostomidae) in North America: insights from regional case studies of a diverse family of non-game fishes. *Biological Conservation*, 121, 317–331.

Correa, C. & Gross, M. R. (2008). Chinook salmon invade southern South America. *Biological Invasions*, 10, 615–639.

Costa-Pierce, B. (1997). *From Farmers to Fishers: Developing Reservoir Aquaculture for People Displaced by Dams*, Vol. 369. Washington, DC: World Bank Publications.

Coutant, C. C. & Whitney, R. R. (2000). Fish behavior in relation to passage through hydropower turbines: a review. *Transactions of the American Fisheries Society*, 129, 351–380.

Crespi, B. J. & Teo, R. (2002). Comparative phylogenetic analysis of the evolution of semelparity and life history in salmonid fishes. *Evolution*, 56, 1008–1020.

Crook, V. (2010). Trade in *Anguilla* species, with a focus on recent trade in European eel, *A. anguilla*. *TRAFFIC report prepared for the European Commission*, 52.

da Silva, L. G. M., Nogueira, L. B., Maia, B. P. & de Resende, L. B. (2012). Fish passage post-construction issues: analysis of distribution, attraction and passage efficiency metrics at the Baguari Dam fish ladder to approach the problem. *Neotropical Ichthyology*, 10, 751–762.

Doyle, M. W. & Havlick, D. G. (2009). Infrastructure and the environment. *Annual Review of Environment and Resources*, 34, 349–373.

Dudgeon, D. (2010). Requiem for a river: extinctions, climate change and the last of the Yangtze. *Aquatic Conservation: Marine and Freshwater Ecosystems*, 20, 127–131.

Duncan, J. P. (2013). *Characterization of Fish Passage Conditions through the Fish Weir and Turbine Unit 1 at Foster Dam, Oregon, using Sensor Fish, 2012*. Pacific Northwest National Laboratory.

Fagan, W. F. (2002). Connectivity, fragmentation, and extinction risk in dendritic metapopulations. *Ecology*, 83, 3243–3249.

FAO [Food and Agriculture Organization of the United Nations]. (2012). Fish-Stat Plus database. www.fao.org/fishery/statistics/software/fishstat/en.

Fausch, K. D., Rieman, B. E., Dunham, J. B., Young, M. K. & Peterson, D. P. (2009). Invasion versus isolation: trade-offs in managing native salmonids with barriers to upstream movement. *Conservation Biology*, 23, 859–870.

Feutry, P., Tabouret, H., Maeda, K., Pecheyran, C. & Keith, P. (2012). Diadromous life cycle and behavioural plasticity in freshwater and estuarine Kuhliidae species (Teleostei) revealed by otolith microchemistry. *Aquatic Biology*, 15, 195–204.

Finer, M. & Jenkins, C. N. (2012). Proliferation of hydroelectric dams in the Andean Amazon and implications for Andes–Amazon connectivity. *PLoS ONE*, 7, e35126.

Finney, B. P., Gregory-Eaves, I., Douglas, M. S. V. & Smol, J. P. (2002). Fisheries productivity in the northeastern Pacific Ocean over the past 2,200 years. *Nature*, 416, 729–733.

Flecker, A. S. (1996). Ecosystem engineering by a dominant detritivore in a diverse tropical stream. *Ecology*, 77, 1845–1854.

Flecker, A. S., McIntyre, P. B., Moore, J. W., *et al.* (2010). Migratory fishes as material and process subsidies in riverine ecosystems. *American Fisheries Society Symposium*, 73, 559–592.

Frankham, R. (1995). Conservation genetics. *Annual Review of Genetics*, 29, 305–327.

Gerrard, J. M., Gerrard, P., Maher, W. J. & Whitfield, D. W. A. (1975). Factors influencing nest site selection of bald eagles northern Saskatchewan and Manitoba. *Blue Jay*, 33, 169–176.

Godinho, A. L. & Kynard, B. (2008). Migratory fishes of Brazil: life history and fish passage needs. *River Research and Applications*, 25, 702–712.

Gotelli, N. J. & Taylor, C. M. (1999). Testing metapopulation models with stream-fish assemblages. *Evolutionary Ecology Research*, 1, 835–845.

Greathouse, E. A., Pringle, C. M., McDowell, W. H. & Holmquist, J. G. (2006). Indirect upstream effects of dams: consequences of migratory consumer extirpation in Puerto Rico. *Ecological Applications*, 16, 339–352.

Gross, M. R., Coleman, R. M. & McDowall, R. M. (1988). Aquatic productivity and the evolution of diadromous fish migration. *Science*, 239, 1291–1293.

Hall, C. A. S. (1972). Migration and metabolism in a temperate stream ecosystem. *Ecology*, 53, 585–604.

Hamann, E. J. & Kennedy, B. P. (2012). Juvenile dispersal affects straying behaviors of adults in a migratory population. *Ecology*, 93, 733–740.

Hanski, I. (1999). *Metapopulation Ecology*, Oxford: Oxford University Press.

Hanson, N., Fogel, M., Fong, D. W. & MacAvoy, S. E. (2010). Marine nutrient transport: anadromous fish migration linked to the freshwater amphipod *Gammarus fasciatus*. *Canadian Journal of Zoology*, 88, 546–552.

Hard, J. J., Gross, M. R., Heino, M., *et al.* (2008). Evolutionary consequences of fishing and their implications for salmon. *Evolutionary Applications*, 1, 388–408.

Harden Jones, F. R. (1968). *Fish Migration*. New York, NY: St. Martin's Press.

Hasler, A. D. (1966). *Underwater Guideposts: Homing of Salmon*, Madison, WI: University of Wisconsin Press.

Hermoso, V., Januchowski-Hartley, S., Linke, S. & Possingham, H. P. (2011). Reference vs. present-day condition: early planning decision influence the achievement of conservation objectives. *Aquatic Conservation: Marine and Freshwater Ecosystems*, 21, 500–509.

Hicks, B. J., Wipfli, M. S., Lang, D. W. & Lang, M. E. (2005). Marine-derived nitrogen and carbon in freshwater-riparian food webs of the Copper River Delta, south-central Alaska. *Oecologia*, 144, 558–569.

Hilderbrand, G. V., Hanley, T. A., Robbins, C. T. & Schwartz, C. C. (1999). Role of brown bears (*Ursus arctos*) in the flow of marine nitrogen into a terrestrial ecosystem. *Oecologia*, 121, 546–550.

Hogan, J. D., McIntyre, P. B., Blum, M. J., Gilliam, J. F. & Bickford, N. (2014). Consequences of alternative dispersal strategies in a putatively amphidromous fish. *Ecology*, 95, 2397–2408.

Hortle, K. G. (2009). Fisheries of the Mekong river basin. In *The Mekong. Biophysical Environment of a Transboundary River*. New York, NY: Elsevier.

Huey, J. A., Crook, D. A., Macdonald, J. I., *et al.* (2014). Is variable connectivity among populations of a continental gobiid fish driven by local adaptation or passive dispersal? *Freshwater Biology*, 59, 1672–1686.

Humphries, P. L. & Winemiller, K. O. (2009). Historical impacts on river fauna, shifting baselines and challenges for restoration. *BioScience*, 59, 673–684.

Irvine, J. R. & Fukuwaka, M. (2011). Pacific salmon abundance trends and climate change. *ICES Journal of Marine Science*, 68(6), 1122–1130.

Jackson, C. R. & Pringle, C. M. (2010). Ecological benefits of reduced hydrologic connectivity in intensively developed landscapes. *BioScience*, 60, 37–46.

Janetski, D. J., Chaloner, D. T., Tiegs, S. D. & Lamberti, G. A. (2009). Pacific salmon effects on stream ecosystems: a quantitative synthesis. *Oecologia*, 159, 583–595.

Janetski, D. J., Chaloner, D. T., Moerke, A. H., *et al.* (2012). Resident fishes display elevated organic pollutants in salmon spawning streams of the Great Lakes. *Environmental Science & Technology*, 46, 8035–8043.

Januchowski-Hartley, S., McIntyre, P. B., Diebel, M. & Doran, P. J. (2013). Restoring aquatic ecosystem connectivity requires expanding inventories of both dams and road crossings. *Frontiers in Ecology and Environment*, 11, 211–217.

Januchowski-Hartley, S., Diebel, M. H., Doran, P. J. & McIntyre, P. B. (2014). Predicting road culvert passability for migratory fishes. *Diversity and Distributions*, 20, 1414–1424.

Jeffres, C. & Moyle, P. (2012). When good fish make bad decisions: coho salmon in an ecological trap. *North American Journal of Fisheries Management*, 32, 87–92.

Jennings, C. A. & Zigler, S. J. (2009). Biology and life history of paddlefish in North America: an update. In *Paddlefish Management, Propagation, and Conservation in the 21st Century*. Bethesda, MD: American Fisheries Society, pp. 1–22.

Johnson, G. E. & Dauble, D. D. (2006). Surface flow outlets to protect juvenile salmonids passing through hydropower dams. *Reviews in Fisheries Science*, 14, 213–244.

Johnson, S. E. & Graber, B. E. (2002). Enlisting the social sciences in decisions about dam removal. *BioScience*, 52, 731–738.

Juanes, F., Gephard, S. & Beland, K. (2004). Long-term changes in migration timing of adult Atlantic 480 salmon (*Salmo salar*) at the southern edge of the species distribution. *Canadian Journal of Fisheries and Aquatic Sciences*, 61, 2392–2400.

Kareiva, P. M. (2012). Dam choices: analyses for multiple needs. *Proceedings of the National Academy of Sciences*, 109, 5553–5554.

Katopodis, C. & Williams, J. G. (2012). The development of fish passage research in a historical context. *Ecological Engineering*, 49, 8–18.

Kemp, P. S. & O'Hanley, J. R. (2010). Procedures for evaluating and prioritising the removal of fish passage barriers: a synthesis. *Fisheries Management and Ecology*, 17, 297–322.

Klingler, G. L., Adams, J. V. & Heinrich, J. W. (2003). Passage of four teleost species prior to sea lamprey (*Petromyzon marinus*) migration in eight tributaries of Lake Superior, 1954 to 1979. *Journal of Great Lakes Research*, 29, 403–409.

Kuby, M. J., Fagan, W. F., ReVelle, C. S. & Graf, W. L. (2005). A multiobjective optimization model for dam removal: an example trading off salmon passage with hydropower and water storage in the Willamette basin. *Advances in Water Resources*, 28, 845–855.

Landsman, S. J, Nguyen, V. M., Gutowsky, L. F. G., *et al.* (2011). Fish movement and migration studies in the Laurentian Great Lakes: research trends and knowledge gaps. *Journal of Great Lakes Research*, 37, 365–379.

Laurance, W. F. & Balmford, A. (2013). A global map for road building. *Nature*, 495, 308–309.

Lavis, D. S., Hallett, A., Koon, E. M. & McAuley, T. C. (2003). History of and advances in barriers as an alternative method to suppress sea lampreys in the Great Lakes. *Journal of Great Lakes Research*, 29, 362–372.

Leggett, W. C. & Carscadden, J. E. (1978). Latitudinal variation in reproductive characteristics of American shad (*Alosa sapidissima*): evidence for population specific life history strategies in fish. *Journal of the Fisheries Research Board of Canada*, 35, 1469–1478.

Lehner, B., Liermann, C. R., Revenga, C., *et al.* (2011). High-resolution mapping of the world's reservoirs and dams for sustainable river-flow management. *Frontiers in Ecology and Environment*, 9, 494–502.

Leveque, C. (1997). *Biodiversity Dynamics and Conservation: The Freshwater Fish of Tropical Africa*. Cambridge University Press.

Levi, T., Darimont, C. T., MacDuffee, M., *et al.* (2012). Using grizzly bears to assess harvest-ecosystem tradeoffs in salmon fisheries. *PLoS Biology*, 10(4) e1001303.

Ligon, F. K., Dietrich, W. E. & Trush, W. J. (1995). Downstream ecological effects of dams: a geomorphic perspective. *BioScience*, 45, 183–192.

Limburg, K. E. & Waldman, J. R. (2009). Dramatic declines in North Atlantic diadromous fishes. *BioScience*, 59, 955–965.

Linke, S., Kennard, M. J., Hermoso, V., *et al.* (2012). Merging connectivity rules and large-scale condition assessment improves conservation adequacy in river systems. *Journal of Applied Ecology*, 49, 1036–1045.

Lucas, M. C. & Baras, E. (2001). *Migration of Freshwater Fishes*. Oxford: Blackwell.

Lytle, D. A. & Poff, N. L. (2004). Adaptation to natural flow regimes. *Trends in Ecology and Evolution*, 19, 94–100.

Mallen-Cooper, M. & Brand, D. A. (2007). Non-salmonids in a salmonid fishway: what do 50 years of data tell us about past and future fish passage? *Fisheries Management and Ecology*, 14, 319–332.

Marsh, P. C. & Douglas, M. E. (1997). Predation on endangered humpback chub (*Gila cypha*) by introduced fishes in the Little Colorado River, Arizona. *Transactions of the American Fisheries Society*, 126, 343–346.

Martin, E. H. & Apse, C. D. (2011). Northeast Aquatic Connectivity: an Assessment of Dams on Northeastern Rivers. *The Nature Conservancy, Eastern Freshwater Program*.

McDowall, R. M. (1988). *Diadromy in Fishes: Migrations between Freshwater and Marine Environments*. Portland, OR: Timber Press.

McDowall, R. M. (1997). The evolution of diadromy in fishes (revisited) and its place in phylogenetic analysis. *Reviews in Fish Biology and Fisheries*, 7, 443–462.

McElroy, B., DeLonay, A. & Jacobson, R. (2012). Optimum swimming pathways of fish spawning migrations in rivers. *Ecology*, 93, 29–34.

McIntyre, P. B., Jones, L., Flecker, A. S. & Vanni, M. J. (2007). Fish extinctions alter nutrient recycling in tropical freshwaters. *Proceedings of the National Academy of Sciences*, 104, 4461–4466.

McIntyre, P. B., Flecker, A. S., Vanni, M., *et al.* (2008). Fish distributions and nutrient recycling in a tropical stream: can fish create biogeochemical hotspots? *Ecology*, 89, 2335–2346.

McKeown, B. A. (1984). *Fish Migration*. London: Croom Helm.

McLaughlin, R. L., Smyth, E. R. B., Castro-Santos, T., *et al.* (2012). Unintended consequences and trade-offs of fish passage. *Fish and Fisheries*, 14, 580–604.

Meck, A. (1916). *The Migrations of Fish.* London: Edward Arnold.

Moilanen, A., Wilson, K. A. & Possingham, H. P. (2009). *Spatial Conservation Prioritization: Quantitative Methods and Computational Tools.* Oxford: Oxford University Press.

Moore, J. W., Schindler, D. E. & Scheuerell, M. D. (2004). Disturbance of freshwater habitats by anadromous salmon in Alaska. *Oecologia*, 139, 298–308.

Moser, M. L., Ocker, P. A., Stuehrenberg, L. C. & Bjornn, T. C. (2002). Passage efficiency of adult pacific lampreys at hydropower dams on the lower Columbia River, USA. *Transactions of the American Fisheries Society*, 131, 956–965.

Myers, G. S. (1949). Usage of anadromous, catadromous and allied terms for migratory fishes. *Copeia*, 1949, 89–97.

Naiman, R. J., Bilby, R. E., Schindler, D. E. & Helfield, J. M. (2002). Pacific salmon, nutrients, and the dynamics of freshwater and riparian ecosystems. *Ecosystems*, 5, 399–417.

Neeson, T. M., Ferris, M. C., Diebel, M. W., *et al.* (2015). Enhancing ecosystem restoration efficiency through spatial and temporal coordination. *Proceedings of the National Academy of Sciences*, 112, 6236–6241.

Null, S. E. & Lund, J. R. (2012). Fish habitat optimization to prioritize river restoration decisions. *River Research and Applications*, 28, 1378–1393.

O'Hanley, J. R. (2011). Open rivers: barrier removal planning and the restoration of free-flowing rivers. *Journal of Environmental Management*, 92, 3112–3120.

Oldani, N. O. & Baigún, C. R. M. (2002). Performance of a fishway system in a major South American dam on the Parana River (Argentina/Paraguay). *River Research and Applications*, 18, 171–183.

Parrish, D. L., Behnke, R. J., Gephard, S. R., McCormick, S. D. & Reeves, G. H. (1998). Why aren't there more Atlantic salmon (*Salmo salar*)? *Canadian Journal of Fisheries and Aquatic Sciences*, 55, 281–287.

Patronski, T., Charbonneau, C., Dryer, P., *et al.* (2009). Fish passage and stream barrier management in the Bad River watershed in northern Wisconsin. http://training.fws.gov/courses/ALC/ALC3159/reports/final-reports/2008FR/fish_passage_final.pdf

Pelicice, F. & Agostinho, A. A. (2008). Fish-passage facilities as ecological traps in large Neotropical rivers. *Conservation Biology*, 22, 180–188.

Pepino, M., Rodriguez, M. A. & Magnan, P. (2012). Fish dispersal in fragmented landscapes: a modeling framework for quantifying the permeability of structural barriers. *Ecological Applications*, 22, 1435–1445.

Perkin, J. S. & Gido, K. B. (2012). Fragmentation alters stream fish community structure in dendritic ecological networks. *Ecological Applications*, 22, 2176–2187.

Pikitch, E. A., Doukakis, P., Lauck, L., Chakrabarty, P. & Ericksen, D. L. (2005). Status, trends and management of sturgeon and paddlefish fisheries. *Fish and Fisheries*, 6, 233–265.

Poff, N. L., Allan, J. D, Bain, M. B., *et al.* (1997). The natural flow regime. *BioScience*, 47, 769–784.

Poff, N. L., Olden, J. D., Merritt, D. M. & Pepin, D. M. (2007). Homogenization of regional river dynamics by dams and global biodiversity implications. *Proceedings of the National Academy of Sciences*, 104, 5732–5737.

Pompeu, P. S., Agostinho, A. A. & Pelicice, F. M. (2012). Existing and future challenges: the concept of successful fish passage in South America. *River Research and Applications*, 28, 504–512.

Powers, S. M., Julian, J., Doyle, M. & Stanley, E. H. (2013). Retention and transport of nutrients in a mature agricultural impoundment. *Journal of Geophysical Research: Biogeosciences*, 118, 91–103.

Pracheil, B. M., Pegg, M. A. & Mestl, G. E. (2009). Tributaries influence recruitment of fish in large rivers. *Ecology of Freshwater Fish*, 18, 603–609.

Pracheil, B. M., Pegg, M. A., Powell, L. A. & Mestl, G. E. (2012). Swimways: protecting paddlefish through movement-centered management. *Fisheries*, 37, 449–457.

Quinn, T. P. (2005). *The Behavior and Ecology of Pacific Salmon and Trout.* Bethesda, MD: American Fisheries Society.

Quinn, T. P. & Adams, D. J. (1996). Environmental changes affecting the migratory timing of American shad and sockeye salmon. *Ecology*, 77, 1151–1162.

Radtke, R. L. & Kinzie, R. A. (1996). Evidence of a marine larval stage in endemic Hawaiian stream gobies from isolated high-elevation locations. *Transactions of the American Fisheries Society*, 125, 613–621.

Rechisky, E. L., Welch, D. W., Porter, A. D., Jacobs-Scott, M. C. & Winchell, P. M. (2013). Influence of multiple dam passage on survival of juvenile Chinook salmon in the Columbia River estuary and coastal ocean. *Proceedings of the National Academy of Sciences*, 110, 6883–6888.

Reidy Liermann, C., Nilsson, C., Robertson, J. & Ng, R. (2012). Implications of dam obstruction for global freshwater fish diversity. *BioScience*, 62, 539–548.

Renaud, C. B. (1997). Conservation status of northern hemisphere lampreys (Petromyzontidae). *Journal of Applied Ichthyology*, 13, 143–148.

Richards, R. A. & Rago, P. J. (1999). A case history of effective fishery management: Chesapeake Bay striped bass. *North American Journal of Fisheries Management*, 19, 356–375.

Rinella, D. J., Wipfli, M. S., Stricker, C. A., Heintz, R. A. & Rinella, M. J. (2012). Pacific salmon (*Oncorhynchus* spp.) runs and consumer fitness: growth and energy storage in stream-dwelling salmonids increase with salmon spawner density. *Canadian Journal of Fisheries and Aquatic Sciences*, 69, 73–84.

Roscoe, D. W. & Hinch, S. G. (2010). Effectiveness monitoring of fish passage facilities: historical trends, geographic patterns and future directions. *Fish and Fisheries*, 11, 12–33.

Ruckelshaus, M. H., Levin, P., Johnson, J. B. & Kareiva, P. M. (2002). The Pacific salmon wars: what science brings to the challenge of recovering species. *Annual Reviews of Ecology and Systematics*, 33, 665–706.

Schindler, D. E., Hilborn, R., Chasco, B., *et al.* (2010). Population diversity and the portfolio effect in an exploited species. *Nature*, 465, 609–612.

Scott, M. L., Friedman, J. M. & Auble, G. T. (1996). Fluvial processes and the establishment of bottom-land trees. *Geomorphology*, 14, 327–339.

Sedell, J. R., Reeves, G. H., Hauer, F. R., Stanford, J. A. & Hawkins, C. P. (1990). Role of refugia in recovery from disturbances: modern fragmented and disconnected river systems. *Environmental Management*, 14,711–724.

Smith, D. M., Welsh, S. A. & Turk, P. J. (2012). Available benthic habitat type may influence predation risk in larval lampreys. *Ecology of Freshwater Fish*, 21, 160–163.

Stanley, E. H. & Doyle, M. W. (2003). Trading off: the ecological effects of dam removal. *Frontiers in Ecology and the Environment*, 1, 15–22.

Stein, B. A., Staudt, A., Cross, M. S., *et al.* (2013). Preparing for and managing change: climate adaptation for biodiversity and ecosystems. *Frontiers in Ecology and Environment*, 11, 502–510.

Stone, R. (2007). The last of the leviathans. *Science*, 316, 1684–1688.

Stuart, I. G. & Mallen-Cooper, M. (1999). An assessment of the effectiveness of a vertical slot fishway for non-salmonid fish at a tidal barrier on a large tropical/subtropical river. *Regulated Rivers: Research and Management*, 15, 575–590.

Taylor, B. W., Flecker, A. S. & Hall, R. O. (2006). Loss of a harvested fish species disrupts carbon flow in a diverse tropical river. *Science*, 31, 833–836.

Theriault, V., Dunlop, E. S., Dieckmann, U., Bernatchez, L. & Dodson, J. J. (2008). The impact of fishing-induced mortality on the evolution of alternative life-history tactics in brook charr. *Evolutionary Applications*, 1, 409–423.

Thiem, J. D., Binder, T. R., Dumont, P., *et al.* (2013). Multispecies fish passage behaviour in a vertical slot fishway on the Richelieu River, Quebec, Canada. *River Research and Applications*, 29, 582–592.

Thorstad, E. B., Fleming, I. A., McGinnity, P., *et al.* (2008). Incidence and impacts of escaped farmed Atlantic salmon *Salmo salar* in nature. NINA Special Report 36. World Wildlife Fund, Inc.

Tiegs, S. D., Levi, P. S., Ruegg, J., *et al.* (2011). Ecological effects of live salmon exceed those of carcasses during an annual spawning migration. *Ecosystems*, 14, 598–614.

Todd, B. D., Bergeron, C. M., Hepner, M. J. & Hopkins, W. A. (2011). Aquatic and terrestrial stressors in amphibians: a test of the double jeopardy hypothesis based on maternally and trophically derived contaminants. *Environmental Toxicology and Chemistry*, 30, 2277–2284.

Tsunagawa, T. & Arai, T. (2009). Migration diversity of the freshwater goby *Rhinogobius* sp. BI, as revealed by otolith Sr:Ca ratios. *Aquatic Biology*, 5, 187–194.

Velez-Espino, L. A., McLaughlin, R. L., Jones, M. L. & Pratt, T. C. (2011). Demographic analysis of trade-offs with deliberate fragmentation of streams: control of invasive species versus protection of native species. *Biological Conservation*, 144, 1068–1080.

Visser, M. E. & Both, C. (2005). Shifts in phenology due to global climate change: the need for a yard-stick. *Proceedings of The Royal Society B – Biological Sciences*, 272, 2561–2569.

Vörösmarty, C. J., Maybeck, M., Fekete, B., *et al.* (2003). Anthropogenic sediment retention: major global impact from registered river impoundments. *Global and Planetary Change*, 39, 169–190.

Vörösmarty, C. J., McIntyre, P. B., Gessner, M. O., *et al.* (2010). Global threats to human water security and river biodiversity. *Nature*, 467, 555–561.

Walter, R. P., Hogan, J. D., Blum, M. J., *et al.* (2012). Climate change and conservation of amphidromous fishes endemic to Hawaiian streams. *Endangered Species Research*, 16, 261–272.

Walters, A. W., Barnes, R. T. & Post, D. M. (2009). Anadromous alewives (*Alosa pseudoharengus*) contribute marine-derived nutrients to coastal stream food webs. *Canadian Journal of Fisheries and Aquatic Sciences*, 66, 439–448.

Wang, C. Y., Wei, Q. W., Kynard, B., Du, H. & Zhang, H. (2012). Migrations and movements of adult Chinese sturgeon *Acipenser sinensis* in the Yangtze River, China. *Journal of Fish Biology*, 81, 696–713.

Ward, J. V. & Stanford, J. A. (1983). The serial discontinuity concept of lotic ecosystems. In *Dynamics of Lotic Ecosystems*. Ann Arbor, MI: Ann Arbor Science, pp. 29–42.

Welcomme, R. L. (1985). River fisheries [Pesca fluvial]. *FAO Fisheries Technical Paper*.

Welcomme, R. L., Cowx, I. G., Coates, D., *et al.* (2010). Inland capture fisheries. *Philosophical Transactions of the Royal Society Biological Sciences*, 365, 2881–2896.

Welsh, A. (2004). Factors influencing the effectiveness of local versus national protection of migratory species: a case study of lake sturgeon in the Great Lakes. *Environmental Science and Policy*, 7, 315–328.

Wilbur, H. M. (1980). Complex life cycles. *Annual Review of Ecology and Systematics*, 11, 67–93.

Wilcove, D. S. & Wikelski, M. (2008). Going, going, gone: is animal migration disappearing? *PLoS Biology*, 6, e188.

Williams, J. G. (2008). Mitigating the effects of high-head dams on the Columbia River, USA: experience from the trenches. *Hydrobiologia*, 609, 241–251.

Williams, J. G., Armstrong, G., Katopodis, C., Lariniere, M. & Travade, F. (2012). Thinking like a fish: a key ingredient for development of effective fish passage facilities at river obstructions. *River Research and Applications*, 28, 407–417.

Willson, M. F. & Halupka, K. C. (1995). Anadromous fish as keystone species in vertebrate communities. *Conservation Biology*, 9, 489–497.

Winemiller, K. O. & Jepsen, D. B. (2004). Migratory Neotropical fish subsidize food webs of oligotrophic blackwater rivers. In *Food Webs at the Landscape Level*. Chicago, IL: University of Chicago Press, pp. 115–132.

Wipfli, M. S., Hudson, J. & Caouette, J. (1998). Influence of salmon carcasses on stream productivity: response of biofilm and benthic macroinvertebrates in southeastern Alaska, USA. *Canadian Journal of Fisheries and Aquatic Sciences*, 55, 1503–1511.

Wollebæk, J., Heggenes, J. & Roed, K. H. (2011). Population connectivity: dam migration mitigations and contemporary site fidelity in arctic char. *BMC Evolutionary Biology*, 11, 207.

Ziv, G., Baran, E., Nam, S., Rodríguez-Iturbe, I. & Levin, S. A. (2012). Trading-off fish biodiversity, food security, and hydropower in the Mekong River Basin. *Proceedings of the National Academy of Sciences*, 109, 5609–5614.

Protecting apex predators

KIRK O. WINEMILLER, PAUL HUMPHRIES
AND BRADLEY J. PUSEY

12.1 INTRODUCTION

In recent years, there has arisen within the popular media a fascination with large, predatory freshwater fishes. Television shows, inter-
net videos, blogs and magazine articles tell of adventures stalking 'river
monsters', 'megafish' and 'monster fish'. 'Extreme fishing' for giant
freshwater fish in exotic locations has become somewhat of a fad. Even
before this modern megafish fascination, large predatory fish species
were icons of freshwater fish faunas. Today, many of these species
also are among the most imperilled. Many features of their life histories and habitats combine to present particularly difficult management
challenges (Chapter 2). This chapter presents an overview of the current understanding of the diversity, ecology and conservation status of
freshwater fishes that are apex predators – fishes that occupy positions
at the top of the food web. Large apex predators, arbitrarily defined here
as fishes with the potential to achieve a total length of at least 1 m, tend
to be strongly impacted by fisheries and other human actions.

Apex predators in freshwater ecosystems merit special attention
from ecologists and conservationists for several reasons. First, they generally rank among the most valuable and heavily exploited fish stocks
(Allan *et al.*, 2005; Stone, 2007). Second, they play important roles
in communities and ecosystems that can profoundly affect ecological
structures and dynamics. Third, they often have life-history traits, such
as migration or brood guarding, that make them highly vulnerable to
impacts (Chapter 2). Given their position atop food webs, these large
predators tend to have lower population densities than fishes at lower

Conservation of Freshwater Fishes, eds G. P. Closs, M. Krkosek and J. D. Olden. Published by
Cambridge University Press. © Cambridge University Press 2016.

trophic levels, even within undisturbed ecosystems. Consequently, apex predators are particularly vulnerable to overexploitation and ecosystem alteration. In this regard, freshwater communities might not differ greatly from terrestrial and marine communities, and top predators oftentimes are the first species to be extirpated when humans begin to exploit renewable natural resources and to impact ecosystems (e.g. wolves in Europe and North America; lions, cheetahs and wild dogs in Africa; thylacines in Tasmania; sharks and groupers of coral reefs; sharks and billfishes in tropical oceans).

What qualifies a species to be categorised as an apex predator? The term implies that the species sits atop a trophic pyramid of population density, biomass, or energy. Some define apex predators as species having no natural enemies, but this of course is nonsense, because all species have predators, even if we exclude parasites from consideration. In the case of large predatory fishes, early life stages generally suffer high predation mortality. In contrast to terrestrial systems and in common with marine systems, freshwater apex predators are frequently included in commercial and artisanal harvests (see Chapter 15). In freshwater systems of South America, Southeast Asia and Australasia, fish apex predators may also form part of the diet of crocodilians. Whether or not we adopt a trophic pyramid perspective, it can be said that apex predators have high trophic positions supported by multiple food chains, each having multiple links. Thus, apex species typically are trophic generalists that exploit a diversity of local prey populations. However, this does not imply that apex predators are not selective feeders, and trophic niche partitioning may occur among a guild of predatory species with high trophic positions in the food web. For example, some large predatory fishes of South American rivers feed primarily near the bottom of the river channel (*Brachyplatystoma* spp.), others feed near the surface or at midwater (*Hydrolycus* spp.), and others feed mostly in littoral habitats within the channel and floodplain habitats (*Pseudoplatystoma* spp., *Cichla* spp.) (Winemiller, 2001; Barbarino Duque & Winemiller, 2003).

Apex predators tend to be large. Given that mouth gape normally limits the range of potential prey sizes that can be consumed by a fish, a large fish can normally exploit a wide range of prey sizes. However, there are some predatory fishes that are not gape-limited, and these species are sometimes quite small and yet occupy high trophic positions. Almost all of these cases are observed within tropical freshwater systems: South American piranhas (Serrasalmidae) that remove chunks of flesh or fins, African distichodontids that remove fins, South American characids

(*Roeboides* spp., *Exodon paradoxus*) and Asian glassfish (Ambassidae) that consume scales, South American cetopsid catfishes that consume chunks of flesh, and South American trichomycterid catfishes that consume mucus or blood (Baskin *et al.*, 1980; Sazima, 1983; Winemiller & Yan, 1989; Roberts, 1990). In temperate freshwaters, several species of lampreys (Petromyzontidae) orally attach themselves to fishes and consume flesh and body fluids from the wound. Although most of these species could be considered apex predators, we will not discuss them in this chapter because, with few known exceptions, their populations have not been as greatly impacted by human actions, and they appear to have greater demographic resilience than large apex predators (discussed below).

12.2 A BRIEF GLOBAL SURVEY OF FISHES THAT ARE APEX PREDATORS IN FRESHWATER ECOSYSTEMS

Large predatory fish are present within virtually every kind of freshwater ecosystem on every continent except Antarctica (Figures 12.1–12.3; Table 12.1; Chapter 1). Possible exceptions are small or ephemeral water bodies, and yet even some of these habitats support piscivorous fishes of impressive size. African lungfishes (*Protopterus* spp.), Asian snakeheads (*Channa* spp.), the South American electric eel (*Electrophorus electricus*), Australasian barramundi (*Lates calcarifer*; Figure 12.3) and sawfish (*Pristis microdon*, now considered a synonym of *P. pristis*; Figure 12.1), and North American gars may occupy shallow floodplain or estuarine habitats for major portions of their life cycles. The majority of species listed in Table 12.1 are normally found in large rivers and/or lakes within major drainage basins of continents, with some species migrating to estuarine or coastal marine habitats during portions of their life cycles.

12.2.1 Africa

Large apex predators of African rivers include the aba (*Gymnarchus niloticus*) and Cornish jack (*Mormyrops deliciosus*), two mormyriform fishes that are nocturnal and navigate by sensing perturbations to weak electric fields that they generate with special cells located along their flanks. The Cornish jack is broadly distributed in the West African, Nilo-Sudanian, Guinean, Congo and Zambezi regions of the continent, and the aba is distributed throughout much of West Africa and the Nilo-Sudanian regions. Most African rivers and lakes are inhabited by large predatory

Figure 12.1 Large predatory fishes that have been severely impacted by over-harvest, dams and habitat alteration. (A) Large-tooth sawfish (*Pristis pristis*) inhabit warm coastal waters and coastal rivers globally, but the size, density and distributions of populations have been severely reduced by over-harvest. (B) Ranking among the largest freshwater fishes, the Chinese paddlefish (*Psephurus gladius*) is listed as critically endangered and possibly extinct. (C) Beluga sturgeon (*Huso huso*), formerly common in the Adriatic, Black and Caspian Seas and major tributary rivers, is severely threatened. (D) The endangered dog-eating catfish (*Pangasius sanitwongsei*), a severely threatened apex predator from the Mekong River, Southeast Asia. (E) The Colorado pikeminnow (*Ptychocheilus lucius*), a large apex predator of the Colorado River Basin in western North America, has been greatly impacted by dams and water withdrawal. Except for the sawfish (an elasmobranch with long life span, slow growth, late maturity and low fecundity), these fishes migrate between spawning and feeding areas, have high fecundity, and are long-lived. [Photos courtesy of (A) K. Winemiller, (B) Wei Qi Wei, (C) Phaedra Doukakis, (D) Zeb Hogan and (E) US Fish & Wildlife Service.]

catfishes of the families Bagridae, Clariidae and Claroteidae (Skelton, 1993), with the largest reported to be the kokuni (*Chrysichthys cranchii*, Claroteidae) at 2 m and 130 kg. The sharptooth catfish (*Clarias gariepinus*) and vundu (*Heterobranchus longifilis*) have been reported to reach similar lengths but not weights (Table 12.1). These clariid catfishes have accessory aerial respiration, broad geographic distributions and broad diets that may include macroinvertebrates, amphibians and vegetation in addition to fish. The largest and probably most well-known large predatory fish in Africa is the Nile perch (*Lates niloticus*), a member of

Figure 12.2 Large predatory fishes that have experienced heavy exploitation resulting in stocks with reduced densities, body size distributions and geographic ranges, even though they continue to sustain fisheries in some areas: (A) Murray cod (*Maccullochella peelii*), southeastern Australia; (B) taimen (*Hucho taimen*) of central and northeastern Asia; (C) piraíba (*Brachyplatystoma filamentosum*) of the Amazon and Orinoco rivers in South America; (D) arapaima (*Arapaima gigas*) of the Amazon; (E) alligator gar (*Atractosteus spatula*) of North America; (F) goonch (*Bagarius yarelli*) of southern and southeast Asia, Borneo and Sumatra. [Photos courtesy of (A) Clayton Sharp, (B) Zeb Hogan, (C & E) K. Winemiller, (D) Caroline Arantes and (F) Sunil Choudhary.]

Latidae that also contains freshwater, estuarine and coastal marine species including barramundi. The Nile perch is entirely restricted to freshwater ecosystems, primarily large rivers and lakes of the Nile, Niger and Congo basins. The Nile perch can reach 2.5 m and 250 kg, and several other species in the genus also attain sizes well in excess of 1 m. The Nile perch is native to Lake Tanganyika in the East African rift valley, where it coexists with about 400 fish species, including over 250 cichlid species, nearly all of which are endemic to the lake. In the 1950s, the colonial

Figure 12.3 Large predatory fishes that have been introduced into non-native habitats to support recreational fishing: (A) African tigerfish (*Hydrocynus vittatus*) stocked in lakes in South Africa and Zimbabwe; (B) giant snakehead (*Channa micropeltes*) introduced into lakes throughout southeast Asia and Indonesia; (C) barramundi (*Lates calcarifer*) extensively stocked in lakes and rivers in Australia and the Indo-Pacific region for sportfishing; (D) peacock cichlid (*Cichla temensis*) targeted for sportfishing in the Amazon and Orinoco river basins of South America and introduced to other tropical regions; (E) northern pike (*Esox lucius*) an apex predator with a circumpolar distribution; (F) channel catfish (*Ictalurus punctatus*) a predatory catfish in rivers and lakes of throughout North America that is stocked extensively in lakes for sportfishing and an important species for aquaculture. [Photos courtesy of (A, B, D & E) K. Winemiller, (C) Chris Errity and (F) Brad Durick.]

Table 12.1 *Examples of large piscivorous fishes found in freshwater ecosystems of continents: maximum size, maximum weight, freshwater residence vs. diadromy, life-history strategy, principal threats to stocks.*

Species, maximum length, maximum weight, residence, life-history strategy – principal impacts

Australia

Bull shark (*Carcharhinus leucus*), > 2 m, 100 kg, diadromous, equilibrium – dams, diversions, inadvertent harvest

Freshwater sawfish (*Pristis pristis*), > 2 m, diadromous, equilibrium – dams, diversion, inadvertent harvest

Longfin eel (*Anguilla reinhardtii*), 2 m, 45 kg, diadromous, periodic – dams, diversion

Barramundi (*Lates calcarifer*), 1.8 m, 60 kg, diadromous, periodic – dams, diversion, harvest

Murray cod (*Maccullochella peelii*), 1.5 m, 100 kg, freshwater, equilibrium – harvest, dams, diversion, pollution

Asia

Giant stingray (*Himantura polylepis*), 2.4 m, 600 kg, freshwater, equilibrium – harvest, pollution

Siberian sturgeon (*Acipenser baerii*), 3 m, 210 kg, diadromous, periodic – harvest, dams

Chinese sturgeon (*Acipenser sinensis*), 4 m, 550 kg, diadromous, periodic – harvest, dams, pollution

Kaluga sturgeon (*Huso dauricus*), 5.6 m, 1000 kg, diadromous, periodic – harvest, dams

Giant featherfin (*Chitala lopis*), 1.5 m, 27 kg, freshwater, equilibrium – harvest, pollution

Taimen (*Hucho taimen*), 2.1 m, 105 kg, freshwater, periodic – harvest, diversions, pollution

Japanese huchen (*Hucho perryi*), 2.1 m, 70 kg, diadromous, periodic – harvest, dams

Northern pike (*Esox lucius*), 1.7 m, 37 kg, freshwater, periodic – harvest

Yellowcheek (*Elopichthys bambusa*), 1.65 m, 54 kg, freshwater, periodic – harvest, diversions, pollution

Mangar (*Luciobarbus esocinus*), 2.3 m, 140 kg, freshwater, periodic – harvest, dams, diversion, pollution

Striped pikecarp (*Luciobarbus striolatus*), 2.0 m, 50 kg, freshwater, periodic – harvest, dams, pollution

Golden mahseer (*Tor putitora*), 2.75 m, 63 kg, freshwater, periodic – harvest, dams, habitat loss, pollution

Redfin mahseer (*Tor tor*), 1.75 m, 78 kg, freshwater, periodic – harvest, dams, habitat loss, pollution

Altai osman (*Oreoleuciscus potanini*), 1.0 m, 10 kg, freshwater and saline inland waters – harvest

Giant pangasius, dog-eating catfish (*Pangasius sanitwongsei*), 3 m, 200 kg, freshwater, periodic – harvest, dams, pollution

Soldatov's catfish (*Silurus soldatovi*), 4 m, 40 kg, freshwater, periodic – harvest

Table 12.1 *(cont.)*

Yangtze catfish (*Silurus meridionalis*), 1.5 m, 30 kg, freshwater, periodic – harvest, dams, pollution
Asian redtail catfish (*Hemibagrus wyckioides*), 1.8 m, 86 kg freshwater, periodic – harvest, pollution
Goonch (*Bagarius yarrelli*), 2 m, 105 kg, freshwater, periodic – harvest, dams, pollution
Wallago (*Wallago attu*), 2.4 m, 150 kg, freshwater. periodic – harvest, dams, pollution
Striped wallago (*Wallago leeri*), 1.6 m, 70 kg, freshwater, periodic – harvest, dams, pollution
Burbot (*Lota lota*), 1.5, 34 kg, freshwater, periodic – relatively low harvest
Great snakehead (*Channa marulia*), 1.8, 30 kg, freshwater, equilibrium – harvest
Giant snakehead (*Channa micropeltes*), 1.8 m, 30 kg, freshwater, equilibrium – harvest

Africa
Marbled lungfish (*Protopterus aethiopicus*), 2 m, 50 kg, freshwater, equilibrium – harvest, diversion
Cornish jack (*Mormyrops anguilloides*), 1.5 m, 15 kg, freshwater, periodic – harvest
Aba (*Gymnarchus niloticus*), 1.7 m, 18 kg, freshwater, equilibrium – harvest, diversion
Mottled eel (*Anguilla bengalensis*), 1.8 m, 23 kg, diadromous, periodic – dams, diversion
Giant mottled eel (*Anguilla marmorata*), 2 m, 28 kg, diadromous, periodic – dams, diversion
Tigerfish (*Hydrocynus vittatus, H. forskahlii, H. brevis*), 1 m, 28 kg, freshwater, periodic – harvest, dams
Goliath tigerfish (*Hydrocynus goliath*), 1.5 m, 70 kg, freshwater, periodic – harvest
Kokuni (*Chrysichthys cranchii*), 2 m, 130 kg, freshwater, periodic – harvest
Semutundu (*Bagrus docmak*), 1.6 m, 50 kg, freshwater, periodic – harvest
Kukumai (*Bathybagrus grandis*), >2 m, 190 kg, freshwater, periodic – harvest
Sharptooth catfish (*Clarias gariepinus*), 2 m, 59 kg, freshwater, periodic – harvest
Vundu (*Heterobranchus longifilis*), 2 m, 55 kg, freshwater, periodic – harvest
Giant Lake Malawi catfish (*Bathyclarias gigas*), 2 m, 90 kg, freshwater, periodic – harvest
Nile perch (*Lates niloticus*), 2.5 m, 250 kg, freshwater, periodic – harvest
Tanganyika lates (*Lates angustifrons*), 2 m, 100 kg, freshwater, periodic – harvest
Giant cichlid (*Boulengerochromis microlepis*), 0.9 m, 5 kg, freshwater, equilibrium – low harvest

South America
Disc ray (*Paratrygon ajereba*), 1.3 m, 60 kg, freshwater, equilibrium – harvest
Arapaima (*Arapaima gigas*), 3 m, 200 kg, freshwater, equilibrium – harvest
Electric eel (*Electrophorus electricus*), 2.8 kg, 41 kg, freshwater, equilibrium – low harvest
Dorado (*Salminus brasiliensis*), 1.4 m, 34 kg, freshwater, periodic – dams, diversions, harvest
Payara (*Hydrolycus armatus, H. scomberoides*), 1.3 m, 30 kg, freshwater, periodic – harvest, dams

Table 12.1 *(cont.)*

Trahira, aymara (*Hoplias amaira, H. macrophthalmus*), 1.2 m, 40 kg, freshwater, equilibrium – harvest

Dourada (*Brachyplatystoma rousseauxii*), 1.9 m, 100 kg, freshwater, periodic – harvest

Piraíba (*Brachyplatystoma filamentosum*), 2.7 m, 200 kg, freshwater, periodic – harvest

Pintado (*Pseudoplatystoma coruscans*), 2.5 m, 150 kg, freshwater, periodic – dams, diversions, harvest

Bagre rayado (*Pseudoplatystoma* spp.), 1.3 m, 70 kg, freshwater, periodic – harvest

Cajaro, piarara (*Phractocephalus hemioliopterus*), 1.8 m, 80 kg, freshwater, periodic – harvest

Peixe-lenha (*Sorubimichthys planiceps*), 2 m, 50 kg, freshwater, periodic – harvest

Manguruyu (*Zungaro jahu*), 2.5 m, 178 kg, freshwater, periodic – dams, diversions, harvest

Jaú (*Zungaro zungaro*), 2 m, 90 kg, freshwater, periodic – harvest

Peacock cichlid (*Cichla temensis*), 1 m, 12 kg, freshwater, equilibrium – harvest

North America

Alligator gar (*Atractosteus spatula*), 3 m, 137 kg, freshwater, periodic – harvest, dams, diversions

Longnose gar (*Lepisosteus osseus*), 2 m, 22 kg, freshwater, periodic – low harvest

Lake sturgeon (*Acipenser fulvescens*), 2.7 m, 140 kg, freshwater, periodic –harvest, pollution

Pallid sturgeon (*Scaphirhynchus albus*), 2 m, 130 kg, freshwater, periodic – dams, diversions, harvest

White sturgeon (*Acipenser transmontanus*), 6.1 m, 830 kg, diadromous, periodic – dams, diversions, harvest

Sheefish (*Stenodus leucichthys*), 1.5 m, 40 kg, diadromous, periodic – harvest

Atlantic salmon (*Salmo trutta*), 1.4 m, 50 kg, diadromous, periodic – dams, diversions, harvest, pollution

Lake trout (*Salvelinus namaycush*), 1.8 m, 46 kg, freshwater, periodic – harvest, pollution, exotics

Muskellunge (*Esox masquinongy*), 1.9 m, 36 kg, freshwater, periodic – harvest, pollution

Northern pike (*Esox lucius*), 1.7 m, 37 kg, freshwater, periodic – harvest

Colorado pikeminnow (*Ptychocheilus lucius*), 2 m, 35 kg, freshwater, periodic – dams, diversions, exotics

Blue catfish (*Ictalurus furcatus*), 1.6 m, 68 kg, freshwater, equilibrium – harvest, dams, diversions

Channel catfish (*Ictalurus punctatus*), 1.3 m, 26.3 kg, freshwater, equilibrium – harvest, dams, diversions

Flathead catfish (*Pylodictis olivaris*), 1.7 m, 63 kg, freshwater, equilibrium – harvest, dams, diversions

Burbot (*Lota lota*), 1.5, 34 kg, freshwater, periodic – relatively low harvest

Largemouth bass (*Micropterus salmoides*), 1 m, 12 kg, freshwater, equilibrium – harvest, pollution

Walleye (*Sander vitreus*), 1 m, 11 kg, freshwater, periodic – harvest, dams, diversions, pollution

Table 12.1 *(cont.)*

Striped bass (*Morone saxatilis*), 1.8 m, 56 kg, diadromous, periodic – dams, diversions, harvest, pollution

Europe

Beluga sturgeon (*Huso huso*), 8 m, 3200 kg, diadromous, periodic – harvest, dams, diversions, pollution

Danube sturgeon (*Acipenser gueldenstaidtii*), 2.3 m, 115 kg, diadromous, periodic – harvest, dams, diversions, pollution

Sturgeon (*Acipenser sturio*), 6 m, 400 kg, diadromous, periodic – harvest, dams, diversions, pollution

Brown trout (*Salmo trutta fario*), 1.2 m, 25 kg, diadromous, periodic – harvest, dams, diversions, pollution

Huchen (*Hucho hucho*), 1.8 m, 70 kg, freshwater, periodic – harvest, dams, diversions, pollution

Northern pike (*Esox lucius*), 1.5 m, 35 kg, freshwater, periodic – harvest

Greek catfish (*Silurus aristotelis*), 1.5 m, 50 kg, freshwater, periodic – harvest, pollution

Wels catfish, sheatfish (*Silurus glanis*), 4.5 m, 327 kg, freshwater, periodic – harvest, dams, diversions, pollution

Burbot (*Lota lota*), 1.5, 34 kg, freshwater, periodic – relatively low harvest

Pikeperch, zander (*Sander lucioperca*), 1.3 m, 20 kg, freshwater, periodic – harvest, dams, diversions, pollution

fisheries department introduced the Nile perch into Lake Victoria, another major rift lake that contains hundreds of endemic haplochromine cichlids. The Nile perch population expanded in Lake Victoria and now supports a major commercial fishery that exports much of the catch to international markets. The major impact of the Nile perch on endemic cichlid fishes and the Lake Victoria ecosystem has been much discussed, and is briefly summarised in Section 12.3.9.

12.2.2 Asia

A great diversity of fishes, including some of the largest fishes in the world, are found in the large river basins of Asia. Two of the largest freshwater fishes in the world occur in Asia: the predatory Chinese paddlefish (*Psephurus gladius*) of the Yangtze River (Wei *et al.*, 1997; Pikitch *et al.*, 2005) and the omnivorous giant Mekong catfish (*Pangasianodon gigas*) (Hogan *et al.*, 2004). Both of these species have been severely impacted by over-harvest and dams to the point that, given so few individuals remain in the wild, they seem destined for extinction (Stone, 2007). The rivers of tropical and subtropical southern Asia are home to numerous piscivorous fishes that measure well over 1 m (Table 12.1), and catfishes of the families Bagridae, Pangasidae, Siluridae and Sisoridae are well represented within this group. Other apex predators of Asian tropical rivers include the snakeheads (Channidae) and giant stingray (*Himantura polylepis*). Several tropical Asian carps (Cyprinidae) are piscivorous, but only *Aaptosyax grypus* of the middle Mekong Basin is reported to reach 1 m, and this species is apparently now rare (Rainboth, 1996). Two species of featherbacks (*Chitala* spp., Notopteridae) are reported to reach or exceed 1 m, and some of these osteoglossiform fishes are broadly distributed in Southeast Asia and support fisheries. These osteoglossiform fishes are found in diverse habitats in rivers, floodplains and lakes where they feed mostly on small fishes and, like the snakeheads, guard their broods for variable periods after the eggs hatch. The mangar (*Luciobarbus esocinus*, Cyprinidae) is the apex predator of large rivers in the Middle East and can attain lengths well over 2 m. In freshwaters of temperate regions of Asia, the largest piscivores are various species of sturgeons (Acipenseridae), pike (*Esox lucius*; Figure 12.3), taimen (*Hucho taimen*, Salmonidae; Figure 12.2), Japanese huchen (*Hucho perryi*), Soldatov's catfish (*Silurus soldatovi*, Siluridae) and burbot (*Lota lota*). Many stocks of these temperate-zone piscivores have been depleted by over-harvest.

12.2.3 Australia and the IndoPacific

Compared to other major landmasses at similar latitudes, Australia has relatively few native freshwater fishes and few that, for our purposes, would be classified as a large apex predator. One such species is the Murray cod (*Maccullochella peelii*; Figure 12.2) of the Murray–Darling Basin. The Murray cod is a large, long-lived carnivore whose diet is dominated by decapod crustaceans. It occupies the main channel of largely medium–large rivers and forms the basis of the most popular recreational fishery in southern inland Australia. Other species in this genus rarely exceed 1 m total length. Two Australian osteoglossiform species (bonytongues), *Scleropages leichardti* and *S. jardinii*, may attain 1 m in length, but terrestrial arthropods and not fishes are more commonly consumed by these inhabitants of pools in low-gradient streams. Fishes comprise about half of the diets of the freshwater ariid catfishes, *Neoarius midgleyi* and *N. paucus*, which uncommonly attain 1 m (Pusey *et al.*, 2004). At least three species of anguillid eel (*Anguilla reinhardtii*, *A. australis* and *A. obscura*) frequently exceed 1 m in length, and *A. reinhardtii* may exceed

2 m, especially when prevented from migrating seaward. Fishes make up less than 30% of the diet of these species, the remainder being comprised of a diverse array of invertebrates (Pusey *et al.*, 2004). A few large diadromous fishes are important apex predators in coastal drainages of Australasia and the IndoPacific, including the barramundi (Figure 12.3), bull shark (*Carcharhinus leucus*) and sawfish (Figure 12.1). These species have diverse diets dominated by fishes. The Papuan black bass (*Lutjanus goldiei*, a species belonging to the snapper family Lutjanidae) may attain lengths of about 1 m (Allen, 1991). New Zealand's longfin eel (*A. dieffenbachia*) grows larger than 1 m, and this diadromous species has a broad diet that sometimes includes fishes.

12.2.4 South America

The neotropical region has the richest freshwater fish fauna on Earth, and therefore it is logical that it also supports great diversity of large piscivores. The arapaima, or pirarucu (*Arapaima gigas*, Arapaimidae; Figure 12.2), is a bonytongue that can grow to over 3 m and 200 kg, which ranks it among the largest non-migratory freshwater fishes. The arapaima is highly valued as a food fish and stocks have been greatly depleted throughout the Amazon and Essequibo basins (discussed below). Other giant predatory fishes of the Amazon, Orinoco and Paraná basins include catfishes from diverse families, peacock cichlids (*Cichla* spp.) and several characiforms, including the trahira (*Hoplias amaira*), dourado (*Salminus brasiliensis*) and payara (*Hydrolycus armatus*). Pimelodid catfishes are the dominant large piscivores in the main channels of large tropical rivers in South America (Barthem & Goulding, 1997; Barbarino Duque & Winemiller, 2003), and at least two of them appear to make long-distance migrations (discussed below). These large catfishes have a periodic life-history strategy, with spawning taking place during the early phase of the annual flood pulse, and eggs and larvae drifting in the water column until they arrive in productive riparian wetlands. Large pimelodids are highly valued food fishes and have been heavily exploited. These stocks tend not to persist in reservoirs (Hoeinghaus *et al.*, 2009), where they are generally replaced at the top of the food web by peacock cichlids. Peacock cichlids (sometimes called peacock bass) are substrate nesters and biparental brood guarders that thrive in floodplain lakes as well as reservoirs (Winemiller, 2001). Rivers of small coastal drainages of South and Central America and the Caribbean lack large piscivorous fishes from freshwater families, but marine invaders frequently occupy the position of apex predator. Some of the common marine invaders in the neotropics include various species of snappers (Lutjanidae) and snooks (Centropomidae) as well as the tarpon (*Megalops atlanticus*, Megalopidae), bull shark and sawfish (*Pristis* spp.).

12.2.5 North America

Many of the largest apex predatory fishes of North America belong to the primitive subclasses Chondrostei (paddlefish and sturgeons) and Neopterygii (gars) (Table 12.1). The largest sturgeons are diadromous, and fishes of impressive size can be found in rivers far inland. The Great Lakes historically supported large sturgeon populations, but these were overfished in the nineteenth century, with much of the harvest taken as by-catch that was burned on the lakeshore or used as fertiliser (Trautman, 1981). Gars (Lepisosteidae) and paddlefish (*Polyodon spathula*) are essentially restricted to freshwater ecosystems of the Mississippi and other major river basins of the eastern and central regions of the continent, although the alligator gar and spotted gar are sometimes abundant in estuaries with high salinity. Sturgeons, gars and paddlefish traditionally have been considered 'trash fish' with little value as sportfish or tablefare; however, in recent years, there have developed sport fisheries for these species in some regions. Sturgeons and paddlefish are harvested in some regions for roe for caviar to address market demand in the face of much reduced sturgeon stocks in Eurasia. Other large piscivorous fishes in North America include the lake trout (*Salvelinus namaycush*), sheefish (*Stenodus leucichthys*), burbot (*Lota lota*), pikes (*Esox lucius, E. masquinongy*) and walleye (*Sander vitreus*) at northern latitudes, and ictalurid catfishes (blue catfish, channel catfish), largemouth bass (*Micropterus salmoides*) and striped bass (*Morone saxatilis*) at lower latitudes. Fish of the latter group have been introduced throughout the continent in support of sportfisheries, and the largemouth bass and channel catfish (Figure 12.3) have been introduced to other regions of the world.

12.2.6 Europe

Some of the high-latitude predatory fishes of North America are also found in Europe. For example, Northern pike and burbot are found within diverse freshwater habitats and have circumpolar distributions. The brown trout (*Salmo trutta*) is broadly distributed in rivers and streams of northern Europe, and due to its popularity as a sportfish, the species has been widely introduced in temperate regions of the Southern Hemisphere. An even larger salmonid is the huchen (*Hucho hucho*) of the Danube River (Table 12.1). Europe has several sturgeon species, all of which are diadromous, including the massive beluga sturgeon (*Huso huso*; Figure 12.1) of the Caspian Sea, a species now critically threatened (discussed further in the next section). Widely distributed in European rivers and lakes, the wels catfish (*Silurus glanis*) is an apex predator that ranks among the world's largest fishes restricted entirely to freshwater. The wels catfish has been widely introduced into reservoirs throughout Europe (Copp *et al.*, 2009), but its congener, the Greek catfish (*Silurus aristotelis*), has a very restricted range and is now considered threatened due to over-harvest and competition and predation from introduced wels catfish (Leonardos *et al.*, 2009).

12.3 THREATS AND CHALLENGES TO LARGE PREDATORY FISHES AND FRESHWATER BIODIVERSITY

Do large apex predators have characteristics that make them more vulnerable to human impacts compared to smaller fishes or fishes feeding at lower trophic positions? Let us first explore life-history strategies, which are syndromes (intercorrelated sets) of attributes associated with reproduction, growth and demography. Life-history strategies have ramifications for predicting how populations respond to impacts, such as harvest, habitat fragmentation from dams, and habitat degradation from pollution and flow regime alteration (Chapter 2). Many large apex predators are periodic strategists *sensu* the triangular life-history model of Winemiller (1989, 2005) and Winemiller and Rose (1992) (Figure 12.4; Table 12.1). Periodic strategists are broadcast spawners with high fecundity and episodic recruitment that reveal varying degrees of migration. This strategy exploits large-scale patterns of environmental variation in space and time that allows for strong recruitment when conditions are encountered that promote early life-stage growth and survival. Populations of long-lived periodic fishes frequently show uneven age distributions, in which a small number of year classes dominate the population. Other large predatory species are equilibrium strategists (Figure 12.4; Table 12.1) that construct nests and guard their broods, with many of these species revealing fewer migratory tendencies. Equilibrium strategists are presumed to have early life-stage survival, growth during the juvenile stage and reproductive success that are more density-dependent compared to periodic strategists. Large species

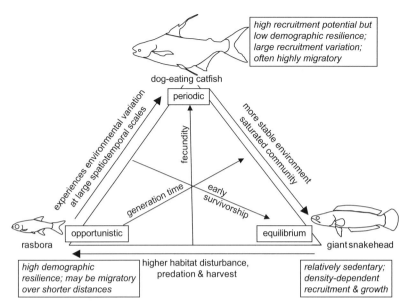

high recruitment potential but low demographic resilience; large recruitment variation; often highly migratory

dog-eating catfish

periodic

experiences environmental variation at large spatiotemporal scales

more stable environment saturated community

fecundity

generation time

early survivorship

opportunistic

equilibrium

rasbora

giant snakehead

high demographic resilience; may be migratory over shorter distances

higher habitat disturbance, predation & harvest

relatively sedentary; density-dependent recruitment & growth

Figure 12.4 Schematic diagram of a triangular continuum of life history strategies based on Winemiller (1989, 2005) and Winemiller and Rose (1992). Vectors inside the triangle show the direction of increase for key life-history variables; vectors along the edges of the triangle show environmental factors that select for end-point strategies. Examples are fishes from Southeast Asian rivers; large apex predators occupy positions near the periodic and equilibrium end-points of the continuum as well as the region between them.

of equilibrium and periodic strategists generally have delayed maturation, which is a major factor reducing the maximum rate of population increase, r_{max}, and limiting population resilience to disturbance. For example, most sturgeons achieve sexual maturation at 10–25 years, and some species have life spans estimated at over 100 years (Pikitch et al., 2005). Consequently, once a local stock has been reduced, it would require many years to rebuild it even after the sources of impact are eliminated.

The third life-history strategy (opportunistic) applies to small species and thus is not represented among the large apex predators. Opportunistic strategists mature at small sizes and have high, fairly continuous reproductive effort, high mortality at all size intervals and short life spans. By virtue of their short generation times and high reproductive efforts, opportunistic fishes have high demographic resilience and these stocks generally can withstand or rebound from high mortality from predation or habitat disturbances. Nonetheless, stocks of relatively small-bodied marine fishes with high economic value have been overfished in many regions of the world (Pinsky *et al.*, 2011), and in freshwater systems, many small-bodied fishes with restricted geographic distributions are threatened by habitat destruction (Winemiller, 2005; Olden *et al.*, 2007; Chapters 2 and 4). Many prey of piscivorous fishes, so-called 'forage fishes', such as anchovies (Engraulidae), minnows (Cyprinidae) and hardyheads/silversides (Atherinidae), are opportunistic strategists. The triangular life history model has been used to explain interannual population variation in relation to hydrology (Mims & Olden, 2012), fish assemblage composition in relation to hydrology (Tedesco *et al.*, 2008), extirpation of native fishes and invasion success by exotics (Olden *et al.*, 2006), resilience to harvest (Winemiller, 2005), and potential for sexual selection (Winemiller, 1992).

Several conservation scenarios involving large apex predators are summarised in the next section. These examples were chosen to illustrate the full range of situations faced by large predatory fishes: some are critically endangered, others have been severely impacted, some are presently stable yet vulnerable to human activities, and others are intensively managed with expanding geographic distributions from stocking. From these examples it is apparent that large apex predators are sensitive to harvest because they take a long time to attain sexual maturity and therefore have low demographic resilience. Froese's (2004) three tenets for sustainable fisheries are particularly relevant for these large, long-lived species: 'let them spawn', 'let them grow' and 'let the megaspawners live'. He suggested three management targets: (1) 100% mature fish in the catch, (2) 100% of individuals with optimum length in catch, and (3) zero harvest of megaspawners. Many of the most critically endangered large apex predators are periodic strategists that migrate long distances between feeding and spawning areas (Chapter 11). Large apex predators that have characteristics tending toward the equilibrium strategy (relatively sedentary, nesting and parental care) have populations sensitive to harvest, but also seem to respond more rapidly to fishing regulations than large periodic-type fishes.

12.3.1 Beluga and Danube sturgeons – critically endangered

The beluga sturgeon (*Huso huso*; Figure 12.1) and Danube, or Russian, sturgeon (*Acipenser gueldenstaedtii*) are two of the largest fishes that inhabit freshwater systems, and both are listed as *critically endangered* by the International Union for the Conservation of Nature (IUCN).

Beluga sturgeons are pelagic and feed mostly on fish, but they have also been reported to feed on birds and even seals. The Danube sturgeon feeds on benthic fishes and macroinvertebrates. Like most sturgeons, these species are diadromous, migrating long distances from coastal waters, such as the Black Sea and Caspian Sea, to spawning and feeding areas in large tributary rivers. High river flows apparently stimulate larger spawning migrations, and there is some evidence of multiple conspecific stocks that have peak migrations during different periods. These species are periodic strategists with high fecundity and are primarily exploited for their roe, which are marketed as Russian caviar, a luxury food item that commands exorbitant prices in the international marketplace (thousands of dollars per kilogram). Management of these sturgeon stocks is also complicated by their international jurisdiction. In the 1960s, the Soviet Union imposed a ban on open-sea fishing within the portion of Caspian Sea that was under its control, approximately four-fifths. Following the collapse of the Soviet Union, five nations now surround the Caspian Sea, and only Russia maintains a quota system which is largely ineffective due to poaching fuelled by the lucrative caviar market. According to Pikitch *et al.* (2005), sturgeon catches in the Caspian Sea peaked during the seventeenth century. In the 1970s, the catch peaked at 27,300 tonnes and declined to 1388 tonnes by 2002. Dams and pollution from industries, including oil production, have also significantly impacted the giant sturgeons of the Caspian Basin. An estimated 90% of spawning habitat on the Volga River is now inaccessible to sturgeons due to dams (Pikitch *et al.*, 2005). The Iron Gates Gorge dams, constructed on the Danube River between Romania and Serbia, created the largest reservoir/hydropower system on the river. Two dams are located 863 and 942 km upstream from the Danube Delta on the Black Sea, and they lack fish-passage structures and therefore have blocked beluga and Danube sturgeon migrations to spawning areas in the middle Danube. Many sturgeons were captured below the first dam during the first five years following its closure, but thereafter catches plummeted. The future survival of these huge apex predators may depend upon stocking programmes. For example, Khodorevskaya *et al.* (1997) estimated that 15–20 million fingerlings would need to be released from hatcheries each year to maintain the Caspian beluga sturgeon stock. However, Doukakis *et al.* (2010) concluded from analyses based on demographic models that stocking is an ineffective means to rebuild depleted beluga sturgeon stocks in the Caspian Sea and

Ural River, which support the only remaining population with significant natural reproduction. Their study concluded that the protection of large spawners has greater potential for recruitment and population growth than stocking, and given that maturation occurs at 9–20 years, they recommended banning harvest on fish younger than 31 years. Moreover, reliance on the production of recruits from natural spawning reduces the risks of unintended degradation of the gene pool.

12.3.2 Dog-eating catfish – critically endangered

The iconic fish of the Mekong River is the giant Mekong catfish (*Pangasianodon gigas*), a huge migratory omnivore that has been driven to the brink of extinction by over-harvest and dams (Hogan *et al.*, 2004; Chapter 1). Much less publicised, but nearly as large and perhaps even more endangered, is the dog-eating catfish (*Pangasius sanitwongsei*), or giant pangasius, a predator that may attain lengths of 3 m and weights of 300 kg. The fish's common name derives from its purported propensity to be caught on hooks baited with dog meat. Unlike the giant Mekong catfish, a robust fish reported to feed on detritus, vegetation and benthic invertebrates, the dog-eating catfish is a piscivore with a rather shark-like appearance (Figure 12.1). The dog-eating catfish was once common in the Chao Phrya and Mekong rivers in Southeast Asia, but now it appears to lack a self-sustaining population in the Chao Phrya and the Mekong population is severely endangered, with few catches reported in recent years (Hogan, 2011). The IUCN lists the species as *critically endangered*. Baran and Myschowoda (2009) identified dams and overfishing as the main impacts on this and other migratory catfish species of the Mekong. The dog-eating catfish would be classified as a periodic strategist in the extreme, with large body size, high fecundity and seasonal spawning without subsequent parental care. Like the giant Mekong catfish, the dog-eating catfish appears to undertake long-distance seasonal migrations between feeding areas in the lower river course in Thailand, Laos and Vietnam and regions upstream in Laos and on the Thailand–Laos border where spawning is suspected to occur during April and May. This region includes the site of the Xayaburi dam, a controversial hydroelectric dam being built in Laos on the mainstem of the Mekong River. This dam will not only block the migration route of the dog-eating catfish and other large migratory fishes, but it will alter the flow regime downstream which will impact many aspects of the aquatic and riparian ecosystems (Dugan *et al.*, 2010; Dudgeon, 2011; Chapter 4).

12.3.3 Murray cod – critically endangered, stocked

The Murray cod (*Maccullochella peelii*; Figure 12.2) can be found, at least historically, from clear upland streams to turbid lowland rivers, but likes cover, especially submerged logs or snags (Lintermans, 2007). It is an equilibrium species, maturing at 4–5 years when about 500 mm in length and breeding in spring/summer within the main channel of rivers. Maximum size can exceed 1.5 m and more than 100 kg and individuals of approximately 50 years old have been recorded. Murray cod can at times move 100–1000 km, although this is not necessary for completion of its life cycle. It spawns once per season, either in depressions in the bank or in snags, and lays between 10,000 and 90,000 eggs that are subsequently guarded by the male. The male continues to guard the newly hatched larvae until they are about a week old, at which time the young fish swim up into the water column and drift downstream for several days. The diet of adults historically included decapods, fish, freshwater mussels, water birds and mammals, but is now dominated by decapods and fish (Ebner, 2006). The species has suffered a decline in abundance and range since the mid 1850s, early on because of commercial fishing (Dakin & Kesteven, 1937; Rowland, 1989), but more recently other threats include recreational fishing, habitat alteration, river regulation, barriers to movement, thermal pollution and alien species. There was a profitable commercial fishery of this species for several decades; however, evidence for the impacts of fishing emerged within three years of the commencement of the fishery in the mid 1800s. By the 1950s, the fishery was no longer commercially viable in much of the Murray system (Rowland, 1989), and the Murray cod is currently listed as *critically endangered* by the IUCN, and as *vulnerable* under the Australian EPBC Act (1999). Recreational fishers are now encouraged to catch-and-release Murray cod, and a 2006–2008 survey showed that while most undersized fish were released (although ~2% of these die later), 68–86% of fish greater than 60 cm were kept (Fulton, 2011).

12.3.4 Taimen – impacted

The taimen (*Hucho taimen*) is the world's largest salmonid (Figure 12.2; Table 12.1) and an apex predator that feeds on diverse fishes and, to a lesser extent, small birds and mammals. Populations have greatly declined throughout its range in rivers draining into the Caspian Sea and Arctic and Pacific Oceans in northeastern Asia (Holcík *et al.*, 1988), and several countries list the species as threatened. According to Vander Zanden *et al.* (2007), this species has several traits that make it vulnerable to impact from exploitation – slow growth, long life span and naturally low population densities. Healthy stocks still exist in some rivers of northern Mongolia where historically there was little tradition of fishing (Jensen *et al.*, 2009). Immigrants to this region have introduced new fishing practices that are impacting stocks, and some areas are being developed for international sportfishing. Reproductive behaviour is similar to other trout species, with fish moving variable distances, often to tributaries, as spring temperatures warm to the appropriate range. The spawning season may be only one to two weeks, during which females release 10,000–30,000 eggs into redds excavated in the streambed. The taimen is estimated to mature at 3–8 years, requires 10 years to reach 1 m, and may live for more than 30 years (Holcík *et al.*, 1988; Jensen *et al.*, 2009). In areas where stocks are exploited, fish larger than 1 m generally are eliminated in less than 20 years (Jensen *et al.*, 2009). Jensen *et al.* (2009) used a model to simulate the effects of fishing on a taimen population in northern Mongolia that had not been exploited historically. The model predicted that a sportfishery that retained captured fish would have a 57% chance of eventually eliminating the population, but that a catch-and-release recreational fishery would reduce fish abundance and mean weight by less than 10%. These authors proposed that international tourism for catch-and-release sportfishing of taimen could be sustainable and provide economic benefits to offset increasing mining interests in the region and their associated environmental impacts.

12.3.5 Arapaima – impacted

The arapaima, or pirarucu (*Arapaima gigas*), is one of the most valuable fishes in the Amazon, and therefore one of the most intensively harvested. This giant boneytongue is particularly vulnerable to harvest using harpoons because of its relatively sedentary nature and the fact that it periodically surfaces to perform aerial respiration. The arapaima is most frequently found in floodplain lakes and may move into flooded forests and savannas during major flood pulses. Although fish of impressive size (Figure 12.2; Table 12.1) are still captured, the abundance and body size of these apex predators has greatly declined in most regions of the Amazon and Essequibo Basins (Arantes *et al.*, 2010). Recent taxonomic research has suggested that there actually are several *Arapaima* species (Stewart, 2013). Adults form pairs during the low-water period and become territorial. As the water level rises, pairs build nests and spawn along the margins of water bodies, and brood guarding lasts for about 3 months (Castello, 2008). Arapaima feed on diverse fishes, and large adults probably have few natural predators other than humans.

In Brazil, the minimum allowable size for harvest is 1.5 m, and fishing is closed from December to May. In an attempt to rebuild stocks, fishing has been banned in three states, but apparently the regulations have been poorly enforced (Castello *et al.*, 2009). In recent years, community-based management has allowed local stocks to recover in several areas. Fairly accurate population censuses can be made based on sightings of fish that surface to gulp air (Castello *et al.*, 2009), and this information facilitates agreement on harvest quotas. Within the Mamirauá Reserve in Brazil's Amazonas State, the overexploited arapaima stock increased several-fold after implementation of new harvest regulations and enforcement based on community-based management (Castello *et al.*, 2009). Castello *et al.* (2011) modelled arapaima population dynamics and found that recruitment was density-dependent in accordance with the Beverton–Holt relationship, and that stocks can sustain annual harvest of up to 25% of adults under enforcement of the minimum size and seasonal closure.

12.3.6 Alligator gar – impacted

Historically, the alligator gar (*Atractosteus spatula*) was a major predator of rivers in the Mississippi River valley from Ohio and Illinois to the Delta and Gulf Coast drainages of North America. It is a member of a primitive order of bony fishes (Lepisosteiformes) that possesses a modified swimbladder that can function as a lung for aerial respiration. The alligator gar can survive in water with degraded quality as well as moderate salinities. Fishes are the most important prey, but given their huge size (Figure 12.2; Table 12.1), alligator gars are capable of consuming a variety of vertebrate prey, including amphibians, turtles and birds. In the USA, alligator gars were long considered 'trash fish' that adversely affected populations of desirable gamefishes, such as largemouth bass and channel catfish (Scarnecchia, 1992). In recent years, the popularity of 'monsterfish' has stimulated sportfisheries for giant alligator gar in regions where they remain abundant. This species also is targeted by bowhunters who take aim when the fish surface to gulp air. In Mexico, gillnets are used to capture gar for commercial fisheries (Garcia de Leon *et al.*, 2001).

Female alligator gars scatter thousands of adhesive eggs over submerged vegetation along river margins, in oxbows and other floodplain habitats, but sometimes in reservoirs (McCarley & Hill, 1979). In some regions, alligator gars appear to move between deep pools of larger rivers where they reside most of the year to tributary streams and inundated floodplain habitats where they spawn. A tagging study indicated that larger gars have greater home ranges (Sakaris *et al.*, 2003). Dams that block longitudinal movement, levees that block lateral movement to riparian zones and water diversion structures that reduce high flow pulses have greatly reduced alligator gar populations to the point the species is considered threatened throughout most of its historic range. In Lake Pontchartrain, Louisiana, the alligator gar has declined by 99% in trawl and seine samples since the 1950s (O'Connell *et al.*, 2007). The species apparently maintains stable populations in the southern portion of its range within the lower Mississippi and several rivers of Texas and Mexico, and most natural resource agencies of states within the gar's historic range have initiated efforts to conserve or rehabilitate populations.

12.3.7 Dourada – vulnerable

The dourada (*Brachyplatystoma rousseauxii*) is a large, predatory pimelodid catfish of the Amazon and Orinoco Rivers and their major tributaries. In the Apure River (Orinoco Basin), the species feeds mostly on nocturnal, weakly electric gymnotiform fishes (Barbarino Duque & Winemiller, 2003). The dourada is believed to be a long-distance migratory species that uses the estuarine region of the Amazon–Solimoes River as a nursery and adult feeding ground during the high-water period (Barthem *et al.*, 1991; Barthem & Goulding, 1997). As seawater invades this region during the low-water period (June–October), large dourada schools move up the Amazon River and its tributaries. The schools are sometimes mixed with other catfish and characiform species. Upstream migrations are not believed to be spawning runs, because mature dourada have never been reported in these schools. Instead, they seem to be associated with dispersal and foraging (Barthem *et al.*, 1991). Mature individuals have been found at locations in the western region of the Amazon during the ascending phase of the annual flood pulse. The downstream movements of larvae within the main channel of the Amazon indicate that spawning may take place in these regions. Barthem *et al.* (1991) estimated that the strong flow of the Amazon during the high-water season (~11 km/h) could transport eggs and larvae a distance of 3500 km in as little as 13 days. They concluded that, during the high-water period, the estuary is the main nursery area for early life stages and also a feeding ground for many adults, and most spawning may occur in tributaries within the western portion of the basin.

The dourada is the most important species landed in the fishery of the Amazon estuary. Landings in the estuary increased every year between 2001 and 2004, from 1269 tons to 2938 tons (Batista *et al.*, 2012). Although the dourada fishery appears sustainable at these levels, given the large size of mature individuals, its population would be impacted by a sufficiently large increase in fishing effort. Given the dourada's migratory nature, its population is vulnerable to dams constructed on tributaries, and fisheries on tributaries with large impoundments already have declined (Barthem *et al.*, 1991). The piraíba (*Brachyplatystoma filamentosum*; Figure 12.2) is another species of giant pimelodid catfish of the Amazon that is migratory, but possibly less so than the dourada. Young piraíba are captured from the Amazon estuary and many other locations throughout the river. The piraíba shows strong evidence of overexploitation in the western Amazon, where it traditionally supported a major fishery, but now comprises a small portion of the total catch (Petrere *et al.*, 2004).

12.3.8 Peacock cichlids – vulnerable, stocked

There are several peacock cichlids native to the Amazon, Orinoco and Essequibo river basins of South America, but only *Cichla temensis* (Figure 12.3) is reported to exceed 1 m. Nonetheless, all *Cichla* species are voracious piscivores that commonly dominate the predator biomass of lentic habitats, especially floodplain lakes. Like all cichlids, peacock cichlids have well-developed parental care that, in this case, involves nest defence and biparental brood guarding for several weeks after larvae become free-swimming. Peacock cichlids have large mouth gapes, and are capable of consuming prey measuring more than half their body length. Research in rivers of the Orinoco Basin has indicated that peacock cichlids have the potential to influence the size structure and composition of fish assemblages during the low-water season (Layman *et al.*, 2005; Montaña *et al.*, 2011). Several *Cichla* species have been introduced into reservoirs in tropical and subtropical regions of the world, but most of the successful introductions have been with *C. ocellaris*. Following its introduction into Lake Gatun, Panama, *Cichla ocellaris* caused major changes in the fish assemblage and food web structure and dynamics of the lake (Zaret & Paine, 1973). *Cichla ocellaris* and *C. piquiti* were introduced from the Amazon Basin into reservoirs within the Paraná Basin of southern Brazil, and they rapidly became established as dominant piscivores within lentic habitats of the Upper Paraná River floodplain (Agostinho *et al.*, 2008). *Cichla ocellaris* is established as a dominant piscivore in canals of south Florida and reservoirs in Puerto Rico and Hawaii, and in Thailand and Malaysia peacock cichlids stocked in pay-to-fish lakes pose an obvious high risk to the native communities and ecosystems should they escape (see Chapter 8).

Despite the fact that peacock cichlids have invaded and frequently dominated fish communities of reservoirs and lakes in many tropical regions, their populations appear to be sensitive to harvest. Based on a comparative study of age and growth of *Cichla* species in Venezuela, Jepsen *et al.* (1999) hypothesised that even low mortality from sport-fisheries can rapidly eliminate large size classes of *C. temensis* and reduce population abundance. Commercial gillnetting in the Aguaro River, Venezuela, virtually eliminated the *C. temensis* population, but the smaller *C. orinocensis* persisted at reduced densities. All peacock cichlids require relatively high water transparency, but *C. temensis* appears to have a stronger requirement for blackwater conditions (low pH from dissolved humic substances) than *C. ocellaris* and *C. orinocensis*. Both *C. temensis* and *C. ocellaris* were stocked into canals of south Florida, but only *C. ocellaris* established strong populations.

12.3.9 Nile perch – vulnerable, stocked

The Nile perch (*Lates niloticus*), a protandrous hermaphrodite, matures as a male at 2 years of age when about 50–55 cm in length and as a female at 4 years when greater than 80–85 cm in length (Hughes, 1992). It may grow to 2 m and ~200 kg, live for 15 years or more, and is highly fecund (~10 million eggs per spawning) (Barlow & Lisle, 1987). It is primarily piscivorous, may consume fish more than 30% its own body length, and is opportunistically able to switch between prey species depending on availability. Juveniles also consume prawns (*Caridina*), and less than 10% of mature females also consume this food resource (Hughes, 1992). Nile perch occur naturally in all major river basins between the

Senegal and Congo rivers in East Africa, inland drainages of Lake Chad and in the Nile River from Alexandria to Lake Albert and into the headwaters of the Blue Nile in Ethiopia. Its downstream distribution in this river is apparently limited by low temperature (< 10°C) (Barlow & Lisle, 1987), and its upstream distribution in the White Nile (and those of other large predatory fishes, such as the tigerfish, *Hydrocynus vittatus*) is blocked by the Murchison Falls below Lake Kyoga. Secret introductions in 1954 or 1955 followed by government-sanctioned mass introductions during the 1960s established its presence in Lake Victoria, above this natural barrier, where it became an important commercial species (replacing a diverse multispecies haplochromine cichlid fishery) (Pringle, 2005). Mass extinctions of endemic cichlids and catfish followed (Goudswaard & Witte, 1997), resulting in a massive simplification of the food web and a trophic cascade combined with increased eutrophication that contributed to reduced water quality (Goldschmidt *et al.*, 1993; Witte *et al.*, 2000). Predation by Nile perch, in combination with intensified fishing pressure, resulted in an evolutionary shift in the life history (smaller size at maturation with greater reproductive effort) of a small cyprinid fish (*Rastrineobola argentea*) in Lake Victoria (Sharpe *et al.*, 2012; Chapter 8). Although the ecology and impact of this species in Lake Victoria has been extensively studied, little is known of its ecology in fluvial ecosystems and lakes, such as Lake Tanganyika, in which it occurs naturally.

12.3.10 Barramundi – vulnerable, stocked

The barramundi (*Lates calcarifer*; Figure 12.3) is common in marine, estuarine and freshwater habitats of the IndoPacific region (Pakistan to China, south to Australia). In Australia, barramundi occur in all rivers of northern Australia (north of 26°S) penetrating as a juvenile many hundreds of kilometres upstream, even into intermittent stream systems, and into floodplain water bodies, where it may reside for 3–4 years before migrating downstream to breed (Pusey *et al.*, 2011). The species is a protandrous hermaphrodite, with sex inversion occurring at about 80 cm at between 4 and 6 years of age. Primary females (i.e. not progressing through an initial male stage) are known to occur, as are totally marine populations. Elsewhere across its range, *L. calcarifer* is more commonly found in estuaries and the near shore environment. Spawning occurs prior to or during the wet season near river mouths. This species has a periodic life-history strategy, with extremely high fecundity (up to 10 million eggs for a fish of 100 cm). Larvae mature quickly (larval duration of 7–20 days) and are delivered by tidal action to supralittoral swamps and estuarine mangrove systems. Temporal variation in recruitment success is tightly related to annual variation in rainfall, with greatest recruitment being associated with years of high rainfall and discharge in lower floodplain sections of rivers (Robins *et al.*, 2005). The barramundi is a generalist predator consuming macrocrustaceans and a wide range of fish species including conspecifics. It is heavily targeted by recreational fishers and forms an important component of the commercial and indigenous fish harvest of northern Australia.

Wild barramundi stocks are commercially exploited in northern Australia, where it forms the basis of a targeted and economically important fishery, and elsewhere throughout its range where it occurs as a component of multispecies fisheries (both commercial and artisanal). Historically, commercial fishing in Australia has severely impacted stock structure and abundance (Russell, 1987), necessitating revision of the number of fishing licences, fishing seasons, gear types, and minimum and maximum size limits (which vary between state jurisdictions). These measures have been largely successful, although rivers subject to exploitation still retain lower abundances than equivalent unexploited rivers (Halliday *et al.*, 2001), and declines have resulted in the establishment of extensive stocking programmes in Queensland (Rimmer & Russell, 2001). Pusey *et al.* (2004) reported that, in one reach of a tropical northeastern Australian river that had been closed to fishing, barramundi abundance was over an order of magnitude greater than reaches open to recreational fishing. Barramundi are stocked in both impounded and non-impounded reaches of Queensland rivers to meet this demand, and in accordance with strict guidelines aimed at maintaining genetic integrity of stocks. There has been little study of the ecological impact of this activity, but Pusey *et al.* (2006) argued that the continual introduction of a large piscivorous fish in areas where they have traditionally been absent (i.e. above natural barriers) is likely to be detrimental and no different from the impact of alien species. Barramundi is vulnerable to man-made barriers to movement (e.g. dams, weirs, road crossings; Pusey *et al.*, 2004).

Barramundi is an increasingly important aquaculture species in Australia and Southeast Asia. Annual production increased twofold to 28,000 tonnes over the period 1997–2006, with a total value of approximately $77 million in 2006. About half of this production occurred in Thailand, with Australia contributing less than 10% (Rimmer, 2008).

Escape of individuals from sea-based aquaculture and their interbreeding with wild fish have been identified as an important threat to the genetic integrity of native stocks in Southeast Asia (Norfatimah *et al.*, 2009).

12.4 INFLUENCE OF LARGE PREDATORY FISHES ON FRESHWATER COMMUNITIES AND ECOSYSTEMS

Ecosystem processes can be affected by the extirpation, reduction, or alteration of size structure of the population of single species. To date, only a handful of studies have documented the consequences of depletion of large apex predators on communities and ecosystems. The most obvious influence of apex predators is on food web dynamics. Food webs are influenced by two kinds of processes: the bottom-up flow of energy and nutrients from resources to their consumers and the top-down regulation of resources by their consumers. A trophic cascade occurs when there are strong top-down interactions. In three-link cascades, predators suppress herbivores, and thereby release plants from herbivory. For example, Carpenter and Kitchell (1988) conducted experiments in which piscivorous largemouth bass were removed from lakes in Wisconsin and the response of prey (minnows), zooplankton and water-column chlorophyll concentrations were monitored. After removal of bass, minnows increased in abundance and reduced zooplankton density, which in turn reduced grazing pressure on phytoplankton. This trophic cascade resulted in greater water-column productivity and turbidity following bass removal.

In longer cascades, apex predators suppress smaller mesopredators, releasing their prey. Without apex predators, smaller mesopredators sometimes become superabundant and able to suppress their prey to an extent not otherwise possible. In some systems, release of mesopredators following the elimination of apex predators from overfishing or other causes severely impacts prey at the next level (Myers *et al.*, 2007; Strong & Frank, 2010). For example, Kennard (1995) manipulated barramundi densities in seasonally flooded lagoons in northeastern Australia. Removal of this predator resulted in significant changes in assemblage structure (i.e. a multispecies response) and habitat use and foraging behaviour of many species. Moderately sized species previously consumed by barramundi assumed a more piscivorous role in its absence. In some cases, mesopredators released from top-down control from apex predators can reverse the direction of food-chain regulation, so that mesopredators begin to suppress apex predators. This occurs, for example, when prey fish feed on the eggs and larvae of their predators.

This can prevent fisheries recovery and lead to persistent alternative ecosystem states (Strong & Frank, 2010).

Heithaus *et al.* (2007) argued that our ability to predict the full ecological ramifications of loss of apex predators in marine communities requires better understanding of both direct (direct mortality) and indirect (risk effects that cause behavioural changes) influences of predators on prey. Based on predictions from an integrative framework, they concluded that both demographic and behavioural responses of prey to apex predators are important for the maintenance of ecological processes. Risk-sensitive foraging by prey can affect the spatial patterns of ecosystem processes like primary productivity. Power *et al.* (1985) found that the standing biomass of benthic algae in a stream was greater in pools where largemouth bass were present, and pools lacking bass had abundant herbivorous minnows and depleted benthic algae. A ramification of these predator effects on prey fish populations is the introduction of predatory sportfishes. Based on a literature review, Jackson (2002) found that largemouth bass have directly and indirectly reduced the diversity of small-bodied fishes, changed competitive interactions within local fish assemblages, reduced energy flow to other predatory fishes, and changed planktonic and benthic communities.

Stocks of apex predators in freshwater and marine systems have collapsed as a result of over-harvesting, and some of these communities seem to have shifted into new states that seem to be irreversible (Post *et al.*, 2002; Frank *et al.*, 2005; Chapter 15). Persson *et al.* (2007) showed, for predators feeding on prey that exhibit food-dependent growth, that the culling of fish prey may promote predator recovery. When they removed older, larger individuals from a preyfish population in a lake with low productivity, it caused an increase in the availability of small-sized preyfish and allowed their predator to recover. The shift they observed in community state was sustained for more than 15 years after the culling of larger preyfish had been discontinued. Because most fish have food-dependent growth, shifts to alternative stable states resulting from overcompensating prey growth may be common in nature and may require modification of traditional management strategies.

Impacts that change fish community structure can affect nutrient recycling in ecosystems, because fish species excrete nitrogen and phosphorus at different rates. McIntyre *et al.* (2007) used models to simulate effects of fish extinctions on nutrient recycling in a Venezuelan river and Lake Tanganyika in central Africa. Their models indicated that loss of fish species targeted by fishermen was followed by rapid declines in nutrient recycling. Compensatory responses by the surviving fish species after the extirpation of the overexploited stocks had strong moderating effects on nutrient dynamics. Nutrient input from decomposing carcasses of anadromous salmon to aquatic and riparian ecosystems of oligotrophic streams has been well documented (Gende *et al.*, 2002). Less attention has been paid to the potential for predators moving from a source habitat into a sink habitat to influence ecological dynamics within the latter. Casini *et al.* (2012) analysed a long-term data set to reveal that the spillover of cod (*Gadus morhua*), a marine apex predator, from its main area of distribution in the Baltic Sea to an adjacent coastal ecosystem resulted in a cascade of changes in the latter involving densities of fishes, zooplankton and phytoplankton. Winemiller and Jepsen (2004) proposed that long-distance migratory fishes in tropical river basins could produce top-down effects on fish communities and aquatic ecosystems on a seasonal or interannual basis. For example, the long-distance migrations of large predatory catfishes in the Amazon River represent a great deal of predation pressure that is seasonally transferred hundreds of kilometres from one location to another across the riverscape.

12.5 POTENTIAL BARRIERS AND SOLUTIONS TO THE CONSERVATION OF FRESHWATER FISH DIVERSITY AND ABUNDANCE

The insidious 'shifting baseline' syndrome (Pauly, 1995; Humphries & Winemiller, 2009) – the fact that our perceptions of what is 'natural' changes with each generation of scientists – taints our expectations about the characteristics that fish stocks should have in the absence of human impacts. Importantly, the shifting baseline syndrome thus influences our approaches to conservation and management. Historical records of species distributions, abundance and body size distributions prior to impacts, such as overfishing, are often lacking (Rowland, 1989; Lichter *et al.*, 2006), and targets for restoration are problematic. However, shifting perceptions of what caused fish declines are equally worrying. Although industrialisation and barriers to movement were important factors in degrading fish stocks in Europe, North America and Australia, for example, overfishing may have pre-dated these effects, sometimes by centuries (Vickers, 2004). For example, concerns about unfettered exploitation of barramundi stocks in the Fitzroy River (Queensland) were expressed in 1880, and legislation was enacted to regulate fisheries in this region around that time (Pusey, unpublished data). Only by acquiring evidence of the historical range of variation of past abundances, distributions and composition of fish assemblages (e.g. from palaeoecological or zooarchaeological sources) can we really be in a position to make sound conservation and management decisions (Romme *et al.*, 2012). Fortunately, palaeoecologists are increasingly teaming up with conservation ecologists to make use of a wealth of formerly unknown data sources (e.g. Lyman, 2006; Frazier, 2007; Miller *et al.*, 2011).

Rivers and lakes are some of the most degraded ecosystems in the world (Palmer *et al.*, 2005; Dudgeon, 2010; Chapter 2) and degradation will continue in the future as human populations and demands for water increase. Thus, more water will be needed for irrigation, more dams will be built, and the quality of water will likely suffer (Palmer *et al.*, 2005), all placing pressure on already depleted fish stocks. Ironically, coincident with the increasing degradation of worldwide freshwater systems, the commercial fishing of apex predators, in industrialised countries at least, is probably as low as it has been for decades, if not centuries. However, pressure from commercial fisheries has, in many cases, been replaced by impacts from recreational fishing (Cooke & Cowx, 2004, 2006; Chapter 15). In addition, seemingly inexorable human population increase places demand on maintenance of critical habitats needed to sustain large apex predators. For example, measures to alleviate poverty in many parts of Southeast Asia have shifted the efforts of formerly artisanal fishermen towards aquaculture, especially mariculture. Such developments are frequently associated with a destruction of coastal mangrove systems that are critical for diadromous fishes, such as barramundi.

Conservation and management of apex predators is well established in many countries, with recreational fishing regulations common. However, most management plans are motivated by the desire to maintain or rehabilitate populations of these species for their own sake or in support of fisheries, and the fish's functional role in the ecosystem is rarely considered (Lipsey & Child, 2007; Cooke *et al.*, 2012). Yet, it is becoming increasingly clear that reintroduction and restoration ecology should be combined to achieve desired outcomes, because apex predators perform important roles in environmental modification and biodiversity regulation, and so increase the likelihood of success of restoration programmes generally (Seddon *et al.*, 2007). This should be carried out, however, recognising that we cannot return to past conditions, and we therefore must conserve and manage with future conditions in mind (Choi, 2007).

Freshwater protected areas may be one solution for the conservation of apex predators (Chapter 14). This concept has been around for a while, and is most developed for marine and estuarine systems, but has great potential in freshwater systems as well (Saunders *et al.*, 2002; Kingsford & Nevill, 2005; Suski & Cooke, 2007). Terrestrial-based protected areas can go some way in protecting freshwater fishes (Lawrence *et al.*, 2011), but in most cases, dedicated freshwater protected areas will be needed. The protection of intact ecosystems (landscape units) is one option, and should be encouraged (Revenga *et al.*, 2005; Abell *et al.*, 2007; Palmer *et al.*, 2010); however, there is the potential to rehabilitate fish populations through appropriate regulations (e.g. catch-and-release), reintroductions, removal of barriers, and even no-take zones that might be critical areas for breeding and recruitment (Baird & Flaherty, 2005; Cooke & Schramm, 2007).

According to Welcomme (2008), the world's river fisheries produce about 5 million tonnes annually, most of which is derived from major floodplains. Most tropical river fisheries are overexploited and show signs of a fishing-down pattern in which body sizes and market values of species comprising the catch decline even while the total biomass of the catch is sustained. River channels are being modified and floodplains are being drained for irrigated agriculture, both of which result in modifications to ecosystem structure and ecological functions. The construction of new infrastructure to meet the growing demand for water reduces flows and disconnects rivers from floodplains (Chapter 4). Brazil, for example, has constructed more than 700 large dams and associated reservoirs on its large rivers. Immediately after reservoir formation, fish species richness is usually high, but decreases as reservoirs age. Flow alteration from dams negatively impacts rheophilic and long-distance migratory fishes in downstream reaches. Management actions taken to minimise the impacts of dams in Brazil, such as fish passages, have failed (Agostinho *et al.*, 2008). Scores of new hydroelectric dams are planned for the Amazon and Mekong Basins (Baran & Myschowoda, 2009; Dugan *et al.*, 2010; Finer & Jenkins, 2012), two of the most biodiverse river basins that, as discussed in this chapter, also support unique large apex predators, some of which are highly migratory.

12.6 FUTURE OUTLOOK FOR THE CONSERVATION OF LARGE APEX PREDATORY FISHES

By virtue of their size alone, large apex predators have high economic value that stimulates intense fishing pressure in most parts of the world. Their predatory nature also makes them prime targets for sportfisheries.

Large apex predators in freshwaters face the same 'double jeopardy' that confronts large apex predators of the marine realm (Myers & Worm, 2003) – high value and long life (Collette *et al.*, 2011). High value results in fishing mortality, and long life imparts low demographic resilience to this additional source of mortality. Based on this review of ecological characteristics, human impacts and current status of stocks of large apex predators from regions around the world, what are realistic expectations for the future of large predatory fishes in freshwater ecosystems? Unfortunately, some species will continue to decline, and some almost certainly will go extinct unless they can be maintained by hatcheries with possible releases to the wild (e.g. beluga sturgeon and other sturgeons, giant Mekong catfishes). These at-risk species are strongly migratory and have periodic life-history strategies. Dams that block access to upstream spawning habitats are a particular threat to these large apex predators. The Yangtze paddlefish, widely considered the largest freshwater fish in the world, is probably extinct, a victim of habitat fragmentation from dams, pollution and over-harvest. The dog-eating catfish and Mekong giant catfish may be facing the same fate for the same reasons (Hogan *et al.*, 2004).

Other species of large apex predators likely will persist with proper management, but they probably will never regain the population abundance, size structure, or geographic extent that was exhibited prior to the industrial revolution (e.g. Murray cod, arapaima, alligator gar, taiman), and this is because there will always be fishing pressure, often intense, that eliminates large individuals so that maximum sizes are rarely achieved and population resilience is compromised. These species span the periodic-to-equilibrium life-history gradient. They are all vulnerable to over-harvest (low demographic resilience) and also to habitat alteration (flow alteration) and habitat fragmentation, but in some settings (e.g. where there is effective community-based management of harvest, where commercial and subsistence fishing pressure is non-existent or low, as in most parts of Europe, Australia, North America), these populations can be sustained with proper management. The community-based management of the arapaima in the Amazon provides a model for successful management of apex predatory fishes in other tropical regions. Even in these cases, the altered size structure of the populations may have consequences for food web interactions that may affect biodiversity and ecosystem processes in ways that are, at present, difficult to forecast. Moreover, size-selective fisheries create new regimes of directional selection that change population life history attributes, such as age at maturation and fecundity (Sharpe *et al.*, 2012; Belgrano & Fowler, 2013).

Finally, some large predatory fishes will continue to be managed for sportfishing, in many cases as exotic species stocked in reservoirs located well beyond their native range (e.g. peacock cichlids, wels catfish, channel catfish, largemouth bass). These species generally have characteristics of the equilibrium life-history strategy, such as nesting, brood guarding and density-dependent recruitment. These traits probably allow them to compete within the relatively stable environment of lakes and reservoirs. Under catch-and-release management scenarios, some of these populations can produce individuals of impressive size that probably approach or equal those recorded from periods prior to major human impacts. In settings where exotic populations of large apex predators become established, the impacts to native communities and ecosystems are often large, yet difficult to predict given the challenges associated with modelling the complexity of food webs. The establishment of exotic apex predators can cause a regime shift that may be difficult or perhaps virtually impossible to reverse.

12.7 DISCUSSION QUESTIONS

1. Why are large fishes that are apex predators so often threatened with extirpation, and what makes these species particularly vulnerable to human impacts?
2. Given that so many fishes that are large apex predators are in decline, what has allowed certain species, such as the wels catfish, to remain common and in some cases expand their geographic distributions?
3. Even if large fishes that are apex predators are not completely extirpated, what are some potential effects of reductions in their population abundance and body size on food webs and ecosystems?
4. Compare and contrast the conservation challenges for freshwater fishes that are large apex predators with marine fishes that are large apex predators.
5. Propose some strategies for restoration of populations of freshwater fishes that are large apex predators.

12.8 REFERENCES

Abell, R., Allan, J. D. & Lehner, B. (2007). Unlocking the potential of protected areas for freshwaters. *Biological Conservation*, 134, 48–63.

Agostinho, A. A., Pelicice, F. M. & Gomes, L. C. (2008). Dams and the fish fauna of the Neotropical region: impacts and management related to diversity and fisheries. *Brazilian Journal of Biology*, 68(Suppl. 4), 1119–1132.

Allan, J. D., Abell, R., Hogan, Z., *et al.* (2005). Overfishing of inland waters. *BioScience*, 55, 1041–1051.

Allen, G. R. (1991). *Field Guide to the Freshwater Fishes of New Guinea*. Publication No. 9. Madang, Papua New Guinea: Christensen Research Institute.

Arantes, C. C., Castello, L., Stewart, D. J., Cetra, M. & Queiroz, H. L. (2010). Population density, growth and reproduction of the arapaima in an Amazonian river-floodplain. *Ecology of Freshwater Fish*, 19, 455–465.

Baird, I. G. & Flaherty, M. S. (2005). Mekong River fish conservation zones in southern Laos: assessing effectiveness using local ecological knowledge. *Environmental Management*, 36, 439–454.

Baran, E. & Myschowoda, C. (2009). Dams and fisheries in the Mekong Basin. *Aquatic Ecosystem Health and Management*, 12, 227–234.

Barbarino Duque, A. & Winemiller, K. O. (2003). Dietary segregation among large catfishes of the Apure and Arauca rivers, Venezuela. *Journal of Fish Biology*, 63, 410–427.

Barlow, C. G. & Lisle, A. (1987). Biology of the Nile perch *Lates niloticus* (Pisces: Centropomidae) with reference to its proposed role as a sport fish in Australia. *Biological Conservation*, 39, 269–289.

Barthem, R. & Goulding, M. (1997). *The Catfish Connection. Ecology, Migration and Conservation of Amazon Predators*. New York, NY: Columbia University Press.

Barthem, R. B., Ribeiro, M. C. L. B. & Petrere, M. (1991). Life strategies of some long-distance migratory catfish in relation to hydroelectric dams in the Amazon basin. *Biological Conservation*, 55, 339–345.

Baskin, J. N., Zaret, T. A. & Mago-Leccia, F. (1980). Feeding of reportedly parasitic catfishes (Trichomycteridae and Cetopsidae) in the Rio Portuguesa Basin, Venezuela. *Biotropica*, 12, 182–186.

Batista, V. S., Isaac, V. J., Fabré, N. N., *et al.* (2012). *Peixes e Pesca no Solimões-Amazonas: Uma Avaliação Integrada*. Brasília: Ibama/ProVárzea.

Belgrano, A. & Fowler, C. W. (2013). How fisheries affect evolution. *Science*, 342, 1176–1177.

Carpenter, S. R. & Kitchell, J. F. (1988). Consumer control of lake productivity. *Bioscience*, 38, 764–769.

Casini, M., Blenckner, T., Möllmann, C., *et al.* (2012). Predator transitory spillover induces trophic cascades in ecological sinks. *Proceedings of the National Academy of Science*, 109, 8185–8189.

Castello, L. (2008). Nesting habitat of pirarucu *Arapaima gigas* in floodplains of the Amazon. *Journal of Fish Biology*, 72, 1520–1528.

Castello, L., Viana, J. P., Watkins, G., Pinedo-Vasquez, M. & Luzadis, V. A. (2009). Lessons from integrating fishers of arapaima in small-scale fisheries management at the Mamirauá Reserve, Amazon. *Environmental Management*, 43, 197–209.

Castello, L., Stewart, D. J. & Arantes, C. C. (2011). Modeling population dynamics and conservation of arapaima in the Amazon. *Reviews in Fish Biology and Fisheries*, 21, 623–640.

Choi, Y. D. (2007). Restoration ecology to the future: a call for a new paradigm. *Restoration Ecology*, 15, 351–353.

Collette, B. B. Carpenter, K. E., Polidoro, B. A., *et al.* (2011). High value and long life – double jeopardy for tunas and billfishes. *Science*, 333, 291–292.

Cooke, S. J. & Cowx, I. G. (2004). The role of recreational fishing in global fish crises. *BioScience*, 54, 857–859.

Cooke, S. J. & Cowx, I. G. (2006). Contrasting recreational and commercial fishing: searching for common issues to promote unified conservation of fisheries resources and aquatic environments. *Biological Conservation*, 128, 93–108.

Cooke, S. J. & Schramm, H. L. (2007). Catch-and-release science and its application to conservation and management of recreational fisheries. *Fisheries Management and Ecology*, 14, 73–79.

Cooke, S. J., Paukert, C. & Hogan, Z. (2012). Endangered river fish: factors hindering conservation and restoration. *Endangered Species Research*, 17, 179–191.

Copp, G. H., Britton, J. R., Cucherousset, J., *et al.* (2009). Voracious invader or benign feline? A review of the environmental biology of European catfish *Silurus glanis* in its native and introduced ranges. *Fish and Fisheries*, 10, 252–282.

Dakin, W. J. & Kesteven, G. L. (1937). The Murray cod (*Maccullochella macquariensis* [Cuv. et Val.]). *New South Wales State Fisheries Bulletin*, 1, 1–18.

Doukakis, P., Babcock, E. A., Pikitch, E. K., *et al.* (2010). Management and recovery options for Ural River beluga sturgeon. *Conservation Biology*, 24, 769–777.

Dudgeon, D. (2010). Prospects for sustaining freshwater biodiversity in the 21st century: linking eco-system structure and function. *Current Options in Environmental Sustainability*, 2, 422–430.

Dudgeon, D. (2011). Asian river fishes in the Anthropocene: threats and conservation challenges in an era of rapid environmental change. *Journal of Fish Biology*, 79, 1487–1524.

Dugan, P. J., Barlow, C., Agostinho, A. A., *et al.* (2010). Fish migration, dams, and loss of ecosystem services in the Mekong Basin. *Ambio*, 39, 344–348.

Ebner, B. (2006). Murray cod, an apex predator in the Murray River, Australia. *Ecology of Freshwater Fish*, 15, 510–520.

Finer, M. & Jenkins, C. N. (2012). Proliferation of hydroelectric dams in the Andean Amazon and impli-cations for Andes–Amazon connectivity. *PLoS ONE*, 7, e35126.

Frank, K. T., Petrie, B., Choi, J. S. & Leggett, W. C. (2005). Trophic cascades in a formerly cod-dominated ecosystem. *Science*, 308, 1621–1623.

Frazier, J. (2007). Sustainable use of wildlife: the view from archaeozoology. *Journal for Nature Conservation*, 15, 163–173.

Froese, R. (2004). Keep it simple: three indicators to deal with overfishing. *Fish and Fisheries*, 5, 86–91.

Fulton, W. (2011). *Sustainability of Recreational Fisheries for Murray Cod in the Murray Darling Basin*. Final Report to Fisheries Research & Development Corporation Project No. 2006/053. Department of Primary Industries, Queenscliff, Victoria.

Garcia de Leon, F. J., Gonzalez-Garcia, L., Herrera-Castillo, J. S., Winemiller, K. O. & Banda-Valdes, A. (2001). Ecology of the alligator gar, *Atractosteus spatula*, in the Vicente Guerrero Reservoir, Tamaulipas, Mexico. *The Southwestern Naturalist*, 46, 151–157.

Gende, S. M., Edwards, R. T., Willson, M. F. & Wipfli, M. S. (2002). Pacific salmon in aquatic and ter-restrial ecosystems. *BioScience*, 52, 917–928.

Goldschmidt, T., Witte, F. & Wanink, J. H. (1993). Cascading effects of the introduced Nile perch on the detritivorous/phytoplanktivorous species in the sublittoral areas of Lake Victoria. *Conservation Biology*, 7, 686–700.

Goudswaard, P. C. & Witte, F. (1997). The catfish fauna of Lake Victoria after the Nile perch upsurge. *Environmental Biology of Fishes*, 49, 21–43.

Halliday, I., Ley, J., Tobin, A., *et al.* (2001). *The Effects of Netfishing: Addressing Biodiversity and Bycatch Issues in Queensland Inshore Waters*. Southern Fisheries Centre, Department of Fisheries, Deception Bay.

Heithaus, M. R., Frid, A., Wirsing, A. J. & Worm, B. (2007). Predicting ecological consequences of mar-ine top predator declines. *Trends in Ecology and Evolution*, 23, 202–210.

Hoeinghaus, D. J., Agostinho, A. A., Gomes, L. C., *et al.* (2009). River impoundment results in a mis-match between embodied energy and market value of a tropical artisanal fishery. *Conservation Biology*, 23, 1222–1231.

Hogan, Z. (2011). *Imperiled Giant Fish and Mainstream Dams in the Lower Mekong Basin: Assessment of Current Status, Threats, and Mitigation*. Unpublished report, University of Nevada, Reno, p. 13.

Hogan, Z. S., Moyle, P. B., May, B., Vander Zanden, M. J. & Baird, I. G. (2004). Imperiled giants of the Mekong: ecologists struggle to understand and protect Southeast Asia's large migratory catfish. *American Scientist*, 92, 228–237.

Holčík, J., Hensel, K., Nieslanik, J. & Skacel, L. (1988). *The Eurasian Huchen, Hucho hucho, Largest Salmon of the World*. Boston, MA: Kluwer Academic Publishers.

Hughes, N. F. (1992). Nile perch, *Lates niloticus*, predation on the freshwater prawn, *Caridina nilotica*, in the Nyanza Gulf, Lake Victoria, East Africa. *Environmental Biology of Fishes*, 33, 307–309.

Humphries, P. L. & Winemiller, K. O. (2009). Historical impacts on river fauna, shifting baselines and challenges for restoration. *BioScience*, 59, 673–684.

Jackson, D. A. (2002). Ecological effects of *Micropterus* introductions: the dark side of black bass. In *Black Bass: Ecology, Conservation, and Management*. American Fisheries Society Symposium 31. Bethesda, MD: American Fisheries Society, pp. 221–232.

Jensen, O. P., Gilroy, D. J., Hogan, Z., *et al.* (2009). Evaluating recreational fisheries for an endan-gered species: a case study of taimen, *Hucho taimen*, in Mongolia. *Canadian Journal of Fisheries and Aquatic Sciences*, 66, 1707–1718.

Jepsen, D. B., Winemiller, K. O., Taphorn, D. C. & Rodriguéz-Olarte, D. (1999). Variation in age struc-ture and growth of peacock cichlids from rivers and reservoirs of Venezuela. *Journal of Fish Biology*, 55, 433–450.

Kennard, M. J. (1995). *Factors Influencing Freshwater Fish Assemblages in Floodplain Lagoons of the Normanby River, Cape York Peninsula: A Large Tropical River.* MPhil thesis, Griffith University, Brisbane.

Khodorevskaya, R., Dovgopol, G., Zhuravleva, O. & Vlasenko, A. (1997). Present status of commercial stocks of sturgeon in the Caspian Sea Basin. *Environmental Biology of Fishes*, 48, 209–219.

Kingsford, R. T. & Nevill, J. (2005). Scientists urge expansion of freshwater protected areas. *Ecological Management and Restoration*, 6, 161–162.

Lawrence, D. J., Larson, E. R., Liermann, C. A. R., *et al.* (2011). National parks as protected areas for US freshwater fish diversity. *Conservation Letters*, 4, 364–371.

Layman, C. A., Winemiller, K. O. & Arrington, D. A. (2005). Describing a species-rich river food web using stable isotopes, stomach contents, and functional experiments. In *Dynamic Food Webs: Multispecies Assemblages, Ecosystem Development and Environmental Change.* Elsevier, Amsterdam, pp. 395–406.

Leonardos, I. D., Tsikliras, A. C., Batzakas, I. & Liousia, V. (2009). Life-history characteristics of the endangered Aristotle's catfish (*Silurus aristotelis* Garman, 1890), Lake Pamvotis, north-west Greece. *Journal of Applied Ichthyology*, 25, 746–751.

Lichter, J., Caron, H., Pasakarnis, T. S., *et al.* (2006). The ecological collapse and partial recovery of a freshwater tidal ecosystem. *Northeastern Naturalist*, 13, 153–178.

Lintermans, M. (2007). *Fishes of the Murray–Darling Basin: An Introductory Guide.* Canberra: Murray-Darling Basin Commission.

Lipsey, M. K. & Child, M. F. (2007). Combining the fields of reintroduction biology and restoration ecology. *Conservation Biology*, 21, 1387–1388.

Lyman, R. L. (2006). Paleozoology in the service of conservation biology. *Evolutionary Anthropology*, 15, 11–19.

McCarley, H. & Hill, L. G. (1979). Reproduction of *Lepisosteus spatula* (Lepisostedae) in Lake Texoma. *The Southwestern Naturalist*, 24, 694–695.

McIntyre, P. B., Jones, L. E., Flecker, A. S. & Vanni, M. J. (2007). Fish extinctions alter nutrient cycling in tropical freshwaters. *Proceedings of the National Academy of Sciences USA*, 104, 4461–4466.

Miller, J. A., Butler, V. L., Simenstad, C. A., Backus, D. H. & Kent, A. J. R. (2011). Life history variation in Upper Columbia River chinook salmon (*Oncorhynchus tshawytscha*): a comparison using modern and ~500-year-old archaeological otoliths. *Canadian Journal of Fisheries and Aquatic Sciences*, 68, 603–617.

Mims, M. C. & Olden, J. D. (2012). Life history theory predicts fish assemblage response to hydrologic regimes. *Ecology*, 93, 35–45.

Montaña, C. G., Layman, C. A. & Winemiller, K. O. (2011). Gape size influences seasonal patterns of piscivore diets in three Neotropical rivers. *Neotropical Ichthyology*, 9, 647–655.

Myers, R. A. & Worm, B. (2003). Rapid worldwide depletion of predatory fish communities. *Nature*, 423, 280–283.

Myers, R. A., Braum, J. K., Shepherd, T. D., Powers, S. P. & Peterson, C. H. (2007). Cascading effects of the loss of apex predatory sharks from a coastal ocean. *Science*, 315, 1846–1850.

Norfatimah, M. Y., Siti Azizah, M. N., Othman, A. S., Patimah, I. & Jamsari, A. F. I. (2009). Genetic variation of *Lates calcarifer* in Peninsular Malaysia based on the cytochrome b gene. *Aquaculture Research*, 40, 1742–1749.

O'Connell, M. T., Shepherd, T. D., O'Connell, A. M. U. & Myers, R. A. (2007). Long-term declines in two apex predators, bull sharks (*Carcharhinus leucas*) and alligator gar (*Atractosteus spatula*), in Lake Pontchartrain, an oligohaline estuary in southeastern Louisiana. *Estuaries and Coasts*, 30, 567–574.

Olden, J. D., Poff, N. L. & Bestgen, K. R. (2006). Life-history strategies predict fish invasions and extirpations in the Colorado River Basin. *Ecological Monographs*, 76, 25–40.

Olden, J. D., Hogan, Z. S. & Vander Zanden, M. J. (2007). Small fish, big fish, red fish, blue fish: size-biased extinction risk of the world's freshwater and marine fishes. *Global Ecology and Biogeography*, 16, 694–701.

Palmer, M. A., Bernhardt, E. S., Allan, J. D., *et al.* (2005). Standards for ecologically successful river restoration. *Journal of Applied Ecology*, 42, 208–217.

Palmer, M. A., Menninger, H. L. & Bernhardt, E. (2010). River restoration, habitat heterogeneity and biodiversity: a failure of theory or practice? *Freshwater Biology*, 55, 205–222.

Pauly, D. (1995). Anecdotes and the shifting baseline syndrome of fisheries. *Trends in Ecology and Evolution*, 10, 430.

Persson, L., Amundsen, P. A., De Roos, A. M., *et al.* (2007). Culling prey promotes predator recovery-alternative states in a whole-lake experiment. *Science,* 316, 1743–1746.

Petrere, M. Jr, Barthem, R. B., Córdoba, E. A. & Gómez, B. C. (2004). Review of the large catfish fisheries in the upper Amazon and the stock depletion of piraíba (*Brachyplatystoma filamentosum Lichtenstein*). *Reviews in Fish Biology and Fisheries,* 14, 403–414.

Pikitch, E. K., Doukakis, P., Lauck, L., Chakrabarty, P. & Erickson, D. L. (2005). Status, trends and management of sturgeon and paddlefish fisheries. *Fish and Fisheries,* 6, 233–265.

Pinsky, M. L., Malin, L., Jensen, O. P., Ricard, D. & Palumbi, S. R. (2011). Unexpected patterns of fisheries collapse in the world's oceans. *Proceedings of the National Academy of Science,* 108, 8317–8322.

Post, J. R., Sullivan, M., Cox, S., *et al.* (2002). Canada's recreational fisheries: the invisible collapse? *Fisheries,* 27, 6–17.

Power, M. E., Matthews, W. J. & Stewart, A. J. (1985). Grazing minnows, piscivorous bass and stream algae: dynamics of a strong interaction. *Ecology,* 66, 1448–1456.

Pringle, R. M. (2005). The origins of the Nile perch in Lake Victoria. *BioScience,* 55, 780–787.

Pusey, B. J., Kennard, M. J. & Arthington, A. H. (2004). *Freshwater Fishes of North-eastern Australia.* Melbourne: CSIRO Publishing.

Pusey, B., Burrows, D., Arthington, A. & Kennard, M. (2006). Translocation and spread of piscivorous fishes in the Burdekin River, north-eastern Australia. *Biological Invasions,* 8, 965–977.

Pusey, B., Kennard, M., Burrows, D., *et al.* (2011). Freshwater fish. In *Aquatic Biodiversity of the Wet–Dry Tropics of Northern Australia: Patterns, Threats And Future.* Darwin: Charles Darwin University Press, pp. 71–92.

Rainboth, W. (1996). *Fishes of the Cambodian Mekong.* Rome: MRC-FAO-DANIDA.

Revenga, C., Campbell, I., Abell, R., de Villiers, P. & Bryer, M. (2005). Prospects for monitoring freshwater ecosystems towards the 2010 targets. *Philosophical Transactions of the Royal Society B,* 360, 397–413.

Rimmer, M. (2008). Production update – marine finfish aquaculture in the Asia-Pacific region. *Aquaculture Asia,* 13, 44–46.

Rimmer, M. & Russell, J. (2001). Stock enhancement of barramundi *Lates calcarifer* (Bloch) in Queensland, Australia. In: *Aquaculture and Fisheries Resource Management.* Conference Proceedings of the Joint Taiwan–Australia Aquaculture and Fisheries Resources and Management Forum, Keelung. Taiwan Fisheries Institute No. 4, pp. 185–192.

Roberts, T. R. (1990). Mimicry of prey by fin-eating fishes of the African characoid genus *Eugnathichthys* (Pisces: Distichodontidae). *Ichthyological Explorations of Freshwaters,* 1, 23–31.

Robins, J. B., Halliday, I. A. & Staunton-Smith, J. (2005). Freshwater-flow requirements of estuarine fisheries in tropical Australia: a review of the state of knowledge and application of a suggested approach. *Marine and Freshwater Research,* 56, 343–360.

Romme, W. H., Wiens, J. A. & Safford, H. D. (2012). Setting the stage: theoretical and conceptual background of historical range of variation. In *Historical Environmental Variation in Conservation and Natural Resource Management.* Hoboken, NJ: John Wiley & Sons, pp. 3–18.

Rowland, S. J. (1989). Aspects of the history and fishery of the Murray cod, *Maccullochella peeli* (Mitchell) (Percichthyidae). *Proceedings of the Linnean Society of New South Wales,* 111, 201–213.

Russell, D. J. (1987). Review of juvenile barramundi (*Lates calcarifer*) wildstocks in Australia. In *Management of Wild and Cultured Seabass/Barramundi.* Proceedings of an International Workshop, Darwin. Canberra: Australian Centre for International Agricultural Research, pp. 44–49.

Sakaris, P. C., Ferrara, A. M., Kleiner, K. J. & Irwin, E. R. (2003). Movements and home ranges of alligator gar in the Mobile–Tensaw Delta, Alabama. *Proceedings of the Annual Conference of the Southeastern Association of Fish and Wildlife Agencies,* 57, 102–111.

Saunders, D. L., Meeuwig, J. J. & Vincent, A. C. J. (2002). Freshwater protected areas: strategies for conservation. *Conservation Biology,* 16, 30–41.

Sazima, I. (1983). Scale-eating in characoids and other fishes. *Environmental Biology of Fishes,* 9, 87–101.

Scarnecchia, D. L. (1992). A reappraisal of gars and bowfins in fishery management. *Fisheries,* 17, 6–12.

Seddon, P. J., Armstrong, D. P. & Maloney, R. F. (2007). Developing the science of reintroduction biology. *Conservation Biology,* 21, 303–312.

Sharpe, D. M. T., Wandera, S. B. & Chapman, L. J. (2012). Life history change in response to fishing and an introduced predator in the East African cyprinid, *Rastrineobola argentea*. *Evolutionary Applications,* 5, 677–693.

Skelton, P. H. (1993). *A Complete Guide to the Freshwater Fishes of Southern Africa*. Halfway House, South Africa: Southern Book Publishers.

Stewart, D. J. (2013). Re-description of *Arapaima agassizii* (Valenciennes), a rare fish from Brazil (Osteoglossomorpha, Osteoglossidae). *Copeia*, 2013, 38–51.

Stone, R. (2007). The last of the leviathans. *Science*, 316, 1684–1688.

Strong, D. R. & Frank, K. T. (2010). Human involvement in food webs. *Annual Review of Environment and Resources*, 35, 1–23.

Suski, C. D. & Cooke, S. J. (2007). Conservation of aquatic resources through the use of freshwater protected areas: opportunities and challenges. *Biodiversity and Conservation*, 16, 2015–2029.

Tedesco, P. A., Hugueny, B., Oberdorff, T., *et al.* (2008). River hydrological seasonality influences life history strategies of tropical river fishes. *Oecologia*, 156, 691–702.

Trautman, M. B. (1981). *The Fishes of Ohio, with Illustrated Keys*. Columbus, OH: Ohio State University Press.

Vander Zanden, M. J., Joppa, L. N., Allen, B. C., *et al.* (2007). Modeling spawning dates of *Hucho taimen* in Mongolia to establish fishery management zones. *Ecological Applications*, 17, 2281–2289.

Vickers, D. (2004). Those dammed shad: would the river fisheries of New England have survived in the absence of industrialization? *William and Mary Quarterly*, 61, 685–712.

Wei, Q., Ke, F., Zhang, J., *et al.* (1997). Biology, fisheries and conservation of sturgeons and paddlefish in China. *Environmental Biology of Fishes*, 48, 241–255.

Welcomme, R. (2008). World prospects for floodplain fisheries. *Ecohydrology and Hydrobiology*, 8, 169–182.

Winemiller, K. O. (1989). Patterns of variation in life history among South American fishes in seasonal environments. *Oecologia*, 81, 225–241.

Winemiller, K. O. (1992). Life history strategies and the effectiveness of sexual selection. *Oikos*, 62, 318–327.

Winemiller, K. O. (2001). Ecology of peacock cichlids (*Cichla* spp.) in Venezuela. *Journal of Aquariculture and Aquatic Sciences*, 9, 93–112.

Winemiller, K. O. (2005). Life history strategies, population regulation, and their implications for fisheries management. *Canadian Journal of Fisheries and Aquatic Sciences*, 62, 872–885.

Winemiller, K. O. & Jepsen, D. B. (2004). Migratory neotropical fish subsidize food webs of oligotrophic blackwater rivers. In *Food Webs at the Landscape Level*. Chicago, IL: University of Chicago Press, pp. 115–132.

Winemiller, K. O. & Rose, K. A. (1992). Patterns of life-history diversification in North American fishes: implications for population regulation. *Canadian Journal of Fisheries and Aquatic Sciences*, 49, 2196–2218.

Winemiller, K. O. & Yan, H. Y. (1989). Obligate mucus feeding in a South American trichomycterid catfish. *Copeia*, 1989, 511–514.

Witte, F., Maku, B. S., Wanink, J. H., *et al.* (2000). Recovery of cichlid species in Lake Victoria: an examination of factors leading to differential extinction. *Reviews in Fish Biology and Fisheries*, 10, 233–41.

Zaret, T. M. & Paine, R. T. (1973). Species introduction into a tropical lake. *Science*, 182(4111), 449–455.

Artificial propagation of freshwater fishes: benefits and risks to recipient ecosystems from stocking, translocation and re-introduction

JOHN M. EPIFANIO AND ROBIN S. WAPLES

We must not think that the highly laudable but expensive art and science of restoration ecology is going to put back ecosystem integrity, even though it 'regreens' with a highly convincing look-alike. – M. J. Samways (1996)

13.1 INTRODUCTION

This book examines emerging and historical perspectives and approaches to the conservation of inland fish *biodiversity* and the freshwater ecosystems upon which this diversity relies. Together, these approaches present enormous opportunities, yet daunting challenges, in efforts to sustain viably functioning freshwater ecosystems with the full complement of ecosystem services and societal values they provide. In this chapter, we focus the reader's attention to a set of interrelated practices aimed at augmenting the production of new recruits to a *targeted population* and ecosystem often as means to restore or conserve fish populations. These practices include *artificial propagation* and *stocking*, direct *translocation* and *re-introduction* (NB: key terms used herein are italicised on first use and defined in Box 13.1. Many of these terms have no widely accepted definitions in practice. Therefore, we provide these as they relate to our usage). Widely used in fisheries management, aquatic conservation and restoration, these activities are the subject of considerable scrutiny and lingering debate as to whether they truly provide a demonstrable ecological benefit as opposed to a suite of more societal benefits (economic, cultural, political, and so on). Moreover, critical examination of whether such benefits justify or balance the attendant

Conservation of Freshwater Fishes, eds G. P. Closs, M. Krkosek and J. D. Olden. Published by Cambridge University Press. © Cambridge University Press 2016.

Box 13.1. Definition of key terms used within Chapter 13

- *Aquaculture (or foodfish aquaculture)* – closed-loop production of fish for direct harvest (that is, no intentional release within the production cycle) for consumption or sale. Within this chapter, we do not focus on issues concerned specifically with aquaculture, although many of them (see Lorenzen *et al.*, 2012) can apply to those programmes where release is intended as well.
- *Artificial propagation* – production of young or adults outside of the natural environment through animal husbandry and culture methods.
- *Biodiversity* – the variety and complexity of organisms living within an ecosystem of concern. The composition, structure and function of biodiversity are hierarchically constructed from genes to communities and occur across multiple spatial and temporal scales (see Chapter 1 for a broader discussion).
- *Captive propagation* – a short-term and near-complete containment of brood and young over a whole life cycle. This approach is commonly used to try to maintain a gene pool for a limited period of time.
- *Domestication* – a genetic change in a captive population that increases its fitness in captivity.
- *Evolutionary Significant Unit (ESU)* – a population or group of populations that represent an important component of the evolutionary legacy of the species as a whole.
- *Inbreeding depression* – a reduction in fitness in offspring bred from parents that are closely related.
- *Introduction* – a release of artificially propagated young or adults into a novel environment intended to expand the species' range of reproducing populations (see Chapter 9 for a more thorough discussion).
- *Mitigation* – a release of artificially propagated young intended to replace or compensate for losses due to temporary stresses or ongoing perturbations such as dams (see Chapter 4 for a more thorough discussion).
- *Non-targeted population (or non-target species)* – population(s) of freshwater fishes (or other species within the community) that can be incidentally and unintentionally affected by the release of artificially propagated, transferred, transplanted or translocated fishes.
- *Outbreeding depression* – a reduction in fitness and survival in offspring bred from parents of genetically divergent lineages.
- *Put-and-take* – a release of artificially propagated fish at a size desired for immediate and near-complete harvest as a recreational management rather than a conservation action. Put-grow-and-take is a variation on this theme where young will experience a growth period prior to near-complete harvest.
- *Recipient ecosystem* – a freshwater water body and interconnected watershed that receives stocked, transplanted, transferred or translocated fishes.
- *Re-introduction* – a release of propagated young or adults to an ecosystem from which the species (or other level of taxonomic complexity) was previously extirpated.

- *Stocking* – an intentional or directed release of individuals propagated, translocated, or re-introduced.
- *Stock (or fish stock)* – discrete, isolated or divergent populations within a fish species. Such heterogeneity may be delineated or structured spatially (e.g. watershed), temporally (e.g. seasonal or annual runs) or behaviourally (e.g. resident vs. adfluvial life histories).
- *Stock transfer* – human-mediated movement of a phylogenetic lineage (or stock) into the distribution or range of another lineage. This kind of activity facilitates the mixing of lineages and the potential of outbreeding effects.
- *Supplementation* – a release of artificially propagated young with the expectation of increasing the abundance of a natural population.
- *Targeted population (or target species)* – the population(s) specifically stocked for conservation or enhancement purposes.
- *Translocation* – human-facilitated movement of individuals from a species from one location to another. The receiving location often harboured a now extant population from the species.

risks to recipient biodiversity and ecosystems are generally lacking. We address both sides of the debate by presenting a general discussion of the kinds, range and magnitude of benefits and risks associated with artificial enhancement practices including those used under the banner of 'conservation', but also more generally. Where space and information permit, we also provide evidence gleaned from specific cases documented from the literature or the authors' experiences. We acknowledge a bias toward examples gleaned from our North American experiences with inland and anadromous species, but suggest that underlying concepts are common and broadly relevant to any fauna regardless of geographical location.

If we take a modern perspective on conservation – that is, one aimed at maintaining viable biodiversity – we must recognise the latter as having a hierarchy of compositional, structural and functional components at multiple levels of taxonomic complexity (Noss, 1990; see also Chapter 1 of this text). Therefore, freshwater biodiversity is not simply the number of species in an ecosystem or its overall community richness. Rather, biodiversity is the variety present within and among populations, the patterns or distribution of such variation (for example, spatially across the landscape or temporally across seasons), the evolutionary processes that created the varieties, ecological and life-history expression of the variation, and finally the role played by the diversity in population and ecosystem sustainability and viability. Therefore, to more fully weigh the full range of risks to biodiversity at multiple levels against societal

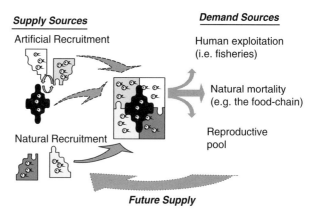

Figure 13.1 The conceptual relationship between supply and demand in fish populations. Conceptually, the 'size' of the pieces and the puzzle as a whole will be influenced by available habitat, food, water quality, etc., that affect the system's carrying capacity.

benefits of any conservation approach – and from artificial propagation, stocking, translocation and re-introduction more specifically – we assert that it is important to consider how these activities affect not only the *targeted population*, but also other *non-targeted populations* that co-occur in *recipient ecosystems*. Inherent in our assertion is the premise that ecological or evolutionary benefits are often assumed, yet rarely documented or achieved.

Conceptually, conservation of biodiversity in fishes is often an exercise to address the 'supply and demand' on living organisms to ensure long-term persistence and viability. For example, harvest exploitation (see Chapter 15) and natural mortality (e.g. predation or naturally occurring diseases) might be viewed as 'demand-side' pressures on biodiversity, along with the effects of pollution, habitat loss or degradation (Chapters 2, 10 and 11), dams (Chapter 4), interactions with exotic species (Chapter 8), and other variables that ultimately reduce natural production and recruitment (Figure 13.1). Conversely, artificial propagation, stocking, re-introduction and translocation can be viewed as 'supply-side' approaches to complement *in situ* habitat improvements or the removal of other demand-side pressures by enhancing production above the carrying capacity inherent in a pristine or in a stressed ecosystem. Supply and demand of living natural resources are not as easily separated as they are for production and consumption of non-living goods or services, because some portion of the production must be reserved as a reproductive template for future supply.

13.2 BRIEF HISTORY OF PROPAGATION AND STOCKING

Histories of artificial propagation and fish stocking chronicle a constantly evolving set of practices, reflective of advances in information about reproductive biology and technology in fish husbandry and culture (Bowen, 1970; Bottom, 1997). Artificial propagation began as a means to produce food-fish largely as an agricultural enterprise – although the production of ornamental fishes for cultural purposes might have occurred alongside food production. While the roots of fish culture go back a thousand years or more, the recent era is about 150 years old and greatly expanded in parallel with the modern industrial revolution. This expansion intensified after World War II, when inland freshwater and diadromous species belonging to diverse taxonomic groups such as trout, pike, perch, bass, eel and salmon (to name but a few) were bred and released widely across several continents to provide protein production and commercial opportunity, compensate for loss of natural supply from human activities, or enhance production in nature. More recently, application toward enhancing recreational outcomes has taken precedence in many economically developed jurisdictions. More recent still, artificial propagation and stocking have been undertaken under the banner of conservation or restoration. As a result, many of the longstanding programmes have changed their focus and goals, in parallel with the evolution of fishery-resource management. Although the many and diverse programmes for artificial propagation and stocking are motivated by an equally diverse set of goals, Cowx (1998) and Utter and Epifanio (2002) suggested that most programmes follow one of a few major categories based on their goals, method of operation, sources of bloodstock, and relationships with natural populations (Figure 13.2).

Translocations, stock transfers and *re-introductions* are related but subtly different activities that ultimately move freshwater fishes within and among ecosystems in which they evolved. *Introductions* – in a general sense – are the movement of a species to a new area beyond its native historical (evolutionary) range. Typically, fish introductions do not serve a conservation purpose (although we acknowledge some conservation benefits where introductions might serve a biocontrol function); instead, they are often intended to meet a recreational or commercial goal. They have a long history in natural resource management, with well-known

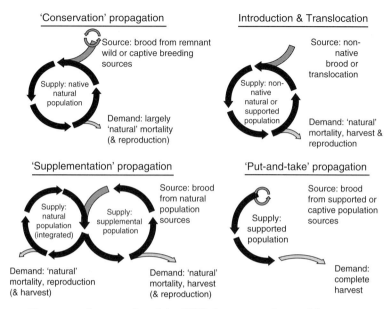

Figure 13.2 Conceptual models of differing propagation models. See Box 13.2 for a description and a set of illustrative examples of these and other approaches. The reader should note that 'natural' mortality may be amplified by anthropogenic influences on the ecosystem.

examples being the global movements and establishment of rainbow (*Oncorhynchus mykiss*) and brown trout (*Salmo trutta*), largemouth bass (*Micropterus salmoides*) and peacock cichlid (*Cichla* species) and white shad (*Alosa sapadissima*) (see Chapter 8 for a more expansive examination of examples and consequences of introductions). A stock transfer involves moving a distinct and divergent lineage (whether defined as a stock, population, *ESU*, genetic management unit, etc.) from within the species' historical range to a different locale presently occupied by another extant discrete stock. In these instances, the chances for the transferred stock to interbreed with a remnant native stock is facilitated, potentially leading to outbreeding effects and erosion of diversity among populations. In much of the twentieth century, stock transfers of fish were commonplace and were carried out in the absence of an appreciation for the genetic, phenotypic and ecological differences that typically occur among natural populations. Stock transfers routinely resulted from ignorance of any genetic-level structure between donor and recipient populations. This activity is typified by the movement of Chinook salmon (*O. tshawytscha*) stocks from one North American river

Box 13.2. Highlighted examples of fish species and the purposes for their propagation, translocation or re-introduction

- *Conservation propagation* – examples include Devil's Hole pupfish (*Cyprinodon diabolis*); humpback chub (*Gila cypha*); razorback sucker (*Xyrauchen texanus*); redspotted sunfish; Laurentian Great Lakes deep-water ciscos and chubs (*Coregonus* species); west slope cutthroat trout (*O. clarki lewisi*); and several sturgeon species worldwide (e.g. *Huso* species, *Acipenser* species).

  Species highlight: desert pupfish (*Cyprinidon macularius*) – endemic to palm oases and small streams in arid deserts of the southwest USA, the desert pupfish has been extirpated from much of its range due to dewatering, predation by introduced crayfish and other stressors. A population of this species is being captively bred and maintained at a hatchery until suitable habitat is identified and secured. The goal is ultimately to release and re-establish viable populations within their historical range.

- *Supplementation* – examples include various Pacific salmon species in North America; Atlantic salmon (*Salmo salar*) in Europe and North America; American (white) shad in North America; pez blanco (*Chirostoma estor estor*) in Mexico.

- *Mitigation* – examples include various Pacific salmon species in the northwest USA ('Mitchell Act' response to hydroelectric dams and river impoundment).

- *Introduction* – examples include Pacific salmonids into the Laurentian Great Lakes, Australia, New Zealand; largemouth bass worldwide; Florida largemouth bass (*M. floridanus*) into northern USA and southern Canada; Nile perch (*Lates niloticus*) into Lake Malawi and other African rift lakes; rainbow trout and brown trout worldwide.

- *Put-and-take* – rainbow trout in recreational waters (both within and beyond its native range) at a legally harvestable size.

  Species highlight: largemouth and smallmouth bass – native to a broad range of southern Canada and the middle USA, black bass species are artificially propagated and released to provide recreational opportunities, especially where reproduction and recruitment in nature cannot support sport-harvest pressure. Fish may be raised and released at a size targeted for immediate harvest or for delayed harvest (put-grow-take).

- *Translocation and re-introduction* – translocation examples include the creation of greenback cutthroat trout refuge populations in waters adjacent to its native range.

  Species highlight: Some of the pitfalls associated with translocations and re-introductions may be viewed by the example of the greenback cutthroat trout. This subspecies of cutthroat trout is native to the southern Rocky Mountain west region of the USA and is an imperilled subspecies and formally protected under federal law. An effort was undertaken to recover this subspecies through a combination of artificial propagation, provisions against harvest and habitat improvement. Establishment of temporary or longer-term *ex situ* populations (a translocation) has been used as a refuge until habitat improvements permit re-introduction into formerly occupied habitats. The goal was to secure a network of viable and self-sustaining populations, but began with artificial production and captive breeding to obtain a demographically sufficient number of founders for the translocated refuge populations. Unfortunately, even with the best of intentions and fairly extensive background investigation, some recent evidence has come to light that the original source of brood may have been hybridised with a different subspecies, thus compromising the primary restoration goal of the programme.

to another or the movement of southern populations of largemouth bass into the range of northerly bass populations across North America. Transfers can be either 'direct' (such that embryos, young or adults are captured from one location and moved to another) or 'indirect' (e.g. adults captured from one location and bred to produce young that are subsequently released into a targeted habitat).

Re-introductions, which are more generally applicable to conservation of biodiversity, involve the return or restoration of a species to a locale within its previously occupied native historical range from which it has been extirpated. Examples include attempts to re-establish formerly abundant alligator gar in the upper reaches of the Mississippi River basin and locally extirpated deep-water whitefishes (*Coregonus* species) to several of the Laurentian Great Lakes in North America. Similarly, a translocation is a conservation-driven action involving movement of a species to a locale or habitat where that species does not presently occur, including off-site refuges, from one that is degraded or incapable of supporting the species. Such a movement is undertaken as a temporary measure (although such measures may be viewed as necessarily longer-term or permanent) for the purposes of preserving the species while the stressors or threats are addressed. The creation of multiple refuge redspotted sunfish (*Lepomis miniatus*) within the fringe range of Illinois (USA) is an example of this approach.

Throughout most of the history and various uses of artificial propagation and stocking, little consideration was given toward conservation or biodiversity criteria. Over the last half century in particular, the use of artificial propagation has grown globally. Arguably the USA and Canada have developed the world's most expansive network of artificial propagation and stocking programmes. Unlike countries or governments that sponsor closed-system *food-fish aquaculture*, the USA and Canada

Table 13.1 *Some indices of the size and scope of the fish artificial propagation enterprise in North America.*[a]

Taxonomic survey	
Number of inland freshwater or anadromous species (Families) propagated for release	95 (17)
Subspecies propagated[b]	10
Documented hybrids propagated intentionally[c]	10
Artificial propagation facilities operated in North America	
by US State Governmental Agencies	426
by Canada Provincial Agencies	> 20
by US Federal Governmental Agencies	70
by Canada Federal Government Agencies	21
by Tribal or First Nation Government Agencies	>35?
by non-governmental agencies (including public utilities), lake associations, or other private entities	'Undocumented'

[a]Information assembled from governmental websites in November 2012. These figures represent a snapshot in time and do not include species or facilities intended for closed-loop aquaculture purposes. Exact counts may change along with evolving programme objectives and funding constraints; contact the authors for specific information and sources.
[b]Primarily described and undescribed cutthroat trout (*Oncorhynchus clarki* subspecies).
[c]Primarily from five families (Salmonidae, Moronidae, Percidae, Esocidae and Centrarchidae).

operate more than 500 artificial production facilities (not including facilities operated by private citizens, companies or non-governmental organisations) that produce resident freshwater or anadromous fishes. The largest share of production and release are presently geared toward anadromous salmon and resident trout (see Epifanio, 2000, for a description).

Global production by artificial propagation is massive. Worldwide, facilities annually produce multiple billions of individual fish (and multiple millions of kilograms) for stocking into the environment (Table 13.1). For example, about 5 billion juvenile Pacific salmon are released annually by nations around the Pacific Rim (Mahnken *et al.*, 1998). Information on other nations is not easily accessible. According to the Food and Agriculture Organization – which collects information on food-fish aquaculture and fishery management – 94 nations report intentionally releasing fish and other species into the freshwater and marine environments for various purposes (FAO, 2005). The scale of the enterprise in North America, let alone globally, warrants a careful examination of just how well the societal benefits from artificial propagation and related activities stack up to the risks. Importantly, these comparisons should be viewed not only for the populations targeted for production, but also for the non-target aquatic biota and communities that are part of recipient ecosystems.

13.3 POTENTIAL BENEFITS OF ARTIFICIAL PROPAGATION

Informed decisions about appropriate uses of artificial propagation (as well as translocations and re-introductions) are best made following a comprehensive assessment of potential benefits and risks. In this and the next section, we illustrate some of the potential benefits and risks, respectively, using an approach modified from Waples and Drake (2004) and Waples et al. (2012). Later, we offer a conceptual framework for addressing artificial propagation and the benefits and risks it introduces to conservation decisions and actions.

Artificial propagation has the potential to provide two general types of benefits: (1) those that accrue to human society, either broadly or to a narrower constituency; and (2) those that accrue to natural populations and associated ecosystems.

Societal benefits can include the following.

1. Increase commercial, recreational and cultural harvest opportunities for food – undoubtedly the most commonly realised societal benefit (Table 13.2 includes purposes and potential benefits for some examples of North American species).
2. Satisfy legal mandates for mitigation or tribal treaty rights – artificial propagation programmes are often tied to legal obligations to produce specified numbers or biomass of fish for stocking to compensate losses to natural populations resulting from human activities such as dam construction or habitat degradation. In the USA, for example, of the approximately 70 federally operated facilities and programmes, many were built and now operate (under laws that may date back to the 1940s) specifically to produce and release migratory salmon to compensate for the construction and operation of dams within the Columbia River Basin (NRC, 1996). Similar mitigation operations operate and produce inland freshwater fish (such as inland trout, sturgeons, eels and other species lost for other dam projects in many large US rivers).
3. Employment opportunities – artificial propagation programmes are labour-intensive and require a significant capital investment in planning, design and construction. These facilities and the programmes

Table 13.2 *Freshwater fish taxa (systematic Family) propagated in North America by state and federal governments. Typical habitats and purposes (i.e. societal benefit) for propagation are identified. Interspecific hybrids are grouped together rather than with Families.*

Common name	Genus	Species	Subspecies	Typical habitat	Purpose(s) for propagation and release
1. Family Acipenseridae					
Shortnose sturgeon	*Acipenser*	*brevirostrum*		FW-W	Conservation, imperilled
Lake sturgeon	*Acipenser*	*fulvescens*		FW-W	Conservation, imperilled
Atlantic sturgeon	*Acipenser*	*oxyrhynchus*		FW-W	Conservation, imperilled
Gulf sturgeon	*Acipenser*	*oxyrhynchus*	*desotoi*	FW-W	Conservation, imperilled
White sturgeon	*Acipenser*	*transmontanus*		FW-W	Conservation, imperilled
Pallid sturgeon	*Scaphirhynchus*	*albus*		FW-W	Conservation, imperilled
2. Family Catostomidae					
White sucker	*Catostomus*	*commersoni*		FW-W	Aquaculture, forage
Rio Grande sucker	*Catostomus*	*plebeius*		FW-W	Conservation, imperilled
Cui-ui sucker	*Chasmistes*	*cujus*		FW-W	Conservation, imperilled
Robust redhorse	*Moxostoma*	*robustum*		FW-W	Conservation, imperilled
Sicklefin redhorse	*Moxostoma*	*sp.* (undescribed)		FW-W	Conservation, imperilled
Razorback sucker	*Xyrauchen*	*texanus*		FW-W	Conservation, imperilled
3. Family Centrarchidae					
Rock bass	*Ambloplites*	*rupestris*		FW-W	Supplementation, recreation
Redbreast sunfish	*Lepomis*	*auritus*		FW-W	Supplementation, recreation
Pumpkinseed	*Lepomis*	*gibbosus*		FW-W	Supplementation, recreation
Bluegill	*Lepomis*	*macrochirus*		FW-W	Supplementation, recreation
Redear sunfish	*Lepomis*	*microlophus*		FW-W	Supplementation, recreation
Redspotted sunfish	*Lepomis*	*miniatus*		FW-W	Conservation, imperilled
Redeye bass	*Micropterus*	*coosae*		FW-W	Supplementation, recreation

Table 13.2 (cont.)

Common name	Genus	Species	Subspecies	Typical habitat	Purpose(s) for propagation and release
Smallmouth bass	*Micropterus*	*dolomieu*		FW-W	Supplementation, recreation
Florida (largemouth) bass	*Micropterus*	*floridanus*		FW-W	Supplementation, recreation
Spotted bass	*Micropterus*	*punctulatus*		FW-W	Supplementation, recreation
Largemouth (northern) bass	*Micropterus*	*salmoides*		FW-W	Supplementation, recreation
Alabama spotted bass	*Micropterus*	sp. (undescribed)		FW-W	Supplementation, recreation
White crappie	*Pomoxis*	*annularis*		FW-W	Supplementation, recreation
Black crappie	*Pomoxis*	*nigromaculatus*		FW-W	Supplementation, recreation
4. Family Cyprinidae					
Goldfish	*Carassius*	*auratus*		FW-W	Aquaculture, forage
Grass carp	*Ctenopharingodon*	*idella*		FW-W	Aquaculture, vegetation control
Guzman beautiful shiner	*Cyprinella*	*formosa*		FW-W	Conservation, imperilled
Koi	*Cyprinus*	*carpio*		FW-W	Aquaculture, ornamental
Humpback chub	*Gila*	*cypha*		FW-W	Conservation, imperilled
Bonytail chub	*Gila*	*elegans*		FW-W	Conservation, imperilled
Chihuahua chub	*Gila*	*nigrescens*		FW-W	Conservation, imperilled
Roundtail chub	*Gila*	*robusta*		FW-W	Conservation, imperilled
Pahranagat roundtail chub	*Gila*	*robusta*	*jordani*	FW-W	Conservation, imperilled
Virgin River chub	*Gila*	*seminuda*		FW-W	Conservation, imperilled
Rio Grande silvery minnow	*Hybognathus*	*amarus*		FW-W	Conservation, imperilled
Plains minnow	*Hybognathus*	*placitus*		FW-W	Conservation, imperilled
Common shiner	*Luxilis*	*cornutus*		FW-W	Aquaculture, forage
Golden shiner	*Notemigonus*	*crysoleucas*		FW-W	Aquaculture, forage
Arkansas River shiner	*Notropis*	*girardi*		FW-W	Conservation, imperilled

Cape Fear shiner	*Notropis*	*mekistocholas*	FW-W	Conservation, imperilled
Suckermouth minnow	*Phenacobius*	*mirabilis*	FW-W	Conservation, imperilled
Northern redbelly dace	*Phoxinus*	*eos*	FW-W	Conservation, imperilled
Southern redbelly dace	*Phoxinus*	*erthrogaster*	FW-W	Conservation, imperilled
Fathead minnow	*Pimephales*	*promelas*	FW-W	Aquaculture, forage
Woundfin minnow	*Plagopterus*	*argentissimus*	FW-W	Conservation, imperilled
Colorado pikeminnow	*Ptychocheilus*	*lucius*	FW-W	Conservation, imperilled
5. Family Cyprinodontidae				
Leon Springs pupfish	*Cyprinodon*	*bovinus*	FW-W	Conservation, imperilled
Comanche Springs pupfish	*Cyprinodon*	*elegans*	FW-W	Conservation, imperilled
Desert pupfish	*Cyprinodon*	*macularius*	FW-W	Conservation, imperilled
6. Family Clupeidae				
River (blueback) herring	*Alosa*	*aestivalis*	FW/A-W	Supplementation, commercial
Hickory shad	*Alosa*	*mediocris*	FW/A-W	Supplementation, commercial, recreational
American shad	*Alosa*	*sapadissima*	FW/A-W	Supplementation, mitigation, commercial, recreational
7. Family Esocidae				
Northern pike	*Esox*	*lucius*	FW-W	Supplementation, recreation, cultural
Muskellunge	*Esox*	*masquinongy*	FW-W	Supplementation, recreation
Chain pickerel	*Esox*	*niger*	FW-W	Supplementation, recreation
8. Family Gadidae				
Burbot	*Lota*	*lota*	FW-C	Conservation, supplementation, commercial, recreation, cultural

Table 13.2 *(cont.)*

Common name	Genus	Species	Subspecies	Typical habitat	Purpose(s) for propagation and release
9. Family Ictaluridae					
Blue catfish	*Ictalurus*	*furcatus*		FW-W	Supplementation, recreation
Brown bullhead	*Ictalurus*	*nebulosu*		FW-W	Supplementation, recreation
Channel catfish	*Ictalurus*	*punctatus*		FW-W	Supplementation, recreation
Flathead catfish	*Pylodictis*	*oliaris*		FW-W	Supplementation, recreation
10. Family Lepisosteidae					
Alligator gar	*Lepistoseus*	*spathula*		FW-W	Conservation, supplementation, recreation
11. Family Moronidae					
White bass	*Morone*	*chrysops*		FW-W	Supplementation
Striped bass	*Morone*	*saxatilis*		FW/A-W	Supplementation, mitigation, recreation
12. Family Petromyzontidae					
Pacific lamprey	*Lampetra*	*tridenta*		FW/A-W&C	Conservation, supplementation, mitigation, cultural
Sea lamprey	*Petromyzon*	*marinus*		FW/A-W&C	Biological control
13. Family Percidae					
Arkansas darter	*Etheostoma*	*cragini*		FW-W	Conservation, imperilled
Yellow perch	*Perca*	*flavescens*		FW-W	Supplementation, recreation
Sauger	*Sander*	*canadense*		FW-W	Supplementation, recreation
Walleye	*Sander*	*vitreum*		FW-W	Supplementation, recreation, cultural

14. Family Poecilidae

Common name	Genus	Species	Subspecies	Code	Notes
Ozark cavefish	Amblyopsis	rosae		FW-W	Conservation, imperilled
Barrens topminnow	Fundulus	julisia		FW-W	Conservation, imperilled
Mosquitofish	Gambusia	affinis		FW-W	Aquaculture, insect control
Big Bend gambusia	Gambusia	gaigei		FW-W	Conservation, imperilled
Pecos gambusia	Gambusia	nobilis		FW-W	Conservation, imperilled
Gila topminnow	Poeciliopsis	occidentalis		FW-W	Conservation, imperilled

15. Family Polyodontidae

Common name	Genus	Species	Subspecies	Code	Notes
Paddlefish	Polyodon	spathula		FW-W	Conservation, supplementation, commercial, recreation

16. Family Salmonidae

Common name	Genus	Species	Subspecies	Code	Notes
California golden trout	Oncorhynchus	aguabonita		FW-C	Conservation, imperilled
Kern River golden trout	Oncorhynchus	aguabonita	gilberti	FW-C	Conservation, imperilled
Apache trout	Oncorhynchus	apache		FW-C	Conservation, mitigation
Yellowstone cutthroat trout	Oncorhynchus	clarki	bouvieri	FW-C	Supplementation, recreation
Lahontan cutthroat trout	Oncorhynchus	clarki	henshawi	FW-C	Conservation, imperilled
Westslope cutthroat trout	Oncorhynchus	clarki	lewisi	FW-C	Supplementation, recreation
Colorado River cutthroat trout	Oncorhynchus	clarki	pleuriticus	FW-C	Conservation, imperilled
Greenback cutthroat trout	Oncorhynchus	clarki	stomias	FW-C	Conservation, imperilled
Bonneville cutthroat trout	Oncorhynchus	clarki	utah	FW-C	Conservation, imperilled
Rio Grande cutthroat trout	Oncorhynchus	clarki	virginalis	FW-C	Conservation, imperilled
Finespotted Snake River cutthroat trout	Oncorhynchus	clarki	(subspp.)	FW-C	Supplementation, conservation
Gila trout	Oncorhynchus	gilae		FW-C	Conservation, mitigation
Pink salmon	Oncorhynchus	gorbuscha		FW/A-C	Supplementation, commercial, recreation, mitigation, conservation, imperilled

Table 13.2 *(cont.)*

Common name	Genus	Species	Subspecies	Typical habitat	Purpose(s) for propagation and release
Chum	*Oncorhynchus*	*keta*		FW/A-C	Supplementation, commercial, recreation, mitigation, conservation, imperilled
Coho salmon	*Oncorhynchus*	*kisutch*		FW/A-C	Supplementation, commercial, recreation, mitigation, conservation, imperilled
Rainbow trout	*Oncorhynchus*	*mykiss*		FW/A-C	Supplementation, mitigation, put & take, conservation, imperilled
Sockeye	*Oncorhynchus*	*nerka*		FW/A-C	Supplementation, commercial, recreation, mitigation, conservation, imperilled
Chinook salmon	*Oncorhynchus*	*tshawytscha*		FW/A-C	Supplementation, commercial, recreation, mitigation, conservation, imperilled
Round whitefish	*Prosopium*	*cylandraceum*		FW-C	Conservation, imperilled
Brown trout (multiple strains)	*Salmo*	*trutta*		FW/A-C	Introduction, supplementation, put & take, recreation
Atlantic salmon	*Salmo*	*salar*		FW/A-C	Conservation, imperilled
Arctic char	*Salvelinus*	*alpinus*		FW/A-C	Supplementation, recreation
Bull trout	*Salvelinus*	*confluentus*		FW/A-C	Conservation, mitigation, imperilled
Brook trout	*Salvelinus*	*fontinalis*		FW/A-C	Supplementation, recreation
Lake trout	*Salvelinus*	*namaycush*		FW-C	Supplementation, recreation
Arctic grayling	*Thymallus*	*arcticus*		FW-C	Supplementation, recreation

17. Family Sciaenidae

Freshwater drum	Aplodinotus	grunniens	FW-W	Aquaculture, forage
18. Interspecific hybrids				
Rainbow × 'unidentified' cutthroat trout hybrid	Oncorhynchus	(mykiss × clarki)	FW-W	Put (grow) & take, recreation
Rainbow × YCT cutthroat trout hybrid	Oncorhynchus	(mykiss × clarki)	FW-W	Put (grow) & take, recreation
Striped × white bass hybrid	Morone	(saxatilis × chrysops)	FW-W	Put (grow) & take, recreation
Walleye × sauger (saugeye)	Sander	(vitreum × canadense)	FW-W	Put (grow) & take, recreation
Lake × brook trout hybrid (splake)	Salvelinus	(namaycush × fontinalis)	FW-W	Put (grow) & take, recreation
Muskellunge × northern pike (tiger)	Esox	(masquinongy × lucius)	FW-W	Put (grow) & take, recreation
'Unidentified' hybrid sunfish	Lepomis		FW-W	Put (grow) & take, recreation
Brown × brook trout (tiger)	a) Salmo ×b) Salvelinus	a) trutta × b) fontinalis	FW-W	Put (grow) & take, recreation
White × black crappie (magnolia)	Pomoxis	(annularis × nigromaculatus)	FW-W	Put (grow) & take, recreation
Bluegill × green sunfish	Lepomis	(macrochirus × cyanellus)	FW-W	Put (grow) & take, recreation

'Typical habitat' key: FW-W, Freshwater–Warmwater species; FW/A-W: Freshwater/Anadromous–Warmwater species; A-W: Anadromous–Warmwater species; FW-C: Freshwater–Coldwater species; FW/A-C: Freshwater/Anadromous–Coldwater species; A-C: Anadromous–Coldwater species.

they support often require a significant labour force to operate and maintain – frequently in remote areas where jobs are scarce.

4. Provide public education – many hatcheries have public exhibits and/ or tours that provide biological information about the species propagated and the rationale for the programmes. For a good example, see the Living Museum of Toyohira Salmon (www.sapporo.world-guides .com/sapporo_museums.html), which provides a wealth of information about chum salmon (*Oncorhynchus keta*) and the very large (approx. 1 billion juveniles released per year) hatchery production programme on the island of Hokkaido, Japan.

The potential for artificial propagation to provide benefits to natural populations and ecosystems is inversely related to the level of risk faced by the target population(s). Populations at high short-term risk of extinction might benefit substantially from captive intervention, but more robust natural populations have little to gain and much to lose from artificial propagation. Potential benefits to natural populations include those that directly result from culture activities (items 1–3 below), as well as those that are indirect consequences of artificial production (items 4–5 below).

Natural population and ecosystem benefits include the following.

1. Reduce short-term extinction risks for endangered populations – an artificial propagation programme that enhances survival during a life-history stage with high natural mortality can help reduce short-term risks. This approach has yielded examples of at least short-term successes (e.g. Redfish Lake sockeye salmon, *Oncorhynchus nerka*); however, long-term consequences of such interventions are largely unknown (e.g. see Metcalf *et al.*, 2007; George *et al.*, 2009 for a cautionary perspective on *O. clarki stomias*).

2. Maintain depressed populations and accelerate recovery – artificial propagation programmes may indefinitely maintain or support depressed populations. The premise is that propagation will facilitate and ultimately achieve self-sustainability. In theory, a hatchery might speed recovery by temporarily increasing abundance, after which the supplementation programme terminates and the population remains stable at the higher population size. For example, otolith marking of hatchery juveniles helped managers demonstrate that supplementation greatly boosted restoration efforts for American shad in the Susquehanna River in the eastern US (Hendricks, 1995). In addition,

Brown *et al.* (2000) examined and failed to observe any changes to the genetic background in American shad in the James River before and after release of captively bred fish, signalling at least an initial level of complementarity between donor and recipient – although we cannot rule out that the releases simply do not survive or interbreed. However, although many such hatchery programmes have demonstrated the ability to propagate fish populations for several decades or 5–10 fish generations, few empirical examples exist of sustained recovery after such programmes terminate. This strategy is most likely to be effective if coupled with efforts to identify and alleviate the factors that caused initial declines and are impeding natural recovery.

3. Reseed vacant or vacated habitat – translocations have been widely used with terrestrial and freshwater species, particularly for restoration purposes.

4. Reduce harvest pressure on natural populations – overfishing (i.e. harvest demand at unsustainable levels) is a serious problem for many species, and existence of a consistent supply of high-quality cultured product at a reasonable price could potentially reduce demand for (and profit from) wild harvests. While this has been suggested in theory, the authors are unaware of any well-documented cases.

5. Provide positive ecosystem effects – fish are often keystone species in aquatic ecosystems and their loss can substantially alter ecosystem function through destabilisation of trophic relationships. For example, studies that track marine-enriched isotopes of carbon and nitrogen have shown that large fractions of riparian vegetation in otherwise unproductive freshwater stream systems derive nutrients from decomposing salmon carcasses. Reintroduction or supplementing native fish populations to levels at which they contribute substantially to ecosystem functions can provide substantial benefits (Merz & Moyle, 2006). Translocation also has a potential to provide ecosystem benefits. Examples include providing forage (such as with minnows or suckers), biological control of weeds or insects (such as with exotic grass carp or mosquito fish covered in Chapter 8), and even control of invasive species (such as the sterile male sea lamprey in the Laurentian Great Lakes). However, each of these examples carries a risk for unanticipated or unintended impacts to the local community or ecosystem.

Artificial propagation programmes aimed at commercial and recreational enhancement rarely seek or achieve demonstrable benefits for natural communities. Therefore, most risk-versus-benefit considerations for commercial artificial propagation must focus on evaluating trade-offs between potential benefits to society and risks to natural populations.

13.4 RISKS ASSOCIATED WITH ARTIFICIAL PROPAGATION

13.4.1 Genetic risks

1. Loss of genetic diversity within populations – genetic variability within populations provides the raw material for evolution. Populations with low levels of genetic variability have less capacity to respond to environmental change. Levels of genetic variability within populations are determined largely by effective population size (N_e), which can be influenced by many factors, including artificial propagation. N_e is typically lower – and often substantially lower – than the population's census size. In general, risks increase when N_e in the captive population is much lower than in the wild and when the fraction of natural spawners that are of captive origin is relatively high. Inbreeding depression and loss of genetic diversity has been documented in hatchery populations of several salmonids (see Waples & Drake, 2004, for citations). Chapter 16 presents more detail on issues related to conserving genetic diversity.

2. Loss of genetic diversity among populations – genetic and phenotypic diversity among populations confers resilience to natural systems on a variety of temporal and spatial scales. The major concern regarding loss of among-population diversity is that widespread escapees from one or a few artificial propagation broodstocks will lead to the replacement of existing patterns of locally adapted populations with a smaller number of relatively homogeneous ones. Diversity among populations typically has evolved over many hundreds to thousands of years and cannot easily be regenerated once lost (Figure 13.3). This diversity plays much the same role in buffering effects of natural fluctuations as the diversification of a portfolio of investments (Schindler *et al.*, 2010).

3. Loss of reproductive fitness – *domestication* is a process by which populations become better adapted to artificial environments. Because selective regimes and mortality patterns differ so greatly between wild and artificial propagation environments, domestication will occur to some extent whenever populations are reared under controlled

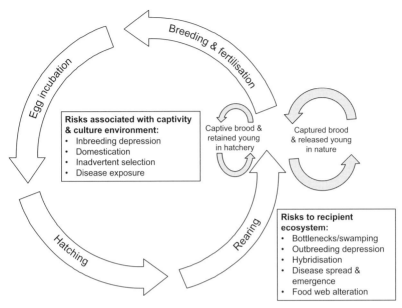

Figure 13.3 Conceptual depiction of risks to culture environments and recipient ecosystems from propagated fishes. The realisation or magnitude of such risks will be specific to the conditions of the propagation (including translocation and re-introduction) enterprise.

conditions. Note that domestication does not require (or generally involve) genetic change to or modification of individual fish; rather, it occurs across generations as certain individuals (and their associated genotypes) survive and reproduce at different rates than they would in the wild. Although some level of domestication is probably essential in a profitable artificial propagation operation, domesticated individuals are generally less fit in the wild, and if they reproduce successfully their progeny can depress fitness and compromise viability of the wild population (Figure 13.4). A related undesirable phenomenon, outbreeding depression, is the loss of fitness that occurs when individuals from two genetically divergent populations interbreed. Outbreeding depression can be facilitated by artificial propagation operations that import non-local stocks that subsequently escape and interbreed with local populations. Use of local broodstocks will not avoid domestication, but it will avoid losses of fitness arising from translocation and subsequent interbreeding of genetically divergent populations. Some artificial propagation operations will involve both

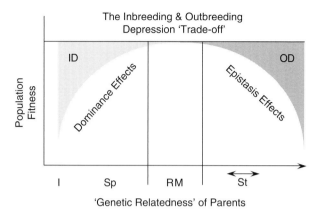

Figure 13.4 Conceptual relationship between genetic relatedness of parents (I, inbred or highly related; RM, random mating within a population; St, divergent stocks; Sp, divergent species) and population fitness. Population fitness is expected to be maximised under random mating of a large population. Fitness may be reduced with increasingly inbred or outbred parents, but due to different causes (accumulation of recessive alleles through inbreeding and breakup of co-adapted, interacting genes through outbreeding). Adapted from Philipp *et al.* 2002.

factors (broodstock derived from a non-local source and strong domestication for market-related reasons). Modern molecular methods for parentage analysis have allowed researchers to quantify effects of domestication on fitness of supplemented fish populations (Christie *et al.*, 2011).

13.4.2 Ecological risks

1. Altered foodweb relationships through changes in competition and predation – stocked fish, especially at high densities or otherwise in excess of the local carrying capacity, can prey on wild populations and compete with natural individuals for food or territory. Released fish that survive to maturity can also compete and interfere with natural adults for spawning sites and access to mates, regardless of whether they are successful in spawning. If the number of stocked fish is sufficiently large, their presence might impose other forms of selection that alter characteristics of wild populations important to survival and reproductive success. Effects on wild populations can occur even if

the stocked individuals do not complete a life cycle in the wild – for example, if they destroy spawning nests of wild individuals or divert mating efforts by wild individuals into unproductive spawnings.

2. Disease and other mortality factors – diseases and parasites are part of the ecological and evolutionary histories of all aquatic species. However, because of high rearing density, hatchery operations can greatly magnify the incidence of naturally occurring diseases and/ or parasites, and movements of eggs, fish or fish products into new geographic areas can expose natural populations to new pathogens to which they have little or no natural resistance. Risks from introduced pathogens and parasites include temporary epidemics and/ or long-lasting population reductions and even extinction of wild populations. Risks posed by disease agents are harder to quantify than those posed by competition or predation, as a single individual transferred to a recipient population can have dramatic consequences. Disease agents also can be spread by water, independent of any releases of cultured individuals. Ultimately in regard to assessment of pathogens, parasites and overall fish health, much of the focus regrettably tends to be on the impact of disease on propagated fish rather than wild fish health (Chapter 7 presents more detail on infectious disease risks).

13.4.3 Other risk factors

1. Power to detect adverse effects is typically low – natural variability in most natural populations is relatively high, which makes it hard to detect deleterious effects of artificial propagation, even in the most ambitious monitoring programme. This means that by the time adverse effects are detected, they might have already occurred for many years and might have long-lasting consequences.

2. Programmatic inertia – experience indicates that, once begun, fish hatchery programmes can be difficult to stop for political or social reasons, even in the face of strong evidence demonstrating their ineffectiveness or adverse impacts on natural populations. Therefore, a key question in considering whether to initiate a new artificial propagation programme is, 'Can the operation be terminated or modified if a biological evaluation indicates its costs and/or detrimental effects exceed its benefits?'

13.5 RISKS ASSOCIATED WITH INTRODUCTIONS, STOCK TRANSFERS, RE-INTRODUCTIONS AND TRANSLOCATIONS

Although the above activities can be motivated by a variety of management or conservation objectives, the magnitude of potential and realised risks and benefits depend on just a few characteristics. First is the complexity and stability of the food web. Introducing a predator into a formerly fishless environment, for example, can have an ecologically catastrophic impact on a resident community (Knapp *et al.*, 2001). Second is the occurrence and history of host–pathogen interactions. Moving a species or population that has co-evolved with a pathogen into an area where a residing fish population is naïve to a pathogen can have profound impacts if (as often occurs) the pathogen hitchhikes along on the introduction. The emergence of epizootic whirling disease (*Myxobolis*) in the US West as a threat to native cutthroat trout subspecies has been traced to voluminous movements of rainbow trout (Allendorf *et al.*, 2001). Third is whether a remnant conspecific population or related species remains in the targeted locale. Releasing divergent but interfertile individuals on top of a remnant population risks causing genomic extinction through interbreeding of stocks or interspecific hybridisation. Here again, the release of rainbow trout within the range, and often on top of, various cutthroat trout species in the western USA has led to the creation of hybrid swarms that effectively eliminated the intact native taxon. Fourth, the degree of ecological dissimilarity between donor and recipient environments can greatly influence the success of a translocation (e.g. Figure 13.5, Table 13.3). Finally, the size of the breeding pool will determine whether any founder effects or genetic bottlenecks might limit viability in the new location.

13.6 A PROPOSED ANALYTICAL FRAMEWORK FOR ADDRESSING BENEFIT AND RISK CONFLICTS

It should be apparent from the previous two sections that artificial propagation can provide a number of benefits, but also creates several risks. This raises a key question: 'Under what conditions do the potential benefits justify the risks?' This question can be answered in part by conducting a formal risk–benefit analysis (e.g. Busack & Currens, 1995). We note, however, that such an analysis is not a trivial exercise and is very difficult to conduct in a comprehensive manner for several reasons.

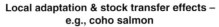

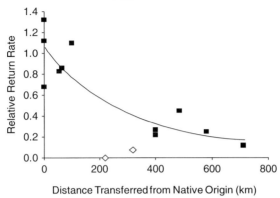

Figure 13.5 Relationship between distance transferred and adult return rate (a measure of reproductive success) for coho salmon. Adapted from Reisenbichler (1988).

Table 13.3 *Comparison of relative fitness (Wr) of four 'Parental' stocks of largemouth (LMB) and Florida (FLB) bass bred and reared sympatrically in geographically distant experimental ponds. Values represent four-year means for proportional contribution to year-classes (1987–1990) for each stock in a pond relative to the stock with the maximum contribution (Wr = 1.00). In each experimental pond, the local stock provided the highest proportional contribution (see diagonal). Adapted from Philipp et al. (2002).*

		Stock	Wr	
Pond location	LMB (Wisconsin)	LMB (Illinois)	LMB (Texas)	FLB (Florida)
Minnesota	1.00	0.00	0.00	0.00
Illinois	0.70	1.00	0.59	0.04
Texas	0.55	0.67	1.00	0.78
Florida	0.01	0.03	0.03	1.00

1. Except for programmes involving critically endangered populations, most benefits will accrue to society (e.g. more fish for food or harvest; jobs; satisfying legal obligations), while most risks will be borne by natural populations. It might be possible to express all potential benefits in terms of dollars, but it is much harder to place a dollar value on loss or harm to species in the wild. This means that a full cost–benefit

analysis would have to include analyses by social scientists as well as natural scientists.

2. Although some of the deleterious effects of artificial propagation are well documented and understood, our ability to predict the consequences of any particular action is limited. This means that any risk–benefit evaluations will include a good deal of uncertainty. This in turn means that managers will need to wrestle with difficult burden-of-proof questions such as, 'Should a proposed project be allowed to go forward unless it can be convincingly demonstrated that serious harm to natural populations would result?' or, 'Should a proposed project NOT be allowed to proceed unless it can be convincingly demonstrated that it will NOT cause serious harm?' Given the large uncertainties involved, the way in which the burden-of-proof issue is addressed can have a profound effect on the type of programmes that are considered to have acceptable risks.

3. Many biological risks are negatively correlated, such that reducing one risk increases another. For example, releasing juvenile fish from a hatchery early in their life cycle reduces opportunities for domestication selection, but at the same time it increases opportunities for ecological interactions (competition, predation, disease transfer) with wild fish. Similarly, if one is trying to enhance a severely depressed natural population with artificial propagation, taking only a few adults for broodstock exposes a smaller fraction of the population to domestication selection and minimises demographic effects on the wild population, but it makes it more likely that the hatchery programme will include only a fraction of the overall diversity. These inherent trade-offs mean that it is impossible to construct a completely risk-free programme (Waples, 1999).

Because risks and benefits alike can vary depending on species, location and methods used, an ideal approach would be to conduct a separate risk–benefit analysis for each programme. General information about risks and benefits can be supplemented by data specific to the species/population in question. A potentially powerful approach for the latter is to conduct a monitoring study using the BACI (before/after, control/impact) study design (Figure 13.6; also see Stewart-Oaten et al., 1992). In a BACI study, pre-impact (baseline) data are collected from one or more target populations that will be affected, as well as one or more control populations that will not be treated. Comparable data are then collected during and after the treatment (in this case, some type of

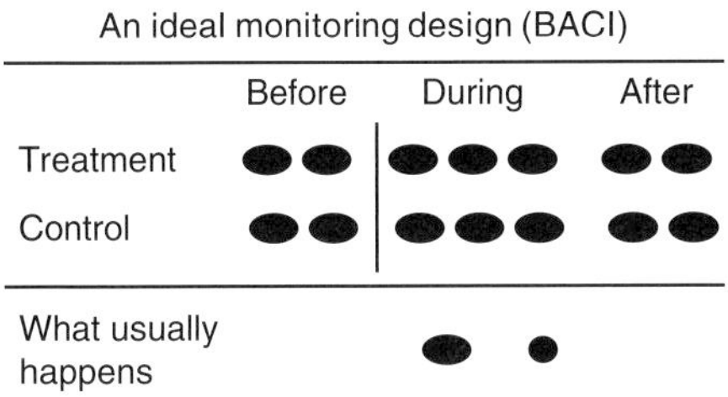

An ideal monitoring design (BACI)

	Before	During	After
Treatment	●●	●●●	●●
Control	●●	●●●	●●

| What usually happens | | ● ● | |

Figure 13.6 Schematic of ideal and actual experimental design for monitoring effects of artificial propagation on natural populations. A rigorous design would follow the BACI (before/after control/impact) protocol, which includes samples before (if possible during), and after the intervention, as well as samples from one or more controlled populations not intentionally affected. In reality, pre-impact data are rarely collected; what monitoring that does occur usually takes place during the treatment, and long-term (post-intervention) consequences are seldom monitored.

artificial propagation). Collecting data both before and after treatment is essential to evaluate effectiveness of the treatment, and monitoring control population(s) is important to rule out changes caused by other factors (such as broad-scale environmental change). In reality, this seldom occurs; rigorous risk–benefit analyses have been conducted for only a tiny fraction of all artificial propagation programmes, and we are not aware of any that have attempted the challenging tasks of formally comparing risks to natural populations and benefits to society.

13.6.1 Addressing 'uncertainty'

Uncertainty is a key consideration in virtually every aspect of supplementation. Although many risks can be identified, not all can be avoided (at least simultaneously), and long-term impacts to natural populations are difficult to predict. Similarly, the benefits of supplementation can be substantial but might never be realised. New programmes under consideration involve additional uncertainties because each new natural population that is brought into a hatchery environment typically displays its own idiosyncratic behaviour. Each new combination of natural source population, hatchery facility and hatchery staff can be expected to result in a programme that differs in important, but unpredictable, ways from other such programmes. In short, it is safer to assume that some unexpected things will occur and to prepare for handling such occurrences than to ignore such chances. Allowance for the likelihood of unexpected developments should be included in the initial risk–benefit analysis.

13.7 'BEST PRACTICES' AND ARTIFICIAL PROPAGATION REFORM

As risks from artificial propagation to recipient ecosystems have become better understood by conservation professionals (Table 13.4), efforts have been undertaken to reform propagation and release practices. Moreover, those governmental and non-governmental agencies charged with biological resource management have searched for a set(s) of best practices to guide the appropriate use of propagation and stocking. For example, one result of the 1993 Symposium on the Uses and Effects of Cultured Fish in Aquatic Ecosystems was a set of 'Considerations for stocked fishes' (Unauthored, 1995). Comparable efforts to develop 'codes of practice' have been developed by the European Inland Fishery Advisory Commission (EIFAC; Coates, 1998; but see also Miller & Kapuscinski, 2003 and George et al., 2009 for other guidelines). Yet, in spite of the periodic attempts to produce a universally acceptable set of practical guidelines, this goal remains elusive for two major reasons. First, as noted above, eliminating all kinds of risk simultaneously is not a reasonable expectation. As a consequence, difficult decisions must be made about various risk–risk or risk–benefit trade-offs. Second, science can provide information about how much risk and how much uncertainty will be associated with different types of artificial propagation programmes, but science alone cannot determine whether these levels of risk and uncertainty are 'acceptable' – that determination requires consideration of societal values and tolerance for different kinds of risk. With this in mind, can we identify practices that generally reduce risks and improve general compatibility of artificial propagation programmes, while others tend to increase the risks or lead to greater risk–benefit disparity? Figure 13.7 provides a conceptual framework that associates the relative risk from several example characteristics about the propagated population to recipient populations and ecosystems.

1. Genetic divergence – outbreeding depression risks are expected to increase with deeper divergence. To reduce the chance for such hazards to be expressed, it is generally prudent for the donor source to be genetically related to recipient and adjacent populations.
2. Background reproductive isolation – as a cause of divergence, breakdown of long-reinforced reproductive isolation (e.g. allopatric, seasonal or behavioural) between donor and recipient populations can lead to reduced fitness from outbreeding depression.

Table 13.4 *Recent-history international conferences addressing concerns related to artificial propagation, re-introduction and translocation.*

Year	Event	Sponsor (location)	Reference
1980	Stock Concept International Symposium (STOCS)	Great Lakes Fishery Commission (McAllister, ON, Canada)	Berst and Simon (1981)
1983	Symposium on the Role of Fish Culture in Fisheries Management	American Fisheries Society (Lake of the Ozarks, MO, USA)	Stroud (1986)
1992	World Fisheries Congress, Theme 3 (Protection Aquatic Biodiversity)	World Council of Fisheries Societies (Athens, Greece)	Philipp et al. (1995)
1993	International Symposium on the Uses and Effects of Fish Culture	American Fisheries Society (Albuquerque, NM; Denver, CO, USA)	Schramm and Piper (1995)
1993	Expert Consultation on the Utilisation and Conservation of Aquatic Genetic Resources	Food and Agriculture Organization (Grottaferrata, Italy)	FAO (1983)
1994	Conference on Evolution and the Aquatic Ecosystem	American Fisheries Society (Monterey, CA, USA)	Nielsen (1995)
1996	Symposium on Stocking and Introduction of Fish in Freshwater and Marine Ecosystems	European Inland Fisheries Advisory Commission (Thonon-le-Bain, France and Hull, UK)	Cowx (1998)
2000	Puget Sound/Coastal Washington & Columbia River Basin Hatchery Reform	Hatchery Scientific Review Group (Seattle, WA, USA)	Hatchery Scientific Review Group (HSRG) (2004, 2011)
2002	2nd International Symposium on Marine Stock Enhancement and Sea Ranching	Japan Fisheries Agency and Japan Sea Farming Association (Kobe, Japan)	Leber et al. (2004)
2003	International Symposium on Propagated Fish in Resource Management	American Fisheries Society, US Fish and Wildlife Service, and Canada's Department of Fisheries and Oceans (Boise, ID, and San Antonio, TX, USA)	Nickum et al. (2004)
2013	Hatcheries and Management of Aquatic Resources Symposia	American Fisheries Society (Nashville, TN, and Little Rock, AR, USA)	Status unknown

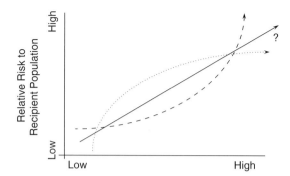

• Phylogenetic Divergence (donor v recipient)
• Background Reproductive Isolation (donor v recipient)
• Life-History Complexity (donor v recipient)
• Relative Abundance (donor v recipient)
• Propagule Pressure of Stock Transfers (Frequency, Magnitude, or Duration)
• Meta-population Complexity & Structure
• Reproductive Success of Mixed Progeny
• Domestication of Donor Stock
• Genetic Bottleneck/Drift within Donor Stock

Figure 13.7 Conceptual relationship between some key characteristics of concern between donor and recipient populations against the relative risk to recipient population. Although a general relationship is expected between these factors and degree of risk, the specific shape of the curves will vary depending on the species and factors involved need to be resolved on an empirical basis.

3. Life-history complexity – matching the complement of life histories (e.g. non-migratory, fluvial and adfluvial types) between donor and recipient populations may reduce the risks of losing elements of functional diversity in recipient waters.

4. Relative abundance – the risk of demographic, ecological, or genetic 'swamping' increases along with the relative abundance of donor to recipient populations. Keeping the proportion of propagated to wild fish at as low a level as possible helps to reduce such risks.

5. Propagule pressure of stock transfers – migration of young, juveniles or adults across stock boundaries can lead to a variety of pathogen, ecological or genetic risks. Such risks may be amplified with frequent transfers, those executed at a massive scale or are ongoing and continual. Where necessary, such transfers should be at a scale appropriate to a specific goal or outcome.

6. Metapopulation complexity and structure – species are generally mosaics of populations, or stocks, with multiple life histories and spatial

or temporal architecture. Matching such complexity and structure reduces risks associated with either mixing stocks or homogenising diversity. Here, it is crucial to avoid substituting a resident life-history form where a migratory form historically occurred (e.g. kokanee for anadromous sockeye salmon, *O. nerka*).

7. Reproductive success of mixed progeny – mixed-source progeny can exhibit high levels of fitness in early generations (F_1) – presumably from hybrid vigour – that declines in subsequent generations. Moreover, even where this fitness breakdown does not occur, native diversity may be supplanted by the hybridised gene pool. To reduce such risks, mixing of brood from divergent stock sources should be generally avoided or undertaken only to meet a specific conservation goal or outcome.

8. Domestication – selection for success by a donor stock in the captive or cultured environment can be in direct conflict with success in the wild. Such risks may be reduced by avoiding captive brood and by introducing more natural settings for animal husbandry.

9. Genetic bottlenecks and random drift – random loss of allelic diversity within populations is generally associated with small brood population (effective) size within a propagated donor stock. Sex ratio biases among breeders, high variance in family size associated with high fecundities, and variable brood size through time (especially in relationship to recipient populations) can magnify such bottleneck risks. To reduce such risks, sex ratios, breeding structure, family sizes and early survival (i.e. the life-table) should mimic closely those in natural populations (see Figure 13.8).

Complementary to a 'best practices' approach, some conservation managers and resource scientists advocate using an Adaptive Management approach (NRC, 1996) coupled with an ecosystem perspective (NPCC, 2009), wherein propagation and release is undertaken as an experimental conservation action. Ultimately, this is an iterative approach that requires formal goal-setting and a decision framework, rigorous effectiveness monitoring (including reference conditions and measurable indicators), and independent evaluation and review of outcomes (see Table 13.5 for examples). This framework is designed to address critical situations where there is considerable uncertainty – as would be the case with critically imperilled species.

A fundamental need for any adaptive management enterprise is an ability to measure a response to the proposed treatment (here,

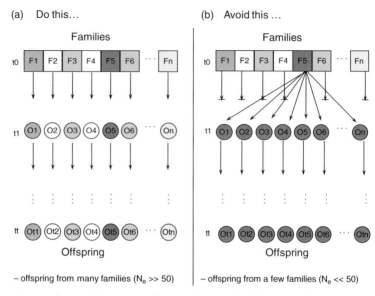

Figure 13.8 To reduce potential genetic bottleneck effects in artificial breeding programmes, it is important to maximise the number of families created to produce offspring.

propagation or translocation, etc.). Effective monitoring and evaluation of stocking for whatever the intended purpose generally requires the ability to distinguish (a) hatchery fish from natural fish, and (b) fish from the target and non-target populations. In some cases, marking all or nearly all of the hatchery fish may be necessary for adequate power to accomplish these objectives. However, such a high level of marking effort may tax both financial and human resources and may incur unacceptably high costs to the natural population in the form of incidental mortality or other adverse effects. Those considering the use of supplementation, translocation or re-introduction should realise that if adequate levels of marking cannot be achieved, risks to natural populations will increase.

13.8 OUR CALL TO INCLUDE FORMALLY ANALYSING A FULLER RANGE OF RISKS AND BENEFITS

As noted earlier, in an ideal scenario, a comprehensive risk–benefit analysis would be conducted *before* making a decision whether to implement

Table 13.5 *Examples of performance standards, corresponding indicators and monitoring and evaluation plans to achieve the artificial propagation programme goals or purposes. These elements are generally measurable, realistic and time specific.*

Performance standards	Performance indicators	Monitoring and evaluation plan
Meet programme production goals	Number of adults collected and individuals reared by life stage	Monthly programme production records
Minimise opportunities for interaction with wild fish through proper broodstock management and marking	Numbers (type) of marked fish	Monthly programme production records
Maintain stock integrity and genetic diversity	Number of broodstock collected, phenotypic characteristics (sex, age, size, etc.), N_e	Spawning guidelines, monthly programme production records
Minimise interactions with wild fish through effective containment methods	Records of documented escapes and losses unaccounted for	Monthly programme production records, marking data
Maximise in-culture survival of broodstock and their progeny	Fish culturists and pathologists will monitor the health of cultured stocks on a monthly basis and recommend preventative actions to maintain fish health	Monthly programme production records
Limit the impact of pathogens associated with culture programme on wild fish	Fish pathologists will diagnose fish health problems and minimise their impact	Fish health monitoring records
	Vaccines will be administered when appropriate to protect fish health	Fish health monitoring records
	A fish health database will be maintained to identify trends in fish health and disease and implement fish health management plans based on findings	
Ensure culture operations comply with state and federal water quality standards through proper environmental monitoring	National Pollutant Discharge Elimination System (NPDES) compliance requirements	Monthly NPDES report records

a proposed artificial propagation programme. If policy-makers find within the analysis good reason to believe that the benefits will outweigh the risks, the programme can be initiated. Collecting pre-impact baseline data in a BACI monitoring study design will provide a basis for using the resulting information specific to the target species/population to modify the programme as appropriate in an adaptive management framework.

This is not to say that good monitoring/evaluation programmes for artificial propagation do not exist. Laikre *et al.* (2010) summarised a number of state-of-the-art genetic monitoring studies that have been conducted on red drum (*Sciaenops ocellatus*) in the Gulf of Mexico and Atlantic Ocean, and impressive studies of the effects of artificial propagation have also been conducted for several other species, including Atlantic cod in Norway (Svasand *et al.*, 2000), brown trout in Denmark (Hansen *et al.*, 2009) and red sea bream in Japan (Kitada & Kishino, 2006). However, even these 'cream-of-the-crop' studies often do not include the important pre-impact data, and virtually none has collected rigorous data on what is arguably one of the most important response variables: the long-term fitness of affected natural populations.

Given this situation, what can be done to improve the ability of managers to make sounder decisions about the appropriate use of artificial propagation? We offer the following suggestions.

1. It is unrealistic to expect a state-of-the-art monitoring and evaluation programme for every implementation. However, the level of monitoring should be commensurate with the risks – programmes expected to create substantial risks could be required to have rigorous monitoring in an adaptive management framework. Moreover, a broader umbrella evaluation effort can assess impacts and response to incremental actions – to target and non-target populations, as well as the receiving ecosystems more generally.

2. Meta-analytical tools could be used to derive as much adaptive management information as possible from a large collection of unplanned, uncoordinated 'experiments'.

3. Think regional: find ways to foster large-scale cooperation/planning by managers to (1) maximise the value of information that is collected for other purposes, (2) add specific monitoring tasks as needed to address critical uncertainties, and (3) identify appropriate controls that are not directly/intentionally 'treated'.

4. Think local: industrial-scale programmes seek to maximise efficiency over effectiveness. They also may suffer from institutional inertia

regarding ongoing operation and costs. Smaller, more temporary operations are often more flexible and able to be adapted or cancelled as well-conceived monitoring data (effectiveness) are collected.

5. Creation of a central and structured database (and associated publication warehouse) for keeping track of stocking, translocations and re-introductions at a variety of jurisdictions levels (George *et al.*, 2009). Such a data warehouse can take advantage of existing assembled data at provincial to global scales.

13.9 CONCLUSION

In the end, making generalisations about specific benefits or risks of any single release, propagation, re-introduction or translocation has proven difficult. Therefore, we avoid offering a cookbook or set of rules that professionals and decision-makers must follow. Surely the success (or failure) of any conservation, rehabilitation or restoration action will rest in how reasonable a facsimile to native biodiversity such activities may achieve across the entire hierarchy of ecosystem complexity, from genes to landscapes. Such a facsimile is achieved by avoiding those disrupting processes within the hatchery or at time of release through a well-conceived 'code of practice' or set of 'best management practices' (including founder or bottleneck effects, inadvertent or domestication selection, interspecific hybridisation or extreme outbreeding, stock transfer or homogenisation, genetic swamping, etc.), or more importantly within the recipient environment (alteration of trophic structure, invasive pathogen or parasites, exceeding carrying capacity, etc.). Robust evaluation of the impact (good or problematic) requires active and rigorously designed monitoring. An adaptive management framework ultimately may permit necessary modifications or even cessation if the conservation actions fail to achieve the desired outcomes. In sum, to be effective for and truly consistent with conservation purposes, it is crucial to look at propagation, re-introduction and translocation broadly through the lens of the freshwater ecosystem rather than narrowly on the fish being targeted, bred or released.

In summary, we challenge current students and future leaders (including managers, scientists and policy-makers) of fish conservation to learn the lessons of the recent and distant past on using artificial propagation, stocking, translocation and re-introduction as a primary component of restoration and conservation. As such, we guide you to be cognisant of both the opportunities and limitations of these practices, to share honest appraisals of successes and failures, and to creatively identify more effective means at achieving conservation of freshwater fishes.

13.10 DISCUSSION QUESTIONS

1. Which of the various propagation 'models' depend on young, artificially produced fish entering and integrating into the wild population? For which models might this be undesirable?
2. Define and describe a set of metrics (or measurable benchmarks) that are appropriate for evaluating whether the release of artificially propagated fish has a realistic conservation versus some other form of societal value.
3. Discuss the importance of inbreeding and outbreeding risks in fish artificial propagation. Which is more important to consider in breeding design?
4. In the USA, artificial propagation is not focused across the trophic pyramid; most propagated fish are top predators. What are the pitfalls and benefits of focusing on a subset of the functional diversity (trophic level) in an aquatic community?
5. How would you go about deciding whether potential societal benefits of a hatchery programme (e.g. substantially increased harvest opportunities and associated jobs and economic activity) justify increased risks to natural populations?

13.11 ACKNOWLEDGEMENTS

The authors thank Ms Kacie Jonasen (INHS) for collecting information on propagation in North America. We also thank Dr Glenn Sandiford and an anonymous reviewer for their editorial recommendations to improve this chapter. The perspectives provided are the sole responsibility of the authors and do not represent those of our employers or funding agencies.

13.12 REFERENCES

Allendorf, F. W., Spruell, P. & Utter, F. M. (2001). Whirling disease and wild trout: Darwinian fisheries management. *Fisheries*, 26, 27–29.

Berst, A. H. & Simon, R. C. (Eds). (1981). Proceedings of the 1980 Stock Concept International Symposium. *Canadian Journal of Fisheries and Aquatic Sciences*. Vol. 38.

Bottom, D. L. (1997). To till the water: a history of ideas in fisheries conservation. In *Pacific Salmon and Their Ecosystems: Status and Future Options*. New York, NY: International Thomson Publishing, pp. 569–598.

Bowen, J. T. (1970). A history of fish culture as related to the development of fishery programs. In *A Century of Fisheries in North America*. Special Publication No. 7. Washington, DC: American Fisheries Society, pp. 71–93.

Brown, B. L., Waters, J. M., Gunter, T. & Epifanio, J. M. (2000). Restoration genetics of American shad in the James River, Virginia (USA): evaluation of captive release. *Conservation Biology*, 14, 294–303.

Busack, C. A. & Currens, K. P. (1995). Genetic risks and hazards in hatchery operations: fundamental concepts and issues. *American Fisheries Society Symposium*, 15, 71–80.

Christie, M. R., Marine, M. L., French, R. A. & Blouin, M. S. (2011). Genetic adaptation to captivity can occur in a single generation. *Proceedings of the National Academy of Sciences*, 109, 238–242.

Coates, D. (1998). Codes of practice for the stocking and introduction of fish. In *Stocking and Introduction of Fish*. Oxford: Fishing News Books, pp. 383–396.

Cowx, I. G. (Ed.). (1998). *Stocking and Introduction of Fish*. Oxford: Fishing News Books.

Epifanio, J. (2000). A status report of coldwater fishery management in the U.S. – an overview of state programs. *Fisheries*, 25, 13–27.

Food and Agriculture Organization (FAO). (1983). *Report of the Expert Consultation on the Utilisation and Conservation of Aquatic Genetic Resources*. FAO Fisheries Report, No. 491. Rome: FAO.

Food and Agriculture Organization (FAO). (2005). World inventory of fisheries. Stocking techniques for increased production. Issues Fact Sheets. In: FAO Fisheries and Aquaculture Department [online]. Rome: FAO (available from www.fao.org/fishery/topic/14885/en).

George, A. L., Kuhajda, B. R., Williams, J. D., *et al.* (2009). Guidelines for propagation and translocation for freshwater fish conservation. *Fisheries*, 34, 529–545.

Hansen, M. M., Fraser, D. J., Meier, K. & Mensberg, K. D. (2009). Sixty years of anthropogenic pressure: a spatio-temporal genetic analysis of brown trout populations subject to stocking and population declines. *Molecular Ecology*, 18, 2549–2562.

Hatchery Scientific Review Group (HSRG). (2004). Hatchery reform: principles and recommendations of the HSRG. Available from www.hatcheryreform.us

Hatchery Scientific Review Group (HSRG). (2011). Hatcheries, conservation, and sustainable fisheries – achieving multiple goals: results of the Hatchery Scientific Review Group's Columbia River basin review. *Fisheries*, 36, 547–561.

Hendricks, M. L. (1995). The contribution of hatchery fish to the restoration of American shad in the Susquehana River. *American Fisheries Society Symposium*, 15, 329–336.

Kitada, S. & Kishino, H. (2006). Lessons learned from Japanese marine finfish stock enhancement programmes. *Fisheries Research*, 80, 101–112.

Knapp, R. A., Corn, P. S. & Schindler, D. E. (2001). The introduction of nonnative fish into wilderness lakes: good intentions, conflicting mandates, and unintended consequences. *Ecosystems*, 4, 275–278.

Laikre, L., Schwartz, M. K., Waples, R. S., Ryman, M. & The GeM Working Group. (2010). Compromising genetic diversity in the wild: unmonitored large-scale release of plants and animals. *Trends in Ecology and Evolution*, 25, 520–529.

Leber, K. M., Kitada, S., Blankenship, H. L. & Svåsand, T. (Eds.). (2004). *Stock Enhancement and Sea Ranching: Developments, Pitfalls and Opportunities*, second edition. Oxford: Blackwell.

Lorenzen, K., Beveridge, M. C. M. & Mangel, M. (2012). Cultured fish: integrative biology and management of domestication and interactions with wild fish. *Biological Reviews*, 87, 639–660.

Mahnken, C. V., Ruggerone, G., Waknitz, F. W. & Flagg, T. A. (1998). A historical perspective on salmonid production from Pacific Rim hatcheries. *North Pacific Anadromous Fish Commission Bulletin*, 1, 38–53.

Merz, J. E. & Moyle, P. B. (2006). Salmon, wildlife, and wine: marine-derived nutrients in human-dominated ecosystems of central California. *Ecological Applications*, 16, 999–1009.

Metcalf, J. L., Pritchard, V. L., Silvestri, S. M., *et al.* (2007). Across the great divide: genetic forensics reveals misidentification of endangered cutthroat trout populations. *Molecular Ecology*, 16, 4445–4454.

Miller, L. M. & Kapuscinski, A. R. (2003). Genetic guidelines for hatchery supplementation programs. In *Population Genetics: Principles and Applications for Fisheries Scientists*. Bethesda, MD: American Fisheries Society, pp. 329–356.

National Research Council (NRC) (1996). *Upstream: Salmon and Society in the Pacific Northwest*. Washington, DC: National Academy Press.

Nickum, M. J., Mazik, P. M., Nickum, J. G. & MacKinlay, D. D. (Eds). (2004). *Propagated Fish in Resource Management*. American Fisheries Society Symposium 44. Bethesda, MD: American Fisheries Society.

Nielsen, J. L. (Ed.). (1995). *Evolution and the Aquatic Ecosystem: Defining Unique Units in Population Conservation*. American Fisheries Society Symposium 17. Bethesda, MD: American Fisheries Society.

Noss, R. F. (1990). Indicators for monitoring biodiversity: a hierarchical approach. *Conservation Biology*, 4, 355–364.

Northwest Power & Conservation Council (NPPC) (2009). Columbia River Basin Fish and Wildlife Program – 2009 Amendments. Document 2009–09. Portland, OR, USA. Available from www.nwcouncil.org/media/115273/2009_09.pdf.

Philipp, D. P., Epifanio, J. M., Marsden, J. E., Claussen, J. E. & Wolotira, Jr, R. J. (Eds). (1995). *Protection of Aquatic Biodiversity*. Proceedings of the World Fisheries Congress, Theme 3. New Delhi: Oxford & IBH Publishing Co. Pvt.

Philipp, D. P., Claussen, J. E., Kassler, T. W. & Epifanio, J. M. (2002). Mixing stocks of largemouth bass reduces fitness through outbreeding depression. *American Fisheries Society Symposium*, 31, 349–364.

Reisenbichler, R. R. (1988). Relation between distance transferred from natal stream and recovery rate for hatchery coho salmon, *Oncorhynchus kisutch*. *North American Journal of Fisheries Management*, 8, 172–174.

Samways, M. J. (1996). The art of unintelligent tinkering. *Conservation Biology*, 10, 1307.

Schindler, D. E., Hilborn, R., Chasco, B., *et al.* (2010). Population diversity and the portfolio effect in an exploited species. *Nature*, 465, 609–612.

Schramm, H. L. & Piper, R. G. (Eds). (1995). *Uses and Effects of Fish Culture in Aquatic Ecosystems*. American Fisheries Society Symposium 15. Bethesda, MD: American Fisheries Society.

Stewart-Oaten, A., Bence, J. R. & Osenberg, C. W. (1992). Assessing effects of unreplicated perturbations: no simple solutions. *Ecology*, 73,1396–1404.

Stroud, R. H. (Ed.). (1986). *Fish Culture in Fisheries Management*. American Fisheries Society Publication Symposium 9. Bethesda, MD: American Fisheries Society.

Svåsand, T., Kristiansen, T. S., Pedersen, T., *et al.* (2000). The enhancement of cod stocks. *Fish and Fisheries*, 1, 173–205.

Unauthored. (1995). Considerations for the use of cultured fishes in fisheries resource management. *American Fisheries Symposium*, 15, 603–606.

Utter, F. & Epifanio, J. (2002). Marine aquaculture: genetic pitfalls and potentialities. *Reviews in Fish Biology and Fisheries*, 12, 59–77.

Waples, R. S. (1999). Dispelling some myths about hatcheries. *Fisheries*, 24, 12–21.

Waples, R. S. & Drake, J. (2004). Risk/benefit considerations for marine stock enhancement: a pacific salmon perspective. In *Stock Enhancement and Sea Ranching: Developments, Pitfalls, and Opportunities*. Oxford: Blackwell, pp 260–306.

Waples, R. S., Hindar, K. & Hard, J. J. (2012). Genetic risks associated with marine aquaculture. U.S. Department of Commerce, NOAA Technical Memorandum NMFS-NWFSC-119, 149 pp.

Freshwater conservation planning

VIRGILIO HERMOSO, SIMON LINKE,
STEPHANIE R. JANUCHOWSKI-HARTLEY
AND MARK J. KENNARD

14.1 INTRODUCTION

Freshwater fishes represent among the most diverse and threatened taxa globally, accounting for more than 25% of total vertebrates (> 30,000 species described), 37% of which are threatened with extinction (Darwall *et al.*, 2008; Chapter 1). The poor conservation status of freshwater biodiversity is directly related to the pressure that these systems experience worldwide (Vörösmarty *et al.*, 2010). Because of their importance to human welfare and development, freshwater ecosystems and biodiversity are subject to higher pressures and threats than are adjacent terrestrial ecosystems (Nel *et al.*, 2007). Water pollution and abstraction coupled with invasive species and habitat modification (e.g. channelling and damming) are the principal threats to the conservation of freshwater biodiversity (Strayer & Dudgeon, 2010; Vörösmarty *et al.*, 2010). These pressures are rapidly growing due to the increase of human population worldwide and the effect of climate change (Dudgeon *et al.*, 2006; Chapter 3).

Although freshwater ecosystems and biodiversity are in urgent need of protection, there has been little emphasis on declaring protected areas for the primary purpose of conserving freshwater biodiversity (although see attempts in South Africa since the early 1970s (Roux & Nel, 2013 for a brief history) or the USA (Moyle & Yoshiyama, 1994)). Instead, uninformed opportunism has reigned, whereby the conservation of freshwater ecosystems has remained peripheral to conservation goals developed for terrestrial ecosystems, unless considered important for terrestrial biodiversity (Nel *et al.*, 2007; Olden *et al.*, 2010). The implementation of

Conservation of Freshwater Fishes, eds G. P. Closs, M. Krkosek and J. D. Olden. Published by Cambridge University Press. © Cambridge University Press 2016.

conservation is constrained by limited budgets and potential conflicts with other human uses. For this reason, it is unfeasible to protect all the areas that contribute to the persistence of biodiversity (Margules *et al.*, 2002), and adequate planning is required. Conservation planning is a discipline that tries to deal with these issues to inform stakeholders and decision-makers on how to best invest limited resources available for conservation. The development of a conservation plan typically draws on knowledge spanning several scientific disciplines, increasingly also from the social sciences.

To be effective for freshwater conservation in general and fish in particular, protected areas must consider some particularities of freshwater ecosystems from the early planning stages (e.g. when deciding where to implement conservation) to the daily management. Freshwater ecosystems pose some unique challenges to the implementation of effective conservation (Abell, 2002), such as the importance of connectivity at maintaining natural processes and facilitating the propagation of threats (Linke *et al.*, 2011). Spatial–temporal connectivity plays a key role in maintaining important ecological processes, such as periodic migrations or dispersal from refuge areas, gene flow or transport of energy and matter, essential for the persistence of populations and species (Chapters 10 and 11). Connectivity in freshwater ecosystems is defined in four dimensions (Ward & Stanford, 1989): longitudinal (upstream–downstream), lateral (interactions between channel and riparian/floodplain systems), vertical (connections between the surface and groundwater systems), all of which are subject to temporal dynamics (Figure 14.1). All of these dimensions have been deeply modified through the construction of barriers such as dams, road crossings or culverts that affect the longitudinal movement of individuals (Pepino *et al.*, 2012), reduce the flow of nutrients and sediments (Stanley & Doyle, 2003) and affect lateral connectivity by diminishing flood pulses (Arthington, 2012).

Freshwater ecosystems are extremely dependent on the surrounding terrestrial landscape (e.g. as a source of material or vulnerability to threats originated in terrestrial systems, such as sediments or pollutants). For this reason, the adequate management for conservation of freshwater biodiversity needs to go beyond the limit of the freshwater ecosystem and extend to the terrestrial realm (sometimes even far distant to account for threatening processes, for example).

An additional key aspect to conservation in freshwater ecosystems is the need for water, often running. Flows define healthy freshwater ecosystems and sustain important features such as temperature,

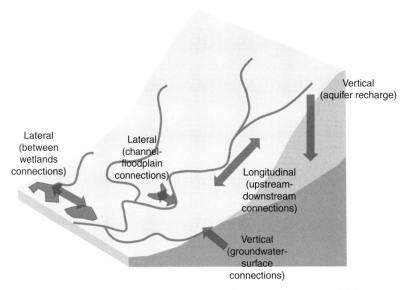

Figure 14.1 Different spatial components of connectivity meaningful for freshwater conservation planning.

geomorphology or habitat diversity and processes like transport of nutrients, key for the maintenance of freshwater biodiversity (Poff *et al.*, 1997; Arthington, 2012). However, as for connectivity, natural flows have been deeply transformed by human intervention: modification of quantity and timing of flow due to river regulation, water abstraction and overexploitation of aquifers (Bunn & Arthington, 2002; Chapter 4).

The implementation of effective conservation programmes requires on-the-ground management actions to address threats to the persistence of biodiversity, rather than simply static protection as is normally done (e.g. change in tenure by declaring an area protected). The variety and magnitude of threats and the peculiar needs of freshwater ecosystems make conservation planning and implementation a challenge that is not well resolved.

In this chapter, we synthesise past progress in the application and implementation of freshwater conservation planning and provide a novel, forward-thinking perspective on freshwater conservation planning for the future. We commence with a summary of past approaches to identify spatial conservation priorities in freshwaters and highlight their strengths and limitations. We introduce the topic of systematic conservation planning, broadly used in the terrestrial and marine realm, but an emerging discipline in the freshwater realm. We highlight the strengths and weaknesses of this approach with some examples taken from different parts of the world and describe how addressing ecosystem processes can improve the adequacy of systematic conservation planning. We summarise the current challenges to effective planning and implementation of conservation plans in freshwater ecosystems and conclude with a section on emerging challenges to achieving effective conservation planning for freshwater biodiversity.

14.2 PAST APPROACHES TO IDENTIFYING FRESHWATER CONSERVATION PRIORITIES

From the early 1800s, when the establishment of parks and reserves began, up until 30 years ago, efforts to conserve natural environments and species were carried out in a relatively *ad hoc* manner. *Ad hoc* assessments of conservation value were based on subjective judgements or on other criteria extraneous to biodiversity conservation, such as scenic value, remoteness, low agricultural production potential or, simply, availability (Pressey *et al.*, 1996). Therefore, these strategies often resulted in the protection of the areas that were easiest to reserve, sometimes with the least need for urgent or immediate protection (Pressey, 1994). Fresh waters tended to be inadequately dealt with in the assessments of terrestrial ecosystem conservation. For example, rivers were often used to define boundaries of terrestrial reserves, like the Kagera River (Rwanda), which shapes the eastern limit of the Akagera National Park declared in 1934. While such incidental representation of rivers did not warrant effective protection (i.e. there could be some species and processes not adequately represented), there remained little consideration or evaluation of these limitations until the early twenty-first century (see Abell *et al.*, 2007).

The effectiveness of existing terrestrial protected areas for mitigating threats to freshwater ecosystems and their dependent species remains relatively poorly understood (Herbert *et al.*, 2010; Januchowski-Hartley *et al.*, 2011). Methods such as gap analyses that evaluate the effectiveness and opportunities of terrestrial protected areas (Roux *et al.*, 2002) for protecting freshwater ecosystems can be used to guide the selection of additional protected areas to achieve both terrestrial and freshwater conservation objectives (Abell *et al.*, 2007; Nel *et al.*, 2009; Lawrence *et al.*, 2011). A case study in the Wet Tropics of Northern Australia (Januchowski-Hartley *et al.*, 2011) evaluated the effectiveness of terrestrial protected areas for protecting fresh waters and freshwater fish species. While there was a large proportion of a terrestrial ecosystem protected in the Wet Tropics (approximately 60%), the terrestrial protected-area network was far from optimal in providing adequate coverage of fresh waters and the fish species they support. Similar findings in other regions of Australia (Hermoso *et al.*, 2011a), South Africa and North America (Roux *et al.*, 2002; Herbert *et al.*, 2010; Lawrence *et al.*, 2011) emphasise the need for freshwater-specific conservation priorities areas.

Alternative methods have evolved to address the challenge of identifying conservation priorities more objectively than *ad hoc* approaches. One such approach is the ranking of areas according to the biodiversity features that they contain, using criteria such as species richness, number of rare or threatened species or landscape diversity (e.g. Williams *et al.*, 1996; Myers *et al.*, 2000). Using these ranking methods, priority areas for conservation are identified based on highest richness, endemism or threatened status (e.g. Rosset *et al.*, 2013). Turner and List (2007) mapped the spatial distribution of 34 native freshwater fish species and used them in conjunction with additional information on the distribution of invasive species to identify conservation priority areas in Arizona (USA). Filipe *et al.* (2004) ranked river reaches in the lower Guadiana River basin (Portugal) for their conservation importance using information on the conservation status of fish species and their spatial distribution. They combined the probability of occurrence of freshwater fish species with their conservation value (a measure of their relative endemicity and local abundance) in a conservation value score to determine conservation priorities. Hermoso *et al.* (2009) also used the probabilities of occurrence of freshwater fish in the Guadiana River basin (Spain) under two alternative scenarios (current vs. reference or pre-disturbance) to identify areas that had not suffered a significant biodiversity loss. Using an overall score of biodiversity loss, they highlighted refuge subcatchments that should be the focus of conservation efforts as they still held freshwater fish assemblages in consonance with that expected under the reference condition (not significantly affected by threats). Although this was an interesting approach to bring a condition assessment into the conservation planning process (sites highlighted as conservation priorities were those in best condition), the approach failed to account for future vulnerability to expanding threats and did not explicitly trade off costs and benefits.

Overall, despite ranking methods being more objective than *ad hoc* approaches, the identification of conservation priority areas based on criteria such as richness does not guarantee the adequate representation of species (Williams *et al.*, 1996). For instance, some biodiversity might not occur in richness hotspots and would in turn not be adequately protected (e.g. some rare species that only occur in low-diversity headwater reaches). In this sense, if protected areas were implemented along the five rivers identified by Turner and List (2007) in Arizona, these areas would only represent 63% of freshwater fishes, and 37% of species would lack adequate protection. To achieve adequate protection for all species requires the protection of additional rivers, which would incur additional costs. An additional drawback of ranking approaches is the inconsistent incorporation of management costs and ecological processes, which are essential to ensure cost-effective conservation recommendations (Carwardine *et al.*, 2008).

Different attempts have been made to incorporate more ecologically sound information into the identification of conservation priority areas. Higgins *et al.* (2005) and Thieme *et al.* (2007) used a stepwise procedure to select areas that fulfil some criteria that account for the viability of populations or connectivity between them. Both studies used a combination of representation criteria (all freshwater species and habitat types were adequately covered) and connectivity and other population viability criteria. More recently, Holland *et al.* (2012) demonstrated the framework to identify freshwater Key Biodiversity Areas (KBAs), following the approach proposed by Darwall and Vie (2005). KBAs are also identified according to ecosystem-based criteria, to account, for example, for the capacity of different areas to maintain populations of endangered species (e.g. 'a site is known or thought to be critical for any life history stage of a species, such as migration routes, spawning or feeding grounds'; see Holland *et al.*, 2012, for more detail on the different criteria proposed). While these modified scoring methodologies refine the set of areas selected for conservation efforts, the stepwise selection of sites used in some cases might undermine the efficiency of conservation recommendations (see systematic conservation planning methods, below). Moreover, priority areas identified through these methods tend to be large (including whole catchments or large rivers) and then more objective ways of prioritising within them would be required to provide more feasible conservation recommendations (see Holland *et al.*, 2012).

A common approach to prioritise among large potentially valuable areas is the use of multi-criteria decision analysis (MCDA). MCDA has been used to rank sites according to a number of criteria by computing an overall scoring value obtained from a weighted sum of all the independent criteria. For example, Ausseil *et al.* (2007) used four criteria (representativeness, area, surrounding naturalness and connectivity) to prioritise palustrine and estuarine wetlands for conservation in New Zealand. The final score was obtained through an additive function that weighted the relative importance of each indicator. Although MCDA can be a helpful tool to ensemble information on multiple criteria that account for patterns and processes, the results obtained are highly dependent on the weights applied to each criterion in the function and the cost of implementing action is not traded off against these benefits. Weightings are inherently subjective, and could result in a high level of uncertainty around the selection of priority areas. Moreover, these methods do not consider the cost when identifying priority areas for conservation as previous scoring approaches. In addition to these scoring and ranking approaches, conservation programmes have also evolved for the protection of particular rare, iconic or threatened species, using the so-called flagship and umbrella species (see Hess *et al.*, 2006). Species-based conservation often assumes that the implementation of conservation actions for one species could also benefit species in the broader community (Roberge & Angelstam, 2004). This is probably the most used practice when setting conservation programmes as the protection of endangered species or communities is often implicit in environmental laws (see the European Habitats Directive (European Council, 1992) or the US Endangered Species Act (ESA, 1973)). Flagship species are typically large-bodied and charismatic with substantial public appeal, whose conservation could indirectly conserve other species that share its habitat (Hess *et al.*, 2006). An example of the use of a flagship species in conservation of freshwater fish is the conservation programme for the piracucu (*Arapaima gigas*). This species was one of Brazil's most important fisheries in the nineteenth century. Due to overexploitation the populations of piracucu declined dramatically, resulting in local extinctions along its distribution range. Conservation programmes that establish collaborations with traditional fishermen have been developed in an effort to try and estimate local abundance to reassess extraction quotas. In addition, government policies established minimum size of catch (1.5 m total length) and closed seasons (December–May) to help recover the exhausted populations. Exploitation of piracucu is only legal for fishers who joined the conservation programme, and has proved to be successful over the last decade (Castello *et al.*, 2009).

The conservation value of flagship species should be considered cautiously, as the identification of these species can be politically or socially driven, and the selection of species as good indicators or umbrella species is not ensured. In some cases the conservation programme aiming to protect the flagship species might only benefit the species under concern and have little or no effect on the remaining community (Verissimo *et al.*, 2011). Moreover, the implementation of conservation programmes targeting flagship species should be cautiously triaged (Bottrill *et al.*, 2008), as they are normally expensive and might compromise the availability of resources for other species. An alternative way of addressing conservation needs for whole communities is by using umbrella species. Umbrella species are those that use such large areas of habitat that protecting them will, by default, protect the requirements of many other species. A good example from fresh waters would be Atlantic salmon (*Salmo salar*) in Norway (Skaala *et al.*, 2014) or other diadromous species such as sturgeons, eels or clupeids (ICES, 2005) that have large ranges that depend on intact connectivity between freshwater, and freshwater and marine, ecosystems. The effective conservation of these salmon populations would require a combination of conservation efforts to maintain migratory pathways and to implement actions to secure good habitat quality in key spawning areas. Other migratory species (e.g. trout) and local resident species would also benefit from the implementation of conservation programmes for these umbrella species. However, conservation plans for umbrella species normally lack of cost-effectiveness analyses. A potential solution to this limitation of current approaches would be to first identify a budget available for a particular programme, and then maximise the benefit gained from the budget by identifying those areas where apart from protecting the umbrella species, the maximum number of other native species would also be adequately protected and in turn achieve the maximum conservation potential for the given budget.

14.3 SYSTEMATIC CONSERVATION PLANNING METHODS/TOOLS

In the early 1980s, conservation scientists began to realise that scoring-based methods, regardless of the criteria considered, were not the most efficient way of designing protected area networks and it often led to the underrepresentation or even completely missing some species. Jamie Kirkpatrick, a plant researcher from Tasmania (Australia) was likely the first one to question whether a ranking system based on single metrics could theoretically lead to the most efficient solution (Kirkpatrick, 1983). Prioritising for gaps in reserve coverage of plant species, he noticed that many of the high scoring areas were home to the exact same species (Kirkpatrick, 1983). If all of these areas were recommended for prioritisation, conservation effort would be duplicated, while areas with few yet important species were not assigned high priority. Therefore, Kirkpatrick applied an iterative procedure, marking the highest scoring area as reserved and removing the species covered by this area to adjust the value of the remaining areas for the species not adequately covered yet. The work of Kirkpatrick and others demonstrated the importance of *complementarity*-based methods for ensuring cost-effective reserve allocation for the conservation of ecosystems and species. *Complementarity* is defined as the gain in representation of biodiversity when a site is added to an existing set of areas.

Systematic conservation planning (SCP; Margules & Pressey, 2000) aims to inform decision-makers on how to achieve conservation goals (e.g. representing all species in a certain area within the set of priority areas) in the most effective and efficient way. This is done by explicitly defining conservation objectives and integrating socioeconomic (e.g. conservation cost) and other ecological (e.g. connectivity) aspects when looking for optimal allocation of priority areas for conservation.

Some of the strengths of systematic conservation planning that help overcome the issues highlighted above are as follows.

1. The use of explicit and quantitative targets or objectives. These can be set and achieved in line with quantitative policy guidelines (e.g. Australia is committed to the protection of representative ecosystems and to the protection of rare and endangered species). For example, a set of targets might be to conserve 15% of each ecosystem type, or 50% of the range of all rare species.
2. The enhancement of efficiency. Different costs are associated with conservation, such as acquiring land to be protected, or implementing management decisions that maintain or restore conservation values (Naidoo *et al.*, 2006). Because conservation usually competes with other human uses, it involves both direct implementation costs and additional socioeconomic implications in the areas under protection. These opportunity costs are a measure of what could have been gained by the next-best use of a resource had it not been put to the current use (e.g. gross economic production of a cropping area if it had to be protected and no longer cultivated). Approaches that incorporate costs when identifying priority areas not only reduce potential conflicts between conservation and economic activities, but also makes

them more feasible and attainable because they are more efficient and therefore cheaper to implement (e.g. Carwardine *et al.*, 2008).

3. Adequacy and persistence. Adequacy refers broadly to the persistence of biodiversity, including the processes on which it depends like population dynamics, movement and migration, patch dynamics, catchment processes and river flows, and many others. Adequacy is difficult to quantify and implement, but systematic methods are being developed to achieve explicit objectives related to adequacy (Linke *et al.*, 2011). Some of these are being adapted specifically for freshwater systems to consider longitudinal and lateral connectivity (see below).

Systematic planning has been extensively applied to conservation problems in marine and terrestrial environments (Margules & Pressey, 2000). The application of systematic methods to conservation planning for freshwater biodiversity started in 2002, when Roux *et al.* (2002) prioritised for river types and processes using a *complementarity*-based algorithm while setting defined targets. Between 2003 and 2007, a handful of studies concerned with comprehensive data collections and extrapolation of data across whole study areas laid the ground for the application of systematic frameworks at the larger scale: while two papers (Fitzsimons & Robertson, 2005; Higgins *et al.*, 2005) dealt with theoretical innovation of basin-wide classifications, other studies developed classification systems that are directly biologically informed (Turak & Koop, 2008).

In terms of direct planning efforts, Thieme *et al.* (2007) prioritised based on detailed maps in a very large Amazonian basin while on-ground data were almost completely lacking. They also trialled a new scheme for multitiered catchment protection, which splits catchments into three zones of mixed protection, from freshwater reserves to management zones across the entire catchment.

Between 2007 and 2009, two studies used systematic frameworks to identify priority areas for conservation in rivers through different approaches. Linke *et al.* (2007) selected protected areas for invertebrates in Victoria (Australia), ensuring protection for entire catchments and complementarity between catchments so all taxa were adequately represented. Moilanen *et al.* (2008) applied for the first time the software ZONATION to the identification of priority areas for freshwater biodiversity in a case study carried out in New Zealand.

The year 2011 posed a major breakthrough in the field of freshwater systematic conservation planning with the publication of a special issue

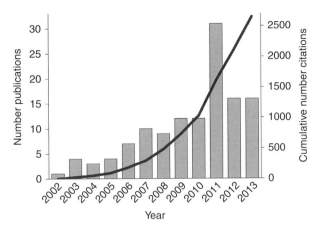

Figure 14.2 Change in number of published journal articles on 'Freshwater conservation planning' published since 2002 (grey bars) and the number of citations (black line). *Source*: Thomsons ISI *Web of Science* search Topic= ('systematic conservation planning' OR 'spatial conservation prioritisation' OR 'conservation planning' OR 'conservation reserve design') AND 'freshwater' (*N* = 125 journal papers, search date 20 February 2014).

on this topic in *Freshwater Biology* (Turak & Linke, 2011). Although the number of scientific publications in international journals had been increasing from 2002, there was a more than twofold increment due to the contribution of this special issue (11 contribution manuscripts; Figure 14.2). This special issue made an important contribution to the field of freshwater conservation planning, showing novel ways of tackling conservation planning in freshwaters, some of which are depicted below, with case studies all around the world. There has also been an exponential increase in the impact of freshwater conservation planning over the last decade as indicated by the number of citations that the mentioned manuscripts have received (Figure 14.2).

Different decision support tools have been developed over the last decade to address systematic conservation planning, but two of them are the most broadly used: MARXAN (Ball *et al.*, 2009) and ZONATION (Moilanen *et al.*, 2005). These tools use different optimisation algorithms to try to find the best way of representing all the species while attending to cost and connectivity constraints (see next section). They were both designed to address conservation planning in marine and terrestrial environments, but they have also been successfully applied in freshwater ecosystems (Moilanen *et al.*, 2008; Hermoso *et al.*, 2011b) and are publicly available (MARXAN: www.uq.edu.au/Marxan; and

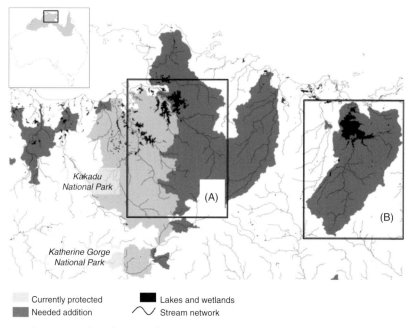

Currently protected   Lakes and wetlands
Needed addition   Stream network

Figure 14.3 Identification of priority areas for the conservation of freshwater biodiversity in northern Australia. The new set of priority areas (dark grey) were identified, using the software MARXAN, to fill the gap in protection of existing terrestrial reserves (light grey) and adequately address some key ecological processes for freshwater biota (e.g. connectivity). Some of these new priority areas included the unprotected headwaters of already protected wetlands (A) and additional important wetlands and streams (B).

ZONATION: www.helsinki.fi/bioscience/consplan/software/Zonation/ index.html). Other optimisation algorithms and spatial prioritisation tools that are freely available include: WORLDMAP (Williams, 2001), ResNet (Kelley *et al.*, 2002), MultCSync (Moffet *et al.*, 2005), ConsNet (Ciarleglio *et al.*, 2009) and C-Plan (Pressey *et al.*, 2009).

Building on the existing network of terrestrial protected areas has been suggested as a practical approach to improve the poor representation of freshwater ecosystems and biodiversity in currently protected areas (Abell *et al.*, 2007). Systematic conservation planning methods have been used to identify sets of additional areas that would require protection to adequately protect freshwater biodiversity. For example, Hermoso *et al.* (2011a) (Figure 14.3) identified additional priority areas for conservation in northern Australia to improve both the representation of freshwater biodiversity and the adequacy of current terrestrial protected areas at addressing freshwater ecosystem processes (see next section). Similarly, Lawrence *et al.* (2011) evaluated the potential of the National Park Service in the USA to protect freshwater biodiversity. They identified the 50 most important parks that would substantially contribute to the conservation of freshwater fish in the USA, but also highlighted the need for further additions in order to ensure the adequate protection of all species.

14.4 ADDRESSING FRESHWATER ECOSYSTEM PROCESSES AND THREATS TO IMPROVE ADEQUACY OF SYSTEMATIC CONSERVATION PLANNING IN A CHANGING WORLD

There are a number of distinctive characteristics of freshwater ecosystems that pose several challenges to conservation practitioners and that might hinder the implementation of effective conservation plans on the ground (Nel *et al.*, 2009; Linke *et al.*, 2011). The important role of connectivity in maintaining key ecological processes is one example of this distinctive character of conservation in freshwater ecosystems (Chapter 10). The connected nature of freshwater systems also means that perturbations originating far away might impact on high conservation priority areas. Moreover, given the poor conservation status of freshwater ecosystems around the world (Strayer & Dudgeon, 2010), the implementation of conservation programmes for freshwater biodiversity often needs to be accompanied by on-the-ground management actions aimed at rehabilitation instead of simply static protection. These issues bring additional layers of complexity to conservation in freshwater environments and reinforce the need for cost-effective planning.

Different approaches have been proposed to address longitudinal (Linke *et al.*, 2007; Moilanen *et al.*, 2008; Hermoso *et al.*, 2011b), lateral (Thieme *et al.*, 2007; Ausseil *et al.*, 2011; Hermoso *et al.*, 2012a) and vertical (Nel *et al.*, 2011) components of connectivity within systematic planning frameworks. The main aim of all these approaches is to obtain spatially clustered priority areas to ensure connectivity (in any of its dimensions). For example, Linke *et al.* (2007) included all the upstream segments to a given priority area to ensure effective protection and fully upstream longitudinal connectivity. This might lead to the selection of large areas which works against the efficiency of solutions. To solve this problem, Hermoso *et al.* (2011b) applied a more flexible approach where only the closest areas (weighted by distance) to a given priority area were selected, without the need of full upstream catchment protection. This method was shown to be very effective at identifying longitudinal corridors to enhance movements within protected areas.

In freshwater ecosystems, spatial connectivity, in all of its dimensions, is limited by the presence of water, which is strongly influenced by temporal dynamics. In this way, functional connectivity (facilitation of movement of organisms through the landscape, *sensu* Taylor *et al.*, 2006) is constrained by the presence of a minimum amount of water in the landscape. This applies not only to obligate freshwater species restricted to movements within river networks, but also to organisms with aerial or terrestrial movements (e.g. spatial proximity of floodplain lakes, important for some birds and turtle species; Haigh *et al.*, 1998; Rea *et al.*, 2009). In climatic regions with pronounced wet to dry seasonality, such as the wet–dry tropics in northern Australia (Kennard *et al.*, 2010), or the Mediterranean areas (Gasith & Resh, 1999), dry periods constrain the presence of water and the connections between the different elements of the system. Rivers and streams that are hydrologically connected during the wet season may cease to flow during the dry season and become reduced to a set of isolated pools. Important dry-season refugia such as floodplain wetlands and lakes are also affected, becoming completely dry or constrained to a reduced area.

Temporal changes in spatial connections can have important implications for population dynamics and community structure (Taylor *et al.*, 2006) and ultimately for population viability (Magalhães *et al.*, 2007). For example, connectivity can affect reproduction and mortality, by allowing or limiting access to potential breeding sites (Taylor *et al.*, 2006). Incorporating the temporal dimension of connectivity (Pringle, 2001) in freshwater conservation planning is therefore a crucial challenge to adequately address the persistence of freshwater biodiversity (Nel *et al.*, 2009). Native biodiversity in dynamic wet–dry seasonal freshwater environments have developed resistant (e.g. featuring desiccation-resistant life-history stages or retreating to refuge habitats offering less harsh conditions) or resilient (e.g. ability to recolonise and recruit after disturbance) strategies to cope with extended dry periods (Bond *et al.*, 2008). For example, Magalhães *et al.* (2007) found that fish assemblages in Mediterranean streams recovered quickly from short-term fluctuations in water level. However, they also warned that exacerbated disruptions in longitudinal connectivity caused by low water levels could negatively influence sensitive species. In an attempt to address the temporal component of connectivity, Hermoso *et al.* (2012b) used long-term series of MODIS flood and post-flood Landsat satellite imagery, which are publicly available, to estimate water residency time (WRT) in the landscape. These estimates of WRT were then used to inform how long hydrological connectivity was effective along an average year and then focus the selection of priority areas on those zones that maintained effective connections for longer periods and that could also play an important refugial role.

Despite the unquestionable value of connectivity for conservation, enhancement of connectivity should not be viewed as a generalised conservation solution, especially when implementing conservation actions in systems that have been highly modified by humans. Artificial barriers (structures such as dams or road culverts) can be an effective way of minimising the propagation of threats and can facilitate species management by excluding both the invasive species and other threats such as pollution that could hinder native species recovery (Rahel, 2013; Chapter 4).

14.4.1 Priority refugia for freshwater biodiversity

Despite droughts being natural phenomena, especially in temporary river systems, the frequency and magnitude of these events is expected to increase in some areas under the effects of climate change (Bates *et al.*, 2008). Global-scale predictions include a two- to threefold increase in the frequency of extreme low flows in many areas (Arnell, 2003) and a reduction in mean annual discharge exacerbated by increasing temperatures and evaporation rates. As a consequence of this change, some currently perennial freshwater ecosystems will become non-perennial and the duration and extent of water scarcity in already wet–dry seasonal ecosystems will increase. Under these conditions it is likely that riverine habitats will become increasingly fragmented for longer periods (Morrongiello *et al.*, 2011), which could compromise the persistence of an important proportion of global freshwater biodiversity. Future persistence of freshwater biodiversity in temporary systems will depend on our capacity to enhance the resilience of these systems to stressful events. This can be achieved by, for example, focusing conservation and rehabilitation efforts on key refuge areas (Crook *et al.*, 2010). Refuge areas maintain individuals that can repopulate a wider range of habitats when more favourable conditions are restored after seasonal or prolonged droughts (Larned *et al.*, 2010). Consequently, refuge areas help sustain freshwater populations over broader geographic ranges (metapopulations, or groups of spatially segregated populations of the same species that interact through exchange of individuals). There is strong evidence that recolonisation can be highly effective at the catchment scale in these types of systems when connectivity is re-established. Balcombe *et al.* (2006) found freshwater fish assemblages to be very similar along a temporary river catchment in Australia (Warrego River) during a period of high connectivity, suggesting efficient dispersal after a dry period when significant dissimilarities in species composition were reported. This hypothesis is further supported by genetic analyses. Carini *et al.* (2006) found low levels of genetic differentiation among different waterholes within the same catchment in two freshwater fish and an invertebrate species, respectively. However, the maintenance of the ecological role these areas play will require the implementation of active management, to ensure the harsh conditions during key periods (e.g. seasonal droughts) will allow freshwater biodiversity to persist. Hermoso *et al.* (2013a) demonstrated how to use systematic conservation planning methods to find priority refuges to maximise the recolonisation capacity after the dry period for freshwater fish in North Australia.

Alteration of natural flow regimes is just one of the several consequences that climate change will have on freshwater ecosystems (Chapter 3). For example, climate change could also lead to the modification of freshwater thermal regimes, which must be considered when identifying refugia for freshwater biodiversity. As water temperature increases it is expected that species will shift their distribution ranges towards more suitable areas where their thermal ranges are not exceeded (Wenger *et al.*, 2011). These areas will be mainly located in higher elevations, which entails a high risk of isolation of local populations. In some cases where the disconnection between populations or lack of further favourable conditions within their current distribution ranges threatens the persistence of a species, more controversial management options such as assisted colonisation (Olden *et al.*, 2011) might need to be considered.

14.4.2 Planning beyond static conservation plans to account for threats

Due to the connected nature of freshwater ecosystems perturbations originating upstream or downstream to the priority area for conservation can easily propagate through hydrological networks and seriously threaten the persistence of the biodiversity (Allan, 2004) within reserved areas (Pringle, 2001; Roux *et al.*, 2008) and threaten the persistence of apparently protected freshwater communities (Januchowski-Hartley

et al., 2011; Linke *et al.*, 2011). For example, non-native species introduced at a point source can spread rapidly upstream and downstream
(Olden & Poff, 2005). Moreover, if threats are not included in the prioritisation process, highly degraded areas can be selected over areas in better condition and ultimately be more costly to manage. For this reason,
it is recommended to account for the spatial distribution of threats while
identifying priority areas for conservation (Linke *et al.*, 2012). Different
methodologies have used condition as either a pre-processing step to
filter out degraded areas or a *post-hoc* analysis in recent studies carried
out in North America, South America and South Africa (Thieme *et al.*,
2007; Khoury *et al.*, 2011; Nel *et al.*, 2011). However, removing degraded
areas after running a planning algorithm undermines the efficiency of
the systematic conservation planning method. An alternative strategy
recently proposed in different studies is to directly integrate condition in
the prioritisation process. This is normally done by applying a penalty to
degraded areas in the spatial prioritisation algorithm (Esselman & Allan,
2011; Heiner *et al.*, 2011; Linke *et al.*, 2012), so the degraded areas are
avoided from the solutions whenever possible. For example, if two alternatives existed to protect a given species, the option in the less-degraded
area would always be recommended. In this way, the cost of rehabilitation could be minimised.

The implementation of these conservation recommendations is also a
challenge, as it would require protection of large portions of catchments
in many cases to ensure that biodiversity is safe from the propagation
of threats along the river network or to maintain migration processes.
This is not affordable from a socioeconomic point of view (e.g. constrain
human uses within protected areas), so alternative approaches need to
be tested in order to incorporate these requirements into more flexible
schemes. For example, Saunders *et al.* (2002) proposed the creation of
multiple-use modules (MUM), structured in a central core conservation
area surrounded by a series of buffer areas in which different human
activities could be allowed. Thieme *et al.* (2007) structured conservation
recommendations for a large area in the Amazon River in three different
management zones: Level 1 (critical management zone), Level 2 (indigenous territories) and Level 3 (threat mitigation zones). The first two
levels consisted of mostly undisturbed catchments, whereas Level 3 zones
were potentially degraded. Each of these levels requires different mixtures of conservation actions and different uses are allowed. Abell *et al.*
(2007) presented a similar hierarchical approach based on three different
management zones. Their schedule was composed of 'freshwater focal

areas', which are key areas for the protection of freshwater biodiversity, similar to protected areas in terrestrial or marine realms; 'critical management zones', as areas that need to be managed to maintain the functionality of a focal area and where uses that do not interfere with the function of this area are allowed; 'catchment management zones' link the entire upstream catchment to a critical management zone where human uses are not constrained but best practices (treat wastewater disposals, maintain riparian buffers in good condition, or by restricting the use of pesticides) are required. The Abell *et al.* (2007) framework has been incorporated in several planning exercises around the globe (e.g. Esselman & Allan, 2011; Heiner *et al.*, 2011; Nel *et al.*, 2011). However, the location and extent of the different management zones and actions were not explicitly accounted for in the optimisation; rather, the zones were a *post-hoc* allocation. An alternative approach to this would be to incorporate management actions from the beginning of planning to influence the spatial distribution and configuration of priority areas. This approach would be in accordance with systematic conservation planning, as the *post-hoc* identification of conservation management actions would not necessarily be the most cost-effective way of achieving conservation objectives for the persistence of biodiversity. To effectively integrate management actions in the planning process will require explicit accountancy of cost and efficacy of different management actions (e.g. Carwardine *et al.*, 2011).

14.4.3 Planning for functional and genetic diversity

Functional and genetic diversity are key ecological and evolutionary determinants of freshwater biodiversity persistence (Hooper *et al.*, 2005; Forest *et al.*, 2007). Functional diversity refers to the number, type and distribution of functions performed by organisms within an ecosystem (Strecker *et al.*, 2011) and is a good indicator of ecosystem health. Genetic diversity describes past and present evolutionary processes and is a warranty of resilience. Conserving genetic diversity is important not only to maintain continuity of evolutionary processes, but also to enhance the capacity of local populations to adapt to future changes.

Different studies have shown incongruent patterns between taxonomic diversity and functional and genetic diversity (Forest *et al.*, 2007; Strecker *et al.*, 2011). For this reason, conservation plans focused on taxonomic diversity might not represent adequately these important components of biodiversity (Devictor *et al.*, 2010). Strecker *et al.* (2011) also demonstrated that functional and genetic diversity decline more rapidly than taxonomic diversity when reducing the area under protection. This suggests that these two components of biodiversity might be more sensitive to loss of quality habitat and ideally should be considered when planning for conservation.

14.5 CURRENT CHALLENGES TO EFFECTIVE PLANNING AND IMPLEMENTATION OF CONSERVATION IN FRESHWATER ECOSYSTEMS

The methodological approaches described in the previous sections pose a great advance towards objective, adequate and cost-effective conservation planning in freshwater ecosystems. However, there are some challenges that conservation planners usually face that might compromise effective planning and implementation of conservation actions to sustain biodiversity over the long term. Some of these challenges that have not been adequately addressed yet in freshwater conservation planning relate to the quality of data used for identifying priority areas, the selection of appropriate conservation targets (e.g. how much representation of a given species is enough as to secure its persistence), or the appropriate spatial extent and scale of the planning exercise. Failing to adequately engage with local stakeholders and landowners can also hinder the implementation of conservation recommendations on the ground.

14.5.1 Effective conservation in data-poor regions: data availability and uncertainty

The reliability of recommendations that arise from conservation planning depends on the quality of data on biodiversity patterns available. Poor-quality or sparse data are potentially subject to high uncertainty and can lead to poor decision-making (Possingham *et al.*, 2007), the misuse of the limited resources available and ultimately the failure of conservation practice. Errors in conservation planning outcomes associated with poor-quality data can compromise effectiveness (e.g. when a species is erroneously thought to be present within a reserve, or commission errors) and efficiency (e.g. when a species is erroneously thought to be absent forcing the selection of additional and unnecessary areas, or omission errors).

Ideally, the distribution of all species across a study region would be known (Grantham *et al.*, 2010). However, data collection is expensive and time-consuming (Gardner *et al.*, 2008) so, in reality, conservation planners have to deal with incomplete data sets that are usually biased spatially (e.g. along road networks) or toward particular taxa (Funk & Richardson, 2002). In the absence of extensive biological surveys, Thieme *et al.* (2007) used environmental classifications to design a reserve network in a large area of the upper Amazon (Madre de Dios River, Peru and Bolivia). Similar approaches to using environmental classifications have also been used to complement scarce biological data-sets elsewhere (see Heiner *et al.*, 2011; Nel *et al.*, 2011). However, given the poor concordance of some of these classifications and biodiversity patterns (some classifications do not portray adequately the spatial dis-tribution of species; see Box 14.1 for more detail), special care should be taken when using them as the only surrogate for biodiversity in conser-vation planning (see Hermoso *et al.*, 2013c for an extended review).

Improved biodiversity surveys have been highlighted as an effect-ive way of increasing certainty in data (Hirzel & Guisan, 2002) and enhancing conservation planning accuracy. For example, intensive bio-diversity surveys can help to detect some of the rarest species in the landscape, which might be the most in need of conservation actions and difficult to detect. However, at the same time, this would be an expen-sive and time-consuming task. In some cases, delaying intervention actions while collecting new data might risk the persistence of some threatened species (Grantham *et al.*, 2009). Moreover, the use of too many indiscriminate data is not always better than using fewer – but more systematically collected – data (Hermoso *et al.*, 2013b). Given that the addition of new high-quality data is constrained by the high cost and time required, it would be recommended to concentrate survey efforts on gathering more data for those species with the highest uncertainties (especially rare species). Methods are already available to help guide data acquisition (see, for example, Gradsec in which sampling is focused on discrete areas selected to contain maximum environmental heterogen-eity to minimise travelling between sites; Hirzel & Guisan, 2002), but have never been tested in a freshwater context.

Despite the great potential of these approaches to solve the problem of data availability (there are numerous satellite imagery and other GIS sources that have been developed in the last few years), some of these methods have been criticised as they do not adequately represent bio-diversity when used alone for identifying priority areas for conservation. For example, Hermoso *et al.* (2013c) demonstrated that coarse-filter sur-rogates do not help identify priority areas for conservation that represent biodiversity any better than the random selection of areas unless the

Box 14.1. Types of biodiversity surrogates used in conservation planning

The lack of information on complete coverage of species distributions is normally tackled by using surrogates for those distributions. The purpose of biodiversity surrogates in conservation planning is to portray species distribution patterns so that conservation priority areas selected to represent the surrogates will adequately represent species. The most commonly used surrogates are classifications that compartmentalise the spatial extent of the area under study into groups or classes, such as vegetation types (Pressey, 2004). The definition of these classes has either been based on *a posteriori* bottom-up classification of biological assemblages (classes represent homogeneous biological communities) or, more commonly, on *a priori* top-down classifications, referred to as coarse-filter surrogates. Coarse-filter surrogates in freshwater environments have included: habitat types (Thieme *et al.*, 2007; Rivers-Moore & Goodman, 2010) and environmental classifications (Januchowski-Hartley *et al.*, 2011). The use of coarse-filter surrogates in spatial prioritisation relies on two main assumptions. The first is that the surrogates have been defined according to factors that strongly influence the distribution of species. The second, as mentioned before for surrogates in general, is that sampling surrogates within a set of priority areas should represent species and other unmapped ecological variation not directly accounted for (Faith & Walker, 1996).

classification used as a surrogate portrays the patterns of biodiversity that it is meant to represent. The level of classification strength recommended in this work as necessary is rarely ever reported in ecological studies in freshwater ecosystems. These results are confirmed by additional studies (e.g. Januchowski-Hartley *et al.*, 2011) that report low effectiveness of coarse-filter surrogates at representing freshwater biodiversity in conservation plans. Therefore, it is recommended to take these approaches carefully in order to avoid costly mistakes when making conservation recommendations.

Alternative methodologies to overcome the problem derived from the use of inaccurate classifications have been proposed. For instance, predictive models have been used to extrapolate available biological information to unsurveyed areas to obtain continuous estimations of species distributions (Leathwick *et al.*, 2005; Hermoso *et al.*, 2011b, 2012a). These models link the known records on species occurrences obtained from field surveys to the environmental characteristics of those sites where they were found. This relationship is then used to estimate the probability of occurrence beyond the surveyed areas and obtain complete coverage of species distributions. However, these estimates are also subject to errors and uncertainties (Hermoso *et al.*, 2013b) that need to be adequately addressed in order to enhance the value of conservation recommendations. More recent modelling techniques focus on estimating alternative ecological features, such as species turnover, and have proven useful for conservation planning (Leathwick *et al.*, 2011; Turak *et al.* 2011).

14.5.2 Setting adequate conservation targets

Setting adequate conservation targets (how much of each species or any other surrogate used wants to be represented in the conservation plan so it is enough as to secure the persistence of the species) requires a greater understanding of the processes that drive species persistence as well as species habitat and resource needs (Linke *et al.*, 2011; Strecker *et al.*, 2011). When setting conservation targets, practitioners face choices that will be very influential on the final conservation plan, as the total areas required and the spatial allocation of priority areas are sensitive to target setting. For example, targets can be set uniformly (e.g. every taxon will have to be represented in a certain length of river or a fixed proportion of their geographic distribution), or tailored to meet the ecological needs of specific taxa. Furthermore, targets can be set for species directly, but can also be spatial components of the region (e.g. geographic area, river length) that serve as surrogates for biodiversity or ecological processes (Desmet *et al.*, 2002).

In freshwater settings to date, both percentages of the whole distribution range (Thieme *et al.*, 2007; Nel *et al.*, 2011) and explicit targets set in terms of river length or area (Hermoso *et al.*, 2011b, 2012b; Linke *et al.*, 2012) have been chosen in conservation planning studies. However, setting percentage conservation targets could lead to overrepresentation of common taxa while underrepresenting rare taxa. For example, a 50% target to represent a species that is everywhere within a planning region will result in a solution requiring 50% of the landscape, while a 50% target of a very rare and endangered species might not be enough to protect species persistence. Because of this imbalance, it is recommended that targets be set for individual species. For example, target considerations can be simple dynamic models that predict extinction risk (Burgman *et al.*, 2001), or estimations of sustainability derived from population viability analysis (Pressey *et al.*, 2003). Ultimately, persistence is one of the key properties when setting conservation targets (Gaston *et al.*, 2002), because 'biodiversity will ultimately succeed at the population level' (Hughes *et al.*, 1997). Therefore, in real-world planning exercises, critical population numbers will have to be assessed, or at least habitat features have to be selected for persistence (Margules *et al.*, 2002). Although taken from the terrestrial realm, a good example illustrating the process are the targets set for the Cape Floristic Region in South Africa (Pressey *et al.*, 2003), where three groups of features were targeted.

14.5.3 On-the-ground implementation of conservation actions

The identification of priority areas is just part of the process of conservation practice. In order to make recommendations produced in conservation assessments, additional resources are needed to facilitate their implementation. Empowerment of individuals and institutions and securing effective action has been recognised as essential for the adequate implementation of conservation (Barmuta *et al.*, 2011). Knight *et al.* (2008) reported that two-thirds of conservation assessments in terrestrial environments published in peer-reviewed literature did not result in effective management action. One reason for this is the frequent disconnect between desktop planning exercises and the lack of involvement of stakeholders throughout the planning process. Important to the success of implementing a plan is consideration of the governance systems that could influence the plan being designed. For example, if the conservation plan is aimed at identifying priorities for fish conservation through the removal of an aquatic invasive species, then it would be essential to identify the bodies governing aquatic species and invasive species in the study region and any relevant legislation that could influence the success of implementing areas or actions selected through the planning process. The management of aquatic invasive species could also require the support of other local stakeholders such as drainage commissioners, water resource managers or landholders depending on the management question (Chapter 8). Understanding the social and political system, which is dependent on the spatial extent of the planning exercise as mentioned above, could assist planners with identifying suitable stakeholder participation strategies.

Critical to the likelihood of conservation recommendations being implemented is the alignment and collaboration between planners and other stakeholder groups who could be influenced by the implementation of the plan. Therefore, identifying stakeholders early on in the planning process is critical, and an important factor that could influence the possibility of implementing the plan and its success once implemented. Across spatial scales, success has been found with earlier stakeholder involvement, openness in negotiations, building capacity and trust, and making and maintaining explicit links with institutions and community groups (see Barmuta *et al.*, 2011 and references therein). Once stakeholders are identified, questions need to be asked about the characteristics of the stakeholders, and how they might be affected by conservation actions (e.g. their willingness to collaborate; Knight *et al.*, 2008). When these factors are incorporated into spatial conservation prioritisations (Moilanen *et al.*, 2009), they can reveal opportunities such as areas where actions are most likely to be successful if implemented, and then assist planners to more strategically invest their resources (Januchowski-Hartley *et al.*, 2012). For example, private landholders expressing unwillingness to engage in conservation programmes on their land have identified barriers to their participation, such as a lack of flexibility of the conservation programme, excessive complexity of the proposed management actions and incompatibility of the conservation programme with their own personal or property goals (Januchowski-Hartley *et al.*, 2012).

Stakeholder involvement and collaboration could require sufficient resources and time to foster and maintain collaboration. However, limited funds often constrain the number of actions a conservation organisation can afford to implement (Carwardine *et al.*, 2008) and the number of properties on which these actions can be implemented. Therefore, to maximise spending efficiency, stakeholder involvement and support of the planning process and final plan selected for implementation could be beneficial for conservation organisations (Knight *et al.*, 2011), reducing unnecessarily high costs that could be avoided through active stakeholder engagement.

14.6 DISCUSSION QUESTIONS

1. Planning in each realm (marine, terrestrial and freshwater) independently overlooks potential conservation synergisms across them (Adams *et al.*, 2014). How can current planning methods be integrated to holistically plan for the protection and management of systems and species that are dependent on multiple realms?
2. Human-related barriers can be an effective way of minimising the propagation of threats (e.g. spread of invasive species) and can facilitate native species management (Rahel, 2013). However, they also pose important conservation problems as they constrain important

ecological processes such as movements and migrations to complete fish life cycles. How can we find optimal allocation for barriers so the ecological impact is minimised while the conservation benefit is maximised?

3. As human populations expand, it is increasingly difficult to justify conservation for biodiversity's sake without also demonstrating some benefits for people. How can we improve the harmony of conservation goals and the exploitation of ecosystem services?

4. Social and political factors can strongly influence the likelihood of conservation plans being implemented. How can we do a better job of considering sociopolitical processes during the planning process and identifying the sociopolitical factors that influence implementation?

14.7 REFERENCES

Abell, R. (2002). Conservation biology for the biodiversity crisis: a freshwater follow up. *Conservation Biology*, 16, 1435–1437.

Abell, R., Allan J. D. & Lehner, B. (2007). Unlocking the potential of protected areas for freshwaters *Biological Conservation*, 134, 48–63.

Adams, V. M., Álvarez-Romero, J. G., Carwardine, J., *et al.* (2013). Planning across freshwater and terrestrial realms: co-benefits and tradeoffs between conservation actions. *Conservation Letters*, 7, 425–440.

Allan, J. D. (2004). Landscape and riverscapes. The influence of land use on river ecosystems. *Annual Reviews of Ecology, Evolution and Systematics*, 35, 257–284.

Arnell, N. W. (2003). Effects of IPCC SRES emission scenarios on river runoff: a global perspective. *Hydrology and Earth Systems Science*, 7, 619–641.

Arthington, A. H. (2012). *Environmental Flows: Saving Rivers in the Third Millennium*. Berkeley, CA: University of California Press.

Ausseil, A. G. E., Dymond, J. R. & Shepherd, J. D. (2007). Rapid mapping and prioritisation of wetland sites in the Manawatu–Wanganui Region, New Zealand. *Environmental Management*, 39, 316–325.

Ausseil, A. G. E., Chadderton, W. L., Gerbeaux, P., Stephens, R. T. T. & Leathwick, J. R. (2011). Applying systematic conservation planning principles to palustrine and inland saline wetlands of New Zealand. *Freshwater Biology*, 56, 142–161.

Balcombe, S. R., Arthington, A. H., Foster, N. D., *et al.* (2006). Fish assemblages of an Australian dryland river: abundance, assemblage structure and recruitment patterns in the Warrego River, Murray–Darling Basin. *Marine and Freshwater Research*, 57, 619–633.

Ball, I. R., Possingham, H. P. & Watts, M. (2009). MARXAN and relatives: software for spatial conservation prioritization. In *Spatial Conservation Prioritisation: Quantitative Methods and Computational Tools*. Oxford: Oxford University Press, pp. 185–195.

Barmuta, L. A., Linke, S. & Turak, E. (2011). Bridging the gap between 'planning' and 'doing' for biodiversity conservation in freshwaters. *Freshwater Biology*, 56, 180–195.

Bates, B. C., Kundzewicz, Z. W., Wu, S. & Palutikof, J. P. (2008). *Climate Change and Water. Technical Paper of the Intergovernmental Panel on Climate Change*. Geneva: IPCC Secretariat.

Bond, N. R., Lake, P. S. & Arthington, A. H. (2008). The impacts of drought on freshwater ecosystems: an Australian perspective. *Hydrobiologia*, 600, 3–16.

Bottrill, M. C., Joseph, L. N., Carwardine, J., *et al.* (2009). Finite conservation funds mean triage is unavoidable. *Trends in Ecology and Evolution*, 24, 183–184.

Bunn, S. E. & Arthington, A. H. (2002). Basic principles and ecological consequences of altered flow regimes for aquatic biodiversity. *Environmental Management*, 30, 492–507.

Burgman, M. A., Possingham, H. P., Lynch, A. J. J., *et al.* (2001). A method for setting the size of plant conservation area. *Conservation Biology*, 15, 603–616.

Carini, G., Hughes, J. M. & Bunn, S. E. (2006). The role of waterholes as 'refugia' in sustaining genetic diversity and variation of two freshwater species in dryland river systems (Western Queensland, Australia). *Freshwater Biology*, 51, 1434–1446.

Carwardine, J., Wilson, K. A., Watts, M., *et al.* (2008). Avoiding costly conservation mistakes: the importance of defining actions and cost in spatial prioritization setting. *PLoS ONE*, 3, e2586.

Carwardine, J., O'Connor, T., Legge, S., *et al.* (2011). *Priority Threat Management to Protect Kimberley Wildlife*. CSIRO Ecosystem Sciences, Brisbane. www.csiro.au/files/files/pzk8.pdf (accessed 31 March 2011).

Castello, L., Viana, J. P., Watkins, G., Pinedo-Vasquez, M. & Luzadis, V. A. (2009). Lessons from integrating fishers of arapaima in small-scale fisheries management at the Mamirauá Reserve, Amazon. *Environmental Management*, 43, 197–209.

Ciarleglio, M., Barnes, J. W. & Sarkar, S. (2009). ConsNet: a new software for the selection of conservation area networks with spatial and multicriteria analysis. *Ecography*, 32, 205–209.

Crook, D. A., Reich, P., Bond, N. R., *et al.* (2010). Using biological information to support proactive strategies for managing freshwater fish during drought. *Marine and Freshwater Research*, 61, 379–387.

Darwall, W. R. T. & Vie, J. C. (2005). Identifying important sites for conservation of freshwater biodiversity: extending the species-based approach. *Fisheries Management and Ecology*, 12, 287–293.

Darwall, W., Smith, K., Allen, D., *et al.* (2008). Freshwater biodiversity – a hidden resource under threat. In *Wildlife in a Changing World –An Analysis of the 2008 IUCN Red List of Threatened Species*. Gland, Switzerland: Vié, IUCN, pp. 43–53.

Desmet, P. G., Cowling, R. M., Ellis, A. G. & Pressey, R. L. (2002). Integrating biosystematic data into conservation planning: perspectives from Southern Africa's Succulent Karoo. *Systematic Biology*, 51, 317–330.

Devictor, V., Mouillot, D., Meynard, C., *et al.* (2010). Spatial mismatch and congruence between taxonomic, phylogenetic and functional diversity: the need for integrative conservation strategies in a changing world. *Ecology Letters*, 13, 1030–1040.

Dudgeon, D., Arthington, A. H., Gessner, M. O., *et al.* (2006). Freshwater biodiversity: importance, threats, status and conservation challenges. *Biological Reviews*, 81, 163–182.

ESA. (1973). *Endangered Species Act of 1973*. Department of the Interior, US Fish and Wildlife Service Washington, DC 20240.

Esselman, P. C. & Allan, J. D. (2011). Application of species distribution models and conservation planning software to the design of a reserve network for the riverine fishes of northeastern Mesoamerica. *Freshwater Biology*, 56, 71–88.

European Council. (1992). *Directive 92/43/EEC of 21 May 1992 on the conservation of natural habitats and of wild fauna and flora*. Brussels.

Faith, D. P. & Walker, P. A. (1996). Environmental diversity: on the best possible use of surrogate data for assessing the relative biodiversity of areas. *Biodiversity and Conservation*, 5, 399–415.

Filipe, A. F., Marques, T. A., Seabra, S., *et al.* (2004). Selection of priority areas for fish conservation in Guadiana River Basin, Iberian Peninsula. *Conservation Biology*, 18, 189–200.

Fitzsimons, J. & Robertson, H. A. (2005). Freshwater reserves in Australia: directions and challenges for the development of a comprehensive, adequate and representative system of protected areas. *Hydrobiologia*, 552, 87–97.

Forest, F., Grenyer, R., Rouget, M., *et al.* (2007). Preserving the evolutionary potential of floras in biodiversity hotspots. *Nature*, 445, 757–760.

Funk, V. A. & Richardson, K. S. (2002). Systematic data in biodiversity studies: use it or lose it. *Systematic Biology*, 51, 303–316.

Gardner, T. A., Barlow, J., Araujo, I. S., *et al.* (2008). The cost-effectiveness of biodiversity surveys in tropical forests. *Ecology Letters*, 11, 139–150.

Gasith, A. & Resh, V. H. (1999). Streams in Mediterranean climate regions: abiotic influences and biotic responses to predictable seasonal events. *Annual Review of Ecology and Systematics*, 30, 51–81.

Gaston, K. J., Pressey, R. L. & Margules, C. R. (2002). Persistence and vulnerability: retaining biodiversity in the landscape and in protected areas. *Journal of Biosciences*, 27, 361–384.

Grantham, H. S., Wilson, K. A., Moilanen, A., Rebelo, T. & Possingham, H. P. (2009). Delaying conservation actions for improved knowledge: how long should we wait? *Ecology Letters*, 12, 293–301.

Grantham, H. S., Pressey, R. L., Wells, J. A. & Beattie, A. J. (2010). Effectiveness of biodiversity surrogates for conservation planning: different measures of effectiveness generate a kaleidoscope of variation. *PLoS ONE*, 5, e11430.

Haigh, S. M., Melhman, D. W. & Oring, L. W. (1998). Avian movements and wetland connectivity in landscape conservation. *Conservation Biology*, 12, 749–758.

Heiner, M., Higgins, J., Li, X. & Baker, B. (2011). Identifying freshwater conservation priorities in the Upper Yangtze River Basin. *Freshwater Biology*, 56, 89–105

Herbert, M. E., Mcintyre, P. B., Doran, P. J., Allan, J. D. & Abell, R. (2010). Terrestrial reserve networks do not adequately represent aquatic ecosystems. *Conservation Biology*, 24, 1002–1011.

Hermoso, V., Linke, S. & Prenda, J. (2009). Identifying priority sites for the conservation of freshwater fish biodiversity in a Mediterranean basin with a high degree of threatened endemics. *Hydrobiologia*, 623, 127–140.

Hermoso, V., Kennard, M., Pusey, B. & Douglas, M. (2011a). Identifying priority areas for the conservation of freshwater biodiversity. In *Aquatic Biodiversity in Northern Australia: Patterns, Threats and Future*. Pusey, B. J. (Ed.). Darwin: Charles Darwin University Press, pp. 133–149.

Hermoso, V., Linke, S., Prenda, J. & Possingham, H. P. (2011b). Addressing longitudinal connectivity in the systematic conservation planning of fresh waters. *Freshwater Biology*, 56, 57–70.

Hermoso, V., Kennard, M. J. & Linke, S. (2012a). Integrating multi-directional connectivity requirements in systematic conservation planning to prioritise fish and waterbird habitat in freshwater systems. *Diversity and Distributions*, 18, 448–458.

Hermoso, V., Ward, D. P. & Kennard, M. J. (2012b). Using water residency time to enhance spatio-temporal connectivity for conservation planning in seasonally dynamic freshwater ecosystems. *Journal of Applied Ecology*, 49, 1028–1035.

Hermoso, V., Ward, D. P. & Kennard, M. J. (2013a). Prioritizing refuge areas for freshwater biodiversity conservation in highly seasonal ecosystems. *Diversity and Distributions*, 19, 1031–1042.

Hermoso, V., Kennard, M. J. & Linke, S. (2013b). Data acquisition for conservation assessments: is the effort worth it? *PLoS ONE*, 8, e59662.

Hermoso, V., Januchowski-Hartley, S. R. & Pressey, R. L. (2013c). When the suit does not fit biodiversity: loose surrogates compromise the achievement of conservation goals. *Biological Conservation*, 159, 197–205.

Hess, G. R., Bartel, R. A., Leidner, A. K., *et al.* (2006). Effectiveness of biodiversity indicators varies with extent, grain, and region. *Biological Conservation*, 132, 448–457.

Higgins, J. V., Bryer, M. T., Khoury, M. L. & Fitzhugh, T. W. (2005). A freshwater classification approach for biodiversity conservation planning. *Conservation Biology*, 19, 432–445.

Hirzel, A. & Guisan, A. (2002). Which is the optimal sampling strategy for habitat suitability modeling. *Ecological Modelling*, 157, 331–341.

Holland, R. A., Darwall, W. R. T. & Smith, K. G. (2012). Conservation priorities for freshwater biodiversity: the Key Biodiversity Area approach refined and tested for continental Africa. *Biological Conservation*, 148, 167–179.

Hooper, D. U., Chapin, F. S., Ewel, J. J., *et al.* (2005). Effects of biodiversity on ecosystem functioning: a consensus of current knowledge. *Ecological Monographs*, 75, 3–35.

Hughes, J. B., Daily, G. C. & Ehrlich, P. R. (1997). Population diversity: its extent and extinction. *Science*, 278, 689–92.

ICES. (2005). *Report of the study group on the status of diadromous fish species (SGSDFS)*. No. ICES CM 2005/I:02, pp. 56.

Januchowski-Hartley, S. R., Hermoso, V., Pressey, R. L., *et al.* (2011). Coarse-filter surrogates do not represent freshwater fish diversity at a regional scale in Queensland, Australia. *Biological Conservation*, 144, 2499–2511.

Januchowski-Hartley, S. R., Moon, K., Stoeckl, N. & Gray, S. (2012). Social factors and private benefits influence landholders' riverine restoration priorities in tropical Australia. *Journal of Environmental Management*, 110, 20–26

Kelley, C., Garson, J., Aggarwal, A. & Sarkar, S. (2002). Place prioritization for biodiversity reserve network design: a comparison of the SITES and ResNet software packages for coverage and efficiency. *Diversity and Distributions*, 8, 297–306.

Kennard, M. J., Pusey, B. J., Olden, J. D., *et al.* (2010). Classification of natural flow regimes in Australia to support environmental flow management. *Freshwater Biology*, 55, 171–193.

Khoury, M. L., Higgins, J. & Weitzell, R. E. (2011). A freshwater conservation assessment of the Upper Mississippi River basin using a coarse- and fine-filter approach. *Freshwater Biology*, 56, 162–179.

Kirkpatrick, J. B. (1983). An iterative method for establishing priorities for the selection of nature reserves: an example from Tasmania. *Biological Conservation*, 25, 127–134.

Knight, A. T., Cowling, R. M., Rouget, M., *et al.* (2008). Knowing but not doing: selecting priority conservation areas and the research-implementation gap. *Conservation Biology*, 22, 610–617.

Larned, S. T., Datry, T., Arscott, D. B. & Tockner, K. (2010). Emerging concepts in temporary-river ecology. *Freshwater Biology*, 55, 717–738.

Lawrence, D. J., Larson, E. R., Liermann, C. A. R., *et al.* (2011). National Parks as protected areas for U.S. freshwater fish diversity. *Conservation Letters*, 4, 364–371.

Leathwick, J. R., Rowe, D., Richardson, J., Elith, J. & Hastie, T. (2005). Using multivariate adaptive regression splines to predict the distribution of New Zealand's freshwater diadromous fish. *Freshwater Biology*, 50, 2034–2052.

Leathwick, J. R,. Snelder, T., Chadderton, W. L., *et al.* (2011). Use of generalised dissimilarity modelling to improve the biological discrimination of river and stream classifications. *Freshwater Biology*, 56, 21–38.

Linke, S., Pressey, R. L, Bailey, R. C. & Norris, R. H. (2007). Management options for river conservation planning: condition and conservation re-visited. *Freshwater Biology*, 52, 918–938.

Linke, S., Turak, E. & Nel, J. (2011). Freshwater conservation planning: the case for systematic approaches. *Freshwater Biology*, 56, 6–20.

Linke, S., Kennard, M. J., Hermoso, V., *et al.* (2012). Merging connectivity rules and large-scale condition assessment improves conservation adequacy in a tropical Australian river. *Journal of Applied Ecology*, 49, 1036–1045.

Magalhães, M. F., Beja, P., Schlosser, I. J. & Collares-Pereira, M. J. (2007). Effects of multi-year droughts on fish assemblages of seasonally drying Mediterranean streams. *Freshwater Biology*, 52, 1494–1510.

Margules, C. R. & Pressey, R. L. (2000). Systematic conservation planning. *Nature*, 405, 243–253.

Margules, C. R., Pressey, R. L. & Williams, P. H. (2002). Representing biodiversity: data and procedures for identifying priority areas for conservation. *Journal of Biosciences*, 27, 309–326.

Moffet, A., Garson, J. & Sarkar, S. (2005). MultCSync: a sotware package for incorporating multiple criteria in conservation planning. *Environmental Modelling and Software*, 20, 1315–1322.

Moilanen, A., Franco, A. M. A., Early, R., *et al.* (2005). Prioritising multiple use landscapes for conservation: methods for large multi species planning problems. *Proceedings of the Royal Society B*, 272, 1885–1891.

Moilanen, A., Leathwick, J. & Elith, J. (2008). A method for freshwater conservation prioritization. *Freshwater Biology*, 53, 577–592.

Moilanen, A., Arponen, A., Stokland, J. & Cabeza, M. (2009). Assessing replacement cost of conservation areas: how does habitat loss influence priorities? *Biological Conservation*, 142, 575–585.

Morrongiello, J. R., Beatty, S. J., Bennett, J. C., *et al.* (2011). Climate change and its implications for Australia's freshwater fish. *Marine and Freshwater Research*, 62, 1082–1098.

Moyle, P. B. & Yoshiyama, R. M. (1994). Protection of aquatic biodiversity in California: a five tiered approach. *Fisheries*, 19, 6–18.

Myers, N., Mittermeier, R. A., Mittermeier, C. G., da Fonseca, G. A. B. & Kent, J. (2000). Biodiversity hotspots for conservation priorities. *Nature*, 403, 853–858.

Naidoo, R., Balmford, A., Ferraro, P. J., *et al.* (2006). Integrating economic cost into conservation planning. *Trends in Ecology and Evolution*, 21, 681–687.

Nel, J. L., Roux, D. J., Maree, G., *et al.* (2007). Rivers in peril inside and outside protected areas: a systematic approach to conservation assessment of river ecosystem. *Diversity and Distributions*, 13, 341–352.

Nel, J. L., Roux, D. J., Abell, R., *et al.* (2009). Progress and challenges in freshwater conservation planning. *Aquatic Conservation: Marine and Freshwater Ecosystems*, 19, 474–485.

Nel, J. L., Reyers, B., Roux, D. J., Impson, N. D. & Cowling, R. M. (2011). Designing a conservation area network that supports the representation and persistence of freshwater biodiversity. *Freshwater Biology*, 56, 106–124.

Olden, J. D. & Poff, N. L. (2005). Long-term trends in native and non-native fish faunas of the American Southwest. *Animal Biodiversity and Conservation*, 28, 75–89

Olden, J. D., Kennard, M. J., Leprieur, F., *et al.* (2010). Conservation biogeography of freshwater fishes: recent progress and future challenges. *Diversity and Distributions*, 16, 496–513.

Olden, J. D., Kennard, M. J., Lawler, J. J. & Poff, N. L. (2011). Challenges and opportunities in implementing managed relocation for conservation of freshwater species. *Conservation Biology*, 25, 40–47.

Pépino, M., Rodríguez, A. & Magnan, P. (2012). Impacts of highway crossings on density of brook charr in streams. *Journal of Applied Ecology*, 49, 395–403.

Poff, N. L., Allan, J. D, Bain, M. B., *et al.* (1997). The natural flow regime: a new paradigm for riverine conservation and restoration. *BioScience*, 47, 769–784.

Possingham, H. P., Grantham, H. & Rondinini, C. (2007). How can you conserve species that haven't been found? *Journal of Biogeography*, 34, 758–759.

Pressey, R. L. (1994). *Ad hoc* reservations: forward and backward steps in developing representative reserve systems. *Conservation Biology*, 8, 662–668.

Pressey, R. L. (2004). Conservation planning and biodiversity: assembling the best data for the job. *Conservation Biology*, 18, 1677–1681.

Pressey, R. L., Possingham, H. P. & Margules, C. R. (1996). Optimality in reserve selection algorithms: when does it matter and how much? *Biological Conservation*, 76, 259–267.

Pressey, R. L., Cowling, R. M. & Rouget, M. (2003). Formulation of conservation targets for biodiversity pattern and process in the Cape Floristic Region, South Africa. *Biological Conservation*, 112, 99–127.

Pressey, R. L., Watts, M. E., Barrett, T. W. & Ridges, M. J. (2009). The C-Plan conservation planning system: origins, applications, and possible futures. In *Spatial Conservation Prioritization: Quantitative Methods and Computational Tools*. Oxford: Oxford University Press, pp. 211–234.

Pringle, C. M. (2001). Hydrologic connectivity and the management of biological reserves: a global perspective. *Ecological Applications*, 11, 981–998.

Rahel, F. J. (2013). Intentional fragmentation as a management strategy in aquatic systems. *Biosciences*, 63, 362–372.

Rea, J. H., Brinton, A. C. & Georges, A. (2009). Temporal and spatial variation in landscape connectivity for a freshwater turtle in a temporally dynamic wetland system. *Ecological Applications*, 19, 1288–1299.

Rivers-Moore, N. A. & Goodman, P. S. (2010). River and wetland classifications for freshwater conservation planning in KwaZulu-Natal, South Africa. *African Journal of Aquatic Sciences*, 35, 61–72.

Roberge, J. M. & Angelstam, P. (2004). Usefulness of the umbrella species concept as a conservation tool. *Conservation Biology*, 18, 76–85.

Rosset, V., Simaika, J. P., Arthaud, F., *et al.* (2013). Comparative assessment of scoring methods to evaluate the conservation value of pond and small lake biodiversity. *Aquatic Conservation: Marine and Freshwater Ecosystems*, 23, 23–36.

Roux, D. J. & Nel, J. L. (2013). Freshwater conservation planning in South Africa: milestones to date and catalysts for implementation. *Water SA*, 39, 151–163.

Roux, D., de Moor, F., Cambray, J. & Barber-James, H. (2002). Use of landscape-level river signatures in conservation planning: a South African case study. *Conservation Ecology*, 6, 6.

Roux, D. J., Nel, J. L., Ashton, P. J., *et al.* (2008). Designing protected areas to conserve riverine biodiversity: lessons from a hypothetical redesign of the Kruger National Park. *Biological Conservation*, 141, 100–117.

Saunders, D. L., Meeuwig, J. J. & Vincent, C. J. (2002). Freshwater protected areas: strategies for conservation. *Conservation Biology*, 16, 30–41.

Skaala, Ø., Helge-Johnsen, G., Lo, H., *et al.* (2014). A conservation plan for Atlantic salmon (*Salmo salar*) and anadromous brown trout (*Salmo trutta*) in a region with intensive industrial use of aquatic habitats, the Hardangerfjord, western Norway. *Marine Biology Research*, 10, 308–322.

Stanley, E. H. & Doyle, M. W. (2003). Trading off: the ecological effects of dam removal. *Frontiers in Ecology and the Environment*, 1, 15–22.

Strayer, D. & Dudgeon, D. (2010). Freshwater biodiversity conservation: recent progress and future challenges. *Journal of the North American Benthological Society*, 29, 344–358.

Strecker, A. L., Olden, J. D., Whittier, J. B. & Paukert, C. P. (2011). Defining conservation priorities for freshwater fishes according to taxonomic, functional, and phylogenetic diversity. *Ecological Applications*, 21, 3002–3013.

Taylor, P. D., Fahrig, L. & With, K. A. (2006). Landscape connectivity: a return to the basics. Connectivity Conservation. In *Connectivity Conservation*. Cambridge: Cambridge University Press, pp. 29–43.

Thieme, M., Lehner, B., Abell, R., *et al.* (2007). Freshwater conservation planning in data-poor areas: an example from a remote Amazonian basin (Madre de Dios River, Peru and Bolivia). *Biological Conservation*, 135, 484–501.

Turak, E. & Koop, K. (2008). A multi-attribute ecological river typology for assessing river condition and conservation planning. *Hydrobiologia*, 603, 83–104.

Turak, E. & Linke, S. (2011). Freshwater conservation planning: an introduction. *Freshwater Biology*, 56, 1–5.

Turak, E., Ferrier, S., Barrett, T., *et al.* (2011). Planning for persistence of river biodiversity: exploring alternative futures using process-based models. *Freshwater Biology*, 56, 39–56.

Turner, D. S. & List, M. D. (2007). Habitat mapping and conservation analysis to identify critical streams for Arizona's native fish. *Aquatic Conservation: Marine and Freshwater Ecosystems*, 17, 737–748.

Verissimo, D., MacMillan, D. C. & Smith, R. J. (2011). Toward a systematic approach for identifying conservation flagships. *Conservation Letters*, 4, 1–8.

Vörösmarty, C. J., McIntyre, P. B., Gessner, M. O., *et al.* (2010). Global threats to human water security and river biodiversity. *Nature*, 467, 555–561.

Ward, J. V. & Stanford, J. A. (1989). The four-dimensional nature of lotic systems. *Journal of the North American Benthological Society*, 8, 2–8.

Wenger, S. J., Isaak, D. J., Luce, C. H., *et al.* (2011). Flow regime, temperature, and biotic interactions drive differential declines of trout species under climate change. *Proceedings of the National Academy of Sciences*, 108, 14175–14180.

Williams, P., Gibbons, D., Margules, C., *et al.* (1996). A comparison of richness hotspots, rarity hotspots and complementarity areas for conserving diversity of British birds. *Conservation Biology*, 10, 155–174.

Williams, P. H. (2001). WORLDMAP Version 4. Priority areas for biodiversity. www.nhm.ac.uk/science/projects/worldmap

Sustainable inland fisheries – perspectives from the recreational, commercial and subsistence sectors from around the globe

STEVEN J. COOKE, VIVIAN M. NGUYEN, JOHN M. DETTMERS,
ROBERT ARLINGHAUS, MICHAEL C. QUIST, DENIS TWEDDLE,
OLAF L. F. WEYL, RAJEEV RAGHAVAN, MARCELA PORTOCARRERO-AYA,
EDWIN AGUDELO CORDOBA AND IAN G. COWX

15.1 INTRODUCTION

Globally, freshwater ecosystems provide varied fishing opportunities (herein termed inland fisheries) represented by three sectors: recreational, commercial and subsistence fisheries. From the depths of the Laurentian Great Lakes to the shallow floodplains of the Ganges River, and from under-ice fisheries in Scandinavia to the rice fields of Southeast Asia, fish and other aquatic life are omnipresent components of fluvial and lacustrine systems. Freshwater fishes generate many ecosystem services that extend beyond their use in fisheries (Holmlund & Hammer, 1999; Cowx & Portocarrereo, 2011). Given the diversity of freshwater fish assemblages, levels of fisheries productivity, cultural norms, density of human population and socioeconomic conditions, it is not surprising that there is immense variation in how, why and the extent to which freshwater fishes and other aquatic animals are exploited. Whether it be sustaining livelihoods through the provision of essential nutrients, generating income, or enabling leisure time with family, inland fisheries are important. Although there are accepted definitions for the three fishing sectors (i.e. UN FAO – see below), ambiguities and exceptions remain that complicate appraisal and management.

Compared with marine waters where industrial-scale commercial fisheries predominate, inland fisheries tend to be smaller in scale and catches generally do not enter the global marketplace. Moreover, whereas

Conservation of Freshwater Fishes, eds G. P. Closs, M. Krkosek and J. D. Olden. Published by Cambridge University Press. © Cambridge University Press 2016.

exploitation pressures are the primary threat facing marine fish populations and marine ecosystems, in inland systems there are multiple threats
including many unrelated to fishing (Arlinghaus *et al.*, 2002). Indeed,
declines in freshwater fish fauna are implicated with broad-scale economic activities such as flow regulation, hydropower, agriculture, urbanisation and pollution (Limburg *et al.*, 2011; Chapters 4 and 9). Reflecting
the multiple threats, freshwater fishes are among the most imperilled
taxa on the globe (Strayer & Dudgeon, 2010; Chapter 2), freshwater biodiversity is in crisis (Dudgeon *et al.*, 2006) and freshwater ecosystems are
among the most altered (Kennish, 2002; Malmqvist & Rundle, 2002).
Despite the many threats to inland fishes and fisheries, they receive disproportionately less interest and attention from the global conservation
community and international political spheres. Indeed, global capture
statistics underrepresented inland fisheries and their contribution to
global production (Welcomme *et al.*, 2010; Welcomme, 2011a,b), partly
because of the diffuse nature of inland fisheries (Beard *et al.*, 2011). By
contrast, it is comparatively easy to generate data for commercial fisheries where products sold on established domestic and export markets can
be readily monitored. In recent years, there have been attempts to better characterise the magnitude and scale of inland fisheries (Welcomme,
2011a,b). Such efforts are aimed at generating a more realistic picture of
how inland fisheries contribute to food security and generation of income,
as well as to identify potential conservation issues (Beard *et al.*, 2011).

Although still imperfect, statistics from inland fisheries reveal a
steady increase from 2.0 million tonnes in 1950 to 11.6 million tonnes
in 2011 (Figures 15.1 and 15.2). The increase is attributed largely to
improvements in monitoring and reporting, rather than real increase in
harvest (Welcomme, 2011a). Inland water catches, however, appear to be
underreported by an average of 70% (World Bank, 2012). For example,
the reported catch for Vietnam in 2003 was 203,000 tonnes, but the
true catch is closer to 1 million tonnes (World Bank, 2012). Although
inland fisheries harvest is only ~10% of that from marine systems, it is
massively important on a regional basis for food security, especially in
remote, rural areas (Welcomme *et al.*, 2010). Indeed, ~38% of inland fish
capture is from the 71 low-income food deficit countries (as defined by
the FAO; Welcomme, 2011a). Globally, inland capture fisheries generate
2.3% of total animal protein sources, although the contribution of fish
to diet varies widely among countries and regions (Welcomme, 2011a).
Inland fish are also targeted by millions of recreational anglers across
the globe (Arlinghaus *et al.*, 2002; Arlinghaus & Cooke, 2009). Relevant
here is that the recreational sector is quite large, they are not accounted

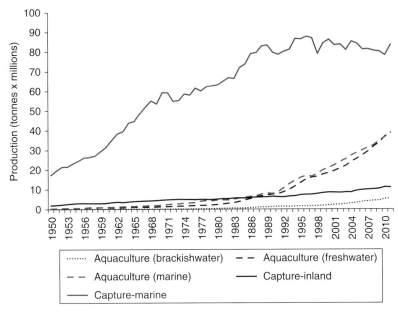

Figure 15.1 Summary of UN FAO statistics illustrating global capture for marine and inland systems as well as global aquaculture production by water body type.

for in FAO statistics, and when economic development occurs, there tends to be a transition from commercial-subsistence fisheries to recreational fisheries (Arlinghaus *et al.*, 2002; Cowx *et al.*, 2010; FAO, 2012), and under these circumstances there is often a reasonable proportion of catch that is released (i.e. catch-and-release). The majority of reported capture (90%) in inland fisheries is for finfish with crustaceans and molluscs locally important in some regions (Welcomme, 2011a,b), although it should be recognised the latter are largely unreported as they pass through informal markets. About 19 million fishers participate in inland fisheries, and inland fisheries support a workforce of 58 million people (World Bank, 2012).

Existing syntheses on inland fisheries have tended to focus on different sectors and their interrelationships (e.g. Welcomme *et al.*, 2010; World Bank, 2012), but not on specific interactions. Given that it is more the rule than the exception for multiple fisheries sectors (two or three) to co-occur in the same region or water body, it would be informative to consider their interactions when characterising inland fisheries and identifying what is needed to achieve sustainable inland fisheries. In the present paper, we adopt a regional case study approach on inland fisheries interactions with a focus on areas where two or three fisheries

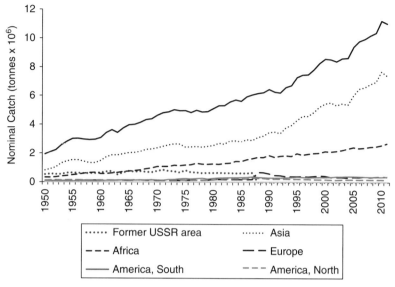

Figure 15.2 Summary of UN FAO statistics for inland fisheries relative to different regions.

sectors co-occur. We preface the case studies with a brief overview of the three fishing sectors acting in inland waters and their characteristics. We then explore the similarities and differences emerging from the case studies to identify general principles related to threats and conservation. We also consider what is needed to chart a course towards a future of sustainable inland fisheries.

15.2 CHARACTERISTICS OF INLAND FISHERIES

Here, we briefly summarise the characteristics (in terms of gear and magnitude) of the three fisheries sectors in inland waters and associated benefits to humans. The definitions for each sector are derived from the FAO Glossary.

15.2.1 Recreational fisheries
Defined as fishing for aquatic animals that do not constitute an individual's primary resource to meet nutritional needs and are not generally sold or otherwise traded on export, domestic, or black markets.

Recreational fisheries in inland waters are described elsewhere in detail (e.g. Arlinghaus *et al.*, 2002; Cowx, 2002; Arlinghaus & Cooke,

2009; FAO, 2012); therefore, we only provide a brief refresher here. Fishing with rod and reel (i.e. angling) is the most common recreational fishing technique (Arlinghaus *et al.*, 2007). However, in some countries, recreational fishers often use other gears such as spear, bow and arrow, rifle, traps, or gill nets (Arlinghaus & Cooke, 2009). Fish are taken home for personal consumption in some jurisdictions, but an increasingly large proportion of fish is released either because of the conservation ethic of the angler, culinary dislike, too many or few fish, or to be compliant with harvest regulations (Arlinghaus *et al.*, 2007). Recreational fisheries can target wild fish or focus on stocked fish (sometimes as put-grow-take fisheries), at times in artificial systems (e.g. ponds). Recreational fisheries constitute the dominant or sole user of many wild freshwater fish stocks in most industrialised countries (Arlinghaus *et al.*, 2002). Globally, approximately 10.6% of the population participate in recreational fishing, but there is insufficient information to partition participation by inland and marine waters (Arlinghaus & Cooke, 2009). Recreational fishing is booming in emerging economies (e.g. India, Brazil) and very popular across the industrialised world, with hundreds of millions of people engaged in the activity (FAO, 2012).

Spatiotemporal closures, bag limits, gear specifications, species/size restrictions and mandatory catch-and-release are various regulatory tools used by recreational fisheries managers (typically in developed countries) to reduce overexploitation and maintain stock structures. Stock enhancement (stocking and introductions), on the other hand, is commonly used to enhance stock of fish in a river or lake system, to create new fisheries, promote angling diversity, improve recruitment and/or maintain productive species (Cowx, 1994; Cooke & Cowx, 2006). Stocking activities can threaten native stocks through competition, predation, reduced genetic diversity, or spread of novel pathogens and parasites (Cowx, 1994). Alternatively, recreational fisheries managers may focus on habitat management to mitigate, rehabilitate, or restore the structure and function of aquatic ecosystems as well as the natural reproduction of recreationally valuable fish populations. In some regions, such as North America, most waters are public access with fisheries managed by natural resource agencies for the public good. In Europe there is a tendency towards private ownership of inland fisheries rights with the expectation that much of the management is the responsibility of the owners (e.g. angling associations; summarised in Arlinghaus *et al.*, 2002).

The socioeconomic benefits of recreational fisheries are numerous and substantial, particularly in inland systems (Arlinghaus *et al.*, 2002; Arlinghaus & Cooke, 2009; Parkkila *et al.*, 2010). Benefits can be viewed from the perspective of the angler and the broader angling community (e.g. tackle manufacturers, guides), as well as the whole of society (Arlinghaus & Cooke, 2009). The whole suite of individual benefits received by anglers is also subsumed within the concept of economic value (Parkkila *et al.*, 2010).

Recreational fishing in inland systems comprises a multibillion dollar global industry that supports economic activity and livelihoods for many (Arlinghaus *et al.*, 2002). There is also the above-mentioned economic value for the individual fisher, where the participation in recreational fishing creates utility over and above the expenditures needed to participate in fishing, often measured as willingness to pay for a fishing experience (Parkkila *et al.*, 2010). In some inland systems, expenditure by recreational fishers represents significant revenues and jobs. Unfortunately, global estimates of the economic value of inland recreational fisheries are lacking, but it is certainly in the hundreds of billions of USD.

15.2.2 Commercial fisheries

Defined as fisheries whose primary aim is to provide resources to meet human nutritional (i.e. essential) needs; in both full-time and part-time commercial fisheries, fishes and other aquatic organisms are sold on domestic and export markets; commercial fisheries include those that supply feed to the aquaculture and agriculture sectors and raw material to other industrial sectors (e.g. biomedical sector).

Commercial fisheries in open marine waters are commonly carried out by large fleet vessels operated by large crews of fishers. Inland water commercial fisheries are usually small-scale (Welcomme, 2001). Globally, there are approximately 1.13 million fishing vessels in inland waters (FAO, 2012). In some areas, fishing gears used in inland fisheries are highly diverse and adapted to exploit the diversity of habitats, species and seasonality. In rivers, fishing gears can be quite complex. For instance, over 114 gears have been described for the lower Mekong River (Deap *et al.*, 2003). In lakes, gears are typically less complex and can include: seines, gillnets, longlines, castnets, trapnets, electrofishing and small trawls (Welcomme, 2011a). Small-scale fishers operate using small craft such as canoes and rafts (with or without outboard motors); whereas industrial fishers require commercial permits and use efficient technologies that have high harvest capacities (De Jesus & Kohler, 2004).

The larger commercial fisheries (e.g. Laurentian Great Lakes, Lake Victoria, the Mekong River) are amongst the most regulated type of the three fishing sectors and contribute much of the data used to estimate global inland fisheries production. Various regulatory constraints help prevent overexploitation of the resources and include spatial and temporal season closures, harvest quotas (catch limits), licensing, gear restrictions and restriction on size and species of harvested fishes (Cooke & Cowx, 2006). Although by-catch does occur in inland systems (Raby et al., 2011), much of it is retained and used, in contrast to many marine systems.

The socioeconomic importance of inland commercial fisheries and their role in the livelihoods of rural households, particularly in developing countries, are significant and often underestimated (Welcomme et al., 2010). Inland commercial fisheries contribute significantly to the livelihoods of millions of people as a source of income for both families of full-time fishers, and for households that live close to water bodies and engage in part-time commercial (i.e. artisanal) fishing. Fishing is necessary to local populations who depend on this activity to access cash quickly and it provides labour for unskilled workers who rely on fishing for income and sustenance (Welcomme et al., 2010). They also contribute to food security, generate revenue with low capital investment, provide monetary resources in times of need, and dissipate risks resulting from crop failures (McKenney & Tola, 2002).

15.2.3 Subsistence fisheries

Defined as fishing for aquatic animals that contribute substantially to meeting an individual's nutritional needs; in pure subsistence fisheries, fishing products are not traded on formal domestic or export markets but are consumed personally or within a close network of family and friends; such fisheries sustain a basic level of livelihood and constitute a culturally significant food-producing and distributing activity.

The difference between subsistence fisheries and the two other sectors is somewhat obscure. In general, if fishing activity supports direct nutritional needs to the family that is difficult to be substituted by alternative food, it would be subsistence rather than recreational. A large number of subsistence fishers participate in inland fisheries, where their products are generally marketed and consumed locally at the point of capture. Subsistence fishers often engage in part-time fishing because they are also involved in other economic activities such as farming or construction when fish catches are low (Welcomme, 2011b), where households may switch from no fishing to subsistence fishing to commercial fishing relative to opportunities and seasonality. Subsistence fishers may harvest individual fish using gears such as rod and reel, spears, dip nets, harpoons or hand lines. They may also catch multiple fishes to provide food for their families and communities using gear similar to that used in small-scale artisanal commercial fisheries (e.g. seines, gillnets, cast nets). An overlap exists between subsistence fisheries and both the recreational and commercial (i.e. small-scale artisanal) fishing sectors. Although recreational fishers engage in fishing for leisure, they may also decide to keep their catch for consumption. Subsistence fishers will also keep their catch for consumption while also trading a portion of the catch at local markets for cash or other products and services. In part, subsistence fisheries are a component of small-scale commercial fisheries with a focus on household consumption rather than commercial activity (World Bank, 2012).

Subsistence fisheries are often informal and catches are rarely sampled or reported because the catch goes directly to domestic consumption, often in cryptic locations (Welcomme, 2001). Even frequently overlooked systems like small lakes, streams and rice paddies support subsistence fisheries (Brummett *et al.*, 2010). Household consumption surveys have been used to estimate fish that avoided passing through formal market chains or recording checkpoints, but because catches are two steps removed from the fishery, information on the origin of the fish consumed is often lost (Welcomme, 2011a). For example, findings from studies of fish consumption in the Mekong Delta (Vietnam) suggested that inland capture fisheries production is five times greater than official reported statistics (World Bank, 2012). Some countries, such as Peru, have expanded their commercial harvest data by 30% to correct for the absence of subsistence reporting (De Jesus & Kohler, 2004). Lymer and Funge-Smith (2009) reassessed Asian inland catch statistics by accounting for countries with large annual increments in fish production. In general, subsistence fisheries are obscure and complex. Understanding and quantifying harvests from these fisheries can be challenging, but critical to assessing global catch estimates and the impacts on livelihoods of river inhabitants.

15.3 CASE STUDIES

What follows are seven case studies from multisector inland fisheries around the globe to provide insight on the interaction among different types of fisheries and its relationship to fish biodiversity conservation. The case studies are focused on larger systems, but the primary themes are likely relevant to smaller systems.

15.3.1 Laurentian Great Lakes fishery

The Laurentian Great Lakes of North America hold 21% of the world's supply of accessible fresh water. The Great Lakes basin straddles Canada and the USA, and is home to more than 30 million people. The Great Lakes serve important roles in terms of transportation, irrigation, recreation and food production. They are ever-changing with respect to the fish assemblages as a result of many intentional (e.g. Pacific salmon) and unintentional (e.g. sea lamprey, round goby) introductions of fish and other organisms (e.g. dreissenid mussels). Recreational and commercial fisheries co-occur on all of the Laurentian Great Lakes with aboriginal subsistence fisheries occurring in discrete locations and often during discrete periods. The economic impact of these fisheries is large, valued at US$7 billion in US waters (American Sportfishing Association, 2008). Commercial fisheries are the dominant form of extraction in Canadian waters, whereas recreational fisheries predominate in US waters. There are, however, viable commercial fisheries in the US, especially for yellow perch (*Perca flavescens*) in lakes Michigan and Erie, and by US tribal governments for lake whitefish (*Coregonus clupeaformis*). An element of the very large recreational fishery on the US side is the charter fisheries for salmonides, walleye (*Sander vitreus*) and yellow perch. In many cases, the recreational and commercial fisheries coexist without extensive conflict because they target different species, the harvest allocation is primarily for either the recreational or the commercial sector, they are separated in space and (or) time, or the harvest rate does not put undue pressure on fish populations. Aboriginal subsistence fisheries, at times, conflict with other sectors. However, this is typically only on a very site-specific basis, usually related to allocation. In the USA, such fisheries are governed by tribal treaty rights, and in Canada, aboriginal subsistence fisheries are protected under the Canadian Constitution. In general, few members of the aboriginal community are fully engaged in subsistence harvesting. Nonetheless, the aboriginal harvest is shared with many throughout the community and has high cultural value.

There is substantial potential for conflict between the recreational and commercial sectors, particularly for walleye and yellow perch, which are prized by commercial and recreational fishers alike. Management of the Lake Erie fishery offers insight into how these fisheries and their potential conflicts are managed in a cooperative, bi-national manner. Cooperative fishery management in the Great Lakes occurs in a non-binding, consensus-based approach as outlined in *A Joint Strategic Plan for Management of Great Lakes Fisheries* (Gaden *et al.*, 2009). As such, all of the jurisdictions with fishery management authority in Lake Erie (Ontario, New York, Pennsylvania, Ohio and Michigan – the Lake Erie Committee) meet, at least annually, in a public forum to decide on management goals, objectives and actions, including the total allowable catch (TAC) for walleye in the quota management zone of western and central Lake Erie, and for yellow perch lakewide. The jurisdictions have agreed to share the TAC based on the amount of habitat for each species in each jurisdiction. For example, the TAC allocated by the Lake Erie Committee (LEC) for walleye in the quota zone for 2012 was 3.49 million fish (LEC, 2012). Within this allocation, Michigan was allocated about 200,000 fish, Ohio about 1.78 million fish, and Ontario around 1.50 million fish. Ontario allocates nearly its entire quota to the commercial sector, whereas Michigan and Ohio allocate all of their harvest to the recreational sector (including charter fisheries). While this practice works well in theory, conflict can arise because the recreational fishery frequently does not harvest its entire assigned quota, leaving commercial fishers questioning why they should not be allowed more of the lake-wide quota. Similarly, recreational anglers who value the opportunity to catch older, larger walleye view the possibility of commercial over-harvest as one that will jeopardise their ability to catch trophy fish.

Although the LEC has generally worked well internally to resolve potential disagreements about quota levels, they have occasionally needed to use dispute resolution procedures (Gaden, 2007). More problematic in recent times has been the distrust between the recreational and commercial sectors about each group's motives and, by extension, the motives of the management agencies associated with each group of fishers. Both managers and resource users recognised the need to directly work together in an open, transparent forum to come to agreement on the future of these fisheries – a process that is currently under way.

Collectively, the recreational, commercial and subsistence fisheries of the Laurentian Great Lakes are reasonably sustainable, given the expected caveats associated with ongoing ecosystem impacts of invasive species, possible future ecosystem shifts due to climate change and changing land-use patterns. Cooperative fishery management across jurisdictional boundaries, stakeholder involvement and strong science help to make this status possible. Nevertheless, the status of exploitable fish populations can change quickly, reinforcing the need for strong science to identify potential population-level changes, managers willing to take appropriate action, and transparent communication among all parties to ensure the future sustainability of these important fisheries.

15.3.2 Mississippi River fishery

The Mississippi River (MSR) is among the world's most important river systems with regard to its size, physico-chemical and biological diversity, and socioeconomics. Its watershed covers about 40% of the surface area of the lower 48 US states, and includes all or portions of 31 US states and 2 Canadian provinces. The MSR and its larger tributaries (Missouri, Ohio and Illinois rivers) played an important role in Native American cultures, and since European settlement, the system has been central to the social and economic development of the USA. While much has changed within the MSR and its watershed, one relative constant has been the importance of fishery resources.

The MSR originates at Lake Itasca, Minnesota, and flows ~820 km to St Anthony Falls, Minnesota. This reach is commonly referred to as the headwater (HW) reach. A number of small dams have been constructed for flood control, power generation, municipal and industrial water supply and recreation. The reach (~1100 km long) from St Anthony Falls, to Alton, Illinois (just upstream of the Missouri River confluence) is the upper Mississippi River (UMR). The UMR has been impounded by 29 locks and dams to provide a minimum channel depth (2.7 m) for navigation, but the reach maintains a high level of habitat diversity relative to downstream reaches. From Alton, the MSR flows unimpounded for ~1800 km to Head of Passes, where it branches into smaller systems that terminate at the Gulf of Mexico. The middle Mississippi River (MMR) is typically defined as the reach from the mouth of the Missouri River to the mouth of the Ohio River, and the lower Mississippi River (LMR) is from the Ohio River to the Head of Passes. A continuous levee system was constructed in the 1930s that disconnected the river from its floodplain. Although discharge and the size of the channel differ between the MMR and LMR, they share similar hydrology, levels of water development and fish assemblages.

The MSR has long provided a subsistence fishery for inhabitants near the MSR. Based on accounts of early explorers and settlers, the MSR provided important resources for Native Americans, including aquatic vertebrates (e.g. fishes, turtles, mammals) and invertebrates (e.g. unionid mussels; Brown *et al.*, 2005). Like most subsistence fisheries, use of fishing the MSR was dependent on the spatial and temporal availability of resources. While a diversity of fishery resources were undoubtedly used throughout the year, accounts indicate that catostomids, acipenserids, paddlefish, esocids and ictalurids were particularly important in subsistence fisheries (Carlander, 1954). Current use of the fishery resources for subsistence is small compared with commercial and recreational fisheries.

The MSR has a long history of supporting commercial fisheries. The HW reach historically supported small, local commercial fisheries, but this is no longer the case. Rather, most commercial fisheries are in the UMR, MMR and LMR. The best records of commercial harvest are from the UMR, where the Upper Mississippi River Conservation Committee (UMRCC; composed of state and federal natural resource managers from Illinois, Iowa, Minnesota, Missouri and Wisconsin) has compiled commercial harvest data for the UMR since the late 1940s. Total landings of fish in the UMR have varied considerably among pools and years. For example, total landings in the UMR were 6.2 million kilograms in 1894 (Carlander, 1954) and 3.1 million kilograms in 2006 (MCQ, unpublished information). Commercial landings also vary spatially, where in 2006, total harvest varied from ~6500 kg in Pool 3 to ~417,000 kg in Pool 13. Similar data are unavailable for recent harvest in the MMR and LMR; however, for those years with comparable data (1954–1976), total landings were within the range of harvest in the UMR (Risotto & Turner, 1985).

Despite substantial spatiotemporal variation in harvest, the dominant species in the commercial catch of the MSR have remained fairly consistent. Specifically, common carp, buffalo species, freshwater drum and ictalurids (primarily channel catfish) have dominated the commercial fishery. In the UMR, these taxa comprised 79–93% of the total catch between 1894 and 1922, and 84–94% between 2000 and 2006. Nevertheless, the proportion of ictalurids has declined during the last 30 years, largely due to consumer concerns about contaminants and the development of aquaculture (Krogman *et al.*, 2011). Commercial harvest of fish during the last 120 years has largely focused on the flesh market; however, harvest of roe in the basin has increased substantially during the past 8–10 years. In particular, harvest of shovelnose sturgeon, paddlefish and bowfin has increased 10-fold in some portions of the system and spurred research focused on the effects of harvest on population dynamics and the sustainability of fisheries (e.g. Columbo *et al.*, 2007; Koch *et al.*, 2009).

Similar to subsistence and commercial fisheries, the MSR has long provided a diverse and productive recreational fishery. Records of recreational angling in the MSR date to as early as the 1840s (Carlander, 1954). Currently, the system supports a number of recreational fisheries, but most are in the upper reaches. Recreational fisheries in the HW reach are intensively monitored by the Minnesota Department of Natural Resources, and most statistics (e.g. angler effort, catch rate, mean length) indicate stable fisheries with low overall effort (Sledge, 1998). The most popular sport fishes in the HW reach include northern pike, smallmouth bass, walleye, channel catfish and panfishes. Popular sport fishes in the UMR include those popular in the HW with the addition of moronids, freshwater drum, sauger and flathead catfish. The recreational fishery of the MMR and LMR has not been extensively monitored. Recreational fisheries in the lower river are less diverse than in upstream reaches due to the general lack of coolwater species that are abundant in the HW and UMR (e.g. smallmouth bass, northern pike, walleye). Important recreational fisheries in the MMR and LMR are dominated by ictalurids and largemouth bass.

Unlike other systems in the mid-western USA where subsistence, commercial and recreational fisheries are often in conflict (e.g. Beard *et al.*, 2003), competition among fisheries in the MSR is low. Catfish fisheries represent an important exception. For example, in Pools 9–19 of the UMR, total commercial harvest of catfishes (primarily channel catfish) showed a significant decline of ~8600 kg year^{-1} between 1955 and 1985 (Pitlo, 1997). The decline was thought to be the result of recruitment overfishing by commercial harvesters. Following an increase in the minimum length of fish that could be harvested (i.e. implemented in 1985), catch rates of age-0 channel catfish and total yield in the commercial catch increased significantly. Increased abundance and larger size structure of channel catfish in the system had an influence on the recreational fishery; channel catfish went from the sixth most abundant species in the creel in 1963 to second in 1994. While some direct and indirect competition undoubtedly occurs between different fisheries in the MSR, the greatest threats to the fisheries are changes in habitat conditions and the introduction of invasive species. The recent invasion of bighead carp and silver carp, particularly to the MMR, LMR and large tributaries, is a serious threat to the entire ecosystem. Unfortunately, the effects of Asian carp are only beginning to emerge, but deleterious effects on native fish populations and fisheries have already been dramatic in some areas (Sampson *et al.*, 2009). Nonetheless, Asian carp also provide potential opportunities for commercial harvest if markets can be established. An Asian carp processing plant opened in 2014 southern Illinois with the intention of exporting 'clean' Asian carp from the Mississippi River to Asia where such fish are desirable, but where local Asian carp populations have high contaminant burdens. However, the plant is temporarily shut down due to the emission of 'unpleasant odours'. Efforts to develop local markets for Asian carp in North America include carp cook-offs and working with chefs to increase the appeal for carp flesh.

The MSR system has long provided important fishery resources. While the various fisheries have not been in a state of conflict, future stressors to the system suggest that minimising habitat degradation and the effects of invasive species will be critical for continued coexistence of subsistence, commercial and recreational fisheries in the MSR.

15.3.3 The Colombian Amazon fishery

The Colombian Amazon has an area of ca. 480,000 km² corresponding to 42% of the National Territory and over 6% of the entire Amazon Basin (UNEP & ACTO, 2009). It is the largest border of the country with a population of about 900,000 people, of which 9% belong to one of 56 ethnic groups (Gutierrez *et al.*, 2004).

Tropical freshwater ecosystems are extremely dynamic and maintain a diverse structural complexity. Large river-level fluctuations affect environmental conditions on a more or less annual cycle, causing habitats and species to change on a seasonal basis (Junk, 1997). These changes in water levels and the transitions from an aquatic to a terrestrial phase not only create and destroy a series of heterogeneous aquatic habitats, but are the key to maintaining important levels of aquatic biodiversity expressed in complex food webs. So far, 753 fish species have been identified in the region (Agudelo *et al.*, 2011), positioning it as the richest region in the country.

Dependence on fishing is substantial and, arguably, the most important activity supporting local and regional livelihoods and economies along the Amazon River. Fishing supports the livelihoods of around 1000 Colombian families along the Amazon River (Amazonian Trapezium located in the south of the Colombian Amazon region) and a minority of these families supply the national fish market with 8600 tonnes per year, representing 37% of total freshwater fish production in Colombia. According to Riaño-Umbarila (2003), fishing is part of the local people's daily routines. Fishing gives people the chance to bring food to their families. The remainder is sold and the money is used to buy basic goods and supplies.

Four types of fisheries take place in the region (Portocarrero-Aya, 2012). Catch-and-release recreational angling takes place mainly during August and September when water levels are low. The complexity of habitats provides a variety of desirable species for anglers from all over Colombia and the world. Recreational fisheries provide only a small part of the income for tourist agencies and guides, primarily in and around Leticia and Bogota, although the area has high potential for supporting recreational fisheries (Agreda-Rudenko, 2008). There are no clear regulations related to the catch-and-release of fishes and the country requires implementation of a code of practice to regulate any expansion of such fisheries.

Subsistence fisheries are probably the most visible service provided by freshwater ecosystems in the region. Heterogeneity in aquatic habitats contributes to high production of fishes, which is reflected in the contribution of fishes as a major source of protein to local communities in areas south of the Colombian Trapezium (88 species; 100–500 g day^{-1}; Fabre & Alonso, 1998; Agudelo *et al.*, 2011). Dependency on the river and its fishery resources makes the human inhabitants vulnerable to changes in the fish species composition and to changes to freshwater habitats. According to Ochoa *et al.* (2006), institutional data on the intensity of subsistence fishing is absent and, therefore, there are no statistical data showing local aspects such as the amount of fish extracted and traded, or the amount of fishing effort imposed on the resource. However, something is known about the species preferred by locals; selection depends on availability, taste and cultural preferences. Species like palometa (*Mylossoma duriventre*), bocachico (*Prochilodus* spp.), yaraqui (*Semaprochilodus* spp.), sabalo (*Brycons* spp.), piranha (*Serrasalmus* spp.), paco (*Piaractus brachypomus*) and gamitana (*Colossoma macropomum*) constitute the main diet of the local residents (Agudelo, 2007). In the region, all fisheries are artisanal, and subsistence fishing predominates over commercial fishing (Ochoa *et al.*, 2006).

Commercial fishing has supported the local and national economy since 1938, when the arrival of 'cold rooms' (to store fish fresh) and airlines were available to accommodate the supply of Amazonian fish to the large cities of Colombia (Salinas, 1994; Agudelo *et al.*, 2011). Around 95% of the fish traded in Leticia (the Amazonian capital) is shipped, either frozen or dried, by plane to Bogota, where 67% is consumed; the remainder supplies other fish markets in Colombia (Ochoa *et al.*, 2006). According to Ochoa *et al.* (2006), the demand for Amazonian fish has increased 155% from 1977 to 2002. Management of commercial fishing is complex because the fisheries include multiple species harvested using a variety of gears that vary in space and time (as a response to changes in the river level), and include diverse social, cultural and economic dynamics. Either way, an ecosystem approach is important for the management of migratory species, especially those representative of the regional fisheries because threats to the fisheries are limited to the effects of fishing. For instance, logging in the headwaters and the floodplain alters habitats that serve as refuge areas, nursery habitats and foraging areas for predators. Mining, an increasing activity in the Putumayo and Caquetá rivers, affects and pollutes the streams used as migratory and spawning corridors for migratory fishes.

Ornamental fisheries are those where fishes are harvested for the ornamental aquaculture trade. Although they may be considered as a subset of commercial fisheries, they constitute an important income to local fishers and traders. In 2009, approximately 2.2 million individual fishes representing 41 species were sold (Ajiaco-Martinez *et al.*, 2012). The city of Leticia is the most important port of collection and export of ornamental fishes, contributing 86% of the total production of the Colombian Amazon (Portocarrero-Aya, 2012). However, a comprehensive understanding of ornamental fisheries remains poor (Agudelo *et al.*, 2011).

Currently, Colombia does not have either a defined policy on fisheries and aquaculture development or strategic guidelines to manage artisanal fisheries. For the Amazon region, only a few regulations banning fishing at certain times or fishing gears, minimum capture sizes, and restricted fishing areas have been expedited. Legal frameworks do not match the social, economic and environmental dynamics of this complex region, and the low budgets assigned to implement management strategies means little can be done to manage the fisheries effectively. Improvements in the institutional structure and increased financial support of the fishing authorities in the region are vital for the survival of the activity and to ensure that inland fisheries are sustainable (Portocarrero-Aya, 2012).

15.3.4 Lower Mekong River fisheries

The Mekong is the longest river in Southeast Asia. It runs for approximately 4800 km from its source on the Tibetan plateau through China, Myanmar, Lao PDR, Thailand, Cambodia and Vietnam to the South China Sea, where it discharges on average 475,000 million m³ per year. The total Mekong Basin catchment covers an area of 795,000 km² and has about 73 million inhabitants. The Lower Mekong Basin (LMB) comprises the four countries – Cambodia, Lao PDR, Thailand and Viet

Nam – and covers 77% of the total basin, with some 55 million people. The Mekong River Basin hosts one of the most diverse freshwater faunas in the world. There are 1200 recorded fish species.

The capture fisheries of the LMB have a centuries-long history of catch technology, local resource knowledge, fish processing, marketing and social organisation (Sverdrup-Jensen, 2002). Much of the fishing technologies are indigenous to the Basin or have been extensively adapted to local conditions, and are manufactured with extensive use of local materials. Most fisheries activities are based on an intimate knowledge of fishes' responses to seasonal environmental changes.

The Mekong fisheries range from individual seasonal fishers in the highlands of Lao PDR to fishing lot owners with large-scale commercial operations in the Great Lake (Tonle Sap) of Cambodia or certain areas of the mainstem rivers; and from full-time specialised traditional fishers in the Khone Falls area to unemployed people in villages in Northeast Thailand who fish to support themselves in times of economic difficulties. The highest degree of commercial specialisation is found in the large floodplains of central and southern Cambodia and the northern part of the Mekong Delta. Large-scale fishing operations are often based on exclusive access to sites obtained through purchase of government licences, although this system is being disbanded in Cambodia and could affect the intensity of fishing and result in social disruption. Throughout the LMB at least 40 million rural dwellers are estimated to be active in the fishery.

All Mekong fisheries are highly seasonal and largely determined by the river flood-pulse from June to October (Poulson *et al.*, 2002). Catches comprise whitefish species during their migration when the water level rises following the onset of the monsoon rains. In the dry season, catches are mainly taken in or around deeper stretches of rivers where many sedentary and migratory fish species take shelter. Seasonal peaks result in a glut of fresh fish far beyond what local demand can absorb. This has stimulated the development of highly effective, low-technology fish processing and marketing systems based on the production of fermented fish products, pastes and sauces. This large, mainly domestic, industry provides seasonal employment and spreads the nutrition and economic benefits of seasonal fish production over the full year.

In addition, there are numerous small-scale fisheries operating in the LMB (Sverdrup-Jensen, 2002). In all four countries of the LMB, monofilament gillnets with their different ways of operation (floating, set, bottom, surface or mid-water) are the most popular gear. A great diversity of basket traps made of widely available natural raw materials, such as bamboo, rattan and vines, adapted to the needs of fishing–farming households, are also used throughout the LMB. In the uplands of the Lao PDR, northern Thailand, northeast Cambodia and the central highlands in Vietnam, mainly small-scale fishing gears, including fishing by hand, are used to retrieve aquatic animals from the wetlands, including rice fields.

Estimates of the total catch from the fisheries in the LMB have increased dramatically in recent years and exceed 2.6 million tonnes annually with a value exceeding US\$1.7 billion. These figures are based on per-capita consumption of all freshwater fish and other aquatic animal products, but exclude the fish produced in aquaculture and in reservoirs (respectively, 260,000 and 240,000 tonnes; Coates, 2002).

The Mekong is threatened by a range of human activities, including agricultural development (especially for rice), deforestation, urban development flood control, alien invasive species and dam development. Some 25,000 reservoirs have been constructed in the LMB, mostly for irrigation, but in recent years there has been a resurgence of interest to promote hydropower. In particular, 11 large dams are proposed on the mainstem. Existing reservoirs have interfered with the migration of fishes to the point where some species have disappeared and it is estimated that between 1.2 and 1.5 million tonnes of fish yield will be lost if the mainstem dams are constructed (Dugan *et al.*, 2010).

15.3.5 Lower Fraser River Pacific salmon fisheries

The Fraser River is one of Canada's largest rivers (drainage area = 217,000 km^2). Starting in the Rocky Mountains, the Fraser River flows for 1370 km before it discharges into the Strait of Georgia near the city of Vancouver, British Columbia (Thompson, 1981). For at least ten millennia the Fraser River Basin has been home to indigenous groups (in BC called First Nations) who depend on the river for subsistence, as well as cultural and spiritual activities. Today, the lower reaches of the Fraser River and its tributaries are highly altered with evidence of fish population declines (e.g. McDaniels *et al.*, 2010; Martins *et al.*, 2011). The Fraser River watershed is home to more than 100 genetically and geographically distinct populations of Pacific salmon, and supports multisector, multispecies fisheries that have great economic, cultural, social and political significance.

Three fishing sectors – commercial, recreational and subsistence (locally called First Nations (FN)) – co-occur, and target adult Pacific salmon during their upriver return migration from the ocean to spawning grounds. Five Pacific salmon species are present in the Fraser River: pink (*Oncorhynchus gorbuscha*), sockeye (*O. nerka*), chinook (*O. tshawytscha*), chum (*O. keta*) and coho salmon (*O. kisutch*) (Williams, 2007). The three fishing sectors are managed by the Department of Fisheries and Oceans Canada (DFO) with the exception of the marine commercial fishery, which has joint management with the Fraser River Panel and the USA. Most inland fisheries are concentrated in the lower Fraser River, with the upper reaches accessible only to FN communities. As such, we focus our case study on the lower Fraser River and its multisector sockeye salmon fishery.

The FN fisheries generally operate on a small scale, using primarily beach seines and gill nets. The fishing sector involves over 72 FN bands who have the legal right to fish for food, social and ceremonial (FSC) purposes with priority to harvest allocations and access over the recreational and commercial fisheries, subject only to conservation needs (English *et al.*, 2011; Cohen, 2012). The FN fishing sector has been considered a subsistence fishery; however, since 2004, 'economic opportunity' fisheries have been negotiated with lower Fraser River FN allowing the fishers to operate commercially under negotiated allocations and catch monitoring procedures. In 2010–11, a total of 379 communal aboriginal commercial licences were issued for salmon. The average harvest rate (2001–09) for the FN fishing sector resulted in just over 700,000 Fraser sockeye harvested annually (35% of total harvest across all fishing sectors), which included catches from the FSC (29%) and the FN commercial fisheries (6%) (reviewed by English *et al.*, 2011).

The Fraser River supports a substantial recreational fishery that directly overlaps and coexists with the FN fishing sector. Not surprisingly, this has resulted in some conflict and animosity. In 1991, a directed recreational fishery on sockeye salmon in the lower Fraser River opened (Roscoe & Pollen, 2010) with fewer than 10,000 sockeye captured and harvested. By 2002, the fishery grew rapidly and catches were estimated to be more than 100,000. From 2001 to 2009, the average annual catch of Fraser River sockeye salmon by recreational anglers accounted for 3% of the total sockeye salmon harvested (average of ~60,000 sockeye salmon; reviewed in English *et al.*, 2011).

The commercial fishery is a competitive, limited-entry fishery using gillnets, seines and troll gear. The fisheries are managed using time and area closures, as well as individual harvest quotas. Catches are monitored using 'sales slip', log books, observers on deck and in-season reporting systems. Historically, the Fraser River sockeye salmon commercial fishing sector operated only in marine waters and in the tidal portion of the Fraser River; however, with the introduction of FN economic opportunity fisheries, commercial fishing now also occurs in-river.

Conservation is the first priority for fisheries management of the lower Fraser River salmon fisheries. The fisheries are managed with the objective of reaching escapement targets and allowing a proportion of the run to be harvested. In 1998, DFO introduced a selective fishing policy applied to all fishing sectors that aimed to minimise or avoid the harvest of species and stocks of conservation concern (i.e. by-catch) as part of developing responsible fisheries. As such, all fishing sectors are subjected to time and area closures, mandatory live release of non-target species, and gear and technique modifications (e.g. minimum mesh size and shortened net sets for gillnetters, barbless hooks for recreational fishing, brailing for seiners). Some incentives have been initiated for commercial fishers by granting priority fish allocations to vessels instigating selective fishing over traditional commercial fisheries.

Issues and conflict among the three fishing sectors generally surround the 'intersectoral allocations'. Harvest allocation for salmon is the priority for First Nations FSC purposes. Once conservation and the FN FSC needs are met, then allocations are granted to recreational and commercial sectors, but are not guaranteed. As a result, feelings of inequality and unfairness can arise from those sectors with lower priority. The recreational fishing sector is granted a maximum average of 5% of the combined recreational and commercial sockeye salmon harvest, resulting in tension between the two sectors as the recreational fishing sector continues to grow (Cohen, 2012). Furthermore, commercial FN fish allocations are negotiated on the premise of balancing the objective of supporting the economic development of FN communities and the interests of all fishery participants. Evidently, the disparities of harvest allocations can create conflict and tension amongst the different fishing sectors.

The DFO is moving towards a 'shared stewardship' management strategy to promote collaboration, participatory decision-making, and shared accountability and responsibility among the fishing sectors and management. In attempts to facilitate communication among all fishing sectors, the DFO produced an Integrated Fisheries Management Plan (IFMP) to provide a planning framework for the conservation and sustainable use of fisheries resources. It incorporates the results of consultations and input from the FN, recreational and commercial advisors as well as other interest groups in the hope of coordinating fishing plans and resolving potential issues amongst the sectors (Cohen, 2012). Continual effort is being made to engage First Nations and stakeholders to participate in co-management.

Despite the complexity of the salmon fisheries in the Fraser River, habitat alterations, changes to hydrological regimes and warming river temperatures play major roles in the declines of salmon populations (e.g. Martins *et al.*, 2011). Fisheries interact with these environmental changes by causing additional physiological stress on live-released non-target fishes, which can result in unobserved mortality or potential in behavioural impairment such as spawning (e.g. Baker & Schindler, 2009; Gale *et al.*, 2011). The issues and conflicts identified here extend downstream into the ocean, resulting in a strong need for management to not only consider the multisector, multispecies of the Fraser River salmon fishery, but also connecting the multienvironments and stressors affecting salmon and the fisheries.

15.3.6 Eel fisheries of Europe

The panmictic (Dannewitz *et al.*, 2005) European eel population (*Anguilla anguilla*) is under threat (International Council for the Exploration of the Sea, 2010). Current recruitment levels are less than 1% of the maximum historic record (Dekker, 2008, 2009; International Council for the Exploration of the Sea, 2010). The eel decline is negatively affecting many small-scale European inland fisheries (Feunteun, 2002; Dekker, 2008; Dorow *et al.*, 2010), because both commercial and recreational fisheries intensively harvest and compete for the resource. However, the eel is not an important fisheries resource in all European countries due to different customs in consuming the species. In addition, different life stages are sought and harvested in different countries.

Managing eel is technically difficult due to the complex life cycle of the species and the multiple nations and fisheries that exploit it throughout Europe. The species transcends freshwater and marine environments and is thus affected by multiple continental and oceanic factors, none of which are under full control of one country or management agency. Moreover, because important sources of eel mortality in the marine environment appear to include the effect of changing nutrient conditions in the spawning grounds and climate change (Knights, 2003; Friedland *et al.*, 2007), there are no local, regional, or national solutions to the problem of the rapidly diminishing eel stock. International collaboration is needed to halt the eel decline, which is particularly hard to organise politically.

Impacts on eel vary across marine and freshwater environments. As mentioned, changing nutrient conditions and altered oceanographic currents likely impact recruitment in the marine environment. In freshwater, the eel is affected by exploitation of the different life stages by commercial and recreational fishing, pollution, predation by piscivorous birds, parasites (especially the introduced nematode *Anguillicola crassus*), and habitat alterations due to hydropower activities and dams (Feunteun, 2002; Starkie, 2003; Dekker, 2009). Uncertainty about the causes of the eel decline pose a significant challenge for identifying effective interventions to conserve this species and hamper stakeholder buy-in and commitment to conservation (Dorow & Arlinghaus, 2012).

A range of political and management actions have been initiated to conserve the eel population. The species has, for example, been red-listed as critically endangered by the International Union for the Conservation of Nature (Freyhof & Kottelat, 2008). In 2007, the European Union (EU) released the so-called eel regulation (EU Council Regulation 1100/2007), requiring member states to develop eel management plans at a river basin scale by the end of 2008 to safeguard a 40% escapement rate (European Commission, 2007). Since then, States whose management plans are not approved or fail to meet targets face immediate reductions in total eel fishing effort by at least 50% or implementation of other measures to reduce eel harvests by 50% (European Commission, 2007). While from a biological perspective the effectiveness of effort control measures is as uncertain as our understanding of the causes of decline, a 50% reduction in fishing mortality would have significant socioeconomic impacts on recreational and commercial eel fisheries across much of Europe (Dekker, 2008; Dorow *et al.*, 2010). Moreover, because the contribution of eel mortality by fishing to the current degraded state is uncertain, some states are reluctant to implement management actions that affect fisheries negatively, instead focusing on alternative actions such as elevated stocking of glass eels into freshwater ecosystems.

Commercial and recreational fisheries can exert substantial fishing mortality on eel (Dekker, 2008, 2009). Fishing-induced mortality may be reduced by directly restricting harvest rates of captured fish (e.g. by implementing a maximum size limit for eel to allow migration to the ocean) or by reducing fishing effort, or indirectly as a response to altered harvest regulations (Beardmore *et al.*, 2011). Direct regulation of effort includes diverse regulations as permit lotteries or spatiotemporal closures. Certain regulatory policies combine these mechanisms to compound their intended conservation benefits. Any drastic regulation of eel fishing, both commercial and recreational, will be accompanied by outright social conflict and result in welfare consequences of many million € per year across Europe (Dorow *et al.*, 2010). Despite the many possible ways by which eel fishing could be regulated, it is currently unclear how management of eel fishing using traditional harvest regulations would help conserve the eel stock. This is because multiple factors other than fishing influence eel populations. Hence, the benefits of increased escapement due to more restrictive eel harvest on the long-term status of eel stocks remain questionable. Overall, the ability of eel management plans using freshwater conservation strategies to meet EU targets for eel fisheries remains largely unknown, *inter alia* because the bottleneck for the freshwater eel population may lie in the marine environment.

15.3.7 Tigerfish in the Zambezi River of Central Africa

The complex riverine, floodplain and lacustrine environments of the Zambezi River system support multigear subsistence and commercial fisheries. Recreational angling is of lesser importance in most water bodies, where food production is of higher importance. In the Zambezi River channels and man-made lakes, however, one of the world's premier freshwater angling species, the tigerfish, *Hydrocynus vittatus* (world record = 16.1 kg; IGFA, 2012), is a major attraction. A variety of large, uniquely endemic cichlid species add to the attraction of angling in these areas.

The upper Zambezi comprises the area upstream of the Victoria Falls and the river is bordered by four countries: Zambia, Zimbabwe, Botswana and Namibia. Over a 120-km extensive floodplain area, known in Namibia as the Caprivi Floodplains, the river forms the Namibia–Zambia border. Recreational catches have only been assessed in the Namibian sector, where catch records of five local angling lodges in Caprivi show that in 2010 some 4000 visitors caught and released in the region of 38,000 tigerfish and 14,000 cichlids. These riverside lodges are dependent on angling tourists for up to 70% of their revenue and in remote, rural floodplain areas, lodges are often the only source of paid employment and are therefore of major local economic importance (Sweeney *et al.*, 2010).

Historically, recreational anglers and the artisanal fisheries were spatially separated to some extent. Recreational angling focused mainly on the main river channels while artisanal and subsistence fishers set their gear, mainly gillnets and fish traps, in backwaters and on the floodplain. In the last two decades, however, improved communications, together with rapidly increasing human populations, has led to the commercialisation of the previously subsistence fisheries. A huge demand for fish in major Zambian urban centres and also in the neighbouring Democratic Republic of Congo put Zambian fisheries under extreme pressure. Migrant fishers, with no interest in long-term sustainability, encroached into Namibian waters where they began to compete with local fishers, creating severe problems for local communities for whom food security from fishing is a vital livelihoods component. An estimated total annual fish yield from the Caprivi Floodplains (excluding the highly productive but ephemeral Lake Liambezi) of approximately 5000 t in 2011 (Tweddle & Hay, 2011a) is indicative of both local dependence on the fishes and the scale of commercial exploitation. The larger, more valuable species were severely depleted through excessive fishing effort and the use of destructive fishing gears such as shore seines and drifting gillnets. The situation deteriorated to the point where larger species were depleted to such an extent that fishers reduced their mesh sizes to target smaller species as well as the juveniles of the larger species.

The increasing effort created two problems for the angling tourism industry. (1) Trophy tigerfish (> 5 kg) and cichlids (> 2.5 kg) were becoming scarcer, and (2) the visual impact of commercial fishing using gillnets was leading to perceptions by tourists that the upper Zambezi River was being overfished. This was beginning to have significant impacts on the tourist industry, with some lodges losing some of their regular customers. The subsistence fishing communities in Caprivi also suffered from the declining fish stocks. The twin assaults on food security for local communities and the tourism angling resource stimulated the development of co-management arrangements between angling lodges, recreational angling club and the local fishing community.

The concept of conservancies has taken root in Namibia (NACSO, 2010, 2011), with much of the Caprivi Floodplain now under recently established (2005–2009) or planned conservancies. These community organisations are supported by government legislation and are successfully assuming responsibility for the management of wildlife and other natural resources. The concept is now being expanded to protect the fish stocks from excessive exploitation. Supported by a Ministry of Fisheries and Marine Resources (MFMR) project, conservancies established fisheries committees to manage the resources for the communities' benefit. The idea of Fish Protection Areas (FPAs), analogous to Marine Protected Areas (MPAs), was adopted. While the major aim of such areas is to act as protected breeding areas for the most important species (food fishers and anglers target the same large cichlids and tigerfish), the conservancies saw extra potential benefits in the form of earning revenue from anglers by allowing the use of non-consumptive catch-and-release angling on payment of fees.

Two pilot FPAs have been established and are functional (Tweddle & Hay, 2011b). In the first, an 11-km long side channel in Sikunga Conservancy, strong support has been forthcoming from lodges and the Nwanyi Angling Club in Katima Mulilo, organiser of the Zambezi Classic Tournament. The second pilot FPA in Impalila Conservancy is the 13-km long Kasaya Channel that links the Zambezi and Chobe Rivers creating Impalila Island. Lodges in the rapidly growing town of Kasane in Botswana on the edge of Chobe National Park and also on the opposite Namibian bank offer angling as a major tourist activity, increasingly based on flyfishing for tigerfish. Conflicts occur between tourist interests and highly commercialised illegal netting in areas adjacent to the FPA. It is in the long-term interests of the local economy that these issues are resolved through negotiated agreements between conservancies and the tourism industry to expand the FPA to accommodate the prime tigerfish angling zone, while ensuring that fees from tourist angling compensate the conservancies for the loss of fishing areas.

In the region, the conflict between commercialisation of the food fishery and tourism is not restricted to the Upper Zambezi. The neighbouring Okavango River system is a major tourism destination and thus foreign exchange earner for Namibia and Botswana, particularly in the world-renowned Okavango Delta, established (in 2014) as a World Heritage Site. In the narrower, northern part of the delta, known as the Panhandle, commercial fishing was developed through government subsidy in the early 1980s (Nengu, 1995). Since that time, there has been intermittent conflict between commercial fishers and the tourism industry (Tweddle *et al.*, 2003; Mosepele & Ngwenya, 2010). The concept of establishing FPAs for the benefit of all stakeholders may be an alternative approach to the either/or zonation currently promoted. Expectations for direct revenue earning from FPAs need to be managed as there is a limit to potential revenue in areas remote from angling tourism lodges. Emphasis in the Namibian MFMR project activities is on the importance of the FPAs to counter severe overexploitation of the most valuable fish species driven by unsustainable commercial demand.

15.3.8 Mahseer in the Western Ghats region of India

The rivers of the Western Ghats, which are exceptional hotspots of freshwater fish diversity and endemism, support poorly known commercial, recreational and subsistence fisheries. High human population density (300–350 individuals km^{-2}; Molur *et al.*, 2011), including the presence of several marginalised tribes and forest-dwelling communities, and their congregation along the banks of major river systems has resulted in a high dependence on freshwater fishes for both food and income. Freshwater commercial and subsistence fisheries in the Western Ghats are primarily small-scale fisheries. There is also a small but significant fishery for endemic (and mostly threatened) aquarium fishes, and a recreational fishery for large cyprinids. These fisheries are *de facto* open access. Exceptions include protected areas, where there are licensed fishing agreements for food fishes, and a prohibition for both aquarium and recreational fisheries.

Decades of unmanaged harvest has led to severe population declines of large food fish species and the precarious state of fisheries in many rivers and reservoirs of the region. A decrease in both the size and abundance of the commercially important species has also resulted in fishers using destructive techniques such as poisoning and explosives to maintain their capture rates (Molur *et al.*, 2011). The large-scale mortality encountered during such fishing, and its biological consequences, has greatly reduced the numbers of small fishes otherwise available to aquarium fish collectors, and larger juveniles and adults harvested as food.

One of the main targets of food, aquarium and recreational fishers is the mahseer, represented by three endangered species, *Tor khudree*, *T. malabaricus* and *T. mussullah*. While medium-sized mahseer are targeted for food, smaller juveniles are collected for the aquarium trade and large adults are the choice for recreational anglers. Intensive harvest of these species, mainly for food, has resulted in population reductions of more than 50% in the last 10 years, and several local fisheries are under a threat of imminent collapse (Raghavan, 2011; Raghavan & Ali, 2011; Raghavan *et al.*, 2011).

The Cauvery is the only river in the Western Ghats where all three forms of fisheries (commercial, subsistence and recreational) coexist. The river is globally renowned for its mahseer recreational fishery (Anon.,

2012a). In the mid 1970s, several stretches of the Cauvery were leased by a local non-governmental organisation (NGO), the Wildlife Association of South India (WASI). This was to control both indiscriminate fishing (outside the protected areas) and poaching (inside protected areas) by migrant settlers who often used explosives and other destructive techniques. Apart from regular stocking of mahseer fingerlings, WASI set up small, seasonal fishing camps in the river to promote responsible 'catch-and-release' angling. The success of WASI encouraged the establishment of four full-time angling camps in the 1980s along the Cauvery to promote fish-based ecotourism and angling. The revenue obtained from these camps helped to control illegal fishing through the establishment of anti-poaching camps and patrols, as well as rehabilitation of former poachers as fishing guides.

Towards the end of 2012, the Government of India decided to expand the size of a protected area near the fishing camps. This expansion led to the closure of all camps and a ban on recreational angling because 'fishing' is considered equivalent to 'hunting', and the latter is prohibited inside protected areas as per the Wildlife Protection Act of India. This move will no doubt negatively impact populations of the endangered mahseer, as wildlife managers and government officials in the region lack resources to efficiently manage illegal fishing (Anon., 2012b). This is because even licensed fisheries inside several protected areas of the Western Ghats are poorly managed (Raghavan *et al.*, 2011).

The Cauvery fishing camps were also involved in feeding several thousand kilograms of cereals to keep mahseer within the protected stretches of river. The loss of supplemental feeding will likely result in fish migrating to areas outside the protected area where fishing is open-access and often destructive in nature. The recreational angling community believes that without anglers and the annual revenue that the sector provides for anti-poaching patrols and camps, the Cauvery mahseer will be exposed to illegal fishers and poaching gangs. Also, several local people including former poachers, who earned an income from recreational fisheries, could revert to their old profession to sustain livelihoods (Anon., 2012b). This could prove disastrous to both the endangered mahseer and the overall ecosystem health of Western Ghats' rivers.

In general, the role of freshwater fish as 'wildlife' is somewhat anomalous in India, as none of the primary wildlife conservation legislation in the country focuses on the conservation of freshwater fishes. Freshwater fishes are viewed as an open-access resource and a free commodity that can be harvested from nature. Legislation concerning conservation, management and sustainable use of freshwater fishes needs to be redefined to ensure a better future for inland fisheries in the Western Ghats and India.

15.4 SYNTHESIS

These case studies highlight regional variation in inland fisheries activities. Beyond painting a picture of the global diversity of inland fisheries, a goal of adopting the case study approach was to identify the common challenges and opportunities that exist with the sustainable exploitation of inland fisheries. Here we provide a synthesis to identify common issues, threats, opportunities and solutions, presented here as a series of recurring themes.

15.4.1 Freshwater fisheries sustain livelihoods

One of the patterns emerging from all of the case studies was that freshwater fisheries sustain livelihoods and generate significant socio-economic benefits. In some regions, the harvested fish contribute essential protein, which is critical for food security. In addition, even in areas where subsistence fisheries are critical, small-scale commercial fisheries, and increasingly the recreational sector, generate wealth that enables fishers (and others, such as those active in the tourism industry) to have the resources needed to sustain their families. In developed nations, recreational and commercial fisheries also generate considerable income that provides direct and indirect employment. In addition, subsistence fisheries in developed countries, which are often associated with aboriginal communities, are of significant cultural value and provide food for ceremonial purposes and sustenance. If freshwater fisheries were to be severely restricted (especially due to conservation concerns arising from other threats) or if they are poorly managed, the consequences could be severe. Unfortunately, a lack of high-quality information makes it difficult to quantify the true value of inland fisheries (Arlinghaus *et al.*, 2002; Beard *et al.*, 2011; World Bank, 2012), which potentially places all three fishing sectors and the livelihoods of millions of people at risk.

15.4.2 Change and adaptation are the norm

Evident in nearly all of the case studies is that the fisheries are not static and there has been a number of changes to which the fishers and managers have to respond. For example, the introduction of non-native fishes, either intentionally (e.g. Pacific salmon in the Laurentian Great Lakes) or unintentionally (e.g. Asian carp in the Mississippi River), has led to both problems and opportunities that have required adaptation by fishers and managers (Chapter 8). Pacific salmon create many recreational fishing opportunities in the Great Lakes, while simultaneously acting as a biological control of the invasive alewife. At the same time, the economic importance of the recreational fishery for Pacific salmon means that fishery managers can choose to manage for a biomass of alewife that may impede efforts to rehabilitate or reintroduce lake trout and Atlantic salmon populations (Dettmers *et al.*, 2012). In the Mississippi River, commercial fishers are beginning to exploit invasive Asian carp as markets begin to develop. In many locales, commercial fisheries in inland waters are declining while recreational fisheries are expanding. This is particularly the case in emerging economies like India (see mahseer case study) and Brazil and South Africa. Additionally, the increasing concern about climate change on Pacific salmon in the Fraser River and European eel will also call for fishers and fisheries managers to adapt fishing practices and consider the additional effects of fisheries on fish populations. Although stock assessment and harvest monitoring are essential for documenting changes and developing appropriate responses, some of the solutions to the issues identified lie way outside the fisheries sector and are thus difficult to be addressed by fisheries management alone (e.g. climate change).

15.4.3 Connectivity is critical

Freshwater fisheries are inherently connected to their surroundings, especially in a watershed context. For example, the Great Lakes Basin fisheries are intrinsically linked to the 30 million people living on the shores of the Great Lakes and their associated activities (see stressors below). Similarly, some of the discussed freshwater resources, such as eel, transcend boundaries among nations and among freshwater and marine environments. Inland fisheries are also thought of in terms of longitudinal connectivity, considering the role of barriers on fish movement as well as connections between freshwater and marine systems. Diadromous species (e.g. European eel case study, Fraser River case study) require holistic management plans that cover the entire life cycle and migration extent (Chapter 11). Less attention, however, is given to lateral connectivity and the importance of the floodplain to fishery production and biodiversity. Maintaining the links with the floodplain is critical given the propensity to channelise rivers (see Mississippi case study) and the importance of floodplains to fishery production globally (e.g. Mekong, Zambezi and Amazon).

15.4.4 Multiple, complex and interacting stressors create challenges

A consistent aspect is the recognition that there are multiple, complex and often interacting stressors that threaten inland fisheries, most of which are external to the fisheries themselves. Unlike marine fisheries, where fisheries exploitation is the primary threat, inland fisheries face a suite of stressors associated with, among others, urbanisation, agriculture, hydropower and industry, which collectively have led to degraded habitats, reductions in water quality, alterations in flows, and loss of connectivity (e.g. Mississippi, Mekong and European eel case study; Arlinghaus *et al.*, 2002). Layered on top of such competing uses for water are invasive species and climate change (Chapters 3 and 8). There have been many calls for incorporating fisheries management into a broader conceptual framework of integrated water(shed) management in inland systems (Arlinghaus *et al.*, 2002; Collares-Pereira & Cowx, 2004; Cowx & Portcarrereo, 2011; FAO, 2012).

15.4.5 Conflict among sectors is common

There are many examples of conflict between recreational and commercial, between subsistence and commercial, and between recreational and subsistence fisheries. In almost all cases, the basis for the conflict is allocation and perceptions of inequitable allocation. In areas with aboriginal peoples (e.g. Fraser River case study), the issues can be particularly salient given the hierarchy of allocation and the primacy of addressing ceremonial needs for First Nations. However, even in the Fraser River, there has been much effort in the last few years for the different sectors to work cooperatively on joint issues. In the Great Lakes Basin, there is conflict between recreational and commercial fisheries, but an inclusive and transparent management process has gone some way to building understanding and alleviating conflict. In the Mississippi River basin, there is little evidence of conflict, mainly due to little competition for the same species (but see the exception with catfish). In the eel case, conflict is severe because the resource is dwindling and both commercial and recreational fisheries compete for the remaining fish. Governance structures that engage all three sectors in a transparent way are likely the best strategy to reduce intersectoral conflict in inland fisheries (Arlinghaus, 2005; FAO, 2012).

15.4.6 Management across jurisdictions is complex

Nearly all of the case studies represent fisheries that occur across jurisdictions, often multiple states, provinces and countries. For example, the Laurentian Great Lakes basin straddles Canada and the USA, which necessitated a number of bi-national agreements and the formation of the Great Lakes Fishery Commission as well as the International Joint Commission. Similarly, the Fraser River Pacific salmon fisheries in Canada fall under a bi-national agreement with the USA, given joint exploitation in marine waters which had led to the Pacific Salmon Treaty and the formation of the Pacific Salmon Commission. Similar situations exist for European eel (governed in a generic way by the European Commission) and on the lower Mekong River (Mekong River Commission). Although there is inherent political complexity associated with multijurisdiction fisheries, the need to cooperate often promotes the development of strong governance structures. Another inherent component of multijurisdictional governance institutions (i.e. regional fisheries management bodies) is the inclusion of stakeholder groups in decision-making processes, which promotes buy-in and compliance (FAO, 2012).

15.5 ACHIEVING A SUSTAINABLE FUTURE FOR INLAND FISHERIES

It is unclear how many inland fisheries are biologically sustainable and the extent to which they contribute to conservation problems. Despite efforts by the FAO to place greater emphasis on inland fisheries (FAO, 2011), the reality is that subsistence fisheries and recreational fisheries, which are so important in inland waters, are difficult to monitor (Welcomme, 2001; Beard *et al.*, 2011). A prerequisite of sustainability for inland fisheries (actually for all fisheries) is knowing the state of the resource to allocate biomass for harvest (Krueger & Decker, 1999), although it should be recognised that many tropical floodplain fisheries are resilient to exploitation and it is loss of the flood pulse through water resources development schemes that is the major threat to these systems (Welcomme *et al.*, 2010). Until such information is known, we are left with only site-specific information, biased towards larger commercial and recreational fisheries that are of high economic value. What is clear is that biodiversity in inland systems is in decline as a result of a variety of stressors (Dudgeon *et al.*, 2006), but what is less clear is whether unsustainable fisheries activities are contributing to the decline (Allan *et al.*, 2005).

The case studies represent some of the best monitored systems with rigorous and long-term stock assessment programmes (e.g. Laurentian Great Lakes). In reality, many millions of smaller inland water bodies and water courses around the globe are subject to no or infrequent stock assessment, and it is unlikely the resources will exist to monitor all the small lakes and rivers in a landscape. It is for that reason that alternatives to stock assessment are needed for data-poor situations, including harnessing local knowledge. In Ontario, Canada, a broad-scale, landscape-level approach to both assessment and management has been implemented to provide information on the hundreds of thousands of water bodies using information from only a sample of fisheries (see Lester *et al.*, 2003). Such approaches may have relevance in other regions. Also needed are rapid appraisal methods that can be applied in developing countries where small-scale fisheries are the norm (Prince, 2010; Beard *et al.*, 2011).

Beyond characterising the state and trends in fisheries, it is also important to better value inland fish stocks (Arlinghaus *et al.*, 2002; Cowx *et al.*, 2004). By valuing inland fisheries it is then possible to make trade-offs relative to the many other uses of inland waters (e.g. agriculture, hydropower; see Welcomme *et al.*, 2010). Indeed, doing so can generate the public support and political will needed to make decisions that benefit biodiversity and livelihoods (Beard *et al.*, 2011), although one has to be cognisant that economists will tend to compare direct economic impact (provisioning services) of other sectors with fisheries and the latter will tend to suffer because the supporting, regulating and cultural services delivered by inland fish are difficult to quantify and therefore often remain unvalued (Cowx & Portocarrereo, 2011). In developed countries, it is often possible to quantify the economic impacts and the net economic value (producer and consumer surplus, see Parkkila *et al.*, 2010) of commercial and recreational fisheries and use these values in allocation schemes. In developing countries, and particularly for subsistence fisheries, ascribing value to fisheries is much more difficult given their diffuse nature. Agricultural household surveys that track food consumption may be one of the best ways to identify the value of such fisheries. There can also be cultural and spiritual value (especially with indigenous peoples), which is even more difficult to quantify. Many inland fisheries, even when commercial, involve local consumption rather than sending inland fish products to export markets, which makes it difficult to track landings. Knowing which inland fisheries products are consumed, and the type of markets in which they are sold or traded, can provide a means to communicate information about sustainable inland fish products to consumers. Sustainable marine seafood awareness campaigns have been somewhat successful, but similar programmes are less common in the inland realm (Cooke et al., 2011), although the Ontario Commercial Fisheries Association is seeking sustainable certification for its fisheries. Such efforts would need to be tied to local markets given that, unlike the marine realm, most inland fisheries products are not exported.

The complexity and uncertainty inherent in decisions about managing water resources require an ecosystem approach that involves stakeholders from all sectors (Beard et al., 2011). An ecosystem approach, which recognises humans as part of the ecosystem, is fundamental (Garcia, 2003). Similarly, rather than thinking about managing fish and fisheries, it is necessary to think about more holistic approaches such as integrated water(shed) management (Heathcote et al., 1998). This is not to say that fisheries management is not important, but instead to recognise that it cannot be done in isolation of the other users and management objectives for a given system (Collares-Pereira & Cowx, 2004). We presented several case studies that provide excellent examples of multijurisdictional, multistakeholder approaches to managing aquatic ecosystems and hope that such examples will become even more common in the future. Engaging the broader public in inland fisheries issues is also needed to achieve the political will to conserve and manage inland fisheries (Cooke et al., 2013).

To achieve a sustainable future for inland fisheries really means, first, addressing a number of critical information gaps (Welcomme et al., 2010; Beard et al., 2011). Rethinking governance structures and management paradigms to better incorporate diverse stakeholder perspectives and approaching water management from a holistic and integrated perspective are needed. Investments in research and management may be costly, but the long-term costs of not having inland fisheries will be disastrous.

15.6 DISCUSSION QUESTIONS

1. What factors lead to shifts in the prominence in different fisheries sectors following industrialisation?

2. How can one strive for sustainable fisheries when no biological assessment data exist?
3. Which management structures are appropriate for multijurisdictional fisheries?
4. Have inland fisheries led to population collapses as have been observed in some marine fisheries?
5. Why do inland fisheries seem to attract less attention than marine fisheries?

15.7 ACKNOWLEDGEMENTS

Cooke is supported by the Canada Research Chairs Program and the Natural Sciences and Engineering Research Council of Canada. The Idaho Cooperative Fish and Wildlife Research Unit is jointly sponsored by the University of Idaho, US Geological Survey, Idaho Department of Fish and Game, and Wildlife Management Institute. The use of trade, firm or product names is for descriptive purposes only and does not imply endorsement by the US Government. Arlinghaus received funding by the German Federal Ministry for Education and Research within the Program for Social–Ecological Research and the project Besatzfisch (www.besatz-fisch.de, grant # 01UU0907). We thank Craig Paukert, Doug Beard and two anonymous referees for reviewing the document.

15.8 REFERENCES

Agreda-Rudenko, A. (2008). *Sinopsis de la Pesca deportiva en Colombia*. Bogotá, Colombia: Proexport Columbia.

Agudelo, C. E. (2007). *La actividad pesquera en la zona suroriental de la amazonia colombiana: una descripción de la captura y comercialización de los bagres transfronterizos*. Instituto de Ciencias y Tecnologías Ambientales. Barcelona, España: Universidad Autónoma de Barcelona.

Agudelo, E., Sanchez Paez, C. L., Rodriguez Fernandez, C. A., Bonilla-Castillo, C. A. & Gómez Hurtado, G. A. (2011). Diagnóstico de la pesquería en la Cuenca del Amazonas. In *II Pesquerías Continentales de Colombia: cuencas del Magdalena-Cauca, Sinú, Canalete, Atrato, Orinoco, Amazonas y vertiente del Pacífico. Serie Editorial Recursos Hidrobiológicos y Pesqueros Continentales de Colombia*. Bogotá, Colombia: Instituto de Investigación de Recursos Biológicos Alexander von Humboldt.

Ajiaco-Martínez, R. E., Ramirez-Gil, H., Sánchez-Duarte, P., Lasso, C. A. & Trujillo, F. (2012). *IV Diagnóstico de la pesca ornamental en Colombia. Serie Editorial Recursos Hidrobiológicos y pesqueros Continentales de Colombia*. Bogotá, Colombia: Instituto de Investigación de Recursos Biológicos Alexander von Humboldt.

Allan, J. D., Abell, R., Hogan, Z., *et al.* (2005). Overfishing of inland waters. *BioScience*, 55, 1041–1051.

American Sportfishing Association. (2008). *Today's Angler: A Statistical Profile of Anglers, their Targeted Species and Expenditures*. Alexandria, VA: American Sportfishing Association.

Anon. (2012a). www.cauveryfishingcamp.com/about_bheemeshwari.html

Anon. (2012b). www.anglingdirectholidays.com/protect-the-mahseer/fishing-holidays-blog.html

Arlinghaus, R. (2005). A conceptual framework to identify and understand conflicts in recreational fisheries systems, with implications for sustainable management. *Aquatic Resources, Culture and Development*, 1, 145–174.

Arlinghaus, R. & Cooke, S. J. (2009). Recreational fisheries: socioeconomic importance, conservation issues and management challenges. In *Recreational Hunting, Conservation and Rural Livelihoods: Science and Practice*. Oxford: Blackwell Publishing, pp. 39–58.

Arlinghaus, R., Mehner, T. & Cowx, I. G. (2002). Reconciling traditional inland fisheries management and sustainability in industrialized countries, with emphasis on Europe. *Fish and Fisheries*, 3, 261–316.

Arlinghaus, R., Cooke, S. J., Lyman, J., *et al.* (2007). Understanding the complexity of catch-and-release in recreational fishing: an integrative synthesis of global knowledge from historical, ethical, social, and biological perspectives. *Reviews in Fisheries Science*, 15, 75–167.

Baker, M. R. & Schindler, D. E. (2009). Unaccounted mortality in salmon fisheries: non-retention in gillnets and effects on estimates of spawners. *Journal of Applied Ecology*, 46, 752–761.

Beard, T. D. Jr, Rasmussen, P. W., Cox, S. & Carpenter, S. R. (2003). Evaluation of a management system for a mixed walleye spearing and angling fishery in northern Wisconsin. *North American Journal of Fisheries Management*, 23, 481–491.

Beard, T. D., Arlinghaus, R., Cooke, S. J., *et al.* (2011). Ecosystem approach to inland fisheries: research needs and implementation strategies. *Biology Letters*, 7, 481–483.

Beardmore, M., Dorow, M., Haider, W. & Arlinghaus, R. (2011). The elasticity of fishing effort response and harvest outcomes to altered regulatory policies in eel (*Anguilla anguilla*) recreational angling. *Fisheries Research*, 110, 136–148.

Brown, A. V., Brown, K. B., Jackson, D. C. & Pierson, W. K. (2005). Lower Mississippi River and its tributaries. In *Rivers of North America*. Amsterdam: Elsevier, pp. 231–281.

Brummett, R. E., Youaleu, J. L. N., Tiani, A-M. & Kenmegne, M. M. (2010). Women's traditional fishery and alternative aquatic resource livelihood strategies in the southern Cameroonian rainforest. *Fisheries Management and Ecology*, 17, 221–230.

Carlander, H. B. (1954). *A History of Fish and Fishing in the Upper Mississippi River*. Rock Island, IL: Upper Mississippi River Conservation Committee.

Coates, D. (2002). *Inland Capture Fishery Statistic of Southeast Asia: Current Status and Information Needs*. Bangkok, Thailand: Food and Agricultural Organisation of the United Nations.

Cohen, B. I. (2012). *The Uncertain Future of Fraser River Sockeye: Volume 1. Commission of Inquiry into the Decline of Sockeye Salmon in the Fraser River, Canada*. Public Works and Government Services Canada.

Collares-Pereira, M. J. & Cowx, I. G. (2004). The role of catchment scale environmental management in freshwater fish conservation. *Fisheries Management and Ecology*, 11, 303–312.

Colombo, R. E., Garvey, J. E., Jackson, N. D., *et al.* (2007). Harvest of Mississippi River sturgeon drives abundance and reproductive success: a harbinger of collapse? *Journal of Applied Ichthyology*, 23, 444–451.

Cooke, S. J. & Cowx, I. G. (2006). Contrasting recreational and commercial fishing: searching for common issues to promote unified conservation of fisheries resources and aquatic environments. *Biological Conservation*, 128, 93–108.

Cooke, S. J., Murchie, K. J. & Danylchuk, A. J. (2011). Sustainable 'seafood' ecolabelling and initiatives in the context of inland fisheries: needs and opportunities to increase food security and protect freshwater ecosystems. *BioScience*, 61, 911–918.

Cooke, S. J., Lapointe, N. W. R., Martins, E. G., *et al.* (2013). Failure to engage the public in issues related to inland fishes and fisheries: strategies for building public and political will to promote meaningful conservation. *Journal of Fish Biology*, 83, 997–1018.

Cowx, I .G. (1994). Stocking strategies. *Fisheries Management and Ecology*, 1, 15–30.

Cowx, I. G. (2002). Recreational fishing. In *Handbook of Fish Biology and Fisheries*, vol. 2. Oxford: Blackwell Science, pp. 367–390.

Cowx, I. G. & Portocarrero, A. M. (2011). Paradigm shifts in fish conservation: moving to the ecosystem services concept. *Journal of Fish Biology*, 79, 1663–1680.

Cowx, I. G., Almeida, O., Bene, C., *et al.* (2004). Value of river fisheries. In *Sustaining Livelihoods and Biodiversity in the New Millennium. Proceedings of the 2nd Large Rivers Symposium*, vol. 1. Phnom Penh, Cambodia: Mekong River Commission, pp. 1–20.

Cowx, I. G., Arlinghaus, R. & Cooke, S. J. (2010). Harmonizing recreational fisheries and conservation objectives for aquatic biodiversity in inland waters. *Journal of Fish Biology*, 76, 2194–2215.

Dannewitz, J., Maes, G. E., Johansson, L., *et al.* (2005). Panmixia in the European eel: a matter of time. *Proceedings of the Royal Society B: Biological Sciences*, 272, 1129–1137.

De Jesus, M. J. & Kohler, C. C. (2004). The commercial fishery of the Peruvian Amazon. *Fisheries*, 29, 10–16.

Deap, L., Degen, P. & van Zalinge, N. (2003). *Fishing Gears of the Cambodian Mekong.* Phnom Penh, Cambodia: Inland Fisheries Research and Development Institute of Cambodia. Fisheries Technical Paper.

Dekker, W. (2008). Coming to grips with the eel stock slip-sliding away. In *International Governance of Fisheries Ecosystems: Learning from the Past, Finding Solutions for the Future.* Bethesda, MD: American Fishery Society, pp. 335–355.

Dekker, W. (2009). A conceptual management framework for the restoration of the declining European eel stock. In *Eels at the Edge: Science, Status and Conservation Concern American Fisheries Society, Symposium 58.* Bethesda, MD: American Fishery Society.

Dettmers, J. M., Goddard, C. I. & Smith, K. D. (2012). Management of alewife using Pacific Salmon in the Great Lakes: whether to manage for economics or the ecosystem? *Fisheries,* 37, 495–501.

Dorow, M. & Arlinghaus, R. (2012). The relationship between personal commitment to angling and the opinions and attitudes of German anglers towards the conservation and management of the European eel *Anguilla anguilla. North American Journal of Fisheries Management,* 32, 466–479.

Dorow, M., Beardmore, B., Haider, W. & Arlinghaus, R. (2010). Winners and losers of conservation policies for European eel, *Anguilla anguilla:* an economic welfare analysis for differently specialised anglers. *Fisheries Management and Ecology,* 17, 106–125.

Dudgeon, D., Arthington, A. H., Gessner, M. O., *et al.* (2006). Freshwater biodiversity: importance, threats, status and conservation challenges. *Biological Reviews,* 81, 163–182.

Dugan, P. J., Barlow, C., Agostinho, A. A., *et al.* (2010). Fish migration, dams, and loss of ecosystem services in the Mekong basin. *Ambio,* 39, 344–348.

English, K. K., Edgell, T. C., Bocking, R. C., Link, M., & Raborn, S. (2011). *Fraser River sockeye fisheries and fisheries management and comparison with Bristol Bay sockeye fisheries.* LGL Ltd. Cohen Commission Technical Report 7.

European Commission. (2007). Council regulation (EC) No. 1100/2007 of 18 September 819 2007 establishing measures for the recovery of the stock of European eel. *Official Journal of the Europian Union,* 248, 17–23.

Fabre, N. N. & Alonso, J. C. (1998). Recursos icticos no Alto Amazonas: sua importancia nas populações ribeirinhas. *Boletim do Museu Paraense Emilio Goeldi Serie Botanica,* 1, 19–55.

Food & Agriculture Organization [FAO]. (2011). *Guidelines for the Ecolabelling of Fish and Fishery Products from Inland Capture Fisheries.* Rome: Food and Agriculture Organization of the United Nations.

Food & Agriculture Organization [FAO]. (2012). *The state of the world fisheries and aquaculture 2012.* Rome: Fisheries and Aquaculture Department, Food and Agriculture Organization of the United Nations.

Feunteun, E. (2002). Management and restoration of European eel population (*Anguilla anguilla*): an impossible bargain. *Ecological Engineering,* 18, 575–591.

Freyhof, J. & Kottelat, M. (2008). *Anguilla anguilla.* In *2008 IUCN Red List of Threatened Species,* ed. IUCN. www.iucnredlist.org

Friedland, K. D., Miller, M. J. & Knights, B. (2007). Oceanic changes in the Sargasso Sea and declines in recruitment of the European eel. *ICES Journal of Marine Science,* 64, 519–530.

Gaden, M. (2007). *Bridging jurisdictional divides: collective action through A Joint Strategic Plan for Management of Great Lakes Fisheries.* PhD dissertation. University of Michigan, Ann Arbor, MI.

Gaden, M., Krueger, C. & Goddard, C. (2009). Managing across jurisdictional boundaries: fishery governance in the Great Lakes and Arctic-Yukon-Kuskokwim regions. In *Pacific Salmon: Ecology and Management of Western Alaska's Populations.* American Fisheries Society Symposium 70. Bethesda, MD: American Fisheries Society, pp. 941–962.

Gale, M. K., Hinch, S. G., Eliason, E. J., Cooke, S. J. & Patterson, D. A. (2011). Physiological impairment of adult sockeye salmon in fresh water after simulated capture-and-release across a range of temperatures. *Fisheries Research,* 112, 85–95.

Garcia, S. M. (2003). *The Ecosystem Approach to Fisheries: Issues, Terminology, Principles, Institutional Foundations, Implementation and Outlook,* Volume 443. Rome: Food and Agriculture Organization of the United Nations.

Gutiérrez, F., Acosta, L. E. & Salazar, C. A. (2004). *Perfiles urbanos en la Amazonia colombiana: un enfoque para el desarrollo sostenible.* Bogotá, Columbia: Editora Guadalupe.

Heathcote, I. W., Edwards, J. R., Greener, H. & Coombs, H. M. (1998). *Integrated Watershed Management: Principles and Practice.* London: Taylor and Francis.

Holmlund, C. M. & Hammer, M. (1999). Ecosystem services generated by fish populations. *Ecological Economics,* 29, 253–268.

IGFA. (2012). IGFA Online World Record Search. http://wrec.igfa.org/WRecordsList.aspx?lc=AllTac kleandcn=Tigerfish

International Council for the Exploration of the Sea. (2010). *Report of the joint EIFAC/ICES Working Group on eels (WGEEL), 9–14 September 2010, Hamburg Germany.* Rep. No. ICES CM 2010/ACOM: 18.

Junk, W. J. (1997). *The Central Amazon Floodplain. Ecology of a Pulsing System.* Berlin: Springer-Verlag.

Kennish, M. J. (2002). Environmental threats and environmental future of estuaries. *Environmental Conservation*, 29, 78–107.

Knights, B. (2003). A review of the possible impacts of long-term oceanic and climate changes and fishing mortality on recruitment of anguillid eels of the northern hemisphere. *Science of the Total Environment*, 310, 237–244.

Koch, J. D., Quist, M. C., Pierce, C. L., Hansen, K. A. & Steuck, M. J. (2009). Effects of commercial harvest on shovelnose sturgeon populations in the upper Mississippi River. *North American Journal of Fisheries Management*, 29, 84–100.

Krogman, R. M., Fischer, J. R., Quist, M. C., Steuck, M. J. & Marron, M. M. (2011). Historical trends in ictalurid catfish commercial harvest in the upper Mississippi River. *American Fisheries Society Symposium*, 77, 127–140.

Krueger, C. C. & Decker, D. J. (1999). The process of fisheries management. In *Inland Fisheries Management in North America*, second edition. Bethesda, MD: American Fisheries Society, pp. 31–59.

Lake Erie Committee. (2012). www.glfc.org/pressrel/LEC_news_release_2012_3-29-12.pdf

Lester, N. P., Marshall, T. R., Armstrong, K., Dunlop, W. I. & Ritchie, B. (2003). A broad-scale approach to management of Ontario's recreational fisheries. *North American Journal of Fisheries Management*, 23, 1312–1328.

Limburg, K. E., Hughes, R. M., Jackson, D. C. & Czech, B. (2011). Human population increase, economic growth, and fish conservation: collision course or savvy stewardship? *Fisheries*, 36, 27–35.

Lymer, D. & Funge-Smith, S. (2009). *An Analysis of Historical National Reports of Inland Capture Fisheries Statistics in the Asia–Pacific Region (1950–2007).* Bangkok: Food and Agriculture Organization of the United Nations, Regional Office for Asia and the Pacific, p. 18.

Malmqvist, B. R. & Rundle, S. (2002). Threats to the running water ecosystems of the world. *Environmental Conservation*, 29, 134–153.

Martins, E. G., Hinch, S. G., Patterson, D. A., *et al.* (2011). Effects of river temperature and climate warming on stock-specific survival of adult migrating Fraser River sockeye salmon (*Oncorhynchus nerka*). *Global Change Biology*, 17, 99–114.

McDaniels, T., Wilmot, S., Healey, M. & Hinch, S. (2010). Vulnerability of Fraser River sockeye salmon to climate change: a life cycle perspective using expert judgments. *Journal of Environmental Management*, 91, 2771–2780.

McKenney, B. & Tola, P. (2002). *Natural Resources and Rural Livelihoods in Cambodia: A Baseline Assessment. Working Paper no. 23.* Phnom Penh, Cambodia: Cambodia Development Resource Institute.

Molur, S., Smith, K. G., Daniel, B. A. & Darwall, W. R. T. (2011). *The Status of Freshwater Biodiversity in the Western Ghats.* Coimbatore, India. Gland, Switzerland, and Cambridge, UK: International Union for Conservation of Nature (IUCN) and Natural Resources, p. 116.

Mosepele, K. & Ngwenya, B. N. (2010). *Socio-economic survey of commercial fishing in the Okavango Delta, Botswana.* Okavango Report Series No. 7, pp. 1–75.

NACSO. (2010). *Namibia's Communal Conservancies: A Review of Progress and Challenges in 2009.* Windhoek, Namibia: NACSO, pp. 1–152.

NACSO. (2011). *Living with Wildlife – The Story of Namibia's Communal Conservancies.* Windhoek, Namibia: NACSO, pp. 1–28.

Nengu, S. M. (1995). Status of fisheries in wetlands. In *Wetlands Management in Botswana: Proceedings of a Conference held in Kasane, Botswana, 14–16 November 1994.* Botswana: Wetlands Coordinating Committee.

Ochoa, G. I., Wood, A. & Zarate, C. G. (2006). *Puerto Nariño: El pueblo que se mira en el río. Retos al desarrollo sustentable en los municipios amazónicos.* Bogotá, Colombia: Instituto Latinoamericano de Servicios Legales Alternativos (ILSA).

Parkkila, K., Arlinghaus, R., Artell, J., *et al.* (2010). *European inland fisheries advisory commission methodologies for assessing socio-economic benefits of European Inland Recreational Fisheries.* European Inland Fisheries Advisory Commission Occasional Paper 46, 112 pp.

Pitlo, J. Jr. (1997). Response of upper Mississippi River channel catfish populations to changes in commercial harvest regulations. *North American Journal of Fisheries Management*, 17, 848–859.

Portocarrero-Aya, M. (2012). *Conservation of the freshwater biodiversity in key areas of the Colombian Amazon*. PhD Dissertation, The University of Hull, Hull, UK.

Poulsen, A. F., Poeu, O., Viravong, S., Suntornratana, U. & Thanh Tung, N. (2002). *Fish migrations in the Lower Mekong River Basin: implications for development, planning and environmental management*. Phnom Penh, Cambodia: Mekong River Commission Technical Paper No. 8.

Prince, J. (2010). Rescaling fisheries assessment and management: a generic approach, access rights, change agents, and toolboxes. *Bulletin of Marine Science*, 86, 197–219.

Raby, G. D., Colotelo, A. H., Blouin-Demers, G. & Cooke, S. J. (2011). Freshwater commercial bycatch: an understated conservation problem. *BioScience*, 61, 271–280.

Raghavan, R. (2011). *Tor khudree*. In *IUCN 2012. IUCN Red List of Threatened Species. Version 2012.2*. www.iucnredlist.org

Raghavan, R. & Ali, A. (2011). *Tor malabaricus*. In *IUCN 2012. IUCN Red List of Threatened Species. Version 2012.2*. www.iucnredlist.org

Raghavan, R., Ali, A., Dahanukar, N. & Rosser, A. (2011). Is the fishery for the Deccan Mahseer, *Tor khudree* (Sykes, 1839) in the Western Ghats Hotspot sustainable? A participatory approach to stock assessment. *Fisheries Research*, 110, 29–38.

Riaño Umbarila, E. (2003). *Organizando se espacio, construyendo su territorio: transformaciones de los asentamientos Ticuna en la ribera del Amazonas colombiano*. Leticia, Colombia: Universidad Nacional de Colombia.

Risotto, S. P. & Turner, R. E. (1985). Annual fluctuation in abundance of the commercial fisheries of the Mississippi River and tributaries. *North American Journal of Fisheries Management*, 5, 557–574.

Roscoe, D. W. & Pollen, C. (2010). *Report cards for three BC Recreational fisheries*. Watershed Watch Salmon Society.

Salinas, Y. (1994). *Aspectos de la Biología Pesquera de los Grandes Bagres (Ostariophysi: Siluriformes, Pimelodiade) en el Sector Colombiano del Río Amazonas*. Bogotá, Colombia: Tesis (Lic. Biología) Universidad Pedagógica Nacional.

Sampson, S. J., Chick, J. H. & Pegg, M. A. (2009). Diet overlap among two Asian carp and three native fishes in backwater lakes on the Illinois and Mississippi rivers. *Biological Invasions* 11, 483–496.

Sledge, T. (1998). *Mississippi River creel survey from St. Cloud to Dayton, May 10 to September 30, 1997*. St. Paul, MN: Minnesota Department of Natural Resources Completion report.

Starkie, A. (2003). Management issues relating to the European eel, *Anguilla anguilla*. *Fisheries Management and Ecology*, 10, 361–364.

Strayer, D. L. & Dudgeon, D. (2010). Freshwater biodiversity conservation: recent progress and future challenges. *Journal of the North American Benthological Society*, 29, 344–358.

Sverdrup-Jensen, S. (2002). *Fisheries in the Lower Mekong Basin: Status and Perspectives*. Mekong River Commission Technical Paper No. 6.

Sweeney, L., Baker, A., Thaniseb, A., *et al.* (2010). *A preliminary economic assessment of the contribution of fishing lodges in the Caprivi Region to the local economy*. Windhoek, Namibia: Namibia Nature Foundation.

Thompson, R. E. (1981). Oceanography of the British Columbia coast. *Canadian Special Publication of Fisheries Aquatic Science*, 56, 291.

Tweddle, D. & Hay, C. J. (2011a). *Data collection and analysis: Report on workshop conducted from 26–27 October 2011, Katima Mulilo, Namibia*. Integrated Co-Management of Zambezi/Chobe River Fisheries Resource Project, Field Document no. MFMR/NNF/WWF/Phase II/5: 1–15.

Tweddle, D. & Hay, C. J. (2011b). *Fish Protection Areas: documentation for their establishment in Sikunga and Impalila Conservancies*. Integrated Co-Management of Zambezi/Chobe River Fisheries Resource Project, Field Document no. MFMR/NNF/WWF/Phase II/6: 1–28.

Tweddle D., Bills R., van de Waal B., *et al.* (2003). Fish diversity and fisheries in the Okavango Delta, Botswana. In *A Rapid Biological Assessment of the Aquatic Ecosystems of the Okavango Delta, Botswana: High Water Survey: RAP 27*. Washington, DC: Conservation International, pp. 97–110 and 210–245.

UNEP & ACTO. (2009). *Environmental outlook in Amazonia – GEO Amazonia*. United Nations Environmental Program, Amazon Cooperation Treaty Organization (ACTO) and Research Center of Universidad del Pacífico (CIUP).

Welcomme, R. L. (2001). *Inland Fisheries: Ecology and Management*. Oxford: Blackwell.

Welcomme, R. L. (2011a). An overview of global catch statistics for inland fish. *ICES Journal of Marine Science*, 68, 1751–1756.

Welcomme, R. (2011b). *Review of the State of the World Fishery Resources: Inland Fisheries*. Fisheries and Aquaculture Circular No. 942, Rev. 2. Rome: Food and Agriculture Organization of the United Nation.

Welcomme, R. L., Cowx, I. G., Coates, D., *et al.* (2010). Inland capture fisheries. *Philosophical Transactions of the Royal Society B: Biological Sciences*, 365, 2881–2896.

Williams, A. (2007). The Pacific Salmon Treaty: a historical analysis and prescription for the future. *Journal of Environmental Law and Litigation*, 22, 153–195.

World Bank. (2012). *Hidden Harvest: The Global Contribution of Capture Fisheries*. https://openknowledge.worldbank.org/handle/10986/11873.

Understanding and conserving genetic diversity in a world dominated by alien introductions and native transfers: the case study of primary and peripheral freshwater fishes in southern Europe

VALERIO KETMAIER AND PIER GIORGIO BIANCO

16.1 INTRODUCTION

This chapter sets out to explore implications for conservation genetics as illustrated by impacts and patterns of introduced species with a special emphasis on southern European freshwater fishes, where alien fish globalisation has severely affected the genetic structure of native populations. Impacts to native species range from loss of genetic identity and fitness due to hybridisation to reduction in population and range sizes – and ultimately extinction – by predation, competition and habitat modification (Olden *et al.*, 2010). Freshwater fishes are amongst the most imperilled faunas worldwide with over 30% of evaluated species considered threatened with extinction (Vié *et al.*, 2009; Chapter 1) in spite of being often acknowledged as invaluable material to elucidate fundamental biological phenomena such as speciation in space and time.

Southern Europe – that part of the European continent south of the western–eastern divide represented by the main mountain ranges of the Pyrenees and the Alps and surrounding the Mediterranean Sea – hosts an astonishing number of endemic plant and animal species including freshwater fishes. The area is thus unanimously recognised as a biodiversity hotspot (Chapter 1). The roots of such diversity are to be found in the complex geological history of the area coupled with the effects of Quaternary ice ages. These events strongly affected central–northern Europe by extirpating most of their flora and fauna, leaving the area with a dramatically depleted biodiversity after each ice age cycle. However,

Conservation of Freshwater Fishes, eds G. P. Closs, M. Krkosek and J. D. Olden. Published by Cambridge University Press. © Cambridge University Press 2016.

the milder climatic conditions of southern Europe – even during ice age peaks – spared most of its species, allowing the persistence of many different lineages (Hewitt, 1999).

The diversity of freshwater fishes in southern Europe (peri-Mediterranean countries; Figure 16.1) is high and encompasses all four divisions used to classify freshwater fishes ecologically on the basis of their salt tolerance and physiological inability to survive in normal marine salt waters: primary, primary-like, secondary and peripheral (see Myers, 1938; Bianco, 1990; and Chapter 2 for a review of these categories). While primary forms are moderately salt-tolerant as they may stand salinity up to 13–15 ppm (Bianco & Nordlie, 2008), secondary and peripheral forms are euryhaline taxa including many diadromous migratory species and recently land-locked migratory taxa. In the peri-Mediterranean area this category includes, amongst others, trout, brook lampreys, killfishes and sticklebacks. Primary and primary-like forms (Bianco & Nordlie, 2008) are the most informative biogeographically as they are unable to use the sea to disperse and are therefore restricted to the hydrographical networks of drainage basins. Human intervention excluded, the current geographical range of such fishes often reflects the geomorphologic evolution of hydrographic networks (Salzburger et al., 2003). Dispersal across river basins takes place through downstream connections and river captures. Downstream connections are influenced by marine regressions while river captures between opposite sides of mountains usually occur during orogenesis (Bianco, 1990). Most of the endemic southern European fish species are primary forms, and the family Cyprinidae is by far the most diverse, accounting for up to 217 endemic species.

The origin of most of the southern European primary fish fauna has been traced back to a peculiar phase of the Mediterranean Sea, the so-called 'Lago Mare' phase, which took place immediately after the Messinian salinity crisis, about 5 million years ago (Hsü et al., 1973). However, the exact age and the effectiveness of this and other Miocene events are still debated (Orszag-Sperber, 2006; Popov et al., 2006; Krijgsman et al., 2010). Reyjol et al. (2007) demonstrated the uniqueness of the peri-Mediterranean fish fauna, its shared biogeographic ancestry and its distinctiveness from other European areas by hierarchically clustering 406 hydrographical networks based on the respective fish faunistic lists (Figure 16.1). Based on the distribution of endemic cyprinid species, Bianco (1990) defined 13 southern European ichthyogeographic districts, each harbouring a unique array of endemic species

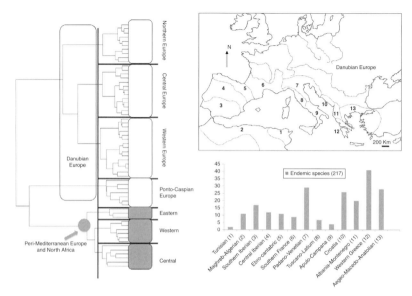

Figure 16.1 Biogeography of European freshwater fishes. The tree on the left is a modification of that presented in Reyjol *et al.* (2007) and shows the relationships among the main European biogeographical regions based on the hierarchical clustering analysis of fish faunistic data of 406 hydrographical networks (see Reyjol *et al.*, 2007 for details on river catchments). The juxtaposition of Danubian (in white) and peri-Mediterranean Europe (in grey) is highlighted; the grey circle shows the monophyly of the peri-Mediterranean area as a whole. On the right, a map of Europe is shown (top) with the 13 ichthyogeographic districts and (bottom) the number of endemic cyprinid species in each of them (numbers in parentheses on the *x*-axis refer to those in the map) (both modified from Bianco, 1990 and updated from Kottelat & Frehyof, 2007 and Bianco, 2014).

(Figure 16.1). The validity of these districts has been repeatedly supported by independent lines of evidence, e.g. molecular evidence (Durand *et al.*, 2003; Ketmaier *et al.*, 2009).

16.2 BIODIVERSITY AT RISK

The unique assemblage of endemic southern European freshwater fishes, which has been generated from long and complex evolutionary processes, is now being negatively impacted by anthropogenic forces. Genetic data are of great help in distinguishing original biogeographic patterns from those of anthropogenic origin and therefore offer valid instruments to set up proper conservation plans and to avoid further

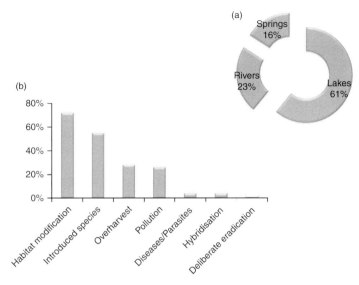

Figure 16.2 Percentages of global decline in freshwater fish diversity from (A) different habitats and (B) attributable to particular causes (after Harrison & Stiassny, 1999).

loss of diversity. Pollution, habitat destruction and dam construction are negatively affecting aquatic diversity on a global and local scale (Bianco, 1995a; Smith & Darwall, 2006). Another more subtle – but not less harmful – risk is the introduction of non-native species. From a genetic perspective, the impact of non-native species includes the loss of genetic identity through hybridisation; this could result in the loss of adaptive potential which could in turn lead to reductions in population and range size and ultimately extinction. Significant invasions are related to growing economies and are widely recognised, especially in developed countries (Bianco, 1995a, 1998, 2005; Copp *et al.*, 2006; Leprieur *et al.*, 2008; Gozlan *et al.*, 2010). The introduction of non-native fish, at a global scale, is recognised as the second most important factor implicated in the extinction of freshwater species, following habitat destruction, with the impacts strongest in lacustrine environments (Olden *et al.*, 2010; Figure 16.2A,B).

The Italian peninsula, due to the lack of strict legislation, can be taken as a paradigmatic example for southern Europe and elsewhere of the negative effects of the introduction of non-native species. About 70% of fish fauna native to the peninsula are either threatened or have been locally extirpated by the introduction of species non-native to the Italian peninsula. Native translocations – namely the introduction of species from one part of a political entity or country in which the species is native to another part of the same country where it is not native (Copp et al., 2006) – also have strong negative effects, and can be as threatening as non-native introductions (Bianco, 2014). A faunal survey carried out on 12 central Italian localities (Bianco & Ketmaier, 2001) showed that none of them were free of non-native or native translocated species, and these were the dominant component (i.e. > 40% in number of species and biomass) in 6 localities.

On a global perspective, there is very little (if any) public awareness that freshwater fish diversity (including the too-often overlooked genetic diversity) is being lost at a rate much higher than that estimated for most terrestrial and marine systems and up to 1000 times higher than the 'natural' background extinction rate (Ricciardi & Rasmussen, 1999). Vié et al. (2009) estimated that about 25% of the evaluated freshwater fish species are threatened with extinction. These evidences are even more dramatic if one considers that species loss is predominant over species replacement when it comes to spatial turnover for native freshwater fishes across Europe (Leprieur et al., 2009).

16.3 HOW CAN GENETICS HELP?

Conservation genetics is truly an interdisciplinary science; it aims to preserve and restore biodiversity by bridging under the same umbrella a variety of disciplines such as population genetics, molecular ecology, evolutionary biology, landscape genetics and systematics. As outlined in the previous section, freshwater fishes are currently faced with a number of threats; as a consequence, many species' populations are declining in abundance and their geographical distribution is becoming increasingly fragmented. Genetics, when coupled with ecological, demographic and systematic studies, can successfully contribute to the design and implementation of plans to preserve and protect threatened or endangered fish species. Scribner et al. (2001) reviewed genetic and statistical methodologies that can be used to detect hybridisation events in fishes due to human activities, such as aquaculture, species introductions and loss or alteration of habitats. In particular, genetic studies can significantly add to (i) the resolution of taxonomic problems in phylogenies and groups where morphology is not definitive; (ii) identifying species boundaries and cryptic species; (iii) describing how genetic diversity is distributed within and amongst geographical populations (i.e. phylogeography); (iv) distinguishing between native versus introduced populations; (v) gaining a better understanding of factors related to fitness (Vrijenhoek, 1998); and (vi) designing programmes for captive breeding. Given the intrinsically fragmented nature of freshwater habitats, a detailed understanding as to how geographic and environmental features shape genetic variation within species – the purview of landscape genetics – is fundamental in any conservation effort.

The last four decades have witnessed an unprecedented boom in molecular techniques that have made the screening of genetic variation in a large number of individuals accessible, even for small- and medium-sized research groups. The early days saw allozymes (i.e. proteins with enzymatic activity) as the markers of choice, but these approaches have now been supplanted by techniques that are able to reveal more variation while being minimally invasive (allozymes generally required the collection of whole animals for analysis). Mitochondrial DNA (mtDNA) has dominated the field for at least two decades and laid the foundation of phylogeography (Avise, 2000). A variety of mtDNA genes – with cytochrome-b playing the foremost role – have proved instrumental in solving taxonomic controversies and assessing phylogenetic relationships due to the evolutionary characteristics of this genomic element (fast rates of evolution, lack of recombination). On the other hand, mtDNA is exclusively of maternal origin (at least in fishes), and hence cannot be used to assess gender-related issues (i.e. asymmetrical gene flow). Its resolving power also tends to decrease as the taxonomic level increases (Avise, 2000). As a general rule, mtDNA allows a relatively straightforward insight into the evolutionary history of a given taxon, whereas nuclear DNA (nDNA) is better suited to track species boundaries, especially in areas where introductions are known. In such circumstances, nDNA should be used to detect hybridisation and introgression events that are likely to go unnoticed in studies based on mtDNA only. The increasing simplicity and affordability of isolating large batteries of highly polymorphic nuclear microsatellite loci has allowed researchers to complement the maternal perspective of mtDNA with bi-parentally inherited nDNA markers. Microsatellites are versatile markers that can reveal population processes, but their main limit is that they are mostly species-specific.

The current blossoming of next-generation sequencing technologies (NGS; high-throughput sequencing technologies that parallelise the sequencing process, producing thousands or millions of sequences simultaneously) is making large-scale genomic resources accessible for non-model organisms. The genomic information that is either already available or will be available in the near future will transform our understanding of the amount, distribution and functional significance of genetic variation in natural populations, turning conservation genetics into conservation genomics (Allendorf *et al.*, 2010). There is a growing list of genomic techniques that can be used to detect variation in natural populations ranging from en-masse single nucleotide polymorphism (SNP) genotyping to whole-genome/transcriptome sequencing. Prior to selecting the NGS approach, researchers should carefully evaluate the trade-off between the economic resources at their disposal and the biological question they aim to target. NGS technologies find their immediate application in estimating neutral populations parameters (i.e. effective population size); here, they simply provide a larger number of markers and, hence, more accurate estimates. Nonetheless, a more exciting and promising avenue of research which is still largely unexplored in conservation is the identification of functionally important genes (candidate genes) implicated in adaptation. This potentially allows for the development of an understanding of the genetic basis of local adaptation and inbreeding depression. Ultimately, by coupling genomics with environmental features, it will be possible to predict viability of populations and their potential to respond to future changes. The flourishing of new methods that harness large numbers of tailored-to-the-scope molecular markers has proceeded in parallel with (and partially stimulated) the development of sophisticated analytical tools (Ketmaier & Caccone, 2013). These can be grouped into two large categories: tools to handle and filter the massive amount of data produced by each NGS run, and the tools to retrieve the information content of the data in a conservation and population genetics framework (Nowrousian, 2010).

16.4 STUDY CASES

We will consider both primary and peripheral divisions of freshwater fish in this review. The primary and peripheral freshwater fishes, although subjected to very different evolutionary histories because of their intrinsic ecological requirements, are the most impacted by human activities since historic times because of their economic importance. They offer very illustrative case studies from which general conclusions in terms of management and conservation of genetic diversity can be derived and eventually exported to other geographical settings. As detailed in the first section of this chapter, freshwater fishes belonging to the primary and primary-like divisions are the most important in studies of terrestrial zoo-geography because their intolerance to salt binds them to inland waters (Myers, 1938; Bianco & Nordlie, 2008). In southern Europe, cyprinids are the most represented and diversified primary freshwater fishes, and hence they are of greatest value in understanding biogeographic struc-ture and relationships. At the other end of the spectrum, peripheral taxa are not limited to inland waters but can migrate (at least theoretically) between fresh and salt waters. Salmonids are typical examples of this division. In southern Europe, however, there are no native migratory salmonid species; reports of trout found at sea are the result of introduc-tion of migratory stock from northern Europe (Snoj, 2002).

16.4.1 Salmonids

Salmonid fishes have received considerable scientific attention over the last 40 years (Verspoor et al., 2005). At the family level, interest in obtaining reliable phylogenies has been stimulated by both fundamen-tal (speciation, evolution of complex life histories) and applied (fisher-ies and conservation) questions (Crête-Lafreniere et al., 2012). For those species that regularly migrate between salt and fresh waters, the marine phase potentially represents an opportunity to cross river boundaries; however, the strong homing behaviour that most migratory salmonids display considerably reduces among-river gene flow (Bernatchez, 2001). The onset of local genetic differences is more pronounced in southern than in northern Europe due to the lack of a migratory behaviour in the former (Bernatchez, 2001). Whilst salmonids have not generated the number of endemic species comparable to that of primary species, nonetheless, they harbour considerable intraspecific diversity with many lineages geographically localised.

Table 16.1 summarises the papers published since 2000 on European salmonids and used for this chapter, with special attention to those explicitly focusing on southern Europe; 18 of the 23 studies considered are based either on mtDNA or nDNA, while the remaining 5 collected information from both genomic compartments. Genetic data revealed a human impact on the population genetic structure of the study species in about half of the cases (52%). When a single locus was used, about 25% of the studies revealed signals of human impact, and this percent-age rises to 75% when multiple markers were analysed simultaneously.

The most severe threat to the genetic integrity of wild salmonid pop-ulations is represented by gene flow from domesticated strains used to

Table 16.1 Genetic studies centred on salmonids reviewed in the present chapter. The geographical setting of the studies is shown (Dan., Danubian; Med., Mediterranean; see Figure 16.1). For each study the molecular approach employed is given and whether genetic data could detect human impact or not.

Taxon	Reference	Geographical area		Molecular marker	Human impact
		Dan.	Med.		
Coregonus lavaretus	Østbye et al.,2005	Yes	No	mtDNA (seq./SSCP, two loci)	No
Thymallus thymallus	Jacobsen et al., 2012	Yes	No	mtDNA (seq., whole genome)	No
	Koskinen et al., 2000	Yes	No	mtDNA PCR-RFLP (six loci)	No
	Koskinen et al., 2002	Yes	Yes	nDNA (17 microsatellite loci)	No
	Weiss et al., 2002	Yes	Yes	mtDNA (seq., one locus)	Yes
	Sušnik et al. 2004	No	Yes	nDNA (15 microsatellite loci)	Yes
	Froufe et al., 2005	Yes	No	mtDNA (seq., two loci)	No
	Gum et al., 2005	Yes	No	mtDNA (seq./RFLP, two loci)	Yes
				nDNA (12 microsatellite loci)	
	Weiss et al., 2013	Yes	Yes	mtDNA (seq./RFLP; one locus)	Yes
				nDNA (13 microsatellite loci)	
Thymallus aeliani	Meraner & Gandolfi, 2012	Yes	Yes	mtDNA (seq., one locus)	Yes
Salmo salar	King et al., 2000	Yes	Yes	mtDNA (RFLP, two loci)	Yes
	Verspoor et al., 2005	Yes	Yes	nDNA (110 enzymatic loci)	Yes
	Ribeiro et al., 2008	No	Yes	nDNA (eight microsatellite loci)	No
Salmo trutta	Aurelle & Berrebi, 2001	Yes	No	mtDNA (seq./SSCP, two loci)	Yes
	Bernatchez, 2001	Yes	Yes	mtDNA (seq., one locus; RFLP, two loci)	Yes
	Bouza et al., 2001	No	Yes	nDNA (34 enzymatic loci)	No
	McMeel et al., 2001	Yes	No	nDNA (RFLP, LDH locus)	No
	Ruzzante et al., 2001	Yes	No	nDNA (seven microsatellite loci)	Yes

Presa et al., 2002	Yes	mtDNA (seq., one locus; RFLP, two loci) nDNA (seq., one locus)	No
Bekkevold et al., 2004	Yes	nDNA (eight microsatellite loci)	Yes
Cortey et al., 2004	No	mtDNA (seq., one locus)	Yes
Ketmaier & Bianco, 2004	No	mtDNA (seq., two loci) nDNA (RFLP, LDH locus)	Yes
Martinez et al., 2007	No	mtDNA (RFLP) nDNA (one enzymatic locus; 12 microsatellite loci)	No
Keller et al., 2011	Yes	nDNA (18 microsatellite loci)	Yes

re-stock declining natural populations, but also by the re-stocking per se, or the translocations of native populations. This practice was initially regarded as an effective measure to counteract the anthropogenically induced population declines due to pollution, habitat destruction, excessive commercial fishing and angling. As evidence based on analysis of molecular markers grew, the negative impact of re-stocking became clear. The mechanisms involved are complex, but they can ultimately be brought back to the loss of phylogeographic histories, the swamping of autochthonous gene pools and possible disruption of local adaptations (Sušnik *et al.*, 2004 and references therein). The remainder of this section describes a few illustrative examples of molecular studies carried out on salmonid species and shows how genetic data could help in implementing conservation strategies.

Studies on the population genetic structure and phylogeographic history of the European whitefish (*Coregonus lavaretus*) has indicated that natural processes rather than human activities dominate the structure of this species. Signs of post-glacial recolonisation events of central and northern Europe from southern (alpine) refugia were clearly found in the species' genetic structure (Bernatchez, 2001). Demographic events such as isolation by distance, population and range size expansion and subsequent fragmentation effectively explain the current geographic distribution of mtDNA haplotypes.

In contrast to the largely natural patterns influencing *Coregonus lavaretus*, molecular work over the past decade on grayling (*Thymallus thymallus*), a popular sport fish throughout most of its geographical range, has indicated that both natural and anthropogenic processes are influencing this species. Grayling are experiencing a dramatic reduction in population sizes at the southern edge of their distribution (i.e. southern France, Switzerland, Austria and northern Italy). The species is subdivided into at least four major mtDNA groups, which differentiated in allopatry during the Pleistocene; Gum *et al.* (2005) proposed these as the basic evolutionary significant units (ESUs) for the conservation and management of grayling throughout Europe. In addition, Sušnik *et al.* (2004 and references therein) and Meraner and Gandolfi (2012) identified a distinct, genetically isolated, endangered Adriatic lineage in northeastern Italy and western Slovenia. This lineage, which is represented by the taxon *Thymallus aeliani* Valenciennes in Cuvier & Valenciennes, 1848 recently rehabilitated (Bianco, 2014), is illustrative of the threats the species is subjected to as a consequence of re-stocking activities. MtDNA haplotypes of Danubian origin were in fact found in

Adriatic populations, confirming human-mediated gene flow between these two clearly distinct lineages as the result of 40 years of re-stocking of the Adriatic lineage with individuals of Danubian origin. Analysis of polymorphic microsatellite loci confirmed that the Danubian gene pool of *Thymallus thymallus* has extensively introgressed into the Adriatic *Thymallus aeliani*, although panmixia has not entirely occurred (Sušnik *et al.*, 2004; Meraner & Gandolfi, 2012; Weiss *et al.*, 2013). According to Allendorf *et al.* (2001), such human-induced, hybridised populations should be managed by putting conservation efforts into maintaining and expanding remaining non-introgressed population segments.

The Atlantic salmon (*Salmo salar*) is one of the most important freshwater fish species both culturally and economically in the Northern Hemisphere. The species includes anadromous stocks, which are of great economic value, along with freshwater resident populations (i.e. land-locked), which occur in many parts of the species' range (Ribeiro *et al.*, 2008). The genetic structure of the species has been described accurately (King *et al.*, 2000). North American and European anadromous populations show strong genetic difference. Northern European populations are the most closely related to the North American ones; this evidence reflects historic (i.e. Quaternary) gene flow between the two regions. Within Europe the species is also highly structured; genetic divergence is more pronounced among populations across the southern part of the continent (France, Spain), although distinct clusters of populations of a more northerly (Baltic lineage) have also been identified. Non-anadromous populations are on the average more genetically fragmented than their anadromous counterparts (King *et al.*, 2000).

Threats to the Atlantic salmon come from two main sources. First, overexploitation and habitat deterioration have caused local declines in population sizes. Second, Atlantic salmon is subject to a number of offshore or coastal fisheries targeting populations from different rivers. Gene flow from farmed salmons into wild populations could also reduce the performance of the wild populations by altering their original genetic make-up. Given the deep phylogeographic break existing between North American and European salmon, stock transfers between continents should be avoided. Such events potentially cause introgression of hatchery-escapee traits into native wild populations and concomitant potential for outbreeding depression and disruption of local adaptation, as well as the possibility of non-native disease transfer (Ruzzante *et al.*, 2001). Farmed Atlantic salmon also seems to have an impact on the frequency of hybridisation events with brown trout. The two species may hybridise where the respective ranges over-lap, but at a low frequency (2%) (Verspoor *et al.*, 2005). The release of genetically marked salmon into natural streams has shown that farmed salmon are more likely to breed with brown trout than wild salmon (Verspoor *et al.*, 2005). However, the species seems capable of restoring viable populations even after a drastic bottleneck. Ribeiro *et al.* (2008) assessed the genetic make-up of the River Eo population (Spain), which experienced a severe decline throughout the 1990s. Genetic variability was found comparable to that of unaffected popu-lations and the estimated effective population size was relatively large. The genetic data were further corroborated by fisheries records, which has attested to a steady increase in the number of salmon returning to the river in recent years, leading the authors to conclude that the River Eo salmon population has not been heavily influenced by ran-dom genetic processes and that a potentially satisfactory recovery may be expected.

The brown trout (*Salmo trutta*) is the most widely distributed fresh-water fish of the Palearctic region and one of the most important eco-nomically. It displays considerable variation in morphology and life history, including anadromous, riverine and lacustrine populations. The species is structured into five major evolutionary lineages, which evolved in isolation during the Pleistocene and have remained largely allopatric ever since (Bernatchez, 2001). Further, within-lineage diver-gence has contributed to the patchy genetic structure of the species. Population structuring is more pronounced at southern latitudes as a consequence of the different impact of glaciations and lack of dispersal at sea. Geographically localised genetic lineages are likely to have experi-enced extensive local adaptation (Bernatchez, 2001). On the other hand, adaptive divergence may be impeded by gene flow owing to human management practices. Throughout Europe, rivers are stocked annu-ally with millions of captive-reared fish (Bianco, 1995a, 1995b, 1998, 2014). Stocked populations easily reproduce and form re-naturalised communities that are often erroneously considered native, resulting in pan-European genetic and biogeographic pollution. These re-stocking activities have often relied on hatchery trout imported from abroad and include translocations across major drainage boundaries. Brown trout that genetically belong to the Atlantic lineage (as defined in Bernatchez, 2001) are widely used in aquaculture and for re-stocking, and there is a wealth of data proving extensive hybridisation between introduced and native lineages (Ruzzante *et al.*, 2001; Keller *et al.*, 2011).

Hybridisation also changes how genetic variation is distributed among populations and locations. In areas not heavily stocked, a higher proportion of genetic variation is distributed among rivers rather than among locations within rivers, reflecting the natural tendency of the species to form local river-wide populations, genetically differentiated from one another (Ryman et al., 1979; Halvorsen & Stabell, 1990). Conversely, where stocking is pronounced, genetic variation is predominantly distributed among locations within rivers rather than among rivers as a consequence of the introduction of individuals carrying non-endemic genetic variants (Ruzzante et al., 2001). Stocking not only impacts the distribution and the genetics of populations, but also seems to have a more global impact on life-history traits as well. Combining mt- and nDNA, Hansen et al. (2000) showed that male hatchery trout contribute more than females to interbreeding with native trout in a Danish river. Also, selection seems to act more intensively against stocked hatchery trout that become anadromous compared to stocked hatchery trout that become resident. Stocked hatchery trout are thought to perform more poorly as sea trout than as resident trout, perhaps as a result of the differences in complexity of the two kinds of life cycles (Hansen et al., 2000). The life history of sea trout is highly complex due to the concomitant actions of various factors (timing of smoltification and migratory behaviour in the sea, amongst others). Stocked hatchery trout that become resident experience a comparatively more stable environment (Hansen et al., 2000). Hansen et al. (2000) also proved that most of the resident trout are males, and this could explain why gene flow from domesticated to wild trout is male-biased. This contrasts strongly with the conclusions of Bekkevold et al. (2004) that gene flow in wild brown trout is mediated mainly by females. However, Bekkevold et al. (2004) also showed that over 40–90 years of stocking with domesticated trout had little effect on sex-specific dispersal behaviour of native Danish trout populations.

Despite the occurrence of massive human-mediated dispersal of non-native genetic variants, local brown trout populations may still harbour native neutral and adaptive variation. The lactate dehydrogenase alleles LDH-C1*90 and LDH-C1*100 are important and useful markers that can be easily applied to readily distinguish the Mediterranean brown trout lineages from the Atlantic farmed brown trout (Atlantic lineage carries the LDH-C1*90 allele only; McMeel et al., 2001). Ketmaier and Bianco (2004) screened variation at that locus and coupled the analysis with mtDNA sequencing on 26 brown trout populations from central and southern Italy, where only the Mediterranean lineage should

Table 16.2 *Genetic studies centred on cyprinids reviewed in the present chapter. The geographical setting of the studies is shown (Dan., Danubian; Med., Mediterranean; see Figure 16.1). For each study the molecular approach employed is given and whether genetic data could detect human impact or not.*

Taxon	Reference	Geographical area		Molecular marker	Human impact
		Dan.	Med.		
Abramis brama	Hayden *et al.*, 2011	Yes	No	mtDNA (seq., one locus)	No
Alburnus spp.	Ketmaier *et al.*, 2009	No	Yes	mtDNA (seq., one locus)	Yes
Barbus barbus	Kotlík & Berrebi, 2001	Yes	Yes	mtDNA (seq., one locus)	No
Barbus graellsii	Bianco & Ketmaier, 2001	No	Yes	mtDNA (seq., one locus)	Yes
Barbus petenyi	Kotlík & Berrebi, 2002	Yes	Yes	mtDNA (seq., one locus)	No
Barbus spp.	Tsingenopoulos *et al.*, 2002	No	Yes	mtDNA (seq., one locus) nDNA (27 enzymatic loci)	Yes
	Livi *et al.*, 2013	No	Yes	mtDNA (seq., two loci) nDNA (three microsatellite loci)	Yes
Luciobarbus spp.	Tsingenopoulos *et al.*, 2003	No	Yes	mtDNA (seq., one locus)	No
Carassius auratus	Gao *et al.*, 2012	Yes	No	mtDNA (seq., two loci)	Yes
Chondrostoma lusitanicum	Mesquita *et al.*, 2001	No	Yes	mtDNA (seq./RFLP; three loci)	No
Chondrostoma spp.	Durand *et al.*, 2003	Yes	Yes	mtDNA (seq., one locus)	No
	Doadrio & Carmona, 2004	Yes	Yes	mtDNA (seq., one locus)	No
Barbus spp. *Chondrostoma* spp. *Luciobarbus* spp.	Mesquita *et al.*, 2007	No	Yes	mtDNA (seq., one locus)	No
Cobitis taenia	Culling *et al.*, 2006	Yes	Yes	mtDNA (seq., one locus)	No
Gobio benacensis	Bianco & Ketmaier, 2006	No	No	mtDNA (seq., one locus)	Yes
Gobio gobio	Schreiber, 2002	Yes	No	nDNA (29 enzymatic loci)	No

Species	Reference			Markers	
Gobio spp.	Mendel *et al.*, 2008	Yes	Yes	mtDNA (seq., one locus) / nDNA (seq., one locus)	Yes
Leuciscus leuciscus	Costedoat *et al.*, 2006	Yes	Yes	mtDNA (seq., three loci) / nDNA (12 enzymatic loci)	No
Squalius lucumonis	Bianco & Ketmaier, 2003	No	Yes	mtDNA (seq., one locus)	Yes
Squalius lucumonis	Tancioni *et al.*, 2013	No	Yes	mtDNA (seq., three loci)	No
Rhodeus amarus	Bryja *et al.*, 2010	Yes	Yes	mtDNA (seq., one locus) / nDNA (12 microsatellite loci)	Yes
Rhodeus spp.	Bohlen *et al.*, 2006	Yes	Yes	mtDNA (seq., one locus)	No
Rhodeus spp.	Zaki *et al.*, 2008	Yes	No	mtDNA (seq., two loci)	No
Rutilus rutilus	Demandt & Björklund, 2007	Yes	No	nDNA (5 microsatellite loci)	Yes
Rutilus rutilus	Larmuseau *et al.*, 2009	Yes	Yes	mtDNA (seq., one locus)	Yes
Rutilus spp.	Ketmaier *et al.*, 2008	Yes	Yes	mtDNA (seq., one locus)	No
Rutilus rutilus	Wyatt *et al.*, 2006	Yes	No	mtDNA (seq., one locus)	No
Abramis brama					
Scardinius erythrophthalmus					
Scardinius dergle	Freyhof *et al.*, 2005	No	Yes	mtDNA (seq., one locus) / nDNA (seq., one locus)	Yes
Telestes muticellus	Stefani *et al.*, 2004	No	Yes	mtDNA (seq., one locus)	No
Telestes muticellus	Zaccara *et al.*, 2007	No	Yes	mtDNA (seq., one locus)	No
Telestes souffia	Salzburger *et al.*, 2003	Yes	Yes	mtDNA (seq., one locus)	No
Telestes souffia	Salducci *et al.*, 2004	No	Yes	mtDNA (seq., one locus) / nDNA (13 enzymatic loci)	No
Telestes souffia	Muenzel *et al.*, 2010	Yes	No	nDNA (10 microsatellite loci)	Yes
Telestes spp.	Ketmaier *et al.*, 2004	Yes	Yes	mtDNA (seq., one locus)	Yes
Scardinius spp.					
Telestes spp.	Dubut *et al.*, 2012	Yes	Yes	mtDNA (seq., one locus) / nDNA (11 microsatellite loci)	Yes

Table 16.2 (cont.)

| Taxon | Reference | Geographical area | | Molecular marker | Human impact |
		Dan.	Med.		
Scardinius spp.	Bianco *et al.*, 2001	No	Yes	nDNA (28 enzymatic loci)	Yes
Scardinius spp.	Ketmaier *et al.*, 2003	No	Yes	nDNA (28 enzymatic loci)	Yes
Squalius spp.	Sanjur *et al.*, 2003	No	Yes	mtDNA (seq., one locus)	No
Squalius alburnoides	Cunha *et al.*, 2004	No	Yes	mtDNA (seq., one locus)	No
Squalius aradensis	Mesquita *et al.*, 2005	No	Yes	mtDNA (seq., one locus) nDNA (six microsatellite loci)	Yes
Squalius cephalus	Seifertová *et al.*, 2012	Yes	Yes	mtDNA (seq., one locus) nDNA (12 microsatellite loci)	No
Squalius spp.	Waap *et al.*, 2011	No	Yes	nDNA (seq., seven loci)	No
Vimba vimba	Hänfling *et al.*, 2009	Yes	Yes	mtDNA (seq., two loci)	Yes

in theory occur (Bernatchez, 2001). Thirteen populations (50%) carried the expected Mediterranean mtDNA haplotype; 4 of these 13 also carried the Mediterranean LDH-C1*100 allele, while in the remaining 9 the Atlantic LDH-C1*90 allele was found. In 5 of the other 13 populations carrying the Atlantic mtDNA haplotype the Mediterranean LDH-C1*100 was present at a very low frequency. Keller *et al.* (2011) showed how stocking potentially impacts candidate loci involved in temperature tolerance, reproductive timing and immune defence in Swiss brown trout. The results showed how the phylogeographic structure of these populations, although impacted by stocking especially in the Mediterranean drainages, was still detectable. Intriguingly, they also uncovered patterns of divergent selection in a marker linked to a quantitative trait locus for temperature tolerance and signs of balancing selection at an immune relevant gene.

16.4.2 Cyprinids

As outlined in previous sections of this chapter, southern European cyprinids have received considerable scientific attention in the last three decades due to their importance for unravelling the biogeographic history of the area. The centre of origin of the family is most probably east–southeast Asia; diversification in the peri-Mediterranean area started in the Oligocene generating most of the extant genera (Perea *et al.*, 2010). At the within-genera level, the 'Lago Mare' phase of the Mediterranean Sea (see Introduction) triggered a burst of speciation events that, in conjunction with Quaternary events, led to the current high number of endemic and allopatric lineages that are range-restricted to the different ichthyogeographic districts (Figure 16.1).

Table 16.2 lists the papers published since 2000 on European cyprinids considered in this chapter. Eleven of the 41 studies are based on both mtDNA and nDNA, while the remaining 30 collected information from one genomic compartment only. Genetic data revealed human impact on the population genetic structure of the study species in about 43% of the cases. When a single locus was used, about 61% of the studies unveiled signals of human impact. This percentage drops to 39% when multiple markers were analysed simultaneously; this evidence is likely an artifact due to the interplay between sampling design (most of the studies on cyprinids do not have a specific conservation angle) and the particular species screened. The remainder of this section describes a few illustrative examples of molecular studies carried out on cyprinid species, and shows how genetic data can help in implementing conservation strategies.

Cyprinids are economically less valuable than salmonids, and are not farmed to the same extent. Indeed, only a few species are farmed for human consumption, although a comparatively larger number of species have a fisheries value. For many years there has been little scientific interest on how to manage and conserve their diversity while their usage as a resource for recreational activities, including recreational and artisanal fishing, has progressively increased (Copp *et al.*, 2005). As a consequence, the political and economic pressures posed by sport-fishing demands have nullified the (very limited) conservation efforts. Attempts to boost fish diversity and production in southern European rivers have been based essentially on transfer of fishes (Bianco, 1995a). However, whereas in salmonids most of the re-stocking of rivers is accomplished with conspecific-farmed animals belonging to genetic lineages other than the native ones, species coming from geographically distant and faunistically distinct zoogeographic regions are often used in introductions of non-salmonid taxa. In these circumstances, introduced taxa do not blend with local communities, but rather interactions are grounded on hetero-specific hybridisation, competition and predation (Bianco, 1995a, 1995b). The limited attention paid to this type of translocation in Europe has had profound impacts in terms of the genetic, biogeographic and faunistic alteration of southern European cyprinids, causing the local extinction of a number of native species (reviewed in Bianco, 1995a, 1995b; Copp *et al.*, 2005). Bianco and Ketmaier (2001) showed that in central Italian rivers non-native species often constitute the predominant biomass; mtDNA and field data revealed for the first time viable populations of the Iberian barbel (*Barbus graellsii*) in Central Italian rivers. Hybridisation events were detected between *Barbus plebejus* and *Barbus caninus* after the successful transfer followed by the invasive establishment of the former within the range of the latter (Tsingenopoulos *et al.*, 2002).

The genetic structure of the Italian bleaks (genus *Alburnus*) carries clear signatures of the effects of the massive translocations of fishes that have been carried out for more than a century along the Italian peninsula. In particular, Ketmaier *et al.* (2009) detected three main mtDNA clusters, two of which contained a mixture of haplotypes from Northern and Southern Italy. In spite of the fact that sampling localities were separated by hundreds of kilometres, allopatric haplotypes were either identical or just a single mutation apart from one another. Conversely, within a single southern Italian river (River Ofanto), a total of 19 haplotypes was found, most of them not phylogenetically related to one other and present at very low frequencies. All this evidence speaks in favour of a scenario based on the artificial introduction of individuals of allochthonous origin from northern to southern Italy (from districts 7 to 9 in Figure 16.1), in close agreement with the historical records of fish transfers within the nation (Bianco, 1995a). In the study, an mtDNA clade of exclusively southern Italian origin was also identified; its range includes a few rivers apparently free of introduction of allochthonous haplotypes that should hence be considered reservoirs of the autochthonous southern Italian bleak.

The Ponto-Caspian gudgeons (genus *Gobio*) offer one of the most remarkable examples of a recent and successful invasion across Europe; this has taken place in the last three decades and has been substantially facilitated by both the translocation of fishes and river regulation (Copp *et al.*, 2005; Mendel *et al.*, 2008). The introduction of the invasive Danubian *G. gobio* is causing the decline of the Italian endemic *G. benacensis*, which is apparently not able to compete with its congeneric (Bianco & Ketmaier, 2006).

The common rudd (*Scardinius hesperidicus*) is yet another illustrative example of how introduced species can outcompete closely related taxa following translocation events. The common rudd, originally distributed north of the Italian Apennines (district 7; Figure 16.1), was extensively transplanted into the original distributional range of the congeneric *S. scardafa*, a species endemic to central Italy (district 8; Figure 16.1) (Bianco, 1995a, 1995b; Ketmaier *et al.*, 2003, 2004). The common rudd has outcompeted and excluded *S. scardafa* throughout the whole of central Italy (Bianco, 1995a, 1995b; Bianco *et al.*, 2001; Bianco & Ketmaier, 2001), with *S. scardafa* now limited to a single population (Bianco, 1995a, 1995b; Ketmaier *et al.*, 2003, 2004). Similar to the pattern in bleaks, the common rudd is genetically homogeneous across geographic barriers that should in theory prevent gene flow. The allochthonous central Italian populations are almost undistinguishable both at the mt- and nDNA levels from the autochthonous northern Italian populations. The analysis of museum samples of *S. scardafa* belonging to the type series used to describe the species in 1837 (prior to any translocations of fishes) confirmed the mtDNA integrity of the single extant central Italian population of this critically endangered species. As a matter of fact, the mtDNA of historical and contemporary samples of *S. scardafa* are identical and contain all the 23 substitutions fixed between the extant *S. scardafa* and the common rudd (Ketmaier *et al.*, 2004).

The examples reported above concern mostly medium- to large-bodied species that are typical of the lower course of rivers and lakes, where demands for sportfishing are more intense and hence the community pressure to stock with introduced taxa is high. It was hence rather unexpected to find evidence of anthropogenic activities in the genetic structure of minnows (*Telestes souffia* complex). Minnows are quite unique among southern European cyprinids because of their specialised ecological niche. The species is found only in small rivers (less than 5 m wide) and within 10 km of the river source (Dubut *et al.*, 2012 and references therein). These fishes are rather small-bodied and not attractive for anglers even though they are sometimes used as bait. This, along with their peculiar ecology, would likely render them less prone to translocation than larger species more commonly targeted by fishermen. Indeed, most of the genetic literature on this and congeneric species (listed in Table 16.2) do describe patterns congruent with a pristine population structure. However, Dubut *et al.* (2012), thanks to a geographically dense sampling scheme, and by combining mt- and nDNA data, were able to detect cases of secondary, human-induced contacts between previously allopatric lineages of *T. souffia* in southern France. In that case, direct translocations were not held responsible; rather channelisation across river drainages to facilitate irrigation appears to be the main mechanism that could have brought previously isolated lineages in contact. Dam construction has been identified by Freyhof *et al.* (2005) as the trigger for hybridisation between two cyprinid species belonging to different genera (*Scardinius dergle* and *Squalius tenellus*), confirming the proneness of members of this family to easily form intergeneric hybrids, a phenomenon generally uncommon in animals.

Habitat alterations can be potentially devastating in species with restricted and fragmented ranges, resulting in demographic crashes potentially leading to severe bottlenecks, through which rare genetic variants are easily lost. The critically endangered *Squalius aradensis* is limited to small southern Portuguese drainages; these habitats exhibit a characteristic Mediterranean hydrological system throughout the year, alternating flooding events in winter and complete drought along many river reaches in summer. Mesquita *et al.* (2005) detected severe bottlenecks in the species; intriguingly, genetic diversity was positively correlated with upstream drainage areas, where severe droughts (and hence fish mortality) are less likely to occur, identifying these areas as genetic reservoirs of greatest relevance for conservation purposes.

16.5 CONCLUSIONS AND FUTURE DIRECTIONS

In both groups of fishes (salmonids and cyprinids) covered in this review, genetic data have clearly revealed the impact of anthropogenic activities, but with some major differences. Genetic data have revealed human impacts on the population genetic structure of salmonids in about half of the cases considered in this chapter (52%). When a single locus was used, about 25% of the studies unveiled signals of human impact. This percentage rises to 75% when multiple markers were analysed simultaneously. In cyprinids, when a single locus was used, about 61% of the studies unveiled signals of human impact, but only in 39% of the cases when multiple markers were analysed simultaneously. It should be reiterated here that most of the genetic studies on cyprinids were designed to qualitatively describe phylogeographic patterns and, to a much smaller extent, to quantitatively assess the genetic consequences of translocations. This is the most parsimonious explanation for the counter-intuitive evidence that increasing the number of markers in cyprinids reveals fewer translocations than a single-marker approach. At the same time, these results also suggest that more markers need to be screened to detect fine alterations in the population genetic structure of salmonids. The fact that translocations in salmonids are often intra-specific while they are interspecific in cyprinids also highlights the need to use a higher number of polymorphic markers in the former to reveal subtle differences.

Mounting evidence suggests that the coupling of human-mediated extirpation of native populations and the introduction of non-native species has resulted in significant changes in freshwater fish faunas at regional and global scales. Fish species are being intentionally and acci-dentally moved around continents at an unprecedented and increasing rate; many species are now distributed in regions they could never have reached naturally. Additionally, human enterprise has greatly acceler-ated the rate at which native fish populations are reduced, eliminated and substituted with non-native ones. We are still missing a global evalu-ation of these changes on freshwater fish biogeography, both at the pre-sent time and a future perspective. To address these questions, experts from different fields (i.e. taxonomists, ecologists and molecular biolo-gists) are required to work together under the same unifying umbrella to face the challenge of reconciling the historical biogeography of native fishes with the current rapid advance of non-native species, against a backdrop of ongoing, global environmental change.

To conclude this chapter, we would like to list a few points that should be kept in mind when it comes to considering freshwater fishes, their diversity and the threats they are currently faced with. We hope that the points below will spark further work on these issues.

In spite of the investment in terms of scientific investigations, protection laws and so on, very little can be done in practice against non-native species once they become established in a certain area. Salmonids and cyprinids present illustrative examples of the genetic consequences of the cross-catchment translocations of congenerics and conspecifics while the threats posed by these actions in terms of loss of phylogeographic histories of native species are still largely ignored by deputy authorities.

The exponential increase of species that are moved around the globe seems unstoppable, as so many of the facts (political, economic) around them often escape the scientific community. As a consequence, scientific efforts aimed at understanding the genetic consequences of these actions are still extremely limited and often directed to a few economically valuable species.

Emphasis should be put on educational programmes and initiatives to increase the general awareness of preserving the original biodiversity of fish communities in terms of native species and the phylogeographic histories written in their genes. These initiatives should aim at reaching both the general audience and primary consumers (i.e. sport anglers); efforts should be made to render the results of genetic studies easily accessible for non-specialists.

Breeding and hatchery practices, when unavoidable, should be routinely based on genetic screening to minimise the potential negative impact of re-stocking activities with allochthonous genetic lineages. Similarly, screening for the occurrence of non-native species and of non-native genetic lineages should be routinely included in any programme aimed at managing a given area; consequently, efforts should be directed to preserve those catchments and/or river sections where only native lineages are present (see the examples of brown trout in central Italy and *Alburnus* in southern Italy).

16.6 DISCUSSION QUESTIONS

1. What can patterns of genetic structure of fish communities tell us about the evolutionary biogeography of freshwater fish communities?

2. Discuss the threat that the introduction of non-native genetic lineages of fish pose to river catchments where only native lineages are present.
3. How can genetics be used to investigate cryptic diversity present in freshwater systems?
4. How are 'genetic bottlenecks' generated, and what threat do they pose to species?
5. How much genetic diversity should we conserve within any particular species?

16.7 REFERENCES

Allendorf, F. W., Leary, R. F., Spruell, P. & Wenburg, J. K. (2001). The problems with hybrids: setting conservation guidelines. *Trends in Ecology and Evolution*, 16, 613–622.

Allendorf, F. W., Hohenlohe, P. A. & Luikart, G. (2010). Genomics and the future of conservation genetics. *Nature Reviews Genetics*, 11, 697–709.

Aurelle, D. & Berrebi, P. (2001). Genetic structure of brown trout (*Salmo trutta*, L.) populations from south-western France: data from mitochondrial control region variability. *Molecular Ecology*, 10, 1551–1561.

Avise, J. C. (2000). *Phylogeography: The History and Formation of Species*. Cambridge, MA): Harvard University Press.

Bekkevold, D., Hansen, M. M. & Mensberg, K. L. (2004). Genetic detection of sex-specific dispersal in historical and contemporary populations of anadromous brown trout *Salmo trutta*. *Molecular Ecology*, 13, 1707–1712.

Bernatchez, L. (2001). The evolutionary history of the brown trout (*Salmo trutta* L.) inferred from phylogeographic, nested clade, and mismatch analysis of mitochondrial DNA variation. *Evolution*, 33, 351–379.

Bianco, P. G. (1990). Potential role of the palaeohistory of the Mediterranean and Paratethys basin on the early dispersal of Europe-Mediterranean freshwater fishes. *Ichthyological Exploration of Freshwaters*, 1, 167–184.

Bianco, P. G. (1995a). Mediterranean endemic freshwater fishes of Italy. *Biology Conservation*, 72, 159–170.

Bianco, P. G. (1995b). Introductions, chief elements of native freshwater fish degradation and use of indices and coefficients in quantifying the situation in Italy. In *Protection of Acquatic Biodiversity*. Proceedings of the World Fisheries Congress. Philipp, D. P. (Ed.). New Delhi: Oxford & IBH, pp. 175–198.

Bianco, P. G. (1998). Freshwater fish transfers in Italy: history, local modification of fish composition, and a prediction on the future of native populations. In *Stocking and Introductions of Fishes*. Cowx, J. (Ed.). Oxford: Blackwell Science, pp. 165–197.

Bianco, P. G. (2005). La transfaunazione e la bio-globalizzazione con particolare riferimento ai pesci d'acqua dolce: un processo inarrestabile. In *Gestione della fauna selvatica e conservazione della biodiversità*. De Filippo, G. & Fulgione, D. (Eds). Rome: Esperienze, pp. 93–104.

Bianco, P. G. (2014). An update on the status of native and exotic freshwater fishes of Italy. *Journal of Applied Ichtyology*, 30, 62–77.

Bianco, P. G. & Ketmaier, V. (2001). Anthropogenic changes in the freshwater fish fauna in Italy with reference to the central region and *Barbus graellsii*, a newly established alien species of Iberian origin. *Journal of Fish Biology*, 59, 190–208.

Bianco, P. G. & Ketmaier, V. (2003). Threatened fishes of the world: *Leuciscus lucumonis* Bianco, 1983 (Cyprinidae). *Environmental Biology of Fishes*, 68, 370.

Bianco, P. G. & Ketmaier, V. (2006). Will the Italian endemic gudgeon, *Gobio benacensis*, survive the interaction with the invasive introduced *Gobio gobio*? *Folia Zoologica*, 54, 42–49.

Bianco, P. G. & Nordlie, F. (2008). The salinity tolerance of *Pseudophoxinus stymphalicus* (Cyprinidae) and *Valencia letourneuxi* (Valenciidae) from western Greece suggests a revision of the ecological categories of freshwater fishes. *Italian Journal of Zoology*, 75, 285–293.

Bianco, P. G., Ketmaier, V. & Busatto, T. (2001). Approccio multidisciplinare all'analisi tassonomica del genere *Scardinius* (Cyprinidae) in Europa. *Quaderni ETP*, 30, 115–120.

Bohlen, J., Šlechtová, V., Bogutskaya, N. & Freyhof, J. (2006). Across Siberia and over Europe: phylogenetic relationships of the freshwater fish genus *Rhodeus* in Europe and the phylogenetic position of *R. sericeus* from the River Amur. *Molecular Phylogenetics and Evolution*, 40, 856–865.

Bouza, C., Castro, J., Sánchez, L. & Martínez, P. (2001). Allozymic evidence of parapatric differentiation of brown trout (*Salmo trutta* L.) within an Atlantic river basin of the Iberian Peninsula. *Molecular Ecology*, 10, 1455–1469.

Bryja, J., Smith, C., Konečný, A. & Reichard, M. (2010). Range-wide population genetic structure of the European bitterling (*Rhodeus amarus*) based on microsatellite and mitochondrial DNA analysis. *Molecular Ecology*, 19, 4708–4722.

Copp, G. H., Bianco, P. G., Bogutskaya, N. G., *et al.* (2006). To be, or not to be, a non-native freshwater fish? *Journal of Applied Ichthyology*, 21, 242–262.

Cortey, M., Pla, C. & Garcia-Marin, J. L. (2004). Historical biogeography of Mediterranean trout. *Molecular Phylogenetics and Evolution*, 33, 831–844.

Costedoat, C., Chappaz, R., Barascud, B., Guillard, O. & Gilles, A. (2006). Heterogeneous colonization pattern of European Cyprinids, as highlighted by the dace complex (Teleostei: Cyprinidae). *Molecular Phylogenetics and Evolution*, 41, 127–148.

Crête-Lafrenière, A., Weir, L. K. & Bernatchez, L. (2012). Framing the Salmonidae family phylogenetic portrait: a more complete picture from increased taxon sampling. *PLoS ONE*, 7, e46662.

Culling, M. A., Janko, K., Boron, A., *et al.* (2006). European colonization by the spined loach (*Cobitis taenia*) from Ponto-Caspian refugia based on mitochondrial DNA variation. *Molecular Ecology*, 15, 173–190

Cunha, C., Coelho, M. M., Carmona, J. A. & Doadrio, I. (2004). Phylogeographical insights into the origins of the *Squalius alburnoides* complex via multiple hybridization events. *Molecular Ecology*, 13, 2807–2817

Demandt, M. & Björklund, M. (2007). Loss of genetic variability in reintroduced roach (*Rutilus rutilus*) populations. *Journal of Fish Biology*, 70, 255–261.

Doadrio, I. & Carmona, J. A. (2004). Phylogenetic relationships and biogeography of the genus *Chondrostoma* inferred from mitochondrial DNA sequences. *Molecular Phylogenetics and Evolution*, 33, 802–815.

Dubut, V., Fouquet, A., Voisin, A., *et al.* (2012). From late Miocene to Holocene: processes of differentiation within the *Telestes* genus (Actinopterygii: Cyprinidae). *PLoS ONE*, 7, e34423.

Durand, J. D., Bianco, P. G., Laroche, J. & Gilles, A. (2003). Insight into the origin of endemic Mediterranean ichthyofauna: phylogeography of *Chondrostoma* genus (Teleostei, Cyprinidae). *Journal of Heredity*, 94, 315–328.

Freyhof, J., Lieckfeldt, D., Bogutskaya, N. G., Pitra, C. & Ludwig, A. (2005). Molecules and morphology: evidence for introgression of mitochondrial DNA in Dalmatian cyprinids. *Molecular Phylogenetics and Evolution*, 37, 347–354.

Froufe, E., Knizhin, I. & Weiss, S. (2005). Phylogenetic analysis of the genus *Thymallus* (grayling) based on mtDNA control region and ATPase 6 genes, with inferences on control region constraints and broad-scale Eurasian phylogeography. *Molecular Phylogenetics and Evolution*, 34, 106–117.

Gao, Y., Wang, S. Y., Luo, J., *et al.* (2012). Quaternary palaeoenvironmental oscillations drove the evolution of the Eurasian *Carassius auratus* complex (Cypriniformes, Cyprinidae). *Journal of Biogeography*, 39, 2264–2278.

Garcia-Castellanos, D., Estrada, F., Jimenez-Munt, I., *et al.* (2009). Catastrophic flood of the Mediterranean after the Messinian salinity crisis. *Nature*, 462, 778–782.

Gozlan, R. E., Britton, J. R., Cowx, I. & Copp, G. H. (2010). Current knowledge on non-native freshwater fish introductions. *Journal of Fish Biology*, 76, 751–786.

Gum, B., Gross, R. & Kuehn, R. (2005). Mitochondrial and nuclear DNA phylogeography of European grayling (*Thymallus thymallus*): evidence for secondary contact zones in central Europe. *Molecular Ecology*, 14, 1707–1725.

Halvorsen, M. & Stabell, O. B. (1990). Homing behaviour of displaced stream-dwelling brown trout. *Animal Behaviour*, 39, 1089–1097.

Hänfling, B., Dümpelmann, C., Bogutskaya, N. G., Brandl, R. & Brändle, M. (2009). Shallow phylogeographic structuring of *Vimba vimba* across Europe suggests two distinct refugia during the last glaciation. *Journal of Fish Biology*, 75, 2269–2286.

Hansen, M. M., Ruzzante, D. E., Nielsen, E. E. & Mensberg, K. L. D. (2000). Microsatellite and mitochondrial DNA polymorphism reveals life-history dependent interbreeding between hatchery and wild brown trout (*Salmo trutta* L.). *Molecular Ecology*, 9, 583–594

Harrison, I. J. & Stiassny, M. L. J. (1999). The quiet crisis: a preliminary listing of freshwater fishes of the World that are either extinct or 'missing in action'. In *Extinctions in Near Time: Causes, Contexts, and Consequences*. New York and London: Plenum Press, pp. 271–331.

Hayden, B., Coscia, I. & Mariani, S. (2011). Low cytochrome b variation in bream *Abramis brama*. *Journal of Fish Biology*, 78, 1579–1587

Hewitt, G. M. (1999). Post-glacial re-colonization of European biota. *Biological Journal of the Linnean Society*, 68, 87–112.

Hsü, K., Ryan, W. B. & Cita, M. B. (1973). Late Miocene desiccation of the Mediterranean. *Nature*, 242, 240–244.

Jacobsen, M. W., Hansen, M. M., Orlando, L., *et al.* (2012). Mitogenome sequencing reveals shallow evolutionary histories and recent divergence time between morphologically and ecologically distinct European whitefish (*Coregonus* spp.). *Molecular Ecology*, 21, 2727–2742.

Keller, I., Taverna, A. & Seehausen, O. (2011). Evidence of neutral and adaptive genetic divergence between European trout populations sampled along altitudinal gradients. *Molecular Ecology*, 20, 1888–1904.

Ketmaier, V. & Bianco, P. G. (2004). Monitoraggio genetico e ibridazione tra popolazioni atlantiche e mediterranee di *Salmo trutta* in Abruzzo e Campania. In: *Ecologia*. Atti del XIII Congresso Nazionale SITE, Aracne. Available at: www.xiiicongresso.societaitalianaecologia.org/articles/Ketmaier-138.pdf

Ketmaier, V. & Caccone, A. (2013). Twenty years of molecular biogeography in the west Mediterranean islands of Corsica and Sardinia: lessons learnt and future prospects. In *Current Progress in Biological Research*. Rijeka, Croatia: InTech.

Ketmaier, V., Bianco, P. G., Cobolli, M. & De Mattheis, E. (2003). Genetic differentiation and biogeography in southern European populations of the genus *Scardinius* (Pisces, Cyprinidae) based on allozyme data. *Zoologica Scripta*, 32, 13–22.

Ketmaier, V., Bianco, P. G., Cobolli, M., *et al.* (2004). Molecular phylogeny of two lineages of Leuciscinae cyprinids (*Telestes* and *Scardinius*) from the peri-Mediterranean area based on cytochrome b data. *Molecular Phylogenetics and Evolution*, 32, 1061–1071.

Ketmaier, V., Bianco, P. G. & Durand, J. D. (2008). Molecular systematics, phylogeny and biogeography of roaches (*Rutilus*, Teleostei, Cyprinidae). *Molecular Phylogenetics and Evolution*, 49, 362–367.

Ketmaier, V., Finamore, F., Largiadèr, C., Milone, M. & Bianco, P. G. (2009). Phylogeography of bleaks (genus *Alburnus*, Cyprinidae) in Italy based on cytochrome b data. *Journal of Fish Biology*, 75, 997–1017.

King, T. L., Spidle, A. P., Eackles, M. S., Lubinski, B. A. & Schill, W. B. (2000). Mitochondrial DNA diversity in North American and European Atlantic salmon with emphasis on the Downeast rivers of Maine. *Journal of Fish Biology*, 57, 614–630.

Koskinen, M. T., Ranta, E., Piironen, J., *et al.* (2000). Genetic lineages and postglacial colonization of grayling (*Thymallus thymallus*, Salmonidae) in Europe, as revealed by mitochondrial DNA analyses. *Molecular Ecology*, 9, 1609–1624.

Koskinen, M. T., Nilsson, J., Veselov, A. J., *et al.* (2002). Microsatellite data resolve phylogeographic patterns in European grayling, *Thymallus thymallus* Salmonidae. *Heredity*, 88, 391–401.

Kotlík, P. & Berrebi, P. (2001). Phylogeography of the barbel (*Barbus barbus*) assessed by mitochondrial DNA variation. *Molecular Ecology*, 10, 2177–2185.

Kotlík, P. & Berrebi, P. (2002). Genetic subdivision and biogeography of the Danubian rheophilic bar *Barbus petenyi* inferred from phylogenetic analysis of mitochondrial DNA. *Molecular Phylogenetics and Evolution*, 24, 10–18.

Kottelat, M. & Freyhof, J. (2007). *Handbook of European Freshwater Fishes*. Cornol, Switzerland: Kottelat and Berlin: Freyhof.

Krijgsman, W., Stoica, M., Vasiliev, I. & Popov, V.V. (2010). Rise and fall of the Paratethys Sea during the Messinian Salinity Crisis. *Earth and Planetary Science Letters*, 290, 183–191.

Larmuseau, M. H. D., Freyhof, J., Volckaert, F. A. M. & Van Houdt, J. K. J. (2009). Matrilinear phylogeography and demographical patterns of *Rutilus rutilus*: implications for taxonomy and conservation. *Journal of Fish Biology*, 75, 332–353.

Leprieur, F., Beauchard, O., Blanchet, S., Oberdorff, T. & Brosse, S. (2008). Fish invasions in the world's river systems: when natural processes are blurred by human activities. *PLoS Biology*, 6, e28.

Leprieur, F., Olden, J. D., Lek, S. & Brosse, S. (2009). Contrasting patterns and mechanisms of spatial turnover for native and exotic freshwater fish in Europe. *Journal of Biogeography*, 36, 1899–1912.

Livi, S., de Innocentiis, S., Longobardi, A., *et al.* (2013). Genetic structure of *Barbus* spp. populations in the Marches Region of central Italy and its relevance to conservation actions. *Journal of Fish Biology*, 82, 806–826.

Martínez, P., Bouza, C., Castro, J., *et al.* (2007). Analysis of a secondary contact between divergent lineages of brown trout *Salmo trutta* L. from Duero basin using microsatellites and mtDNA RFLPs. *Journal of Fish Biology*, 71, 195–213.

McMeel, O. M., Hoey, E. M. & Ferguson, A. (2001). Partial nucleotide sequences, and routine typing by polymerase chain reaction-restriction fragment length polymorphism, of the brown trout (*Salmo trutta*) lactate dehydrogenase, LDH-C1*90 and *100 alleles. *Molecular Ecology*, 10, 29–34.

Mendel, J., Lusk, S., Vasil'eva, E. D., *et al.* (2008). Molecular phylogeny of the genus *Gobio* Cuvier, 1816 (Teleostei: Cyprinidae) and its contribution to taxonomy. *Molecular Phylogenetics and Evolution*, 47, 1061–1075.

Meraner, A. & Gandolfi, A. (2012). Phylogeography of European grayling, *Thymallus thymallus* (Actinopterygii, Salmonidae), within the Northern Adriatic basin: evidence for native and exotic mitochondrial DNA lineages. *Hydrobiologia*, 693, 205–221.

Mesquita, N., Carvalho, G., Shaw, P., Crespo, E. & Coelho, M. M. (2001). River basin-related genetic structuring in an endangered fish species, *Chondrostoma lusitanicum*, based on mtDNA sequencing and RFLP analysis. *Heredity*, 86, 253–264.

Mesquita, N., Hänfling, B., Carvalho, G. R. & Coelho, M. M. (2005). Phylogeography of the cyprinid *Squalius aradensis* and implications for conservation of the endemic freshwater fauna of southern Portugal. *Molecular Ecology*, 14, 1939–1954.

Mesquita, N., Cunha, C., Carvalho, G. R. & Coelho, M. M. (2007). Comparative phylogeography of endemic cyprinids in the south-west Iberian Peninsula: evidence for a new ichthyogeographic area. *Journal of Fish Biology*, 71, 45–75.

Muenzel, F. M., Salzburger, W., Sanetra, M., Grabherr, B. & Meyer, A. (2010). Genetic structure of the vairone *Telestes souffia* in the eastern part of Lake Constance, central Europe. *Journal of Fish Biology*, 77, 1158–1164.

Myers, G. S. (1938). Freshwater fishes of West Indian zoogeography. *Smithsonian Reports*, 1937, 339–364.

Nowrousian, M. (2010). Next-generation sequencing techniques for eukaryotic microorganisms: sequencing-based solutions to biological problems. *Eukaryotic Cell*, 9, 1300–1310.

Olden, J. D., Kennard, M. J., Leprieur, F., *et al.* (2010). Conservation biogeography of freshwater fishes: recent progress and future challenges. *Diversity and Distributions*, 16, 496–513.

Orszag-Sperber, F. (2006). Changing perspectives in the concept of 'Lago-Mare' in Mediterranean Late Miocene evolution. *Sedimentary Geology*, 188–189, 259–277.

Østbye, K., Bernatchez, L., Næsje, T. F., Himberg, K. J. M. & Hindar, K. (2005). Evolutionary history of the European whitefish *Coregonus lavaretus* (L.) species complex as inferred from mtDNA phylogeography and gill-raker numbers. *Molecular Ecology*, 14, 4371–4387.

Perea, S., Böhme, M., Zupancic, P., *et al.* (2010). Phylogenetic relationships and biogeographical patterns in Circum-Mediterranean subfamily Leuciscinae (Teleostei, Cyprinidae) inferred from both mitochondrial and nuclear data. *BMC Evolutionary Biology*, 10, 265.

Popov, S. V., Shcherba, I. G., Ilyina, L. B., *et al.* (2006). Late Miocene to Pliocene palaeogeography of the Paratethys and its relation to the Mediterranean. *Palaeogeography, Palaeoclimatology and Palaeoecology*, 238, 91–106.

Presa, P., Pardo, B. G., Martinez, P. & Bernatchez, L. (2002). Phylogeographic congruence between mtDNA and rDNA ITS markers in brown trout. *Molecular Biology and Evolution*, 19, 2161–2175.

Reyjol, Y., Hugueny, B., Pont, D., *et al.* (2007). Patterns in species richness and endemism of European freshwater fish. *Global Ecology and Biogeography*, 16, 65–75.

Ribeiro, Â., Morán, P. & Caballero, A. (2008). Genetic diversity and effective size of the Atlantic salmon *Salmo salar* L. inhabiting the River Eo (Spain) following a stock collapse. *Journal of Fish Biology*, 72, 1933–1944.

Ricciardi, A. & Rasmussen, J. B. (1999). Extinction rates of North American freshwater fauna. *Conservation Biology*, 13, 1220–1222.

Ruzzante, D. E., Hansen, M. M. & Meldrup, D. (2001). Distribution of individual inbreeding coefficients, relatedness and influence of stocking on native anadromous brown trout (*Salmo trutta*) population structure. *Molecular Ecology*, 10, 2107–2128.

Ryman, N., Allendorf, F. & Ståhl, G. (1979). Reproductive isolation with little genetic divergence in sympatric populations of brown trout (*Salmo trutta*). *Genetics*, 92, 247–262.

Salducci, M. D., Martin, J. F., Pech, N., *et al.* (2004). Deciphering the evolutionary biology of freshwater fish using multiple approaches insights for the biological conservation of the vairone. *Conservation Genetics*, 5, 63–77.

Salzburger, W., Brandstätter, A., Gilles, A., *et al.* (2003). Phylogeography of the vairone (*Leuciscus souffia*, Risso 1826) in Central Europe. *Molecular Ecology*, 12, 2371–2386.

Sanjur, O. I., Carmona, J. A. & Doadrio, I. (2003). Evolutionary and biogeographical patterns within Iberian populations of the genus *Squalius* inferred from molecular data. *Molecular Phylogenetics and Evolution*, 29, 20–30.

Schreiber, A. (2002). Differences in levels of heterozygosity in populations of the common gudgeon (*Gobio gobio*, Cyprinidae) among adjacent drainages in Central Europe: an effect of postglacial range dynamics? *Heredity*, 89, 163–170.

Scribner, K. T., Page, K. & Bartron, M. L. (2001). Hybridization in freshwater fishes: a review of case studies and cytonuclear methods of biological inference. *Reviews in Fish Biology and Fisheries*, 10, 293–323.

Seifertová, M., Bryja, J., Vyskočilová, M., Martínková, N. & Šimková, A. (2012). Multiple Pleistocene refugia and post-glacial colonization in the European chub (*Squalius cephalus*) revealed by combined use of nuclear and mitochondrial markers. *Journal of Biogeography*, 39, 1024–1040.

Smith, K. G. & Darwall, W. R. T. (2006). *The Status and Distribution of Freshwater Fish Endemic to the Mediterranean Basin*. Gland, Switzerland and Cambridge: IUCN.

Snoj, A., Marceta, B., Sušnik, S., *et al.* (2002). The taxonomic status of the 'sea trout' from the north Adriatic Sea, as revealed by mitochondrial and nuclear DNA analysis. *Journal of Biogeography*, 29, 1179–1185.

Stefani, F., Galli, P., Zaccara, S. & Crosa, G. (2004). Genetic variability and phylogeography of the cyprinid *Telestes muticellus* within the Italian peninsula as revealed by mitochondrial DNA. *Journal of Zoological Systematics and Evolutionary Research*, 42, 323–331.

Sušnik, S., Berrebi, P., Dovc, P., Hansen, M. M. & Snoj, A. (2004). Genetic introgression between wild and stocked salmonids and the prospects for using molecular markers in population rehabilitation: the case of the Adriatic grayling (*Thymallus thymallus* L. 1785). *Heredity*, 93, 273–282.

Tancioni, L., Russo, T., Cataudella S., *et al.* (2013). Testing species delimitations in four Italian sympatric Leuciscine fishes in the Tiber River: a combined morphological and molecular approach. *PLoS ONE*, 8, e60392.

Tsigenopoulos, C. S., Ráb, P., Naran, D. & Berrebi, P. (2002). Multiple origins of polyploidy in the phylogeny of southern African barbs (Cyprinidae) as inferred from mtDNA markers. *Heredity*, 88, 466–473.

Tsigenopoulos, C. S., Durand, J. D., Unlu, E. & Berrebi, P. (2003). Rapid radiation of the Mediterranean *Luciobarbus* species (Cyprinidae) after the Messinian salinity crisis of the Mediterranean Sea, inferred from mitochondrial phylogenetic analysis. *Biological Journal of the Linnean Society*, 80, 207–222.

Verspoor, E., Beardmore, J. A., Consuegra, S., *et al.* (2005). Population structure in the Atlantic salmon: insights from 40 years of research into genetic protein variation. *Journal of Fish Biology*, 67, 3–54.

Vié, J. C., Hilton-Taylor, C. & Stuart, S. N. (2009). *Wildlife in a Changing World – An Analysis of the 2008 IUCN Red List of Threatened Species*. Gland, Switzerland: IUCN.

Vrijenhoek, R. C. (1998). Conservation genetics of freshwater fish. *Journal of Fish Biology*, 53, 394–412.

Waap, S., Amaral, A. R., Gomes, B. & Coelho, M. M. (2011). Multi-locus species tree of the chub genus *Squalius* (Leuciscinae: Cyprinidae) from western Iberia: new insights into its evolutionary history. *Genetica*, 139, 1009–1018.

Weiss, S., Persat, H., Eppe, R., Schlötterer, C. & Uiblein, F. (2002). Complex patterns of colonization and refugia revealed for European grayling *Thymallus thymallus*, based on complete sequencing of the mitochondrial DNA control region. *Molecular Ecology*, 11, 1393–1407.

Weiss, S. J., Kopun, T. & Bajec, S. S. (2013). Assessing natural and disturbed population structure in European grayling *Thymallus thymallus*: melding phylogeographic, population genetic and jurisdictional perspectives for conservation planning. *Journal of Fish Biology*, 82, 505–521.

Wyatt, P. M. W., Pitts, C. S. & Butlin, R. K. (2006). A molecular approach to detect hybridization between bream *Abramis brama*, roach *Rutulus rutilus* and rudd *Scardinius erythrophthalmus*. *Journal of Fish Biology*, 69, 52–71.

Zaccara, S., Stefani, F. & Delmastro, G. B. (2007). Phylogeographical structure of vairone *Telestes muticellus* (Teleostei, Cyprinidae) within three European peri-Mediterranean districts. *Zoologica Scripta*, 36, 443–453.

Zaki, S. A. H., Jordan, W. C., Reichard, M., Przybylski, M. & Smith, C. (2008). A morphological and genetic analysis of the European bitterling species complex. *Biological Journal of the Linnean Society*, 95, 337–347.

(17)

Maintaining taxonomic skills; the decline of taxonomy – a threat to fish conservation

MARIA J. COLLARES-PEREIRA, PAUL H. SKELTON
AND IAN G. COWX

17.1 INTRODUCTION

Species, however defined, are the generally accepted unit of biology, including in conservation biology where it is more conventional to consider elements in terms of landscape units, ecosystems or communities (Mace, 2004). Here we consider how a general decline in systematics and taxonomy, the science dealing with the identification, description, naming and classifying of organisms, might affect fish conservation (Reid, 2010).

The link between species and conservation has many facets and implications, all of which have changed over time as the paradigms of both systems themselves have changed. The one constant that a 'species' has for conservation is that of linkage and communication, of being the instrument through which conservation issues are brought into a common understanding for different audiences. The iconic nature of certain species linked to conservation illustrates this point – giant pandas, rhinos, elephants, tigers, the blue whale or, more relevant to this chapter, species like golden trout, Atlantic salmon, sturgeon, Tennessee darter, coelacanth, Australian lungfish and Asian arowana – all have clear conservation connotations. In every case, the icon connects directly to an environment or ecosystem and threats to its existence, and critically to the human dimensions of its conservation needs. Were it not that fish and fisheries are of major economic and cultural importance (Welcomme *et al.*, 2010), this situation might not be considered that serious, but fishes are key components of aquatic ecosystems and are both

Conservation of Freshwater Fishes, eds G. P. Closs, M. Krkosek and J. D. Olden. Published by Cambridge University Press. © Cambridge University Press 2016.

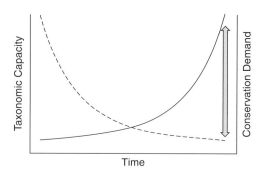

Figure 17.1 Conceptual relationship between a declining taxonomic capacity and a rising conservation demand. The arrow indicates the increasing 'gap' between taxonomic capacity and conservation demand.

an essential food resource as well as a source of other products. They also play a major cultural role, from aesthetics to pets and leisure, in the lives of millions of people (Reid *et al.*, 2013).

Their taxonomy is thus critical to recognise the threats to their existence and the steps required to conserve them. It is our contention, therefore, that conservation is all about 'human' concerns and actions, and, as such, taxonomy is a critical component in the process. A 'taxonomic impediment' (or insufficient taxonomic expertise) has been raised as a problem for understanding the extent of the biodiversity crisis that currently faces society. As represented in Figure 17.1, taxonomic capacity has been declining and conservation demand clearly increasing, particularly under obligations to international directives and conventions (e.g. Convention on Biological Diversity, Rio+20, the European Habitats and Water Framework Directives), and drives to protect landscapes and habitats.

That there is a global 'crisis' for biodiversity, and especially for freshwater environments, is without doubt (Dudgeon *et al.*, 2006, 2011; Chapter 1), even though terrestrial and aquatic environments are irrevocably interlinked. The taxonomy of vertebrate taxa is generally well advanced, but the taxonomy of fishes, by far the most numerous vertebrates, is far from complete (Nelson, 2006). Considering fresh waters, fish assemblages vary greatly in taxonomic composition and species richness within distinct local and regional scales due to specific micro- and macro-ecological patterns and processes, but they have been differentially studied in the several biogeographic realms (Lundberg *et al.*, 2000). Fish taxonomists are not equally distributed geographically and some ichthyofaunae are better documented (North America, Europe and Australia-New Guinea) than others (South and Central America, Africa and Tropical Asia) (Lundberg *et al.*, 2000). New ichthyological discoveries occur frequently, especially in (i) still unexploited water bodies and remote tropical areas, and in (ii) the frame of wider revisions of species-rich (freshwater) groups (e.g. ostariophysans, atherinomorphs and percomorphs) using molecular data (Lundberg *et al.*, 2000). Actually, the rise of molecular DNA tools and genomics has boosted interest in fish systematics and increasingly exposed the extensive taxonomic shortfall that faces the discipline. This chapter examines the origins of this demise in taxonomic skills and interest, and its implications for global fish conservation practices.

17.2 PARADIGM SHIFTS IN CONSERVATION

Fishes, and freshwater fishes in particular, are among the most imperilled taxa on the globe (e.g. Leidy & Moyle, 1997; Ricciardi & Rasmussen, 1999; Collares-Pereira *et al.*, 2002; Strayer & Dudgeon, 2010; IUCN, 2012), and freshwater ecosystems are among the most degraded (Kennish, 2002; Malmqvist & Rundle, 2002; Chapter 1; Box 17.1). Generally, with increasing industrialisation, the focus of fisheries exploitation has changed from commercial extractive activities towards recreational fisheries as social demands for participatory leisure activities increase. This trend continues, but management objectives increasingly rely on non-fishery-related conservation objectives when anthropogenic changes become visible within a society. This is further motivated by changing societal drivers towards improving habitat and environmental quality, under initiatives such as the EU Water Framework Directive or the US Water Act and the recognition that biodiversity and biodiversity change directly interact with, and have an impact on, water availability, food security, human health, climate change and resilience to natural hazards such as floods and droughts (Dudgeon *et al.*, 2006). As a consequence, numerous international conventions and policy drivers (e.g. 1992 United Nations Convention on Biological Diversity (CBD, 2000); EU Habitats Directive (EUHD, 1992); Millennium Ecosystem Assessment (MEA, 2005)) advocate protection of biodiversity and to achieve significant reduction in rates of biodiversity loss within relatively short timeframes (5–10 years; Mace *et al.*, 2010).

This change in management philosophy is coupled with a shift in emphasis of conservation efforts from protection of individual species

Box 17.1. Impacts of human activities on freshwater fish communities

Habitat degradation (e.g. eutrophication, acidification, sedimentation, increased turbidity, removal of riparian vegetation, channelisation), contamination by toxic substances such heavy metals, altered hydrology (dams, flow regulation, water abstraction), introduction of non-native species and transfer of diseases and pathogens are amongst the most insidious impacts of human activities on freshwater fish communities (see Chapter 1; e.g. Cowx, 2002; Dudgeon *et al.*, 2006; Gozlan *et al.*, 2010; Welcomme *et al.*, 2010). Overexploitation through commercial and recreational fishing has also been implicated in such declines (e.g. Cooke & Cowx, 2004; Allan *et al.*, 2005; Arlinghaus *et al.*, 2010; Cowx *et al.*, 2010), but despite this, non-fishing-related anthropogenic disturbances are probably the most important drivers of global loss of fish biodiversity (Richter *et al.*, 1997; Mack *et al.*, 2000; Arlinghaus *et al.*, 2002; Cowx *et al.*, 2010; Gozlan *et al.*, 2010; Chapter 1).

Most freshwater ecosystems have been systematically degraded in the global quest for economic development and this is now accelerating as the emerging economies, such as Brazil, Russia, India, China and many of the countries of southeast Asia and Africa, are trying to sustain economic growth and wealth generation. Unfortunately, the location of these countries overlaps with hotspots of biodiversity (Figure 17.2; Vorösmarty *et al.*, 2010; Chapter 1), especially freshwater biodiversity, and they are suffering in the face of such development (note that ~70% of threatened species are from Africa and Asia).

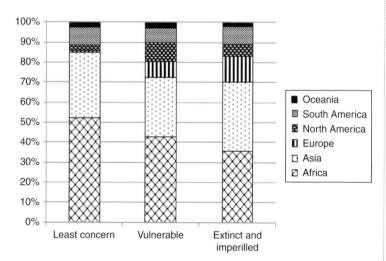

Figure 17.2 Distribution of conservation status of freshwater fish species by region reported in IUCN Red List (IUCN, 2012).

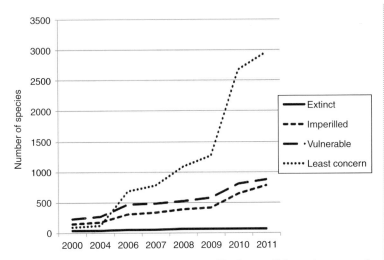

Figure 17.3 Trends in conservation status of freshwater fish species reported in IUCN Red List (IUCN, 2012).

Another emerging issue is one of perception, namely that while the number of freshwater fishes recorded as threatened has progressively increased in recent years (Figure 17.3), so has the number of species of least concern. This is largely due to improved reporting and communication of the conservation status of species and through the global IUCN network. This improved reporting has resulted in the number of species of least concern depressing the proportion of threatened species reported. For example, in 2006, 55% of 1783 freshwater fish assessed were extinct or in decline, but this dropped to 37% of 5719 species assessed in 2011. This has important implications for conservation initiatives because it potentially dissipates the impact and focus on this threatened species group. Furthermore, the definition of 'least concern' hides the emerging problem of cryptic species that have been grouped within existing species descriptions. As a consequence, there is a need to undertake taxonomic and genetic assessment and revision of species groups to account for cryptic species and reassess the IUCN categories.

and later associated habitats, towards understanding and protecting ecosystem functioning and processes (Figure 17.4). The latter is recognised as fundamental to maintaining ecosystem integrity and the continued delivery of services, including biodiversity, from the associated water bodies (Cowx & Portocarrereo, 2011). This has resulted in the promotion of ecosystem-based approaches to conservation (Dudgeon, 2010; Atkins et al., 2011), which require an understanding of the way the ecological

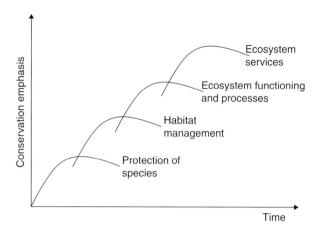

Figure 17.4 Conceptual shift in focus of conservation efforts with time.

system functions, while at the same time understanding the way society manages exploitation of the target ecosystem and the potential adverse and/or advantageous effects of its activities, including mitigation and/or compensation. The greatest problem encountered is valuing the contribution of the aquatic resources, including biodiversity, within the context of wider economic development and sustainable livelihoods (Cowx *et al.*, 2010). This has subsequently resulted in the emerging concept of ecosystem services, which provides a mechanism for integrating economic arguments into sustainable conservation practices. The concept of ecosystem services has received wide acceptance (MEA, 2005) because it enables ecosystem and species conservation to be readily considered alongside other options by planners and decision-makers.

Unfortunately, the shift towards ecosystem-based approaches to conservation has reduced the importance of identifying the impacts of development on individual species of conservation concern to evaluating delivery of goods and services, e.g. fisheries, habitat quality or water quality, and with it less need for species identification (or taxonomy). This problem is particularly prevalent in environmental impact statements that have become less focused on biodiversity and more on mitigating the impact on ecosystem services for societal well-being. This can be seen in the many hydropower dam developments being promoted globally where protection of the fisheries production is considered more important than protecting biodiversity; see for example the ongoing debate about the Xayaburi Dam on the lower Mekong that has been heavily orientated around sediment management and maintaining fisheries productivity (Dugan *et al.*, 2010; MRC, 2011). Such a strategy is particularly problematic because many of the species of conservation concern are small-sized and of no realised economic value (many potentially are good indicators of environmental health, hold bio-prospecting secrets and opportunities in the aquarium trade, for example).

The shift towards the ecosystem approach is exacerbated by the underlying origins of the field of taxonomy. Taxonomic classification and in many cases the identification and recording of new species is embedded in western-driven societal norms and is not inherent in the cultures of the emerging economies. Consequently, recent reporting has often been based on existing knowledge of species with known distributions and less on the identification of new species and their extant conservation status.

17.3 EROSION OF TAXONOMY CAPACITY

One of the traditional bastions of taxonomic skills and information is museums, often linked to universities and learned societies such as the Linnaean Society of London. These institutions have left a legacy of research and exploration, but their heritage is being somewhat eroded with modern developments in taxonomic studies, changing emphasis of conservation objectives, and constraints on funding and resources. However, institutions can forge new ways to raise their prominence in supporting conservation actions once again, by promoting the value of taxonomic work and the use of collections (Box 17.2), engaging with broad stakeholders for taxonomic support and quality control, and embracing new technologies and widening communication.

Natural history museums, one of the most valuable long-term sources of scientific data, have great challenges ahead as key institutions storing and centralising biological collections (including the type-specimens used for the description of new species), and promoting outreach activities related to conservation of biodiversity (Cotterill, 2002). Despite this central role, the research components of museums is being progressively eroded and replaced by public programmes because of a lack of financial support from central governmental sources and the consequent need to raise funds and attract more and more visitors to maintain the fundamental infrastructure of museums. As a consequence, this trend has been changing their role from well-recognised institutions of research and scholarship to centres for public attraction. Also, some have the wrong perception of museums being 'old' and left

Box 17.2. What's in a jar? The value of natural science collections

Museum collections are filled with jars, tubs and vats housing valuable specimens that showcase the diversity and breadth of the fishes for a particular focal region or global picture in the case of certain museum collections (Figure 17.5). They also provide a chronology of exploration and scientific thinking in natural history. Although their primary purpose is for taxonomy, and good taxonomy leads to good science, there is a myriad of other uses. These JARs can represent:

Figure 17.5 Fish collection, MUHNAC (Lisbon). Photo by Luis Filipe Lopes.

Journeys through time

- records of species distribution and change in conservation status
- a taxonomic log (e.g. type-specimens) linked to evolution of species descriptions
- a link to cultural history and discovery

Accurate data for today

- quality control and defensible records – voucher specimens that underpin science
- data for impact assessment and management
- support for ecological and DNA project

Resources for the future

- a source for taxonomic review and discovery
- a bank of previous exploration, energy and investment
- a call for new investigative technology

behind with respect to technology (Guerra-Garcia *et al.*, 2008), which might have obstructed their inclusion in more applied research projects embracing modern taxonomic tools.

Universities and higher education institutions no longer have the facilities and conditions to fill this gap, and certainly no longer have the capacities to take up the training function. They have experienced a strong reduction of courses on basic taxonomic skills to cater for changing student interests from classical (morphological, osteological, ecological, physiological) studies to genetic (molecular) approaches, and to general issues related to the glamorisation of natural history (what has been called the 'Attenborough effect'). As a result, university managers who are very much driven at present by market forces to encourage students to register have been pushing teachers to change the content of courses to make them more attractive for students. Thus, the last generation of taxonomists that was inspired by almost three centuries of dedicated protagonists to describe and understand biodiversity will not be easily replaced because some less popular, although relevant, areas are being effectively erased from tertiary-level academic curricula. The problem is exacerbated because, like museum researchers, most university and/or research centre academic staff are nowadays ranked on their research outputs using tools such as the contrived Impact Factor (IF) of the journals in which they publish plus their subsequent citations (Fischer *et al.*, 2012). Taxonomic studies rarely attract large-scale external funding, and publications do not achieve high citation rates or are rarely published in high IF journals, thus driving those potentially interested in taxonomy into other fields of study that attract funding and kudos (Krell, 2002). The knock-on effect is that investment in taxonomy (jobs and resources) by major research institutions is negligible and appointments are unlikely to be made.

The shift towards ecosystem-based (top-down) approaches to conservation has obscured the importance of identifying and mitigating the impacts at fine scales (bottom-up) at the species conservation level grounded with sound taxonomy and identification in favour of evaluating delivery of goods and services, e.g. fisheries, habitat quality or water quality. The direct consequence has been the general decline of taxonomy and of taxonomists themselves, both at universities (for research and training) and museums (for research and collections maintenance), compromising the production of new knowledge and a more proactive public awareness (Skelton, 2002). The fact that both research and teaching are actually market-driven, shifting interests away from fundamental science towards applied and higher-impact science in societal terms, is also having serious consequences regarding available funds.

The lack of charismatic megafauna among fishes has also contributed to the demise of traditional fish taxonomists by lowering the success in getting funds for research. This makes it even more difficult for the few remaining taxonomists to cover fieldwork expenses and pursue their research. In addition, with some recorded exceptions in South America (Carvalho *et al.*, 2005, 2007), there are now few new positions open for taxonomists, and existing positions at museums are frequently not being renewed (but see Costello *et al.*, 2013a,c). Consequently, taxonomy is now the domain of a few dedicated specialists (Guerra-Garcia *et al.*, 2008), including scientific societies and non-governmental organisations (NGOs), although the loss of these skills has been officially recognised by some governments (e.g. Brazil, USA), and they are investing in the next generation to fill the skills gaps (Carvalho *et al.*, 2007, 2008). This is because knowing which species lives in a place is widely recognised as crucial for managing stocks, undertaking fundamental research and developing strategic conservation actions. Therefore, even for countries still with high investment in data collection, there are growing concerns over the loss of skills to identify the species accurately as well as over how to optimise data sampling and storage while assuring data quality. Interestingly, some recent opinions (Costello *et al.*, 2013a,b,c, and references therein) have claimed that the field of taxonomy has never been so strong and that taxonomists were never so numerous and productive. These questionable statements are based on a bibliometric analysis of species descriptions, assembling all types of publications and authorship including a wider range of people who are not 'taxonomists' in a strict sense but use current online access information.

Information acquisition, storage and accuracy have also been contributing to the demise of taxonomy. Data are now available in computerised databases for numerous applications, namely: (i) answering questions related to the mechanisms that shape diversity patterns; (ii) devising conservation and management actions from regional to global perspectives; and (iii) informing on pathogenic agents and agricultural pests, thus impacting human health and crops production, respectively. In addition to the prestigious Ichthyological Collection of the California Academy of Sciences (http://research.calacademy.org/) and its online resources, the most commonly used source of information on fish is FishBase, a worldwide online database (www.fishbase.org/). It is connected to other global database initiatives such as the Encyclopedia of Life (EoL, www.eol.org) and the Global Biodiversity Information Facility (GBIF, www.gbif.org/). However, despite the enormous role of these so-called 'cyber tools', there is a growing concern about provision of data quality and accuracy in web platforms (Carvalho *et al.*, 2007, 2008). The taxonomy of many fish families, especially of those living in highly fluctuating environments (at both spatial and temporal scales; e.g. Cyprinidae, Cobitidae, Characidae, Salmonidae), requires considerable training, even more so if taxonomists work in remote places. Nevertheless, to lower the negative impacts of the biodiversity crisis, the current challenge to taxonomists is to help promote the quality of input data in the current electronic age by being more proactive in identifying the main problems and offering expert-validation processes. Such a global effort is being implemented for marine species through WoRMS (World Register of Marine Species) which provides a standardisation protocol that is highly valuable for biodiversity research and management (Costello *et al.*, 2013b). Having the technology, as stressed by Costello *et al.* (2013a), the question relies on getting the involvement of the people that are qualified to ensure data assemblage control in open-access online databases.

A typical example of this issue is emerging through the preparation of 'River Basin Management Plans' (RBMPs) under the EU Water Framework Directive. This requires, in many cases, governmental institutions to prepare assessments of the ecological status of rivers, lakes, estuaries and coastal waters. The procedure relies excessively on existing, and not strictly validated, local databases. Among several other situations, the frequent existence of natural hybrids in riverine assemblages is not even reported in most RBMPs (e.g. Moreira *et al.*, 2002). This is worth raising with respect to the highly speciose family Cyprinidae (e.g. Scribner *et al.*, 2001) when attempting to meet obligations under national and international conventions (e.g. IUCN or Bern, EU Habitats Directive, EU Water Framework Directive); having accurate taxonomic studies is of paramount importance. Although several hybrid occurrences correlate with anthropogenic disturbances such as fish translocations, habitat changes and aquaculture activities, it is important to distinguish these from natural processes (Scribner *et al.*, 2001). Moreover, the existence of hybrid complexes in fish assemblages living in pristine conditions has broadened our understanding of the evolutionary role of hybridisation (e.g. Mallet, 2007, 2008; Collares-Pereira *et al.*, 2013). However, this raises the question, 'Do we need to set appropriate conservation guidelines to deal with natural hybridisation and introgression (*sensu* genome invasion)?' With respect to conservation issues, the problems raised by hybrid populations as defended by Kraus (1995) and Allendorf *et al.* (2001), amongst others, can be a cause of concern depending on the situation itself, because the process may lead directly or indirectly to local species extinction and/or contribute to generate evolutionary novelties.

17.4 EVOLUTION OF NEW TOOLS

The perception of biodiversity and the growth of knowledge on taxonomy over time have also been driven by paradigm shifts moving from phenotypes to gene sequence observations and descriptions (Figure 17.6). Currently, global surveys on organisms, populations and species are available through a wide range of 'omics' (chromosomics, genomics, transcriptomics, proteomics, metabolomics, and beyond) and of bioinformatic technologies. It is possible to dissect the distinct levels of biological organisation using enormous quantities of data generated by sophisticated algorithms. The high-throughput next-generation sequencing technologies (NGST) are now impacting both evolutionary and ecological research and applications to non-model organisms are enabling old questions to be addressed, namely how identical genotypes can yield distinct phenotypes under specific environmental conditions, and what is orchestrating the plasticity and dynamics of phenotypes as regards inheritance processes (see e.g. Rokas & Abbot, 2009; Richards *et al.*, 2010).

The whole issue of biodiversity is now made more complicated because of the recognition of an unsuspected genetic variability at the intra- and interpopulation levels (Bernardo, 2011; Reid *et al.*, 2013), of epigenetic transgenerational inheritance (Jablonka & Raz, 2009; Mattick, 2012) and horizontal transfer processes (Koonin, 2011). The discovery of cryptic taxa is another issue disrupting settled taxonomies and is increasingly prevalent with the rise of molecular technologies (Bernardo, 2011). Moreover, hybridisation with or without genome introgression has been increasingly found in fish, frequently linked to polyploidy (Collares-Pereira *et al.*, 2013). Almost all known hybrid complexes lie between non-sexual and sexual reproductive modes, if not both simultaneously. As they are not strictly clonal or hemiclonal, this has begun to question the most commonly used species concept which invokes reproductive isolation between sexual species – the Biological

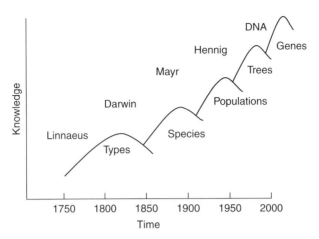

Figure 17.6 The growth of systematic knowledge over time as driven by paradigm shift. The paradigms are indicated by the major driver (above the curve) and the dominant or characteristic element (below the curve).

Species Concept (BSC; e.g. Sokal & Crovello, 1970; Kraus, 1995; de Queiroz, 2007; Mallet, 2007, 2008).

As a consequence, classical taxonomy, i.e. based on studying phenotypes, is now being replaced, or at least firmly complemented, by molecular approaches, which have the added advantage of increasing the probability of being accepted in high-IF journals (Boero, 2010). Indeed, the current view is that such journals prefer to publish data on 'modern' (i.e. molecular) analyses and that 'traditional' taxonomic papers do not score well. As mentioned previously, this is having serious consequences in the way academic profiles are considered, not only at universities but also at museums and research centres; thus, precise measures have to be taken to provide scientific recognition to those working in fields which are undervalued in the IF era, and one way may be to minimise comparative analyses by not using absolute values of IFs (Guerra-Garcia et al., 2008). Nonetheless, before putting a 'fin' in a tube, it is important to identify and give a name to the sample by inspecting the phenotype of the fish, and many molecular taxonomists only have very basic identification skills. Therefore, although molecular taxonomy can have a role in accurately determining biodiversity, adding precise sequences to databases (e.g. GenBank) demands that nominal species are correctly identified. To prove the identity, the use of vouchers is essential (Pleijel et al., 2008) and these supply a critical link between data and taxa. They provide the taxonomic identity of the specimens sequenced, linking phenotypes to genotypes, and such taxonomic services should also be internalised in research project applications.

The Fish Barcode of Life (FISH-BOL) represents the first effort to assemble a global sequence library for this diverse group of vertebrates, and because it usually ensures an accurate taxonomic identification of reference specimens, it has been considered a powerful demonstration of the immense value of collections, museums and taxonomists to both science and society (Ward *et al.*, 2009). The Consortium for the Barcode of Life (CBOL) is an international network of natural history museums, herbaria and biological repositories with experts in genomics, taxonomy and computer science (www.barcodeoflife.org). Barcoding is an approach for rapid species identification of organisms, and for animals and many other eukaryotes uses a short section of around 670 bp of the mitochondrial COI (cytochrome *c* oxidase I) gene as a standard. Despite some serious criticisms, namely concerning sampling and scale issues (reviewed in Bergsten *et al.*, 2012), this tool, which was developed in concert with genomic progress, is not a substitute for taxonomy. It has also been used for other purposes, e.g. to study molecular phylogenics and to discover new species (Ward *et al.*, 2009), but every barcode generates information that has to be linked with a known, described specimen stored in a publicly available collection (Carvalho *et al.*, 2007). The Barcode of Life system is robust so that even in the absence of a formally described taxonomic name for a barcoded organism it provides a barcode index number (BIN) that can substitute until the conventional taxonomy catches up (Ratnasingham & Hebert, 2013).

Life is a continuum of hierarchical levels starting from genes and genomes (i.e. sets of co-adapted genes) to populations and communities. Therefore, discussing the unit of conservation, as well as discussing the unit of selection, can be done either under a holistic perspective (the community within a watershed landscape) or a reductionist approach (the species or the population). In both views, there is an applied value, and proposals have been put forward with regard to fish conservation (e.g. Collares-Pereira & Cowx, 2004; Filipe *et al.*, 2004; Reid *et al.*, 2013).

The concepts of evolutionary significant units (ESUs, historically isolated and thus independently evolving sets of populations) and of management units (MUs) were developed and have been thoroughly discussed (see e.g. Moritz, 1994; Crandall *et al.*, 2000) over the last 20 years. An integrative conceptual framework – the Adaptive Evolutionary Conservation (AEC) – was proposed by Fraser and Bernatchez (2001), but no single approach is applicable to all situations and demographic contexts. The units have to be built on concepts of reproductive isolation, adaptive variation and ecological exchangeability under an integrative scaffold (Crandall *et al.*, 2000). However, identifying species-level lineages within phylogeographic studies without adding a precise name to them should be avoided (illustrated by the examples of fishes of the Cape Floristic Region, Section 17.5). If conservation is a serious goal, a full and comprehensive knowledge of existing diversity for end-users and managers is required. Good field guides are a mainstay of public (i.e. non-scientific) understanding of biodiversity, as is demonstrated by the many and often competing guides to the birds and mammals of different regions. As aquatic organisms, freshwater fishes are not as popular as birds or mammals for the general public, but in some well-researched regions such as North America and Europe, field guides to the freshwater fishes are available and are used by conservators in carrying out their obligations. The availability of a comprehensive field guide to the freshwater fishes of Southern Africa (Skelton, 2001) has been an important factor in conservation circles in the region. The drift in taxonomic understanding with the contents of such a field guide is a problem that encourages other public guides to fill the gap where they can – e.g. by Garrow and Marr (2012) in the Cape Floristic Region, where knowledge of the fish diversity has outpaced the taxonomy.

However, the Linnean system used to name a species is by definition unstable, as a taxon name that has been used for a long time can always be replaced based on new findings (reviewed in Mallet & Willmott, 2003). Thus, taxonomic changes can potentially impact conservation efforts, e.g. by the splitting of taxa that might increase protection of the units, but outcomes are unpredictable (Morrison *et al.*, 2009). An inaccurate taxonomy could easily complicate conservation by diverting or diluting appropriate attention from a problem area. For example, the listing of threatened species by the IUCN depends upon sound taxonomy and an inappropriately listed species by name could result in misguided action or neglect by conservation or resource management authorities. Within the Cape Floristic Region, the Eastern Cape redfin *Pseudobarbus afer* (Peters, 1864) has been considered a widespread polymorphic entity, in which a described species '*Barbus*' *senticeps* Smith, 1936 has been sunk in synonymy. However, the latter species is valid and is narrowly restricted to a single river system that is seriously impacted by human agencies (dams, agricultural abstraction, alien predator fish invasion). A more accurate taxonomy would ensure a more focused conservation attention to the problem. Despite this kind of uncertainty, taxonomists have to find an equilibrium that is acceptable to set sustainable and effective conservation actions. Indeed, a breakdown of species into distinct conservation units risks ending up with most aquatic systems being designated as sites of conservation interest, preventing any further development opportunities and the establishment of effective conservation programmes (e.g. Cowx & Collares-Pereira, 2002; Filipe *et al.*, 2004). In addition, Morrison *et al.* (2009) showed that name changes have the least effect for charismatic organisms, and that taxonomic progress may be detrimental, namely when it reveals the hybrid nature of a species. In such a situation, considering that 'hybrids' have been by definition excluded from conservation acts and laws, a wiser option could be to re-evaluate such policies after distinguishing 'natural' from 'anthropogenic' hybrids (Allendorf *et al.*, 2001; Mallet, 2008).

Identifying 'species' and intraspecific conservation units at distinct scales using genetic, phenotypic and environmental data is crucial in conservation and management efforts, as well as involving people that recognise phenotypes. Thus, both ecologists and geneticists have to be more proactive and interdisciplinary in their conceptual approaches, and the increasingly powerful analytical and database management tools can be used to reinvigorate 'modern' taxonomy (i.e. that combines all types of data – morphology, DNA, biology; embracing technology as digital photography, electronic keys, science communication, etc.). For this, multiscale approaches incorporating many factors to depict both historical and contemporary constraints and to understand ecological landscapes have to be implemented. In addition, a scenario of effective collaboration between researchers from different disciplines instead of competition for resources and funds is essential (e.g. Carvalho *et al.*, 2008; Boero, 2010) as regards fish conservation biology.

17.5 INTEGRATING MODERN TAXONOMY INTO CONSERVATION

The severe loss of taxonomic capacity and skills constitutes 'a threat to the world of fishes' and, therefore, to conservation itself. Despite the previous arguments about paradigm shifts in conservation emphasis and the emerging developments in molecular tools, effective conservation entails the successful integration of three essential components: scientific research and knowledge, informed policy (legislation) and management (implementation of legislated mandates) (Meffe, 2002). The first component, scientific research and knowledge, concerns information about the ecology of the threatened entity that is to be conserved, and all ecological knowledge rests on the identification, description and naming of the species in the ecosystem under consideration, namely taxonomy (Bernardo, 2011). Inaccurate identification of the species simply detracts from the understanding of the ecosystem and its spatiotemporal relationships with the landscape and the broader context of the specific need for conservation of that threatened entity. Thus, in agreement with Likens and Lindenmayer (2012), an integration of approaches is essential for effective conservation of biodiversity.

A few real-life examples will illustrate this point of connection between taxonomy and conservation, with the examples being taken from African freshwater fishes where discovery and conservation have marched hand-in-hand over the past few decades.

- One of the better-known conservation dramas involving freshwater fishes is the ever-unfolding saga of human interference in the fishes of Lake Victoria, East Africa, and its enormous impact on the life, ecosystem and humanity itself of that vast lake. The saga is complex and mired in controversy (Chapman *et al.*, 2008), so only certain aspects emphasising the taxonomic–conservation link will be raised here. Briefly, Lake Victoria is a large (68,800 km^2) tropical lake positioned between the two arms of the East African Rift system. It may have existed for about 1.6 my but has dried out completely on several occasions, the most recent time around 17–14 kabp (Danley *et al.*, 2012). Biologically, the lake is most renowned for a 'superflock' of haplochromine cichlids that, prior to the introduction of Nile perch in the 1950s, was estimated to have included several hundred species (Goudswaard *et al.*, 2008). The interactions between species are often complex and unpredictable, resulting in unexpected population fluxes over time. Following their introduction, there was a delay and then rapid rise in Nile perch stocks which coincided with an equally rapid decline and demise of many *Haplochromis* species, and more latterly there has been a proliferation of the planktivorous cyprinid *Rastrineobola argentea* (Pellegrin, 1904) (Goudswaard *et al.*, 2008) associated with a decline in Nile perch due to overfishing. As many as 150 of the haplochromine species have now been described, but the description of species has slowed to a trickle and no longer keeps pace with the revelations of diversity as indicated by molecular DNA studies. The biological knowledge of these species is essential to understand the ecological processes of extinction, recovery and evolution of species in a dynamic evolutionary arena. Without effective taxonomy to link the species to the ecology, fisheries management of the resource is weakened.

- The Twee River redfin, *Barbus erubescens* Skelton, 1974, is a riverine minnow species only found in a restricted montane tributary system, the Twee River, of the west-coast Olifants River system in the Western Cape of South Africa (Skelton, 1974, 2011). The species has been isolated from downstream species by a barrier waterfall. The investigation leading to its description was prompted by the marked population decline of what proved to be its sister species, the Clanwilliam redfin (*Barbus calidus* Barnard, 1938), which occurs more widely in the system. The causes of the decline of the Clanwilliam redfin are primarily the impacts of introduced alien bass (smallmouth bass *Micropterus dolomieu* Lacepede, 1802 and largemouth bass *M. salmoides* Lacepede, 1802), as well as habitat deterioration arising from water abstraction for agricultural and domestic purposes. Dams and weirs have been built on the mainstream and in tributaries that interrupt connectivity within the system. The Twee River subsystem has been subject to similar impacts as described above, and the Twee River redfin is now, in spite of longstanding conservation attention, Critically Endangered and at serious risk of extinction (Impson *et al.*, 2007; Marr *et al.*, 2009). Without the taxonomic focus that indicated its species status this species would have been considered a population of the more widely distributed *Barbus calidus* with lesser conservation importance. The many environmental issues within the Twee River catchment also would not have received due attention.

- The case history of the Maloti minnow (*Pseudobarbus quathlambae*) is likewise an interplay of taxonomy and conservation (Skelton, 1987, 2002, 2011). The species was described by Barnard (1938) as *Labeo quathlambae* from specimens collected during a trout water survey in the Umkomazana River in KwaZulu-Natal, South Africa. The species was not reported again for over 30 years and was feared to be extinct (Jubb, 1966). Further taxonomic attention indicated that the species was unique in the African landscape and Greenwood and Jubb (1967) reassigned it as a monotypic genus *Oreodaimon quathlambae*, the generic name ('mountain spirit') reflecting its 'extinct' status as at that time. In 1971, fisheries officers conducting a survey of trout waters in Lesotho discovered a small population of *O. quathlambae* in the Tsoelikana stream, a high altitude source tributary of the Orange River system (Pike & Tedder, 1973). Other populations in Orange River tributaries in Lesotho were subsequently reported by Rondorf (1976), Skelton and Masapha (1989) and Rall (1993). Systematic studies on the species indicated that its phylogenetic relationships were with the redfin minnows of the Cape Fold Mountains (Skelton, 1980; Swartz et al., 2009) and it was placed within the genus *Pseudobarbus* (Skelton, 1988). The Lesotho Highlands Water Project (LHWP), a major water and hydroelectric generation project in the Lesotho highlands, brought intense conservation spotlight on the Maloti minnow populations from the late 1980s. In Phase 1 of the LHWP, two populations of the minnow – in the Mohale and the Motsuku streams, respectively, the latter discovered as a result of the LHWP – would be affected by the building of dams and water transfer tunnels. A Maloti Minnow Conservation Project was instituted (Rall, 1999; Tweddle et al., 2009) that included an extensive survey of potential suitable rivers in Lesotho as part of the conservation action plan, and genetic analyses were conducted on specimens from all known sites. This study indicated that the Mohale population, hitherto protected from trout and other fish species by a large barrier waterfall, would be directly drowned by the impounded waters and connected by the transfer tunnel to a source of alien trout and other indigenous fish species from the Katse Dam. Molecular DNA studies indicated that the Mohale population was sufficiently distinct to be considered an ESU, and even a full species according to the genetic distance (Swartz et al., 2009). Specimens were caught from the Mohale catchment and transferred to three sanctuary streams elsewhere in Lesotho in a desperate attempt to prevent the possible extinction of this currently undescribed species (J. Rall, personal communication). This safety-net action was necessary because alternative actions to prevent the invasion of the entire, or at least the inflowing riverine sections, of the Mohale impoundment where the species might have survived indefinitely were not taken. Since then, the Mohale impoundment has been invaded by smallmouth yellowfish (*Labeobarbus aeneus* (Burchell, 1822)) and there are plans awaiting approval for the establishment of a trout cage-aquaculture facility in the impoundment (Advance Africa, 2013).

- Detailed systematic and ecological studies have been made on other species of the redfin minnow *Pseudobarbus* lineage as several species are listed as Vulnerable, Endangered or Critically Endangered by the IUCN (2012). Swartz et al. (2007, 2009) indicated that several undescribed species exist in the complex, and the reconfiguration of these taxa severely complicates their conservation needs.

The Eastern Cape redfin (*Pseudobarbus afer* (Peters, 1864)) is a complex of four species, with one being confined to the eastern sector of its range in the Algoa Bay area (Swartz *et al.*, 2007). Taken together, the 'species' would not be considered threatened according to the IUCN categories. Independently, however, two of the four species are Endangered, and one, restricted to a single river system, is Critically Endangered.

- A similar situation exists with regard to the redfin minnow species *Pseudobarbus burchelli* (Smith, 1841), originally considered Near Threatened. Chakona (2011) studied the population genetics, phylogeny and conservation biology of this species as well as species of the genus *Galaxias* and *Sandelia* in the Breede River system in detail. He revealed four species within the current *P. burchelli* framework, nine species within the Breede River complex of *Galaxias*, and two *Sandelia capensis* (Cuvier, 1829) species, with possibly others within the lineage (there are seven minor branches within the phylogenetic lineage). Three of the four *P. burchelli* complex redfin species are restricted in range and Critically Endangered, and the remaining one is Near Threatened. Neither the *Galaxias* nor the *Sandelia* species have been evaluated for IUCN status, but it is likely that some of the lineages will be Endangered or Critically Endangered.

- Taxonomy has played a vital part in the conservation of the large *Labeobarbus* species flock in Lake Tana (Getahun, 2011). Lake Tana at the source of the Blue Nile is a 3150 km², high-altitude lake in Ethiopia, physically isolated by the Tssisat Falls from the lower reaches of the river. The lake has a fish fauna of 28 species with a unique 'flock' of at least 17 large *Labeobarbus* species, as well as other fishes, including *Oreochromis niloticus* (Linnaeus, 1758) and *Clarias gariepinus* (Burchell, 1822), which constitute the major component of the fisheries of the lake. Until as recently as 1994 the *Labeobarbus* flock was considered a single species (*Labeobarbus intermedius* (Rüppell, 1835); Nagelkerke *et al.*, 1994). Detailed studies, including morphology, genetics, biology and ecology, showed conclusively that there were sufficient differences to recognise the 17 species in the flock, and spawning ecology, which involved riverine migration for at least 7 of the species, was an important criterion for some. Overfishing is a threat to the *Labeobarbus* resource in Lake Tana (Getahun, 2011). The implications of such a complex taxonomy are important in terms of adaptive management of the fisheries resource that, until the flock was recognised, was considered a single-species fishery.

These selected case histories of conservation biology involving African ichthyofauna highlight the importance of taxonomy in recognising species flocks, strains and distinct taxonomic species.

17.6 MAINTAINING TAXONOMIC SKILLS FOR CONSERVATION

The development of modern communication and database technologies enables research on biodiversity and management, namely (i) to identify hotspots of threatened fish diversity, and (ii) to assess patterns of diversity and priority areas of concern (e.g. Filipe *et al.*, 2009). While permitting coordination, storing and access to information on biodiversity as never before, they have the potential to accurately address at least some taxonomic issues. However, there is a need for concerted measures among data providers, developers and users, as well as for taxonomic authority certifications provided by the invaluable Ichthyological Collection of the California Academy of Sciences and its online resources, and by natural history museums. Uncertainties and voids in collecting and merging data from distinct sampling surveys (currently also threatened by the increased restrictions to sample and the very severe animal welfare rules) exist, and the new possibilities of building local and regional databases reinforce the need for accurate validations.

Given the issues raised in terms of importance of taxonomic skills and their critical linkages to robust conservation science and management, there is a fundamental requirement to promote this field of study and raise its profile. The following courses of action, although not exclusive, are recommended in the first instance to address the erosion of taxonomic skills.

There is a need to reintroduce taxonomic studies into educational courses at all academic levels, but starting at primary schools to interest children at an early age. Modern media technologies and interfaces now allow a multitude of mechanisms to interact with society and glamorise the many facets of the natural world, and visual aids and interactive tools should be developed to support this need. Beyond this, taxonomic skills should be reintroduced into secondary and tertiary education courses and become integral to the wider learning experience. Dull identification and systematic courses need to be replaced with interactive, hands-on experiences where the student is engaged and understands the fundamentals of species anatomy, physiology and behaviour and how these fit into ecosystem functioning. Overall taxonomic skills need marketing in the broader environmental framework linked to potential job opportunities and research underpinning sustainable development. One area where taxonomic skills can be enhanced is through field courses where students can learn taxonomy *in situ*. However, the escalating demands of meeting 'health and safety' requirements is suppressing this possibility, and efforts must be found to overcome such restrictions. Another possibility is to involve experts in taxonomy in online advanced courses aimed at training wider audiences on taxonomic specificities and tools within an international context (Wheeler, 2003).

An additional area that needs adjustment is academic recognition. Society needs to recognise the importance of taxonomic skills and output and suitably reward those engaged in the work. Whether this is through dedicated grants such as the UK Department for Environment Food and Rural Affairs Darwin project scheme or better recognition of taxonomic research outputs in academic ranking, mechanisms must be put into place and resources made available to deliver robust science that meets the requirements of management. There is a strong argument regarding the need to cross-subsidise training and resourcing in some less-popular, but nonetheless critically important, areas to achieve this objective.

Many sources of data used by taxonomists have already been or are targeted to be digitalised in distinct platforms. Following Tautz *et al.* (2003), a DNA-based system has to be anchored on traditional taxonomy and make use of the latest technological tools. The combination of molecular (not single sequences, but information on wider genomic regions) and morphological (both morphometric and meristic characters) data should give natural history museums a new impetus for preserving biological diversity. According to Deans *et al.* (2012), this is an exciting field to be engaged in. Moreover, these authors argue that this is the best time to be a taxonomist, because conditions exist to complete the taxonomic revolution (or evolution, see Carvalho *et al.*, 2008) that started with computer innovations. However, the way biodiversity is described has to be changed, with taxonomists being committed to undertake such a transformation by adopting the new generation of supportive digital tools (Deans *et al.*, 2012; Costello *et al.*, 2013a). The good news is that interest in taxonomy is being revitalised, and is open to strategic marketing by the institutions themselves taking advantage of the international networking schemes already in place. All institutions linked to taxonomy can forge new ways to raise their prominence again, by promoting the dynamic value and use of collections, engaging with broad stakeholders for taxonomic support and quality control, and embracing new technologies and communication. Crucially, more funding should be immediately allocated to taxonomic research, museum collections, and training and hiring of professional taxonomists.

To conclude, Greenwood (1992) considered the question 'Are the major fish faunas well-known?' He indicated that they were not, and focused on the issues threatening these faunas. He regarded the threats as fundamentally twofold: in the first instance, through impacts from anthropogenic actions, and secondly through our lack of knowledge of these faunas as a consequence of a paucity of systematic ichthyologists and taxonomists: 'Since taxonomy and systematics are basic to all ichthyological research, I would consider the latter situation to be one of the major non-environmental threats to the world of fishes' (Greenwood, 1992, p. 135). This chapter endorses this realisation, but offers a way forward to address the concern of declining taxonomic skills. Let us hope that in the next 20 years we no longer lament a further loss of fish biodiversity because of the demise of such a fundamental skills base; the removal of the current caveats might, on the contrary, ensure that biodiversity and conservation fuel the discipline of taxonomy itself.

17.7 DISCUSSION QUESTIONS

1. Do you know where to get historic data and taxonomic advice or collaboration to study and identify fish species in your region?
2. Are any species in your region faced with the threat of extinction and, if so, do you know what are the threats to their survival?
3. Have any of these threatened species been the target of a recent genetic and taxonomic revision to assess for cryptic species?
4. Was a systematic/taxonomic revision the first step upon embarking on an ecological or fisheries-based research programme in your region?
5. What do you need to do to undertake a genetic study of the fish diversity in your region?

17.8 ACKNOWLEDGEMENTS

The authors would like to thank all the editors for the invitation and, in particular, Gerry Closs for his permanent support and critical revisions, which have significantly contributed to raise the focus of this chapter.

17.9 REFERENCES

Advance Africa. (2013). Available at: www.advanceafrica.co.za/projects.htm.
Allan, J. D., Abell, R., Hogan, Z., *et al.* (2005). Overfishing of inland waters. *BioScience*, 55, 1041–1051.

Allendorf, F., Leary, R. F., Spruell, P. & Wenburg, J. K. (2001). The problems with hybrids: setting conservation guidelines. *Trends in Ecology and Evolution,* 16, 613–622.

Arlinghaus, R., Mehner, T. & Cowx, I. G. (2002). Reconciling traditional inland fisheries management and sustainability in industrialised countries, with emphasis on Europe. *Fish and Fisheries,* 3, 261–316.

Arlinghaus, R., Cooke, S. J. & Cowx, I. G. (2010). Providing context to the global code of practice for recreational fisheries. *Fisheries Management and Ecology,* 17, 146–156.

Atkins, J. P., Burdon, D., Elliott, M. & Gregory, A. J. (2011). Management of the marine environment: integrating ecosystem services and societal benefits with the DPSIR framework in a systems approach. *Marine Pollution Bulletin,* 62, 215–226.

Barnard, K. H. (1938). Description of a new species of fresh-water fish from Natal. *Annals of the Natal Museum,* 8, 525–528.

Bergsten, J., Bilton, D. T., Fujisawa, T., *et al.* (2012). The effect of geographical scale of sampling on DNA Barcoding. *Systematic Biology,* 61, 851–869.

Bernardo, J. (2011). A critical appraisal of the meaning and diagnosability of cryptic evolutionary diversity, and its implications for conservation in the face of climate change. In *Climate Change, Ecology and Systematics.* Hodkinson, T. R., Jones, M. B., Waldren, S. & Parnell, J. A. N. (Eds). Cambridge University Press, pp. 380–438.

Boero, F. (2010). The study of species in the Era of Biodiversity: a tale of stupidity. *Diversity,* 2, 115–126.

Carvalho, M. R., Bockmann, F. A., Amorim, D. S., *et al.* (2005). Revisiting the taxonomic impediment. *Science,* 307, 353.

Carvalho, M. R., Bockmann, F. A., Amorim, D. S., *et al.* (2007). Taxonomic impediment or impediment to taxonomy? A commentary on systematic and the cybertaxonomic-automation paradigm. *Evolutionary Biology,* 34, 140–143.

Carvalho, M. R., Bockmann, F. A., Amorim, D. S. & Brandão, C. R. F. (2008). Systematics must embrace comparative biology and evolution, not speed and automation. *Evolutionary Biology,* 35, 150–157.

CBD (2000). *Convention on Biological Diversity.* Available at http://69.90.183.227/doc/legal/cbd-un-en.pdf/

Chakona, A. (2011). *Comparative biogeography and ecology of freshwater fishes in the Breede and associated river systems, South Africa.* PhD thesis, Rhodes University, Grahamstown, South Africa.

Chapman, L. J., Chapman, C. A., Kaufman, L., Witte, F. & Balirwa, J. (2008). Biodiversity conservation in African inland waters: lessons of the Lake Victoria region. *Verhandlungen des Internationalen Verein Limnologie,* 30, 16–34.

Collares-Pereira, M. J. & Cowx, I. G. (2004). The role of catchment scale environmental management in freshwater fish conservation. *Fisheries Management and Ecology,* 11, 303–312.

Collares-Pereira, M. J., Cowx, I. G. & Coelho, M. M. (Eds). (2002). *Conservation of Freshwater Fishes: Options for the Future.* Oxford: Fishing News Books, Blackwell Science.

Collares-Pereira, M. J., Matos, I., Morgado-Santos, M. & Coelho, M. M. (2013). Natural pathways towards polyploidy in animals – the *Squalius alburnoides* fish complex as a model system to study genome size and genome reorganization in polyploids. In: *Trends in Polyploidy Research in Animals and Plants.* Stock, M. & Lamatsch, D. (Eds). *Cytogenetic and Genome Research,* 140, 97–116.

Cooke, S. J. & Cowx, I. G. (2004). Considering recreational fisheries impacts in global fish crises. *BioScience,* 54, 857–859.

Costello, M. J., May, R. M. & Stork, N. E. (2013a). Can we name Earth's species before they go extinct? *Science,* 339, 413–416.

Costello, M. J., Bouchet, P., Boxshall, G., *et al.* (2013b). Global coordination and standardisation in marine biodiversity through the World Register of Marine Species (WoRMS) and related databases. *PLoS ONE,* 8, e51629.

Costello, M. J., Wilson, S. & Houlding, B. (2013c). More taxonomists describing significantly fewer species per unit effort may indicate that most species have been discovered. *Systematic Biology,* 62, 616–624.

Cotterill, F. P. D. (2002). The future of natural science collections into the 21st century. Conferencia de Clausura. In *Actas del I Simposio sobre el Patrimonio Natural en las Colecciones Públicas en España (Vitoria, 25–27 September 2001).* Vitoria: Departamento de Cultura, Diputacion Foral de Alava, pp. 237–282.

Cowx, I. G. (2002). Analysis of threats to freshwater fish conservation: past and present challenges. In *Conservation of Freshwater Fish: Options for the Future.* Collares-Pereira, M. J., Cowx, I. G. & Coelho, M. M. (Eds). Oxford: Fishing News Books, Blackwell Science, pp. 201–220.

Cowx, I. G. & Collares-Pereira, M. J. (2002). Freshwater fish conservation: options for the future. In *Conservation of Freshwater Fish: Options for the Future*. Collares-Pereira, M. J., Cowx, I. G. & Coelho, M. M. (Eds). Oxford: Fishing News Books, Blackwell Science, pp. 443–452.

Cowx, I. G. & Portocarrero, M. A. (2011). Paradigm shifts in fish conservation: moving to the ecosystem services concept. *Journal of Fish Biology*, 79, 1663–1680.

Cowx, I. G., Arlinghaus, R. & Cooke, S. J. (2010). Harmonising recreational fisheries and conservation objectives for aquatic biodiversity in inland waters. *Journal of Fish Biology*, 76, 2194–2215.

Crandall, K. A., Bininda-Emonds, O. R. P., Mace, G. M. & Wayne, R. K. (2000). Considering evolutionary processes in conservation biology. *Trends in Ecology and Evolution*, 15, 290–295.

Danley, P. D., Husemann, M., Ding, B., *et al.* (2012). The impact of geologic history and paleoclimate on the diversification of East African cichlids. *International Journal of Evolutionary Biology*, Article ID 574851.

Deans, A. R., Yoder, M. J. & Balhoff, J. P. (2012). Time to change how we describe biodiversity. *Trends in Ecology and Evolution*, 27, 78–84.

De Queiroz, K. (2007). Species concepts and species delimitation. *Systematic Biology*, 56, 879–886.

Dudgeon, D. (2010). Prospects for sustaining freshwater biodiversity in the 21st century: linking ecosystem structure and function. *Current Opinion in Environmental Sustainability*, 2, 422–430.

Dudgeon, D., Arthington, A. H., Gessner, M. O., *et al.* (2006). Freshwater biodiversity: importance, threats, status and conservation challenges. *Biological Reviews*, 81, 163–182.

Dudgeon, D., Paugy, D., Leveque, C., Rebelo, L.-M. & McCartney, M. P. (2011). Background. In *The diversity of life in African Freshwaters: Under Water, Under Threat. An analysis of the status and distribution of freshwater species throughout mainland Africa*. Darwall, W. R. T., Smith, K. G., Allem, D. J., *et al.* (Eds). Cambridge, United Kingdom and Gland, Switzerland: IUCN, pp. 1–31.

Dugan, P. J., Barlow, C., Agostinho, A. A., *et al.* (2010). Fish migration, dams, and loss of ecosystem services in the Mekong basin. *Ambio*, 39, 344–348.

EUHD (1992). *Habitats Directive 92/43/EEC*. Available at http://ec.europa.eu/environment/nature/legislation/habitatsdirective/indx_en.htm/

Filipe, A. F., Marques, T., Seabra, S., *et al.* (2004). Selection of priority areas for fish conservation in Guadiana River Basin, Iberian Peninsula. *Conservation Biology*, 18, 189–200.

Filipe, A. F., Araujo, M. B., Doadrio, I., Angermeir, P. & Collares-Pereira, M. J. (2009). Biogeography of Iberian freshwater fishes revisited: the roles of contemporary *versus* historical constraints. *Journal of Biogeography*, 36, 2096–2110.

Fischer, J., Ritchie, E. G. & Hanspach, J. (2012). Academia's obsession with quantity. *Trends in Ecology and Evolution*, 27, 473–474.

Fraser, D. J. & Bernatchez, L. (2001). Adaptive evolutionary conservation: towards a unified concept for defining conservation units. *Molecular Ecology*, 10, 2741–2752.

Garrow, C. & Marr, S. (2012). *Swimming on the Edge of Extinction*. Grahamstown, South Africa: NISC.

Getahun, A. (2011). A unique species flock in Lake Tana – the *Labeobarbus* complex. Species in the spotlight. In *The Diversity of Life in African Freshwaters: Under Water, Under Threat. An Analysis of the Status and Distribution of Freshwater Species throughout Mainland Africa*. Darwall, W. R. T., Smith, K. G., Allem, D. J., *et al.* (Eds). Cambridge, UK and Gland, Switzerland: IUCN, pp. 82–84.

Goudswaard, P. C., Witte, F. & Katunzi, E. F. B. (2008). The invasion of an introduced predator, Nile perch (*Lates niloticus*, L.) in Lake Victoria (East Africa): chronology and causes. *Environmental Biology of Fish*, 81, 127–139.

Gozlan, R. E., Britton, J. R., Cowx, I. G. & Copp, G. H. (2010). Current knowledge on nonnative freshwater fish introductions. *Journal of Fish Biology*, 76, 751–786.

Greenwood, P. H. (1992). Are the major fish faunas well-known? *Netherlands Journal of Zoology*, 42, 131–138.

Greenwood, P. H. & Jubb, R. A. (1967). The generic identity of *Labeo quathlambae* Barnard (Pisces, Cyprinidae). *Annals of the Cape Provincial Museums (Natural History)*, 6, 17–37.

Guerra-Garcia, J. M., Espinosa, F. & Garcia-Gomez, J. C. (2008). Trends in taxonomy today: an overview about the main topics in taxonomy. *Zoology Baetica*, 19, 15–49.

Impson, N. D., Marriott, M. S., Bills, I. R. & Skelton, P. H. (2007). Conservation biology and management of a critically endangered cyprinid, the Twee River redfin, *Barbus erubescens* (Teleostei: Cyprinidae), of the Cape Floristic Region, South Africa. *African Journal of Aquatic Science*, 32, 27–33.

IUCN. (2012). *The IUCN Red List of Threatened Species. Version 2012.2*. Available at: www.iucnredlist.org.

Jablonka, E. & Raz, G. (2009). Transgenerational epigenetic inheritance: prevalence, mechanisms, and implications for the study of heredity and evolution. *Quarterly Review of Biology*, 84, 131–176.

Jubb, R. A. (1966). *Labeo quathlambae*, a rare freshwater fish now feared to be extinct. *Piscator*, 67, 78–80.

Kennish, M. J. (2002). Environmental threats and environmental future of estuaries. *Environmental Conservation*, 29, 78–107.

Koonin, E. V. (2011). *The Logic of Chance: The Nature and Origin of Biological Evolution*. Upper Saddle River, NJ: FT Press Science.

Kraus, F. (1995). The conservation of unisexual vertebrate populations. *Conservation Biology*, 9, 956–959.

Krell, F. T. (2002). Why impact factors don't work for taxonomy. *Nature*, 415, 957.

Leidy, R. A. & Moyle, P. B. (1997). Conservation status of the world's fish fauna: an overview. In *Conservation Biology for the Coming Decade*. Fiedler P. A. & Karieva, P. M. (Eds). New York, NY: Chapman and Hall, pp. 187–227.

Likens, G. E. & Lindenmayer, D. B. (2012). Integrating approaches leads to more effective conservation of biodiversity. *Biodiversity Conservation*, 12, 3323–3341.

Lundberg, J. G., Kottelat, M., Smith, G. R., Stiassny, M. L. J. & Gill, A. C. (2000). So many fishes, so little time: an overview of recent ichthyological discovery in continental waters. *Annals of the Missouri Botanical Garden*, 87, 26–62.

Mace, G. M. (2004). The role of taxonomy in species conservation. *Philosophical Transactions of the Royal Society London B*, 359, 711–719.

Mace, G. M., Cramer, W., Diaz, S., *et al.* (2010). Biodiversity targets after 2010. *Current Opinions in Environmental Sustainability*, 2, 3–8.

Mack, R. N., Simberloff, D., Lonsdale, W. M., *et al.* (2000). Biotic invasions: causes, epidemiology, global consequences, and control. *Ecological Applications*, 10, 689–710.

Mallet, J. (2007). Hybrid speciation. *Nature*, 446, 279–283.

Mallet, J. (2008). Hybridization, ecological races and the nature of species: empirical evidence for the ease of speciation. *Philosophical Transactions of the Royal Society London B*, 363, 2971–2986.

Mallet, J. & Willmott, K. (2003). Taxonomy: renaissance or Tower of Babel? *Trends in Ecology and Evolution*, 18, 57–59.

Malmqvist, B. & Rundle, S. (2002). Threats to the running water ecosystems of the world. *Environmental Conservation*, 29, 134–153.

Marr, S. M., Sutcliffe, L. M. E., Day, J. A., Griffiths, C. L. & Skelton, P. H. (2009). Conserving the fishes of the Twee River, Western Cape, South Africa: revisiting the issues. *African Journal of Aquatic Science*, 34, 77–85.

Mattick, J. S. (2012). Rocking the foundations of molecular genetics. *Proceedings of National Academy of Sciences USA*, 109, 16400–16401.

MEA. (2005). *Millennium Ecosystem Assessment. Ecosystems and Human Well-being: Synthesis*. Washington, DC: Island Press.

Meffe, G. K. (2002). Connecting science to management and policy in freshwater fish conservation. In *Conservation of Freshwater Fish: Options for the Future*. Collares-Pereira, M. J., Cowx, I. G. & Coelho, M. M. (Eds). Oxford: Fishing News Books, Blackwell Science, pp. 363–372.

Moreira, I., Ferreira, M. T., Cortes, R. M., Pinto, P. & Almeida, R. (2002). *Ecossistemas Aquáticos e Ribeirinhos*. Lisboa: Instituto da Água – Ministérios das Cidades, Ordenamento do Território e Ambiente. (In Portuguese.)

Moritz, C. (1994). Defining 'Evolutionary Significant Units' for conservation. *Trends in Ecology and Evolution*, 9, 373–375.

Morrison III, W. R., Lohr, J. L., Duchen, P., *et al.* (2009). The impact of taxonomic change on conservation: does it kill, can it save, or is it just irrelevant? *Biological Conservation*, 142, 3201–3206.

MRC (Mekong River Commission). (2011). *Procedures for Notification, Prior Consultation and Agreement; Proposed Xayaburi Dam Project – Mekong River Prior Consultation Project Review Report*, Mekong River Commission Secretariat. Vientiane Laos PDR.

Nagelkerke, L. A. J., Sibbing F. A., van den Boogaart, J. G. M., Lammens, E. H. R. R. & Osse, J. W. M. (1994). The barbs (*Barbus* spp.) of Lake Tana: a forgotton species flock? *Environmental Biology of Fishes*, 39, 1–22.

Nelson, J. S. (2006). *Fishes of the World*, fourth edition. Hoboken, NJ: John Wiley & Sons, Inc.

Pike, T. & Tedder, A. J. (1973). Rediscovery of *Oreodaimon quathlambae* (Barnard). *The Lammergeyer*, 19, 9–15.

Pleijel, F., Jondelius, U., Norlinder, E., *et al.* (2008). Phylogenies without roots? A plea for the use of vouchers in molecular phylogenetic studies. *Molecular Phylogenetics and Evolution*, 48, 369–371.

Rall, J. L. (1993). *An ecological study of the Maluti Minnow* (Pseudobarbus quathlambae) *in the catchment area of phase 1B of the Lesotho Highlands Water Scheme.* MSc thesis, Rand Afrikaans University, Johannesburg, South Africa.

Rall, J. L. (1999). *Development of a Conservation Program on* Pseudobarbus quathlambae *in the catchment area of Phase 1B of the Lesotho Highlands Water Project.* PhD thesis, Rand Afrikaans University, Johannesburg, South Africa.

Ratnasingham, S. & Hebert, P. D. N. (2013). A DNA-based registery for all animal species: the Barcode Index Number (BIN) system. *PLoS ONE,* 8, e66213.

Reid, G. McG. (2010). Taxonomy and the survival of threatened animal species – a matter of life and death. In *Systema Naturae 250 The Linnean Ark.* Polaszek, A. (Ed.). Boca Raton, FL: CRC Press, pp. 29–52.

Reid, G. McG., Contreras MacBeath, T. & Csatádi, K. (2013). Global challenges in freshwater fish conservation related to public aquariums and the aquarium industry. *International Zoo Yearbook,* 47, 6–45.

Ricciardi, A. & Rasmussen, J. B. (1999). Extinction rates of North American freshwater fauna. *Conservation Biology,* 13, 1220–1222.

Richards, C. L., Bossdorf, O. & Pigliucci, M. (2010). What role does heritable epigenetic variation play in phenotypic evolution? *BioScience,* 60, 232–237.

Richter, B. D., Braun, D. P., Mendelson, M. A. & Master, L. L. (1997). Threats to imperiled freshwater fauna. *Conservation Biology,* 11, 1081–1093.

Rokas, A. & Abbot, P. (2009). Harnessing genomics for evolutionary insights. *Trends in Ecology and Evolution,* 24, 192–200.

Rondorf, D. W. (1976). New locations of *Oreodaimon quathlambae* (Barnard, 1938) (Pisces, Cyprinidae) populations. *South African Journal of Science,* 72, 150–151.

Scribner, K. T., Page, K. S. & Bartron, M. L. (2001). Hybridization in freshwater fishes: a review of case studies and cytonuclear methods of biological inference. *Reviews in Fish Biology and Fisheries,* 10, 293–323.

Skelton, P. H. (1974). A new *Barbus* species (Pisces, Cyprinidae) from the Olifants River system, Western Cape Province, South Africa. *Special Publications of the Rhodes University Department of Ichthyology,* 13, 1–12.

Skelton, P. H. (1980). *Systematics and biogeography of the redfin* Barbus *species (Pisces, Cyprinidae) from Southern Africa.* PhD thesis, Rhodes University, Grahamstown, South Africa.

Skelton, P. H. (1987). *South African Red Data Book – Fishes.* South African National Scientific Programmes Report No. 137, pp. 1–199.

Skelton, P. H. (1988). A taxonomic revision of the redfin minnows (Pisces, Cyprinidae) from Southern Africa. *Annals of the Cape Provincial Museums (Natural History),* 16, 201–307.

Skelton, P. H. (2001). *A Complete Guide to the Freshwater Fishes of Southern Africa.* Cape Town, South Africa: Struik Publishers.

Skelton, P. H. (2002). An overview of the challenges of conserving freshwater fishes in South Africa. In *Conservation of Freshwater Fish: Options for the Future.* Collares-Pereira, M. J., Cowx, I. G. & Coelho, M. M. (Eds). Oxford: Fishing News Books, Blackwell Science, pp. 221–236.

Skelton, P. H. (2011). The Twee River redfin – a Critically Endangered minnow from South Africa. Species in the spotlight. In *The Diversity of Life in African Freshwaters: Under Water, Under Threat. An Analysis of the Status and Distribution of Freshwater Species throughout Mainland Africa.* Darwall, W. R. T., Smith, K. G., Allem, D. J., *et al.* (Eds). Cambridge, UK and Gland, Switzerland: IUCN, pp. 85–86.

Skelton, P. H. & Masapha, S. (1989). The fauna and physico-chemical characteristics of riverine aquatic systems. Drakensberg/Maloti Mountain Catchment Conservation Programme, Final Report.

Sokal, R. R. & Crovello, T. J. (1970). The Biological Species Concept: a critical evaluation. *American Naturalist,* 104, 127–153.

Strayer, D. L. & Dudgeon, D. (2010). Freshwater biodiversity conservation: recent progress and future challenges. *Journal of the North American Benthological Society,* 29, 344–358.

Swartz, E. R., Skelton, P. H. & Bloomer, P. (2007). Sea-level changes, river capture and the evolution of populations of the Eastern Cape and fiery redfins (*Pseudobarbus afer* and *Pseudobarbus phlegethon,* Cyprinidae) across multiple river systems in South Africa. *Journal of Biogeography,* 34, 2086–2099.

Swartz, E. R., Skelton, P. H. & Bloomer, P. (2009). Phylogeny and biogeography of the genus *Pseudobarbus* (Cyprinidae): shedding light on the drainage history of rivers associated with the Cape Floristic Kingdom. *Molecular Phylogenetics and Evolution,* 51, 75–84.

Tautz, D., Arctander, P., Minelli, A., Thomas, R. H. & Vogler, A. P. (2003). A plea for DNA taxonomy. *Trends in Ecology and Evolution*, 18, 70–74.

Tweddle, D., Bills, R., Swartz, E., *et al.* (2009). The status and distribution of freshwater fishes. In *The Status and Distribution of Freshwater Biodiversity in Southern Africa*. Darwall, W. R. T., Smith, K. G., Tweddle, D. & Skelton, P. (Eds). Gland: IUCN and Grahamstown, South Africa: SAIAB, pp. 21–37.

Vorösmarty, C. J., McIntyre, P. B., Gessner, M. O., *et al.* (2010). Global threats to human water security and river biodiversity. *Nature*, 467, 555–561.

Ward, R. D., Hanner, R. & Hebert, P. D. N. 2009. The campaign to DNA barcode all fishes, FISH-BOL. *Journal of Fish Biology*, 74, 329–356.

Welcomme, R. L., Cowx, I. G., Coates, D., *et al.* (2010). Inland capture fisheries. *Philosophical Transactions of the Royal Society B*, 1554, 2881–2896.

Wheeler, Q. D. (2003). Transforming taxonomy. *The Systematist*, 22, 3–5.

Synthesis – what is the future of freshwater fishes?

GERARD P. CLOSS, MARTIN KRKOSEK
AND JULIAN D. OLDEN

Global ecosystems are increasingly shaped and dominated by the human enterprise, and overwhelming evidence indicates that the abundance and diversity of freshwater fishes are declining at unprecedented rates (Olden *et al.* 2010; Chapters 1 and 2). More often than not, this phenomenon is directly or indirectly attributed to various human activities (Angermeier, 2007; Dudgeon, 2010, 2011; Vörösmarty *et al.*, 2010; Chapter 1). Current trends suggest that many, if not all, regions of high freshwater fish diversity will face greater threats in the near future (Dudgeon, 2011; Chapter 1). In a global environment where human demands placed upon freshwater ecosystems are increasing (Angermeier, 2007; Chapter 2), compelling arguments must be advanced to ensure the future of freshwater fishes.

18.1 WHY SHOULD WE CONSERVE FRESHWATER FISH DIVERSITY?

Fishes play a central role in the structure and function of freshwater ecosystems. Many fishes represent critical keystone species within their communities, and their loss can result in considerable and unpredictable changes to both aquatic and terrestrial ecosystems. Loss of species signifies the erosion of natural capital and loss of ecological function, which ultimately reduces the stability, value and support that such systems provide for sustainable human economies (Angermeier, 2007; Chapter 2). The decline and loss of freshwater fish diversity should be a clear warning of the extent to which we are shifting ecosystems away from their natural

Conservation of Freshwater Fishes, eds G. P. Closs, M. Krkosek and J. D. Olden. Published by Cambridge University Press. © Cambridge University Press 2016.

state, and into ecological domains that are potentially very different to those in which humans have evolved (Chapters 3 and 5). Put another way, we are increasingly moving into unfamiliar ecological territory.

Freshwater fish also contribute significantly to the quality of human life in a diverse range of ways. Across many regions of the world, freshwater fish remain critical for income, nutrition and cultural heritage (Chapter 15). Water-based recreational activities, including fishing and nature watching, continue to be important individual and social activities for many millions of people around the planet. Natural ecosystems also play a central role in global tourism, and contribute greatly to the 'sense of place' of different locations – as natural systems homogenize, much of the uniqueness of different locations around the planet degrades as well. Losing access to such activities and values due to ecosystem degradation represents a significant loss in the quality of human life (Angermeier, 2007). From a scientific perspective, freshwater fish contribute significantly to our understanding of the evolution and maintenance of biological diversity (Chapters 16 and 17). As the most diverse and biogeographically distinctive group of vertebrate taxa (Abell *et al.*, 2008; Chapter 1), freshwater fish provide unique insights into the relationships between biogeographic patterns and evolutionary processes due to the constraints imposed by landscape on their movement and dispersal (Chapters 2, 16 and 17). The biogeography of freshwater fish reflects the movement of continents, the rise and decay of mountain ranges, climatic variation, and the co-evolution of multiple species and their relationships over deep evolutionary time scales (Olden *et al.*, 2010; Chapters 16 and 17). Understanding the relationships between these patterns and processes can reveal much about the evolution of life.

The diversity of freshwater fish also informs on ecological constraints and the solutions that can be evolved to cope with particular environmental challenges. As the most morphologically, behaviourally and ecologically diverse groups of vertebrates, fish exhibit a particularly diverse and fascinating range of evolutionary adaptation (Chapter 2). At first sight, it is almost difficult to believe that the largest and smallest freshwater species, the massive beluga sturgeon (maximum length over 7 m) and the tiny cyprinid *Paedocypris progenetica* (maximum length of 7.9 mm) are both fish – yet they are. The former migrates for hundreds of kilometres along the rivers of Eurasia to spawn and feed, whilst the latter lives out its life within a few metres in the acidic swamps of Indonesia. The staggering differences in life history between these two species demonstrate diverse adaptations to aquatic environments;

evolutionary 'solutions' that have also rendered them highly vulnerable to extinction for different reasons (Chapter 1).

18.2 CAN HUMANS COEXIST WITH DIVERSE FRESHWATER FISH COMMUNITIES?

The conservation and protection of freshwater fish requires changes that involve some degree of perceived or real cost at either an individual or community level (Angermeier, 2007; Chapter 2). Those costs can be weighed against measurable positive benefits, although not all ecosystem services are easily measurable (e.g. value of fisheries landings versus aesthetic value). Whilst the sheer scale of the global threat to freshwater fish is daunting (Chapter 1), there is cause for optimism that humans can sustainably coexist with diverse freshwater fish communities: the great centres of freshwater fish biodiversity are relatively intact, there is a high level of understanding of the diversity and composition of the global freshwater fish fauna, relatively few freshwater fish species have yet gone extinct over the last century despite the litany of threats, and (until relatively recently in human history) there has been a sustainable coexistence of humans and freshwater fishes across many regions of the world. Throughout this book, there are multiple examples of actions and management approaches that are generating real gains towards a sustainable coexistence between humans and diverse communities of freshwater fishes. This provides hope not despair when we envision the future of fishes.

A key cause for optimism is that the great centres of global freshwater fish diversity are still relatively intact: together the regions of South and Central America, Indo-Burma, Indonesia and continental Africa support over 10,000 species, representing approximately 70% of the global diversity of recognised freshwater fish species (Chapter 1). Out of this 10,000 species, slightly less than 10% (949 species) are recognised as threatened; but (perhaps surprisingly) only 30 species are considered extinct. Given the greater number of extinctions in the well-described and studied fauna of North America (39 extinctions) and Europe (15 extinctions), plus the lack of current detailed information on the state of freshwater fauna and the extensive changes in land use that are occurring in parts of West Africa, the Amazon and Orinoco basins, and Southeast Asia, it is likely that the numbers of threatened and extinct species across many of these relatively poorly studied regions are underestimates. Nonetheless, the figures still point to the staggering diversity that is concentrated into these areas, and highlights their importance for maintaining the overall richness of the global diversity of freshwater fishes. Focusing efforts in these relatively discrete areas should help optimise the conservation of the global fauna. Whilst the abundance and distributions of many species have been diminished to varying degrees, the resilience and persistence of most species, and the ongoing distinctiveness of regional assemblages, provide some hope for achieving a sustainable and biodiverse future.

The conservation of freshwater fish should be bolstered by gains in understanding of the global abundance and distribution of species, and the key factors that influence the structure and integrity of communities (Chapters 1 and 2). Whilst this knowledge could hardly be described as complete (Pelayo-Villami *et al.*, 2015), the core patterns of freshwater fish diversity at regional and catchment levels, and the associated threats, are well characterised (Abell *et al.*, 2008; Chapter 1). Similarly, despite rudimentary knowledge of the ecological requirements of most species, our broad level of understanding of the negative impacts of altered flow regimes, habitat degradation, introduced species and overfishing is extensive and documented in this book. Even without knowing the specific ecological requirements for each species within an assemblage, our knowledge is such that many degraded systems could be restored to a functioning community, were there opportunity, resources and motivation to do so (Koehn & Lintermans, 2012). Based on the evidence presented in this text, there is sufficient knowledge of freshwater fish to support sustainable management across most areas; ignorance cannot be claimed as an excuse for a continued degradation. Although the resilience of freshwater systems provides some scope to accommodate human activities, clearly there are limits to the impacts that can be absorbed.

18.3 MAJOR CHALLENGES FOR SCIENTISTS AND MANAGERS

Freshwater fishes present some unique and particularly difficult challenges for management. Freshwater fish are the most diverse group of vertebrates on the planet, with significant proportions of that diversity concentrated in relatively few regions around the planet. Any impacts that impinge on freshwater fish diversity in those species-rich regions have disproportionate impacts on global diversity. Worryingly, large-scale development of the river basins of tropical South America or the Indo-Burma region may be incompatible with the survival of many freshwater fishes, such as has occurred in parts of the Yangtze and Ganges river basins (Chapter 1). Given current and proposed plans for agricultural, urban and industrial development within these regions, such a scenario is a grim but, unfortunately, likely reality (Chapters 2 and 11). Climate change, combined with extensive sea-level rise (Chapter 3), also presents significant threats to these biodiverse regions; for example, large areas of the lowland regions of the Amazon basin would be submerged by a rise in sea level. The intense competition and demands for limited water resources caused by a growing human population is also generating a momentum of change across the globe (Vörösmarty et al., 2010), the trajectory of which is difficult to challenge or alter.

The major diversity of freshwater fishes is in itself a significant challenge. There are so many freshwater fish species, and genetically distinct populations within large river networks, that identifying and protecting all significant populations is a daunting task, particularly in a landscape densely populated by humans. Some losses are inevitable and management decisions typically involve trade-offs among ecological and human variables that are spatially distributed. Freshwater fish are remarkable for their diversity of behaviour, colouration, morphology and life-history strategies, particularly amongst small, closely related species. Small fish species can be particularly difficult to protect due to their high diversity, often spatially restricted distributions, and highly specialised life histories, that collectively result in disproportionate level of extinction threat (Olden et al., 2007; Chapter 2). Considerable, often relatively cryptic, within and between-species diversity is present across many river systems, requiring specialist morphological and genetic taxonomic skills to recognise the often subtle differences between closely related species or genetically distinctive populations (Chapters 16 and 17). Such taxonomic skills are in limited supply across many of the regions where the most diverse fauna occurs, thus increasing the risk of unrecognised loss of genetic and species diversity (Pelayo-Villami et al., 2015; Chapter 17). Even in regions where taxonomic skills are high, ensuring the genetic integrity of populations is retained across multiple tributaries is challenging, particularly when extensive modifications to land use are occurring, and individuals of the same or closely related species, or potential predators or competitors, are being translocated widely (Chapters 8, 13 and 16).

River systems are represented by dendritic networks of narrow habitat that can integrate and concentrate the effects of the surrounding terrestrial landscapes at broad spatial scales. Fish movement or migration along these narrow corridors can be all too easily blocked by a variety of natural and anthropogenic barriers (Chapters 4 and 11). The intimate connectivity of freshwater systems to both the terrestrial landscapes within which they are embedded, and across aquatic habitats from the headwaters to estuaries (Chapters 2 and 9), creates a uniquely high level of exposure and vulnerability to human activities. When that high level of vulnerability to human-mediated impact is combined with the spatial constraints on fish movement, then the risk to freshwater fish is elevated (Olden *et al.*, 2010). Given the diversity of ways humans can directly and indirectly impact and alter freshwater systems, including chemical and nutrient pollution (Chapters 5 and 6), altered flow regimes and barriers to migration (Chapter 4), perhaps it is not surprising that the degradation of freshwater communities is often a consequence of 'multiple stressors' (Chapter 6). Such degradation is difficult to address and remedy given its origins are often spatially diffuse, multifactorial, and can span multiple jurisdictions, management agencies, or nationalities.

18.4 OPPORTUNITIES FOR EFFECTIVE CONSERVATION

The highly connected nature of freshwater systems requires a whole-ecosystem or watershed-scale approach for effective conservation. River and lake catchments form the natural landscape subunit around which conservation planning should be organised (Chapter 14). The loss of freshwater fish diversity and abundance is frequently associated with poor environmental management in the surrounding landscape (Chapter 2), including barriers to migration (Chapters 4, 10 and 11), altered flow regimes (Chapter 4), pollution (Chapters 5 and 6), the establishment of introduced species (Chapters 8 and 16), and erosion that results in excessive inputs of nutrients and sediment (Chapters 6 and 9). Poor management that results in these degrading processes diminishes natural capital and ecosystem function and resilience (Chapters 3 and 12) and reduces scope for sustainable resource use (Chapter 15). By contrast, the maintenance of near-natural freshwater ecosystems that collect water from managed urban or agricultural landscapes is an impressive and powerful indicator of effective and sustainable landscape management. It requires the engagement of multiple regulatory agencies, a navigation of trade-offs among stakeholders, and the support of the local communities.

Sustainable management of freshwater fish communities requires a systematic conservation planning approach that involves cross-community relationships and effective communication (Chapter 14).

Such planning requires the integration of knowledge of basic fish biology and ecology with management and use of fresh waters by the whole community. As outlined in the book chapters, our collective knowledge of the biology and ecology of freshwater fish is such that most of the adverse impacts of human activities on freshwaters can be mitigated or even avoided with appropriate management (Chapters 2 and 14). Working from the premise that the coexistence of near-natural assemblages of freshwater fish and humans is both possible and desirable, systematic conservation planning starts from integrating large-scale perspectives of species distribution and assemblage composition, genetic structure and patterns of connectivity between populations, and ecological processes (Chapter 14). The distribution of species and key populations across regions and catchments must be determined, and critical habitats and migration pathways identified and then protected or restored (Chapters 4, 10–12, 16 and 17). Recognition of the critical importance of maintaining high water quality requires effective cross-catchment management of direct and diffuse pollution to avoid multiple-stressor effects (Chapters 5 and 6). Environmental flows that mimic aspects of natural flow regimes and provide enough water to sustain habitat and water quality are also essential (Chapters 4–6 and 14). Intact riparian habitats play a crucial multifaceted role in stabilising banks, taking up excess nutrients, filtering sediments, providing structure and sources of organic matter, and moderating key environmental parameters including temperature, light and evaporation (Chapter 9). At finer spatial scales, recognition and active management of the intimate relationship between freshwater habitats and the surrounding terrestrial landscape is crucial (Chapter 9); management of terrestrial landscapes should include assessments of potential impacts and mitigation of land use on downstream waterways as a matter of course (Chapter 14).

The central role of fish ecologists in engaging and encouraging communities towards sustainable management of freshwaters cannot be understated (Angermeier, 2007; Monroe et al., 2009). Freshwater fish are a potent symbol of healthy freshwater ecosystems, and the presence or restoration of fish in rivers and streams is a very tangible goal that can motivate positive action. Community recognition of decline or loss of significant fisheries or fish species has motivated attempts to restore river systems around the world, albeit with varying levels of success (Roni et al., 2008; Koehn & Lintermans, 2012; Chapter 14). In an increasingly urbanised world, declines in ecological literacy and community connection with the natural world represent a key challenge in educating communities of our collective anthropogenic impact (Angermeier, 2007). However, freshwater ecosystems are an ever-present reminder of our intricate links with nature – even in the most urbanised environments, the rain still falls and the water flows somewhere. The abundance and diversity of freshwater fish represent a quantifiable measure of failures in management but also successes in conservation and restoration.

18.5 CONFRONTING AN IMMENSE AND IMMINENT THREAT

The scale of the threat to freshwater fishes (Chapter 1) is clearly immense, with significant loss of diversity likely within the next 20 years if current trends in the degradation of freshwaters continue. The potential for catastrophic losses also clearly exists if development results in severe degradation of global hotspots of diversity (e.g. Dudgeon, 2011; Ziv *et al.*, 2012). As ecologists who understand the total dependence of human societies on the services that ecosystems provide, it does seem somewhat perverse that we must argue for the conservation of the natural resources upon which our very survival depends. However, given the increasing demands that modern human societies place on freshwater ecosystems, and the level of disconnect between increasingly urbanised communities and the natural world (Angermeier, 2007; Chapter 2), the need to argue for the conservation of freshwater fish communities will be an ongoing challenge.

Species extinction is final, so while we can 'save' a species on multiple occasions, a single failure is irrevocable. This tenuous nature of conservation achievement is highlighted by the probable extinction of the Chinese paddlefish (*Psephurus gladius*). The creation of the Upper Yangtze River Rare and Endemic Fishes Reserve to protect the habitat of threatened paddlefish, sturgeon and many other 'rare and precious' fish species at one point might have been a sign of hope for the unique fauna of this river system (Dudgeon, 2010). However, changing priorities have resulted in the subsequent construction of dams within the reserve and adjustments to reserve boundaries to accommodate further dams, severely compromising the value of the reserve as a tool for conservation (Dudgeon, 2010, 2011). No adult paddlefish have been seen since 2003, and it now seems likely the species is extinct (Dudgeon, 2011). Interviews with fishers living along the Yangtze River also indicate a lack of awareness of the Yangtze freshwater megafauna, suggesting that the local community memory of the river fishes is rapidly being lost (Turvey *et al.*, 2010). More broadly, this fading memory is a vivid illustration of the 'shifting ecological baseline' (Humphries & Winemiller, 2009; Chapter 12), and represents a large risk for maintaining efforts to conserve freshwater fishes.

The management of freshwaters is further complicated by complex jurisdictional management of river basins and biogeographic regions, which often traverse multiple national or regional government boundaries, each with potentially different political and social cultures; and contrasting conservation and developmental priorities that change over time (Abell *et al.*, 2008; Dudgeon, 2011; Koehn & Lintermans, 2012; Ziv *et al.*, 2012). Furthermore, even within a particular government boundary, the management of a watershed is likely to span disjoint regulatory agencies. Such cross-jurisdictional barriers to management will be an ongoing and ever-present challenge to fisheries management that fisheries biologists must work with, and continue to advocate for collaborative and integrated management arrangements. However, such challenges also represent opportunities for interagency and international collaboration and cooperation.

It is imperative that we maintain a past, present and future perspective of the condition of freshwater fish communities. Critical to this is maintaining momentum in freshwater fish research that documents and advances understanding of past and present distributions and diversity (Chapters 1, 2, 16 and 17), addresses the processes that maintain and degrade abundance and diversity (Chapters 2–12), and then integrates that knowledge into environmental planning for sustainability (Chapters 3 and 13–15). Protecting freshwater biodiversity clearly requires the engagement of fisheries biologists to inform systematic conservation planning (e.g. identification of core goals, critical refugia and reference areas, and planning for future threats); for long-term success, community awareness, engagement and empowerment is critical for resisting species loss and the incremental erosion of ecosystem integrity (Angermeier, 2007; Dudgeon, 2010, 2011; Koehn & Lintermans 2012; Chapters 2 and 14). At present, that level of community awareness and engagement may be inadequate (Angermeier, 2007); however, there are also encouraging examples of community-based restoration projects focused on rebuilding local fish stocks – particularly in the case of salmon and trout.

The growing imperilment of global freshwater fishes represents a staggering challenge to biodiversity conservation and the ecosystem services humans derive from healthy ecosystems. A primary aim of this book is to provide a synthesis of the state of freshwater fishes that is needed to advance fundamental research on freshwater fish, inform conservation research and application, and communicate the scale and urgency of the threats facing freshwater fish. It is the task of the reader to educate and advocate for conservation, restoration and development options that allow for a sustainable coexistence between humans and freshwater organisms, including fish. By informing managers and policy-makers, as well as local communities, choices may be made that will result in freshwater ecosystems and human communities that will be ecologically, economically and socially more resilient and prosperous.

18.6 REFERENCES

Abell, R., Thieme, M. L., Revenga, C., *et al.* (2008). Freshwater ecoregions of the world: a new map of biogeographic units for freshwater biodiversity conservation. *Bioscience*, 58, 403–414.

Angermeier, P. L. (2007). The role of fish biologists in helping society build ecological sustainability. *Fisheries*, 32, 9–20.

Dudgeon, D. (2010). Requiem for a river: extinctions, climate change and the last of the Yangtze. *Aquatic Conservation: Marine and Freshwater Ecosystems*, 20, 127–131.

Dudgeon, D. (2011). Asian river fishes in the Anthropocene: threats and conservation challenges in an era of rapid environmental change. *Journal of Fish Biology*, 79, 1487–1524.

Humphries, P. & Winemiller, K. O. (2009). Historical impacts on river fauna, shifting baselines, and challenges for restoration. *Bioscience*, 59, 673–684.

Koehn, J. D. & Lintermans, M. (2012). A strategy to rehabilitate fishes of the Murray–Darling basin, south-eastern Australia. *Endangered Species Research*, 16, 165–181.

Monroe, J. B., Baxter, C. V., Olden, J. D. & Angermeier, P. L. (2009). Freshwaters in the public eye: understanding the role of images and media in aquatic conservation. *Fisheries*, 34, 581–585.

Olden, J. D., Hogan, Z. S. & Vander Zanden, M. J. (2007). Small fish, big fish, red fish, blue fish: size-biased extinction risk of the world's freshwater and marine fishes. *Global Ecology and Biogeography*, 16, 694–701.

Olden, J. D., Kennard, M. K., Leprieur, F., *et al.* (2010). Conservation biogeography of freshwater fishes: past progress and future directions. *Diversity and Distributions*, 16, 496–513.

Pelayo-Villamil, P., Guisande, C., Vari, R. P., *et al.* (2015). Global diversity patterns of freshwater fishes – potential victims of their own success. *Diversity and Distributions*, 21, 345–356.

Roni, P., Hanson, K. & Beechie, T. J. (2008). Global review of the physical and biological effectiveness of stream habitat rehabilitation techniques. *North American Journal of Fisheries Management*, 28, 856–890.

Turvey, S. T., Barrett, L. A., Hao, Y., *et al.* (2010). Rapidly shifting baselines in Yangtze fishing communities and local memory of extinct species. *Conservation Biology*, 24, 778–787.

Vörösmarty, C. J., McIntyre, P. B., Gessner, M. O., *et al.* (2010). Global threats to human water security and river biodiversity. *Nature*, 467, 555–561.

Ziv, G., Baran, E., Nam, S., Rodriguez-Iturbe, I. & Levin, S. A. (2012). Trading-off fish biodiversity, food security, and hydropower in the Mekong river basin. *Proceedings of the National Academy of Sciences*, 109, 5609–5614.

Index